Social Statistics for a Diverse Society

THE PINE FORGE PRESS SERIES IN RESEARCH METHODS AND STATISTICS

edited by Kathleen S. Crittenden

Through its unique modular format, this Series offers an unmatched flexibility and coherence for undergraduate methods and statistics teaching. The two "core" volumes, one in methods and one in statistics, address the primary concerns of undergraduate courses, but in less detail than found in existing texts. The smaller "satellite" volumes in the Series can either supplement these core books, giving instructors the emphasis and coverage best suited for their course and students, or be used in more advanced, specialized courses.

Social Statistics for a Diverse Society, Second Edition,
by Chava Frankfort-Nachmias and Anna Leon-Guerrero

Experimental Design and the Analysis of Variance *by Robert Leik*

How Sampling Works *by Richard Maisel and Caroline Hodges Persell*

Program Evaluation *by George McCall*

Investigating the Social World: The Process and Practice of Research, Second Edition, *by Russell K. Schutt*

Multiple Regression: A Primer *by Paul Allison*

A Guide to Field Research *by Carol A. Bailey*

Designing Surveys: A Guide to Decisions and Procedures
by Ronald Czaja and Johnny Blair

OTHER PINE FORGE PRESS TITLES OF RELATED INTEREST

Adventures in Social Research: Data Analysis Using SPSS® for Windows™
by Earl Babbie and Fred Halley

Exploring Social Issues Using SPSS® for Windows 95™ Versions 7.5, 8.0, or Higher
by Joseph Healey, John Boli, Earl Babbie, and Fred Halley

Adventures in Criminal Justice Research, Revised Edition: Data Analysis for Windows 95/98™ Using SPSS® Versions 7.5, 8.0, or Higher
by George Dowdall, Kim Logio, Earl Babbie, and Fred Halley

Social Statistics for a Diverse Society

SECOND EDITION

Chava Frankfort-Nachmias
University of Wisconsin–Milwaukee

Anna Leon-Guerrero
Pacific Lutheran University

Pine Forge Press

Thousand Oaks, California ■ London ■ New Delhi

For information, address:

 Pine Forge Press
A Sage Publications Company
2455 Teller Road
Thousand Oaks, California 91320
(805) 499-4224
e-mail: sales@pfp.sagepub.com

Sage Publications Ltd.
6 Bonhill Street
London EC2A 4PU
United Kingdom

Sage Publications India Pvt. Ltd.
M-32 Marker
Greater Kailash I
New Delhi 110 048 India

Production, Scratchgravel Publishing Services; *Copy Editor,* Margaret C. Tropp; *Interior Designer,* Lisa Mirski Devenish; *Artist,* Natalie Hill; *Typesetter,* Scratchgravel Publishing Services; *Cover Designer,* Paula Shuhert and Graham Metcalfe

Printed in the United States of America
99 00 01 02 03 10 9 8 7 6 5 4 3 2 1

Library of Congress Cataloging-in-Publication Data
Nachmias, Chava.
    Social statistics for a diverse society / Chava Frankfort-Nachmias, Anna Leon-Guerrero. — 2nd ed.
        p.   cm. — (Pine Forge Press series on research methods and statistics)
    Includes index.
    ISBN 0-7619-8668-5 (alk. paper)
    1. Social sciences—Statistical methods. 2. Statistics.
I. Leon-Guerrero, Anna. II. Title. III. Series.
HA29 .N25   2000
519.5—dc21   99-6747
                                                                CIP

To all our friends

About the Authors

Chava Frankfort-Nachmias is Associate Professor of Sociology at the University of Wisconsin, Milwaukee, where she teaches courses in research methods, statistics, and gender. She is the author of *Research Methods in the Social Sciences* (with David Nachmias) and numerous publications on ethnicity and development, urban revitalization, science and gender, and women in Israel. She was the recipient of the University of Wisconsin System teaching improvement grant on integrating race, ethnicity, and gender into the social statistics and research methods curriculum.

Anna Leon-Guerrero is Associate Professor of Sociology at Pacific Lutheran University, Washington. She received her Ph.D. in sociology from the University of California, Los Angeles. She teaches courses in statistics, social theory, and social problems. Her areas of research and publications include family business and strategic management, job retention and social welfare policy, and social service program evaluation.

About the Publisher

Pine Forge Press is a new educational publisher, dedicated to publishing innovative books and software throughout the social sciences. On this and any other of our publications, we welcome your comments. Please call or write us at:

Pine Forge Press
A Sage Publications Company
2455 Teller Road
Thousand Oaks, CA 91320
(805)499-4224
E-mail: sales@pfp.sagepub.com

Visit our World Wide Web site, your direct link to a multitude of on-line resources:

http://www.pineforge.com

Brief Contents

Detailed Contents

6 Relationships Between Two Variables: Cross-Tabulation 201

7 Measures of Association for Nominal and Ordinal Variables 253

Series Foreword

The Pine Forge Press Series in Research Methods and Statistics, consisting of core books in methods and statistics in the social sciences and a series of satellite volumes on specialized topics, allows instructors to create a customized curriculum. The authors of the core volumes are all seasoned researchers and distinguished teachers, and the more specialized texts are written by acknowledged experts in their fields. To date, the series offers core texts in research methods and introductory statistics courses and satellite volumes focusing on sampling, field methods, survey research, experimental design and analysis of variance, and multiple regression. A volume on evaluation research is coming soon.

In this second edition of *Social Statistics for a Diverse Society,* Chava Frankfort-Nachmias and Anna Leon-Guerrero carry on the tradition of the popular first edition. The book provides a core introduction to statistics that is strongly grounded in important social issues. Although the typical treatment chooses examples to illustrate particular statistical techniques, the authors organize statistical topics around the tasks of describing the variety of social groupings in society and asking questions about issues of social stratification and inequality. Two "integration and review" chapters and an extensive set of computational and SPSS® exercises, all using real-world data, continue to focus on statistics as a tool for understanding the major social divisions in contemporary society.

This book is available packaged with the student version of SPSS version 9.0. Two satellite volumes are designed to be used with this text: Paul Allison's *Multiple Regression: A Primer,* and Robert Leik's *Experimental Design and the Analysis of Variance.*

Kathleen S. Crittenden
Series Editor

Preface

You may be reading this introduction on the first day or sometime during the first week of your statistics class. You probably have some questions about statistics and concerns about what your course will be like. Math, formulas, calculations? Yes, those will be part of your learning experience. However, there is more.

Throughout our text, we emphasize the relevance of statistics in our daily and professional lives. In fact, statistics is such a part of our lives that its importance and uses are often overlooked. How Americans feel about a variety of political and social topics—safety in schools, gun control, abortion, affirmative action, or our president—are measured by surveys and polls and reported daily by the news media. Consider how news programs are already predicting the Republican and Democratic front runners for the next presidential election based on early poll results. The latest from a health care study on women was just reported on a morning talk show. And that outfit you just purchased—it didn't go unnoticed. The study of consumer trends, specifically focusing on teens and young adults, helps determine commercial programming, product advertising and placement, and ultimately, consumer spending.

Statistics is not just a part of our lives in the form of news bits or information. And it isn't just numbers either. Throughout this book, we encourage you to move beyond being just a consumer of statistics and begin to recognize and utilize the many ways that statistics can increase our understanding of our world. As social scientists, we know that statistics can be a valuable set of tools to help us analyze and understand the differences in our American society and the world. We use statistics to track demographic trends, to assess differences among groups in society, and to make an impact on social policy and social change. Statistics can help us gain insight into real-life problems that affect our lives.

Teaching and Learning Goals

The following three teaching and learning goals continue to be the guiding principles of our book, as they were in the first edition.

The first goal is to introduce you to social statistics and demonstrate its value. Although most of you will not use statistics in your own student research, you will be expected to read and interpret statistical information presented by others in professional and scholarly publications, in the workplace, and in the popular media. This book will help you understand the concepts behind the statistics so that you will be able to assess the circumstances in which certain statistics should and should not be used.

Our second goal is to demonstrate that substance and statistical techniques are truly related in social science research. A special quality of this book is its integration of statistical techniques with substantive issues of particular relevance in the social sciences. Your learning will not be limited to statistical calculations and formulas. Rather, you will become proficient in statistical techniques while learning about social differences and inequality through numerous substantive examples and real-world data applications. Because the world we live in is characterized by a growing diversity—where personal and social realities are increasingly shaped by race, class, gender, and other categories of experience—this book teaches you basic statistics while incorporating social science research related to the dynamic interplay of social variables.

Many of you may lack substantial math background, and some of you may suffer from the "math anxiety syndrome." This anxiety often leads to a less-than-optimum learning environment, with students trying to memorize every detail of a statistical procedure rather than attempting to understand the general concept involved. Hence, our third goal is to address math anxiety by using straightforward prose to explain statistical concepts and by emphasizing intuition, logic, and common sense over rote memorization and derivation of formulas.

Distinctive and Updated Features of Our Book

The three learning goals we emphasize are accomplished through a variety of specific and distinctive features throughout this book.

A Close Link Between the Practice of Statistics, Important Social Issues, and Real-World Examples A special quality of this book is its integration of statistical technique with pressing social issues of particular concern to society and social science. We emphasize how the conduct of social science is the constant interplay between social concerns and methods of inquiry. In addition, the examples throughout the book—most taken from news stories, government reports, scholarly research, and the NORC General Social Survey—are formulated to emphasize to students like you that we live in a world in which statistical arguments are common. Statistical concepts and procedures are illustrated with real data and research, providing a clear sense of how questions about important social issues can be studied with various statistical techniques.

A Focus on Diversity A strong emphasis on race, class, and gender as central substantive concepts is mindful of a trend in the social sciences toward integrating issues of diversity in the curriculum. This focus on the richness of social differences within society is manifested in the application of statistical tools to examine how race, class, gender, and other categories of experience shape our social world and explain social behavior.

Reading the Research Literature   In your student career and in the workplace, you may be expected to read and interpret statistical information presented by others in professional and scholarly publications. The statistical analyses presented in these publications are a good deal more complex than most class and textbook presentations. To guide you in reading and interpreting research reports written by social scientists, most chapters include a section presenting excerpts of published research reports utilizing the statistical concepts under discussion.

Integration and Review Chapters   Two special review chapters are included. The first is Chapter 9, a review of descriptive statistical methods (Chapters 2–8), and the second, Chapter 15, reviews inferential statistics (Chapters 10–14). These review chapters provide an overview of the interconnectedness of the statistical concepts in this book and help test your abilities to cumulatively apply the knowledge from previous chapters. Both chapters include flowcharts that summarize the systematic approach utilized in the selection of statistical techniques as well as exercises that require the use of several different procedures.

Tools to Promote Effective Study   Each chapter closes with a list of main points and key terms discussed in that chapter. Boxed definitions of the key terms also appear in the body of the chapter, as do learning checks keyed to the most important points. Key terms are also clearly defined and explained in the index/glossary, another special feature in our book. Answers to all the odd-numbered problems in the text are included in the back of the book. Complete step-by-step solutions are in the manual for instructors, available from the publisher upon adoption of the text.

Emphasis on Computing   SPSS® for Windows® is used throughout the book, although the use of computers is not required to learn from the text. Real data are used to motivate and make concrete the coverage of statistical topics. These data, from the General Social Survey, are included in a disk packaged with every copy of the text. At the end of each chapter, we feature a demonstration of a related SPSS procedure, along with a set of exercises.

Highlights of the Second Edition

We have made a number of important changes to this book in response to the valuable comments that we have received from the many instructors adopting the first edition and from other interested instructors (and students).

- *Clearer and more concise presentation of topics.* We have carefully edited the discussion of statistical procedures, reducing the redundancy of statistical procedures and clarifying examples, while at the same time preserving the book's easily understood style. For examples, please refer to our discussion in Chapter 7 of lambda or our estimation procedures in Chapter 12.

- *Revisions to "Testing Hypotheses" Chapters.* We have combined Chapters 13 and 14 of the first edition into a single new Chapter 13, Testing Hypotheses about Two Samples. We've eliminated the Z test for one sample, focusing on the two-sample t-test for means and Z-test for proportions. In order to be consistent with the probability of the obtained test statistic as reported in SPSS output (versus setting the critical region of rejection as explained in the first edition), we eliminated the discussion on critical values. In the new Chapter 13, we discuss setting an appropriate alpha level and estimating the probability based on the Z or t distribution tables. The six-step model for hypothesis testing has been revised to a five-step model. We apply the five-step model to our discussion of two-sample tests and the chi-square test. (See Chapters 13 and 14.)

- *Real-world examples and exercises.* A hallmark of the first edition was its extensive use of real data from a variety of sources for chapter illustrations and exercises. Throughout the second edition, we have updated the majority of exercises and examples based on General Social Survey or U.S. Census data. New end-of-chapter exercises have been added to several chapters, highlighting current research in criminal justice, public health, teen behavior, and social attitudes.

- *SPSS version 9.0.* Packaged with this text, on an optional basis, is SPSS Student Version 9.0. SPSS demonstrations and exercises have been updated, using version 9.0 format. Appendix E, How to Use a Statistical Package, has also been updated to highlight 9.0 features. Please telephone the publisher at (805) 499-4224, or access their web site at www.pineforge.com to learn how to order the book packaged with the Student Version of SPSS v. 9.0.

- *General Social Survey updates.* As a companion to the second edition's SPSS demonstrations and exercises, we have created two datasets: one featuring GSS survey items on family and women's issues and the second highlighting attitudes toward racial and ethnic diversity in the United States. SPSS exercises at the end of each chapter utilize certain variables from both data modules. There is ample opportunity for instructors to develop their own SPSS exercises using these data.

- *Supplemental tools on important topics.* The second edition's discussion of inferential statistics remains focused on Z, t and chi-square. The Pine Forge Press Series in Research Methods and Statistics, of which this book is a part, includes additional supplementary volumes by Paul Allison on regression and by Robert Leik on analysis of variance. These supplements were written to closely coordinate with this text and are available from the publisher.

Acknowledgments

We are both grateful to Steve Rutter, the president of Pine Forge Press. His deep understanding of the changing world of college teaching and his remarkable involvement in every phase of this project have made it all possible.

Our largest single debt is to our series editor, Kathleen S. Crittenden. We are grateful for her detailed comments on many drafts of this book (and its revision) and her countless contributions at every step of the way. With her constant support and unflagging patience, she has seen both of us through the completion of this book and its revision.

Many manuscript reviewers recruited by Pine Forge provided invaluable feedback. For their comments to the first edition, we thank:

Catherine W. Berheide, *Skidmore College*
Terry Besser, *University of Kentucky*
Lisa Callahan, *Russell Sage College*
Ashley "Woody" Doane, *University of Hartford*
James Ennis, *Tufts University*
Kristin Esterberg, *University of Missouri–Kansas City*
Gary Gorham, *North Dakota State University*
Barbara Hart, *University of Texas, Tyler*
Colleen Johnson, *University of Memphis*
Barbara R. Keating, *Mankato State University*
Alice Kemp, *University of New Orleans*
John Light, *Vermont Alcohol Research Center*
Thomas J. Linneman, *University of Washington*
Joan Morris, *University of Central Florida*
Chandra Muller, *University of Texas*
Edward Nelson, *California State University, Fresno*
Jeff Pounders, *Ouachita Baptist University*
Josephine A. Ruggiero, *Providence College*
Valerie Schwebach, *Rice University*
Judith Stull, *Lasalle University*
Ira M. Wasseman, *Eastern Michigan University*
Janet Wilmoth, *Purdue University*
Cathy Zimmer, *North Carolina State University*

For their comments and contributions to the second edition, we are grateful to:

Catherine W. Berheide, Skidmore College
Marit Berntson, University of Minnesota
Terry Besser, Iowa State University
Lisa Callahan, Russell Sage College
Raymond G. Devries, St. Olaf College
Ashley A. Woody, University of Hartford
Detis Duhart, Virginia Commonwealth University
David Elesh, Temple University

James Ennis, Tufts University
Joseph Eyerman, Florida State University
Kristin Esterberg, University of Massachusetts, Lowell
Sandra Gill, Gettysburg College
Gary Goreham, North Dakota State University
Barbara Hart, University of Texas, Tyler
A. Leigh Ingram, University of Colorado, Denver
Colleen Johnson, University of Memphis
Barbara R. Keating, Minnesota State University, Mankato
Debra Kelley, Longwood College
Alice Kemp, University of New Orleans
John Light, Vermont Alcohol Research Center
Thomas J. Linneman, College of William and Mary
Andrew London, Kent State University
Ginger Macheski, Valdosta State University
Joan Morris, University of Central Florida
Chandra Muller, University of Texas
Edward Nelson, California State University, Fresno
Jeff Pounders, Ouachita Baptist University
Larry Raffalovich, State University of New York, Albany
Lawrence Rosen, Temple University
Pamela Rosenberg, Gettysburg College
Randi Rosenblum, CUNY, Hunter College
Josephine A. Ruggiero, Providence College
Greg Sanders, Greenville College
Meg Sandifer, University of St. Thomas
Mary Kay Schleiter, University of Wisconsin, Parkside
Valerie Schwebach, Rice University
Pamela Stone, CUNY, Hunter College
Judith Stull, Lasalle University
Ira M. Wasserman, Eastern Michigan University
Janet Wilmoth, Purdue University
Cathy Zimmer, North Carolina State University

We are grateful to Greg and Anne Draus at Scratchgravel Publishing Services and Peggy Tropp for guiding the book through the production process. We would also like to acknowledge Sherith Pankratz, Jean Skeels, and the rest of the Pine Forge Press staff for their assistance and support throughout this project. Thanks also to Joan Saxton Weber for her work on the SPSS version 9.0 appendix.

Chava Frankfort-Nachmias would like to thank and acknowledge the following: I am intellectually indebted to Elizabeth Higginbotham and Lynn Weber Cannon for an instructive and inspiring SWS workshop on integrating race, class, and gender in the sociological curriculum. At that workshop the idea to work on this book began to emerge. I was also greatly influenced by the pioneering work of Margaret L. Andersen and Patricia Hill

Collins, who developed an interdisciplinary and inclusive framework for transforming the curriculum.

My profound gratitude goes to friends and colleagues who have stood by me, cheered me on, and understood when I was unavailable for long periods due to the demands of this project. Special thanks to Stacey Oliker and Eleanor Miller for supporting my work and offering personal encouragement throughout. I am indebted to Shani Beth Halachmi and Marilyn Kraar, my closest friends, for their support and encouragement. I am also grateful to Joanne Spiro, who encouraged me to begin this project, and to Carole Warshaw, who helped me see my way through a difficult period.

I wish to express my indebtedness to Chris Roerden for editing an earlier draft and to Lura Harrison for careful editing of the first edition manuscript.

Indispensable assistance in preparing the first edition manuscript was provided at every step by my research assistant, Pat Pawasarat, who is also the co-author of two chapters in this book. I would like to express my gratitude to Pat for her thoroughness and patience. I also thank Mark Rodeghier, who wrote the first edition exercises and helped with many of the examples from the General Social Survey. I also thank Helen Miller for her work on the first edition instructor's manual, James Harris for help in preparing the final version of the first edition manuscript, and Lisa Amoroso for help in preparing the first edition instructor's manual.

I am grateful to my students at the University of Wisconsin–Milwaukee, who taught me that even the most complex statistical ideas can be simplified. The ideas presented in this book are the products of many years of classroom testing. I thank my students for their patience and contributions.

Finally, I thank Marlene Stern, whose love and support made it possible for me to finish the book, and my daughters, Anat and Talia, for their extraordinary patience, sense of humor, and faith in me.

Anna Leon-Guerrero expresses her thanks to the following: I wish to thank my statistics teaching assistants and students. My passion for and understanding of teaching statistics grow with each semester and class experience. I am grateful for the teaching and learning opportunities we have shared.

My work on this second edition could not have been possible without the contribution and efforts of Marcus Womack (for his SPSS graphics support), Sharon Raddatz (for her technical and manuscript support), and Kerri Fletcher (for her assistance on the GSS datasets). In addition, I would like to express my gratitude to friends and colleagues for their encouragement and support throughout this project. My love and thanks to my husband, Brian Sullivan.

Chava Frankfort-Nachmias
University of Wisconsin–Milwaukee

Anna Leon-Guerrero
Pacific Lutheran University

Social Statistics for a Diverse Society

1 The What and the Why of Statistics

■ ■ ■ ■ Introduction

Are you taking statistics because it is required in your major—not because you find it interesting? If so, you may be feeling intimidated because you know that statistics involves numbers and math. Perhaps you feel intimidated not only because you're uncomfortable with math, but also because you suspect that numbers and math don't leave room for human judgment or have any relevance to your own personal experience. In fact, you may even question the relevance of statistics to understanding people, social behavior, or society.

In this course, we will show you that statistics can be a lot more interesting and easy to understand than you may have been led to believe. In fact, as we draw upon your previous knowledge and experience and relate materials to interesting and important social issues, you'll begin to see that statistics is not just a course you have to take but a useful tool as well.

There are two major reasons why learning statistics may be of value to you. First, you are constantly exposed to statistics every day of your life. Marketing surveys, voting polls, and the findings of social research appear daily in newspapers and popular magazines. By learning statistics you will become a sharper consumer of statistical material. Second, as a major in the social sciences, you may be expected to read and interpret statistical information presented to you in the workplace. Even if conducting research is not a part of your job, you may still be expected to understand and learn from other people's research or to be able to write reports based on statistical analyses.

Just what *is* statistics anyway? You may associate the word with numbers that indicate birth rates, conviction rates, per-capita income, marriage and divorce rates, and so on. But the word **statistics** also refers to a set of procedures used by social scientists. They use these procedures to organize, summarize, and communicate information. Only information represented by numbers can be the subject of statistical analysis. Such information is called **data**; researchers use statistical procedures to analyze data to answer research questions and test theories. It is the latter usage—answering research questions and testing theories—which this textbook explores.

Statistics   A set of procedures used by social scientists to organize, summarize, and communicate information.

Data   Information represented by numbers, which can be the subject of statistical analysis.

■ ■ ■ ■ The Research Process

To give you a better idea of the role of statistics in social research, let's start by looking at the **research process**. We can think of the research process as

Figure 1.1 **The Research Process**

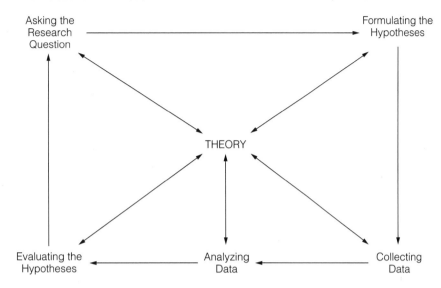

a set of activities in which social scientists engage so they can answer questions, examine ideas, or test theories.

As illustrated in Figure 1.1, the research process consists of five stages:

1. Asking the research question
2. Formulating the hypotheses
3. Collecting data
4. Analyzing data
5. Evaluating the hypotheses

Each stage affects the *theory* and is affected by it as well. Statistics are most closely tied to the data analysis stage of the research process. As we will see in later chapters, statistical analysis of the data helps researchers test the validity and accuracy of their hypotheses.

> ***Research process*** A set of activities in which social scientists engage to answer questions, examine ideas, or test theories.

■ ■ ■ ■ Asking Research Questions

The starting point for most research is asking a *research question*. Consider the following research questions taken from a number of social science journals:

"Does cost control influence the quality of health care?"

"Has sexual harassment become more widespread during the past decade?"

"Does social class influence voting behavior?"

"What factors influence the economic mobility of female workers?"

These are all questions that can be answered by conducting **empirical research**—research based on information that can be verified by using our direct experience. To answer research questions we cannot rely on reasoning, speculation, moral judgment, or subjective preference. For example, the questions "Is racial equality good for society?" or "Is an urban lifestyle better than a rural lifestyle?" cannot be answered empirically because the terms *good* and *better* are concerned with values, beliefs, or subjective preference and, therefore, cannot be independently verified. One way to study these questions is by defining *good* and *better* in terms that can be verified empirically. For example, we can define *good* in terms of economic growth and *better* in terms of psychological well-being. These questions could then be verified by conducting empirical research.

You may wonder how to come up with a research question. The first step is to pick a question that interests you. If you are not sure, look around! Ideas for research problems are all around you, from media sources to personal experience or your own intuition. Talk to other people, write down your own observations and ideas, or learn what other social scientists have written about.

Take, for instance, the issue of gender and work. As a college student about to enter the labor force, you may wonder about the similarities and differences between women's and men's work experiences and about job opportunities when you graduate. Here are some facts and observations based on research reports and our own personal experiences: In 1997 women who were employed full-time earned $431 per week on average; men who were employed full-time earned $579 per week on average.[1] Women's and men's work are also very different. Women continue to be the minority in many of the higher-ranking and higher-salaried positions in professional and managerial occupations. For example, in 1996 women made up 26.4 percent of physicians, 29.0 percent of lawyers and judges, 16.7 percent of architects, and 37.8 percent of managers in marketing and advertising.[2] In comparison, among all those employed in technical, sales, and administrative support positions, 64.2 percent were women.[3] These observations may prompt us to ask research questions such as: Are women paid, on average, less than men for the same types of work? How much change has there been in women's work over time? Does the fact that women and men work in gender-segregated jobs relate to their differences in earnings?

---

[1]U.S. Bureau of Labor Statistics, Women's Bureau, *20 Facts on Women Workers*, May 1998.
[2]*Statistical Abstract of the United States*, 1997.
[3]Ibid.

Empirical research   Research based on evidence that can be verified by using our direct experience.

Learning Check.   *Identify one or two social science questions amenable to empirical research. You can almost bet that you will be required to do a research project sometime in your college career. Get a head start and start thinking about a good research question now.*

■ ■ ■ ■ ■   **The Role of Theory**

You may have noticed that each preceding research question was expressed in terms of a *relationship*. This relationship may be between two or more attributes of individuals or groups, such as gender and income or gender segregation in the workplace and income disparity. The relationship between attributes or characteristics of individuals and groups lies at the heart of social scientific inquiry.

Most of us use the term *theory* quite casually to explain events and experiences in our daily life. We may have a "theory" about why our boss has been so nice to us lately or why we didn't do so well on our last history test. In a somewhat similar manner, social scientists attempt to explain the nature of social reality. Whereas our theories about events in our lives are commonsense explanations based on educated guesses and personal experience, to the social scientist a theory is a more precise explanation that is frequently tested by conducting research.

A **theory** is an explanation of the relationship between two or more observable attributes of individuals or groups. The theory attempts to establish a link between what we observe (the data) and our conceptual understanding of why certain phenomena are related to each other in a particular way. For instance, suppose we wanted to understand the reasons for the income disparity between men and women; we may wonder whether the types of jobs men and women have and the organizations in which they work have something to do with their wages.

One explanation for gender inequality in wages is *gender segregation in the workplace*—the fact that American men and women are concentrated in different kinds of jobs and occupations. For example, in 1990 one-third of all women who were in the labor force worked in only 10 of the 503 occupations listed by the census; only 11 percent worked in occupations that were at least 75 percent male.[4]

[4]Barbara Reskin and Irene Padavic, *Women and Men at Work* (Thousand Oaks, CA: Pine Forge Press, 1994), p. 45.

What is the significance of gender segregation in the workplace? In our society, people's occupations and jobs are closely associated with their level of prestige, authority, and income. The jobs in which women and men are segregated are not only different but also unequal. Although the proportion of women in the labor force has markedly increased, women are still concentrated in occupations with low pay, low prestige, and few opportunities for promotion. Thus, gender segregation in the workplace is associated with unequal earnings, authority, and status. In particular, women and men's segregation into different jobs and occupations is the most immediate cause of the pay gap. Women receive lower pay than men do even when they have the same level of education, skills, and experience as men in comparable occupations.

> *Theory* An elaborate explanation of the relationship between two or more observable attributes of individuals or groups.

■ ■ ■ ■ Formulating the Hypotheses

So far we have come up with a number of research questions about the income disparity between men and women in the workplace. We have also discussed a possible explanation—a theory—that helps us make sense of gender inequality in wages. Is that enough? Where do we go from here?

Our next step is to test some of the ideas suggested by the gender segregation theory. But this theory, even if it sounds reasonable and logical to us, is too general and does not contain enough specific information to be tested. Instead, theories suggest specific concrete predictions about the way that observable attributes of people or groups are interrelated in real life. These predictions, called **hypotheses**, are tentative answers to research problems. Hypotheses are tentative because they can be verified only after they have been tested empirically.[5] For example, one hypothesis we can derive from the gender segregation theory is that wages in occupations in which the majority of workers are female are lower than the wages in occupations in which the majority of workers are male.

Not all hypotheses are derived directly from theories. We can generate hypotheses in many ways—from theories, directly from observations, or from intuition. Probably the greatest source of hypotheses is the professional literature. A critical review of the professional literature will familiarize you with the current state of knowledge and with hypotheses that others have studied.

> *Hypothesis* A tentative answer to a research problem.

[5]Chava Frankfort-Nachmias and David Nachmias, *Research Methods in the Social Sciences* (New York: St. Martin's Press, 1996), p. 62.

Let's restate our hypothesis:

Wages in occupations in which the majority of workers are female are lower than the wages in occupations in which the majority of workers are male.

Notice that this hypothesis is a statement of a relationship between two characteristics that vary: *wages* and *gender composition* of occupations. Such characteristics are called variables. A **variable** is a property of people or objects that takes on two or more values. For example, people can be classified into a number of *social class* categories, such as upper class, middle class, or working class. Similarly, people have different levels of education; therefore, *education* is a variable. *Family income* is a variable; it can take on values from zero to hundreds of thousands of dollars or more. *Wages* is a variable, with values from zero to thousands of dollars or more. Similarly, *gender composition* is a variable. The percentage of females (or males) in an occupation can vary from 0 to 100. (See Figure 1.2 for examples of some variables and their possible values.)

> *Variable* A property of people or objects that takes on two or more values.

Each variable must include categories that are both *exhaustive* and *mutually exclusive*. Exhaustiveness means that there should be enough categories composing the variables to classify every observation. For example, the common classification of the variable *marital status* into the categories "married," "single," "divorced," and "widowed" violates the requirement of

Figure 1.2 **Variables and Value Categories**

Variable	Categories
Social class	Upper class Middle class Working class
Religion	Christian Jewish Muslim
Monthly income	$ 1,000 $ 2,500 $10,000 $15,000
Gender	Male Female

exhaustiveness. As defined, it does not allow us to classify same-sex couples or heterosexual couples who are not legally married. (We can make every variable exhaustive by adding the category "other" to the list of categories. However, this practice is not recommended if it leads to the exclusion of categories that have theoretical significance or a substantial number of observations.)

Mutual exclusiveness refers to the need to classify every observation into one and only one category. For example, we need to define *religion* in such a way that no one would be classified into more than one category. For instance, the categories "Protestant" and "Methodist" are not mutually exclusive because Methodists are also considered Protestant and, therefore, could be classified into both categories.

> **Learning Check.** *Review the definitions of* exhaustive *and* mutually exclusive. *What other categories could be added to the variable* religion *in order to be exhaustive and mutually exclusive? What other categories could be added to* social class? *to* income?

Social scientists can choose which level of social life to focus their research on. They can focus on individuals or on groups of people such as families, organizations, and nations. These distinctions are referred to as **units of analysis**. A variable will be a property of whatever the unit of analysis is for the study. Variables can be properties of individuals, of groups (such as the family or a social group), of organizations (such as a hospital or university), or of societies (such as a country or a nation). For example, in a study that looks at the relationship between individuals' level of education and their income, the variable *income* refers to the income level of an individual. On the other hand, a study that compares how differences in corporations' revenues relate to differences in the fringe benefits they provide to their employees uses the variable *revenue* as a characteristic of an organization (the corporation). The variables *wages* and *gender composition* in our example are characteristics of occupations. Figure 1.3 illustrates different units of analysis frequently employed by social scientists.

> *Unit of analysis* The level of social life on which social scientists focus. Examples of different levels are individuals and groups.

> **Learning Check.** *Remember that research question you came up with? Can you formulate a hypothesis you could test? Remember that the variables must take on two or more values and you must determine the unit of analysis.*

Figure 1.3 **Examples of Units of Analysis**

Individual as unit of analysis:
How old are you?
What are your political views?
What is your occupation?

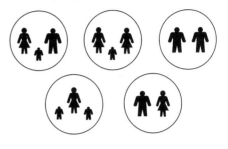

Family as unit of analysis:
How many children are in the family?
Who does the housework?
How many wage earners are there?

Organization as unit of analysis:
How many employees are there?
What is the gender composition?
Do you have a diversity office?

City as unit of analysis:
What was the crime rate last year?
What is the population density?
What type of government runs things?

Independent and Dependent Variables: Causality

Hypotheses are usually stated in terms of a relationship between an *independent* and a *dependent variable*. The distinction between an independent and a dependent variable is an important one in the language of research. Social theories often intend to provide an explanation for social patterns or causal relations between variables. For example, according to the gender segregation theory, gender segregation in the workplace is the primary explanation (although certainly not the only one) of the male/female earning gap. Why should jobs where the majority of workers are female pay less than jobs that employ mostly men? One explanation is that "societies undervalue the work women do, regardless of what those tasks are, because women do them. . . . Physical strength, for example, in which men tend to excel over women, commands premium pay in metal working industries. But manual dexterity, allegedly more common in women than men, does not raise workers' pay in assembly-line jobs."[6] Similarly, "In the United States, where most dentists are male, dentists are near the top of the income hierarchy; in Europe, where most dentists are female, dentists' incomes are much closer to the average income. In general, the more women in an occupation, the lower its average pay."[7]

In the language of research, the variable the researcher wants to explain (the effect) is called the **dependent variable**. The variable that is expected to "cause" or account for the dependent variable is called the **independent variable**. Therefore, in our example, *gender composition of occupations* is the independent variable, while *wages* is the dependent variable.

Dependent variable　The variable to be explained (the "effect").

Independent variable　The variable expected to account for (the "cause" of) the dependent variable.

Cause-and-effect relationships between variables are *not* easy to infer in the social sciences. In order to establish that two variables are causally related, you need to meet three conditions: (1) the cause has to precede the effect in time, (2) there has to be an empirical relationship between the cause and the effect, and (3) this relationship cannot be explained by other factors.

Let's consider the decades-old debate about controlling crime through the use of prevention versus punishment. Some people argue that special counseling for youths at the first sign of trouble and strict controls on access to firearms would help reduce crime. Others argue that overhauling federal and state sentencing laws to stop early prison releases is the solution. Here's an example from the state of Washington, which in 1994 adopted a new measure—"three strikes and you're out"—imposing life prison terms on three-time felons. Let's suppose that two years after the measure was introduced the crime rate in Washington declined somewhat. Does the observation that the incidence of crime declined mean that the new measure caused this reduction? Not necessarily! Perhaps the rate of crime had been going down for other reasons, such as improvement in the economy, and the new measure had nothing to do with it. To demonstrate a cause-and-effect relationship, we would need to show three things: (1) the enactment of the "three strikes and you're out" measure was empirically associated with a decrease in crime; (2) the reduction of crime actually occurred *after* the enactment of this measure; and (3) the relationship between the reduction in crime and the "three strikes and you're out" policy is not due to the influence of another variable (for instance, the improvement of overall economic conditions).

[6]Reskin and Pavadic, pp. 118–119.
[7]Ibid., p. 119.

Independent and Dependent Variables: Guidelines

Because of the limitations in inferring cause-and-effect relationships in the social sciences, be cautious about using the terms *cause* and *effect* when examining relationships between variables. However, using the terms *independent variable* and *dependent variable* is still appropriate even when this relationship is not articulated in terms of direct cause and effect. Here are a few guidelines that may help you to identify the independent and dependent variables:

1. The dependent variable is always the property you are trying to explain; it is always the object of the research.
2. The independent variable usually occurs earlier in time than the dependent variable.
3. The independent variable is often seen as influencing, directly or indirectly, the dependent variable.

The purpose of the research should help determine which is the independent variable and which is the dependent variable. In the real world, variables are neither dependent nor independent; they can be switched around depending on the research problem. A variable defined as independent in one research investigation may be a dependent variable in another.[8] For instance, *educational attainment* may be an independent variable in a study attempting to explain how education influences political attitudes. However, in an investigation of whether a person's level of education is influenced by the social status of his or her family of origin, *educational attainment* will be the dependent variable. Some variables, such as race, age, or ethnicity, because they are primordial characteristics that cannot be explained by social scientists, are never considered dependent variables in a social science analysis.

Learning Check. *Identify the independent and dependent variables in the following hypotheses:*

- *Children who attended preschool day-care centers earn better grades in first grade than children who received home preschool care.*
- *People who attend church regularly are more likely to oppose abortion than people who do not attend church regularly.*
- *Elderly women are more likely to live alone than elderly men.*
- *Individuals with postgraduate education are likely to have fewer children than those with less education.*

What are the independent and dependent variables in your hypothesis?

---

[8]Frankfort-Nachmias and Nachmias, p. 56.

■ ■ ■ ■ Collecting Data

Once we have decided on the research question, the hypothesis, and the variables to be included in the study, we proceed to the next stage in the research cycle. This step includes measuring our variables and collecting the data. As researchers, we must decide how to measure the variables of interest to us, how to select the cases for our research, and what kind of data collection techniques we will be using. A wide variety of data collection techniques are available to us, from direct observations to survey research, experiments, or secondary sources. Similarly, we can construct numerous measuring instruments. These instruments can be as simple as a single question included in a questionnaire or as complex as a composite measure constructed through the combination of two or more questionnaire items. The choice of a particular data collection method or instrument to measure our variables depends on the study objective. For instance, suppose we decide to study how social class position is related to attitudes about abortion. Since attitudes about abortion are not directly observable, we need to collect data by asking a group of people questions about their attitudes and opinions. A suitable method of data collection for this project would be a *survey* that uses some kind of questionnaire or interview guide to elicit verbal reports from respondents. The questionnaire could include numerous questions designed to measure attitudes toward abortion, social class, and other variables relevant to the study.

How would we go about collecting data to test the hypothesis relating the gender composition of occupations to wages? We want to gather information on the proportion of men and women in different occupations and the average earnings for these occupations. This kind of information is routinely collected by the government and published in sources such as bulletins distributed by the U.S. Department of Labor's Bureau of Labor Statistics and the *Statistical Abstract of the United States*. The data obtained from these sources could then be analyzed and used to test our hypothesis.

Levels of Measurement

The statistical analysis of data involves many mathematical operations, from simple counting to addition and multiplication. However, not every operation can be used with every variable. The type of statistical operations we employ will depend on how our variables are measured. For example, for the variable *gender*, we can use the number 1 to represent females and the number 2 to represent males. Similarly, 1 can also be used as a numerical code for the category "one child" in the variable *number of children*. Clearly, in the first example the number is an arbitrary symbol that does not correspond to the property "female," whereas in the second example the number 1 has a distinct numerical meaning that does correspond to the property "one child." The correspondence between the properties we measure and the numbers representing these properties determines the type of statistical operations we can use. The degree of correspondence also leads to different ways of measuring—that is, to distinct *levels of measurement*. In this section, we will discuss three levels of measurement: *nominal, ordinal,* and *interval-ratio.*

Nominal Level of Measurement   At the **nominal** level of measurement, numbers or other symbols are assigned to a set of categories for the purpose of naming, labeling, or classifying the observations. *Gender* is an example of a nominal level variable. Using the numbers 1 and 2, for instance, we can classify our observations into the categories "females" and "males," with 1 representing females and 2 representing males. We could use any of a variety of symbols to represent the different categories of a nominal variable; however, when numbers are used to represent the different categories, we do not imply anything about the magnitude or quantitative difference between the categories. Because the different categories (for instance, males versus females) vary in the quality inherent in each but not in quantity, nominal variables are often called *qualitative*. Other examples of nominal level variables are political party, religion, and race.

Ordinal Level of Measurement   Whenever we assign numbers to rank-ordered categories ranging from low to high, we have an **ordinal** level variable. *Social class* is an example of an ordinal variable. We might classify individuals with respect to their social class status as "upper class," "middle class," or "working class." We can say that a person in the category "upper class" has a higher class position than a person in a "middle class" category (or that a "middle class" position is higher than a "working class" position), but we do not know the magnitude of the differences between the categories; that is, we don't know how much higher "upper class" is compared with "middle class."

Many attitudes we measure in the social sciences are ordinal level variables. Take, for instance, the following question used to measure attitudes toward same-sex marriages: "Same-sex partners should have the right to marry each other." Respondents are asked to mark the number representing their degree of agreement or disagreement with this statement. One form in which a number might be made to correspond with the answers can be seen in Table 1.1. Although the differences between these numbers represent

Table 1.1   **Ordinal Ranking Scale**

Rank	Value
1	Strongly agree
2	Agree
3	Neither agree nor disagree
4	Disagree
5	Strongly disagree

higher or lower degrees of agreement with same-sex marriage, the distance between any two of those numbers does not have a precise numerical meaning.

Interval-Ratio Level of Measurement   If the categories (or values) of a variable can be rank-ordered, and if the measurements for all the cases are expressed in the same units, then an **interval-ratio** level of measurement has been achieved. Examples of variables measured at the interval-ratio level are *age, income*, and *SAT scores*. With all these variables we can compare values not only in terms of which is larger or smaller, but also in terms of *how much* larger or smaller one is compared with another. In some discussions of levels of measurement you will see a distinction made between interval-ratio variables that have a natural zero point (where zero means the absence of the property) and those variables that have zero as an arbitrary point. For example, weight and length have a natural zero point, whereas temperature has an arbitrary zero point. Variables with a natural zero point are also called *ratio variables*. In statistical practice, however, ratio variables are subjected to operations that treat them as interval and ignore their ratio properties. Therefore, no distinction between these two types is made in this text.

Nominal measurement   Numbers or other symbols are assigned to a set of categories for the purpose of naming, labeling, or classifying the observations.

Ordinal measurement   Numbers are assigned to rank-ordered categories ranging from low to high.

Interval-ratio measurement   Measurements for all cases are expressed in the same units.

Cumulative Property of Levels of Measurement   Variables that can be measured at the interval-ratio level of measurement can also be measured at the ordinal and nominal levels. As a rule, properties that can be measured at a higher level (interval-ratio is the highest) can also be measured at lower levels, but not vice versa. Let's take, for example, *gender composition of occupations,* the independent variable in our research example. Table 1.2 shows the percentage of women in four major occupational groups as reported in the 1997 *Statistical Abstract of the United States.*

The variable *gender composition* (measured as the percentage of women in the occupational group) is an interval-ratio variable and, therefore, has the properties of nominal, ordinal, and interval-ratio measures. For example, we can say that the transportation group differs from the service workers group (a nominal comparison), that service occupations have more women than transportation occupations (an ordinal comparison), and that

Table 1.2 **Gender Composition of Four Major Occupational Groups**

Occupational Group	% Women in Occupation
Executive, administrative, and managerial	44
Administrative support	79
Transportation workers	10
Service workers	59

Source: Statistical Abstract of the United States, 1997.

service occupations have 49 percentage points more women (59 minus 10) than transportation occupations (an interval-ratio comparison).

The types of comparisons possible at each level of measurement are summarized in Table 1.3 and Figure 1.4. Notice that where differences can be established at each of the three levels, only at the interval-ratio level can we establish the magnitude of the difference.

> **Learning Check.** *Make sure you understand these levels of measurement. As the course progresses, your instructor is likely to ask you what statistical procedure you would use to describe or analyze a set of data. To make the proper choice, you must know the level of measurement of the data.*

Levels of Measurement of Dichotomous Variables A variable that has only two values is called a *dichotomous variable*. Several key social factors, such as gender, employment status, and marital status, are dichotomies; that is, you are male or female, employed or unemployed, married or not married. Such variables may seem to be measured at the nominal level: you fit in either one category or other. No category is naturally higher or lower than the other, so they can't be ordered.

Table 1.3 **Levels of Measurement and Possible Comparisons**

Level	Different or Equivalent	Higher or Lower	How Much Higher
Nominal	Yes	No	No
Ordinal	Yes	Yes	No
Interval-ratio	Yes	Yes	Yes

Figure 1.4 **Levels of Measurement and Possible Comparisons: Education Measured on Nominal, Ordinal, and Interval-Ratio Levels**

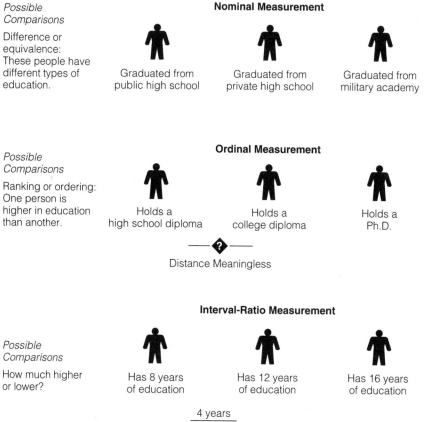

Possible Comparisons

Difference or equivalence: These people have different types of education.

Nominal Measurement

Graduated from public high school

Graduated from private high school

Graduated from military academy

Possible Comparisons

Ranking or ordering: One person is higher in education than another.

Ordinal Measurement

Holds a high school diploma

Holds a college diploma

Holds a Ph.D.

?

Distance Meaningless

Possible Comparisons

How much higher or lower?

Interval-Ratio Measurement

Has 8 years of education

Has 12 years of education

Has 16 years of education

4 years

Distance Meaningful

However, because there are only two possible values for a dichotomy, we can measure it at the ordinal or the interval-ratio level. For example, we can think of "femaleness" as the ordering principle for gender, so that "female" is higher and "male" is lower. Using "maleness" as the ordering principle, "female" is lower and "male" is higher. In either case, with only two classes, there is no way to get them out of order; therefore, gender could be considered at the ordinal level.

Dichotomous variables can also be considered to be interval-ratio level. Why is this? In measuring interval-ratio data, the size of the interval between the categories is *meaningful*: the distance between 4 and 7, for example, is the same as the distance between 11 and 14. But with a dichotomy, there is only one interval. Therefore, there is really no other distance to which we can compare it:

Mathematically, this gives the dichotomy more power than other nominal level variables (as you will notice later in the text).

This is why researchers dichotomize some of their variables, turning a multicategory nominal variable into a dichotomy. For example, you may see race (originally divided into many categories) dichotomized into "white" and "nonwhite." Though this is substantively suspect, it may be the most logical statistical step to take.

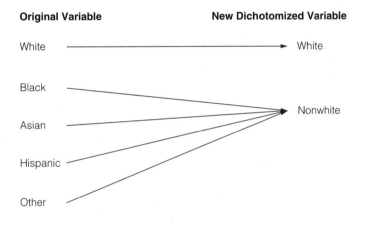

Be sure that the two categories capture a distinction that is important to your research question (for example, a comparison of the number of white versus nonwhite U.S. Senators).

Discrete and Continuous Variables

The statistical operations we can perform are also determined by whether the variables are continuous or discrete. *Discrete* variables have a minimum-sized unit of measurement, which cannot be subdivided. The number of children per family is an example of a discrete variable because the minimum unit is 1 child. A family may have 2 or 3 children, but not 2.5 children. The variable *wages* in our research example is a discrete variable because currency has a minimum unit (1 cent), which cannot be subdivided. One can have $101.21 or $101.22 but not $101.21843. Wages cannot differ by less than 1 cent—the minimum-sized unit.

Unlike discrete variables, *continuous* variables do not have a minimum-sized unit of measurement; their range of values can be subdivided into increasingly smaller fractional values. *Length* is an example of a continuous variable because there is no minimum unit of length. A particular object may be 12 inches long, it may be 12.5 inches long, or it may be 12.532011 inches long. Although we cannot always measure all possible length values with absolute accuracy, it is possible for objects to exist at an infinite number of lengths.[9] In principle, we can speak of a tenth of an inch, a ten-thousandth of an inch, or a ten-trillionth of an inch. The variable *gender composition of occupations* is a continuous variable because it is measured in proportions or percentages (for example, the percentage of women in medicine), which can be subdivided into smaller and smaller fractions.

This attribute of variables—whether they are continuous or discrete—affects subsequent research operations, particularly measurement procedures, data analysis, and methods of inference and generalization. However, keep in mind that, in practice, some discrete variables can be treated as if they were continuous, and vice versa.

Learning Check. *Name three continuous and three discrete variables. Determine whether each of the variables in your hypothesis is continuous or discrete.*

■ ■ ■ ■ **Analyzing Data and Evaluating the Hypotheses**

Following the data collection stage, researchers analyze their data and evaluate the hypotheses of the study. The data consist of codes and numbers used to represent our observations. In our example, each occupational group would be represented by two scores: (1) the percentage of women and (2) the average wage. If we had collected information on 100 occupations, we would end up with 200 scores, 2 per occupational group. However, the typical research project includes more variables; therefore, the amount of data the researcher confronts is considerably larger. We now must find a systematic way to organize these data, analyze them, and use some set of procedures to decide what they mean. These last steps make up the *statistical analysis* stage, which is the main topic of this textbook. It is also at this point in the research cycle where statistical procedures will help us *evaluate* our research hypothesis and assess the theory from which the hypothesis was derived.

Descriptive and Inferential Statistics

Statistical procedures can be divided into two major categories: *descriptive statistics* and *inferential statistics*. Before we can discuss the difference between these two types of statistics we need to understand the terms *population* and *sample*. A **population** is the total set of individuals, objects, groups, or events in which the researcher is interested. For example, if we were interested in looking at voting behavior in the last presidential election, we

[9]Ibid., p. 58.

would probably define our population as all citizens who voted in the election. If we wanted to understand the employment patterns of Hispanic women in our state, we would include in our population all Hispanic women in our state who are in the labor force.

Although we are usually interested in a population, quite often, because of limited time and resources, it is impossible to study the entire population. Imagine interviewing all the citizens of the United States who voted in the last election, or even all the Hispanic women who are in the labor force in our state. Not only would that be very expensive and time consuming, but we would probably have a very hard time locating everyone! Fortunately, we can learn a lot about a population if we carefully select a subset from that population. A subset selected from a population is called a **sample.** Researchers usually collect their data from a sample and then generalize their observations to the larger population.

> *Population* The total set of individuals, objects, groups, or events in which the researcher is interested.
>
> *Sample* A relatively small subset selected from a population.

Descriptive statistics includes procedures that help us organize and describe data collected from either a sample or a population such as a census. Occasionally data are collected on an entire population. **Inferential statistics**, on the other hand, is concerned with making predictions or inferences about a population from observations and analyses of a sample. For instance, the General Social Survey (GSS), from which numerous examples presented in this book are drawn, is conducted every year by the National Opinion Research Center (NORC) on a representative sample of about 1,500 respondents. The survey, which includes several hundred questions, is designed to provide social science researchers with a readily accessible database of socially relevant attitudes, behaviors, and attributes of a cross-section of the U.S. adult population. NORC has verified that the composition of the GSS samples closely resembles census data. But because the data are based on a sample rather than on the entire population, the average for the sample will not equal the average of the population as a whole. For example, in the 1996 GSS respondents were asked what they think is the ideal number of children for a family. Researchers found the average to be 2.8 children. This average will probably differ from the average of the population from which the GSS sample was drawn. The tools of statistical inference help determine the accuracy of the sample average obtained by the researchers.

> *Descriptive statistics* Procedures that help us organize and describe data collected from either a sample or a population.

> *Inferential statistics* The logic and procedures concerned with making predictions or inferences about a population from observations and analyses of a sample.

Evaluating the Hypotheses

At the completion of these descriptive and inferential procedures we can move to the next stage of the research process: the assessment and evaluation of our hypotheses and theories in light of the analyzed data. At this next stage new questions might be raised about unexpected trends in the data and about other variables that may have to be considered in addition to our original variables. For example, we may have found that the relationship between gender composition of occupations and earnings can be observed with respect to some groups of occupations but not others. Similarly, the relationship between these variables may apply for some racial/ethnic groups but not for others.

These findings will provide evidence to help us decide how our data relate to the theoretical framework that guided our research. We may decide to revise our theory and hypothesis to take account of these later findings. Recent studies are modifying what we know about gender segregation in the workplace. These studies suggest that race, as well as gender, shapes the occupational structure in the United States and helps explain disparities in income. This reformulation of the theory calls for a modified hypothesis and new research, which will start the circular process of research all over again.

Statistics provides an important link between theory and research. As our example on gender segregation demonstrates, the application of statistical techniques is an indispensable part of the research process. The results of statistical analyses help us evaluate our hypotheses and theories, discover unanticipated patterns and trends, and provide the impetus for shaping and reformulating our theories. Nevertheless, the importance of statistics should not diminish the significance of the preceding phases of the research process. Nor does the use of statistics lessen the importance of our own judgment in the entire process. Statistical analysis is a relatively small part of the research process, and even the most rigorous statistical procedures cannot speak for themselves. If our research questions are poorly conceived or our data are flawed due to errors in our design and measurement procedures, our results will be useless.

■ ■ ■ ■ **Looking at Social Differences**

By the middle of the next century, if current trends continue unchanged, the United States will no longer be a predominantly European society. Due mostly to renewed immigration and higher birth rates, the United States is being transformed into a "global society" in which nearly half the population will be African American, Asian American, Hispanic American, or Native American.

Is the increasing diversity of American society relevant to social scientists? What impact will such diversity have on the research methodologies we employ?

In a diverse society stratified by race, ethnicity, class, and gender, less partial and distorted explanations of social relations tend to result when researchers, research participants, and the research process itself reflect that diversity. Such diversity shapes the research questions we ask, how we observe and interpret our findings, and the conclusions we draw.

How does a consciousness of social differences inform social statistics? How can issues of race, class, gender, and other categories of experience shape the way we approach statistics? A statistical approach that focuses on social differences uses statistical tools to examine how variables such as race, class, and gender, as well as other categories of experience such as age, religion, and sexual orientation, shape our social world and explain social behavior. Numerous statistical procedures can be applied to describe these processes, and we will begin to look at some of those options in the next chapter. For now, let's preview briefly some of the procedures that can be employed to analyze social differences.

In Chapter 2, we will learn how to organize information using descriptive techniques, such as frequency distributions, percentage distributions, ratios, and rates. These statistical tools can also be employed to learn about the characteristics and experiences of groups in our society that have not been as visible as other groups. For example, in a series of special reports published by the U.S. Census Bureau over the past few years, these descriptive statistical techniques have been used to describe the characteristics and experiences of Native Americans, Hispanics, and the elderly in America.

In Chapter 3, we illustrate how graphic devices can highlight diversity. In particular, graphs help us to explore the differences and similarities among the many social groups coexisting within American society and emphasize the rapidly changing composition of the U.S. population. Using data published by the U.S. Census Bureau, we discuss various graphic devices that can be used to describe differences and similarities among elderly Americans. For instance, by employing a simple graphic device called a bar chart, we depict variations in the living patterns of the elderly and show that in every age category elderly females are more likely than elderly males to live alone. Another graphic device, called a time series chart, shows changes over time in the percentages of divorced European American, African American, and Hispanic women.

Whereas the similarities and commonalties in social experiences can be depicted using measures of central tendency (Chapter 4), the differences and diversity within social groups can be described using statistical measures of variation. For instance, we may want to analyze the changing age composition in the United States, or compare the degree of racial/ethnic or

Box 1.1   A Tale of Simple Arithmetic: How Culture May Influence How We Count

A second-grade schoolteacher posed this problem to the class: "There are four blackbirds sitting in a tree. You take a slingshot and shoot one of them. How many are left?"

"Three," answered the seven-year-old European with certainty. "One subtracted from four leaves three."

"Zero," answered the seven-year-old African with equal certainty. "If you shoot one bird, the others will fly away."*

*Working Woman, January 1991, p. 45.

religious diversity in the fifty states. Measures such as the standard deviation and the index of qualitative variation (IQV) are calculated for these purposes. For example, using IQV, we can demonstrate that Vermont and Maine are the least diverse states, whereas California is the most diverse (Chapter 5).

In Chapters 6 through 9, we review several methods of bivariate analysis, which are especially suited for examining the association between different social behaviors and variables such as race, class, ethnicity, gender, or religion. We use these methods of analysis to show not only how each of these variables operates independently in shaping behavior, but also how they interlock in shaping our experience as individuals in society.[10]

We will learn about inferential statistics in Chapters 11 through 15. Working with sample data, we examine the relationship between class, sex, or ethnicity and several social behaviors and attitudes. Inferential statistics, like the t-test and chi-square statistics, will help us decide how closely our samples resemble the population from which they are drawn.

Finally, a word of caution about all statistical applications. Whichever model of social research you use—whether you follow a traditional one or integrate your analysis with qualitative data, whether you focus on social differences or any other aspect of social behavior—remember that any application of statistical procedures requires a basic understanding of the statistical concepts and techniques. This introductory text is intended to familiarize you with the range of descriptive and inferential statistics widely applied in the social sciences. Our emphasis on statistical techniques should not diminish the importance of human judgment and your awareness of the person-made quality of statistics. Only with this awareness can statistics become a useful tool for viewing social life.

[10]Patricia Hill Collins, "Toward a New Vision: Race, Class and Gender as Categories of Analysis and Connection" (Keynote address at Integrating Race and Gender into the College Curriculum, a workshop sponsored by the Center for Research on Women, Memphis State University, Memphis, TN, 1989).

Box 1.2 Are You Anxious About Statistics?

Some of you are probably taking this introductory course in statistics with a great deal of suspicion and very little enthusiasm. The word *statistics* may make you anxious because you associate statistics with numbers, formulas, and abstract notations that seem inaccessible and complicated. It appears that statistics is not as integrated into the rest of your life as are other parts of the college curriculum.

Statistics is perhaps the most anxiety-provoking course in any social science curriculum. This anxiety often leads to a less than optimum learning environment, with students often trying to memorize every detail of a statistical procedure rather than trying to understand the general concept involved.

After many years of teaching statistics, I have learned that what underlies many of the difficulties students have in learning statistics is the belief that it involves mainly memorization of meaningless formulas.

There is no denying that statistics involves many strange symbols and unfamiliar terms. It is also true that you need to know some math to do statistics. But although the subject involves some mathematical computations, you will not be asked to know more than four basic operations: addition, subtraction, multiplication, and division. The language of statistics may appear difficult because these operations (and how they are combined) are written in a code that is unfamiliar to you. Those abstract notations are simply part of the language of statistics; much like learning any foreign language, you need to learn the alphabet before you can "speak the language." Once you understand the vocabulary and are able to translate the symbols and codes into terms that are familiar to you, you will feel more relaxed and begin to see how statistical techniques are just one more source of information.

For me, the key to enjoying and feeling competent in statistics is to frame anything I do in a familiar language and in a context that is relevant and interesting. Therefore, you will find that this book emphasizes intuition, logic, and common sense over rote memorization and derivation of formulas. I have found that this approach reduces statistics anxiety for most students and improves learning.

Another strategy that will help you develop confidence in your ability to do statistics is working with other people. This book encourages collaboration in learning statistics as a strategy designed to help you overcome statistics anxiety. Over the years I have learned that students who are intimidated by statistics do not like to admit it or talk about it. This avoidance mechanism may be an obstacle to overcoming statistics anxiety. Talking about your feelings with other students will help you realize that you are not the only one who suffers from fears of inadequacy about statistics. This sharing process is at the heart of the treatment of statistics anxiety, not because it will help you realize that you are not the "dumbest" one in the class after all, but because talking to others in a "safe" group setting will help you take risks and trust your own intuition and judgment. Ultimately, your judgment and intuition lie at the heart of your ability to translate statistical symbols and concepts into a language that makes sense and to interpret data using newly acquired statistical tools.*

*This discussion is based on Sheila Tobias' pioneering work on mathematics anxiety. See especially Sheila Tobias, *Overcoming Math Anxiety* (New York: Norton, 1978), Chapters 2 and 8.

MAIN POINTS

- Statistics are procedures used by social scientists to organize, summarize, and communicate information. Only information represented by numbers can be the subject of statistical analysis.

- The research process is a set of activities in which social scientists engage to answer questions, examine ideas, or test theories. It consists of the following stages: asking the research question, formulating the hypotheses, collecting data, analyzing data, and evaluating the hypotheses.

- A theory is an elaborate explanation of the relationship between two or more observable attributes of individuals or groups.

- Theories offer specific concrete predictions about the way observable attributes of people or groups would be interrelated in real life. These predictions, called hypotheses, are tentative answers to research problems.

- A variable is a property of people or objects that takes on two or more values. The variable the researcher wants to explain (the "effect") is called the dependent variable. The variable that is expected to "cause" or account for the dependent variable is called the independent variable.

- Three conditions are required to establish causal relations: (1) the cause has to precede the effect in time, (2) there has to be an empirical relationship between the cause and the effect, and (3) this relationship cannot be explained by other factors.

- At the nominal level of measurement, numbers or other symbols are assigned to a set of categories to name, label, or classify the observations. At the ordinal level of measurement, categories can be rank-ordered from low to high (or vice versa). At the interval-ratio level of measurement, measurements for all cases are expressed in the same unit.

- A population is the total set of individuals, objects, groups, or events in which the researcher is interested. A sample is a relatively small subset selected from a population.

- Descriptive statistics includes procedures that help us organize and describe data collected from either a sample or a population. Inferential statistics is concerned with making predictions or inferences about a population from observations and analyses of a sample.

KEY TERMS

data	*independent variable*
dependent variable	*inferential statistics*
descriptive statistics	*interval-ratio measurement*
empirical research	*nominal measurement*
hypothesis	*ordinal measurement*

population

research process

sample

statistics

theory

unit of analysis

variable

SPSS DEMONSTRATION

Introduction to Data Sets and Variables

We'll be using a set of computer data and exercises at the end of each chapter. All computer exercises will be based on the program SPSS 8.0 for Windows. There are two versions of the program: a standard version with no limits on the number of variables or cases, and a student version with a limit of 1,500 cases and 50 variables. Confirm with your instructor which SPSS version is available at your university.

The GSS96.SAV contains a selection of variables and cases from the complete 1996 General Social Survey. Those of you with the student version of SPSS 9.0 will work with two separate GSS files: GSS Module A features gender and family issues, and GSS Module B highlights race and poverty issues.

Appendix F describes the General Social Survey 1996 in greater detail. Appendix E explains the basic operation and procedures for SPSS for Windows. We strongly recommend that you refer to this appendix before beginning the SPSS exercises.

When you begin to use a data set, you should take the time to review your variables. What are the variables called? What do they measure? What do they mean? There are several ways to do this.

In order to review your data, you must first open the data file. Files are opened in the SPSS by clicking on *File*, then *Open*. After switching directories and drives to the appropriate location of the files (which may be on a hard disk or on a floppy drive), you select one data file and click on *OK*. This routine is the same each time you open a data file. SPSS automatically opens each data file in the SPSS Data Editor Window. We'll use GSS96A.SAV for this demonstration.

One way to review the complete list of variables in a file is to click on the *Utilities* choice from the main menu, then on *Variables* in the list of submenu choices. A dialog box should open (as depicted in Figure 1.5). The SPSS variable names, which are limited to eight characters or less, are listed in the scroll box (left column). When a variable name is highlighted, the descriptive label for that variable is listed, along with any missing values and, if available, the value labels for each variable category. Variables will be listed in alphabetical order.

A second way to review all variables is to click on *Utilities*, then on *File Info*. This choice tells SPSS to put the variable definition information in the Output–SPSS Viewer window (Figure 1.6). You can scroll up and down in the Output window to see the variables, and you can print the complete

Figure 1.5

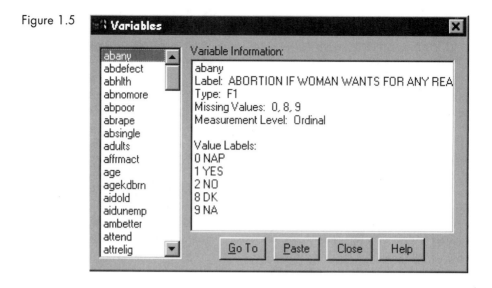

Output window to keep a printed version of the information. Variables will be listed in the order they appear in the Data Editor window.

SPSS PROBLEM

Based on the *Utilities-Variables* option, review the variables from the GSS96 (or Module A or B). Can you identify three nominal and three ordinal variables? Based on the information in the dialog box, you should be able to identify the variable name, variable label, and category values.

CHAPTER EXERCISES

1. In your own words, explain the relationship of data (collecting and analyzing) to the research process. (Refer to Figure 1.1.)

2. Construct potential hypotheses or research questions to relate the variables in each of the following examples. Also, write a brief statement explaining why you believe there is a relationship between the variables as specified in your hypotheses.
   a. Gender and educational level
   b. Income and race
   c. The crime rate and the number of police in a city
   d. Life satisfaction and age
   e. A nation's military expenditures as a percentage of its gross national product (GNP) and that nation's overall level of security
   f. Care of elderly parents and race

Figure 1.6

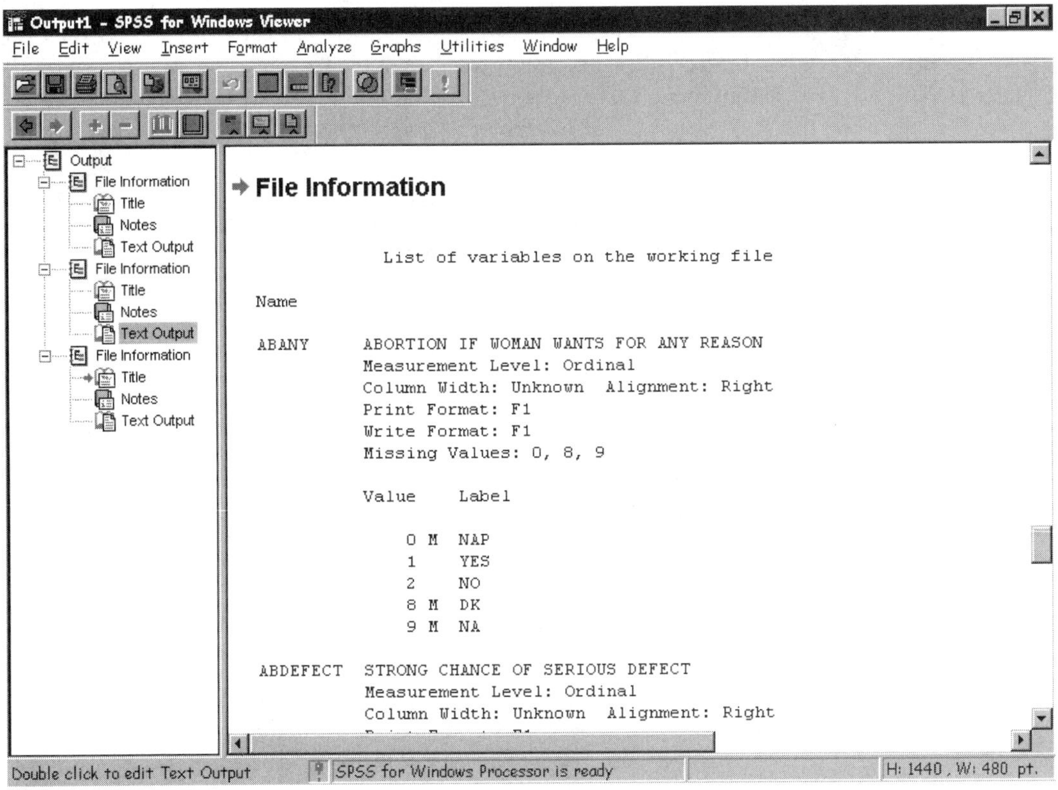

```
██ Output1 - SPSS for Windows Viewer                                    _ ₈ ×

 File  Edit  View  Insert  Format  Analyze  Graphs  Utilities  Window  Help

 [toolbar icons]

 [toolbar icons]

□ [回] Output                ▲
 └ [回] File Information       ⇒ File Information
    └ [图] Title
       [图] Notes
       [图] Text Output                    List of variables on the working file
 └ [回] File Information
    └ [图] Title              Name
       [图] Notes
       [图] Text Output       ABANY     ABORTION IF WOMAN WANTS FOR ANY REASON
 └ [回] File Information                 Measurement Level: Ordinal
    ⇒[图] Title                          Column Width: Unknown  Alignment: Right
       [图] Notes                        Print Format: F1
       [图] Text Output                  Write Format: F1
                                         Missing Values: 0, 8, 9


                           Value     Label

                               0 M   NAP
                               1     YES
                               2     NO
                               8 M   DK
                               9 M   NA


                           ABDEFECT  STRONG CHANCE OF SERIOUS DEFECT
                                     Measurement Level: Ordinal
                                     Column Width: Unknown  Alignment: Right

 Double click to edit Text Output   [?] SPSS for Windows Processor is ready    H: 1440 , W: 480 pt.
```

3. Determine the level of measurement for each of the following variables.
   a. The number of people in your family
   b. Place of residence, classified as urban, suburban, or rural
   c. The percentage of university students who attended public high school
   d. The rating of the overall quality of a textbook, on a scale from "Excellent" to "Poor"
   e. The type of transportation a person takes to work (for example, bus, walk, car)
   f. The highest educational degree earned
   g. The U.S. unemployment rate

4. For each of the variables in Exercise 3 that you classified as interval-ratio, identify whether it is discrete or continuous.

5. Why do you think men and women, on average, do not earn the same amount of money? Develop your own theory to explain the difference. Use three independent variables in your theory, with annual income as

your dependent variable. Construct hypotheses to link each independent variable with your dependent variable.

6. For each of the following examples, indicate whether it involves the use of descriptive or inferential statistics. Justify your answer.
   a. Estimating the number of unemployed people in the United States
   b. Asking all the students at a college their opinion about the quality of food at the cafeteria
   c. Determining the incidence of breast cancer among Asian women
   d. Conducting a study to determine the rating of the quality of a new automobile, gathered from 1,000 new buyers
   e. The average salaries of various categories of employees (for example, tellers and loan officers) at a large bank
   f. Any change in the number of immigrants coming to the United States from Asian countries between 1990 and 1999

7. Identify three social problems or issues that can be investigated with statistics. (One example of a social problem is hate crimes.) Which one of the three issues would be the most difficult to study? Why? Which would be the easiest?

8. Construct measures of political participation at the nominal, ordinal, and interval-ratio levels. (*Hint:* You can use such behaviors as voting frequency or political party membership.) Discuss the advantages and disadvantages of each.

9. Variables can be measured according to more than one level of measurement. For the following variables, identify at least two levels of measurement. Is one level of measurement better than another? Explain.
   a. Individual age
   b. Annual income
   c. Religiosity
   d. Student performance
   e. Social class
   f. Attitude toward affirmative action

10. In a 1998 study Canadian researchers concluded, "Parental education plays a significant role in children's ability to match or improve upon their parents' educational attainment."[11]
    a. Identify the dependent and independent variables in this statement.
    b. Review the chapter discussion on cause-and-effect relationships (see pages 9–11). Are all three conditions met by the relationship identified in this study?
    c. What other variables may be involved in the relationship between the dependent and independent variables? Explain.

[11]*Canadian Social Trends* 49 (Summer 1998): 15.

2 Organization of Information: Frequency Distributions

Introduction

Frequency Distributions

Proportions and Percentages

Percentage Distributions

Comparisons

Statistics in Practice: Labor Force Participation of Native Americans

The Construction of Frequency Distributions
Frequency Distributions for Nominal Variables
Frequency Distributions for Ordinal Variables
Frequency Distributions for Interval-Ratio Variables

Cumulative Distributions

Box 2.1 Real Limits, Stated Limits, and Midpoints of Class Intervals

Rates

Statistics in Practice: Marriage and Divorce Rates over Time

Reading the Research Literature: Statistical Tables
Basic Principles
Tables with a Different Format

Conclusion

MAIN POINTS
KEY TERMS
SPSS DEMONSTRATIONS
SPSS PROBLEMS
CHAPTER EXERCISES

■ ■ ■ ■ **Introduction**

As social researchers we often have to deal with very large amounts of data. For example, in a typical survey, by the completion of your data collection phase, you will have accumulated thousands of individual responses represented by a jumble of numbers. To make sense out of these data you will have to organize and summarize them in some systematic fashion. The most basic method for organizing data is to classify the observations into a frequency distribution. A **frequency distribution** is a table that reports the number of observations that fall into each category of the variable we are analyzing. Constructing a frequency distribution is usually the first step in the statistical analysis of data.

> *Frequency distribution* A table reporting the number of observations falling into each category of the variable.

■ ■ ■ ■ **Frequency Distributions**

Let's begin with an example of a frequency distribution of a variable described in a study on Native Americans. When Columbus first encountered "the original inhabitants of the Americas," people he later described as "Indios," nothing was known about their numbers, where they lived, or the characteristics of their social structure.[1] During the past 200 years we have gathered a wealth of information about other immigrant groups who settled in North America. In comparison, until a few years ago we knew little more about Native Americans than Columbus did 500 years ago.

In 1980 the U.S. Bureau of the Census went to great lengths to collect data on the Native American population. Because so little is known about their contemporary experiences, this information presents an opportunity for significantly advancing the current state of knowledge. A summary and analysis of these data appeared in a book published in 1989, *American Indians: The First of This Land*.[2] In its exposition, the book relies heavily on some of the techniques we will discuss in this chapter.

Native Americans are not one group but many extremely diverse groups. Today there are about 200 different Native American tribes characterized by distinct lifestyles and cultural practices. Therefore, the question of who is a Native American is not a simple matter: Native American identity also depends on personal perception of Indian race and ethnicity. Table 2.1 shows the frequency distribution of the variable *identity categories of Native Americans*. These categories are based on several different patterns of

---

[1]C. Matthew Snipp, *American Indians: The First of This Land*. New York: Russell Sage Foundation, 1989, 1.

[2]Ibid.

Table 2.1   **Frequency Distribution for Categories
of Native American Identity**

Identity	Frequency (f)
Native American	947,500
Native American of multiple ancestry	269,700
Native American of Indian descent	5,537,600
Total (N)	6,754,800

Source: Adapted from C. Matthew Snipp, *American Indians:
The First of This Land* (New York: Russell Sage Foundation, 1989),
p. 51. Based on 1980 census data.

self-identified race and ethnic background. The first category includes persons who identify their race and their ethnic background as Native American. This group is referred to as "Native American." The second category includes persons who report their race as Native American but include non-Indian ancestry in their ethnic background. The designation for these individuals is "Native American of multiple ancestry." A third category contains persons who cite a non-Indian race yet claim Native American ancestry for their ethnic background.[3] These individuals are classified as "Native Americans of Indian descent."

Notice that the frequency distribution is organized in a table, which has a number (2.1) and a descriptive title. The title indicates the kind of data presented: "Categories of Native American Identity." The table consists of two columns. The first column identifies the variable (categories of Native American identity) and its categories (Native American, Native American of multiple ancestry, and Native American of Indian descent). The second column, headed "Frequency (f)," tells the number of cases in each category as well as the total number of cases ($N = 6,754,800$). Notice also that the source of the table is clearly identified as a source note in the table. It shows that the table was adapted from a book by C. Matthew Snipp, *American Indians: The First of This Land,* and that the data come from the 1980 census. In general, the source of data for a table should appear as a source note in the table unless it is clear from the general discussion of the data.

What can you learn from the information presented in Table 2.1? The table shows that, as of 1980, approximately 6.8 million persons (6,754,800) reported that their race and/or ethnic ancestry was Native American. Out of this group, the majority, about 5.5 million persons (5,537,600) cite a non-Indian race yet claim Native American ancestry for their ethnic background (Native American of Indian descent), another 947,500 report both their race and ethnic background as Native American, and the remaining 269,700 include persons who identify themselves as Native American but are of multiple ancestry (Native American of multiple ancestry).

[3]Ibid., pp. 50–51.

> **Learning Check.** *You will see frequency distributions throughout this book. Take the time to familiarize yourself with the parts in this basic example; they will get more complicated as we go on.*

■ ■ ■ ■ Proportions and Percentages

Frequency distributions are helpful in presenting information in a compact form. However, when the number of cases is large, the frequencies may be difficult to grasp. Even though there is nothing wrong in concluding that out of 6.8 million Native Americans 5.5 million cite a non-Indian race yet claim Native American ancestry for their ethnic background, most of us find it difficult to think of the relative sizes of such large numbers. To standardize these raw frequencies, we can translate them into *relative frequencies*—that is, *proportions* or *percentages*.

A **proportion** is a relative frequency obtained by dividing the frequency in each category by the total number of cases. To find a proportion (P) divide the frequency (f) in each category by the total number of cases (N):

$$P = \frac{f}{N} \tag{2.1}$$

where

f = frequency
N = total number of cases

Thus, the proportion of Native American respondents who in 1980 identified themselves as simply "Native American" is

$$\frac{947,500}{6,754,800} = .14$$

The proportion who identified themselves as "Native American of multiple ancestry" is

$$\frac{269,700}{6,754,800} = .04$$

and the proportion who identified themselves as "Native American of Indian descent" is

$$\frac{5,537,600}{6,754,800} = .82$$

Proportions should always sum to 1.00 (allowing for some rounding errors). Thus, in our example the sum of the three proportions is

$$.14 + .04 + .82 = 1.00$$

To determine a frequency from a proportion, we simply multiply the proportion by the total N:

$$f = P(N) \tag{2.2}$$

Thus, the frequency of Native American respondents who in 1980 identified themselves as simply "Native American" can be calculated as

$.14(6,754,800) = 945,672$

Note that the obtained frequency differs somewhat from the actual frequency of 947,500. This difference is due to rounding of the proportion. If we use the actual proportion instead of the rounded proportion, we obtain the correct frequency:

$.140270622(6,754,800) = 947,500$

Learning Check. *Compare group A with group B in Figure 2.1 and answer the following questions: Which group has the greatest number of women? Which group has the largest proportion of women?*

Figure 2.1 **Numbers and Proportions**

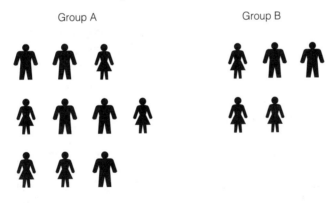

We can also express frequencies as percentages. A **percentage** is a relative frequency obtained by dividing the frequency in each category by the total number of cases and multiplying by 100. In most statistical reports, frequencies are presented as percentages rather than proportions. Percentages express the size of the frequencies as if there were a total of 100 cases.

To calculate a percentage, simply multiply the proportion by 100:

$$\text{Percentage } (\%) = \frac{f}{N}(100) \tag{2.3}$$

or

$$\text{Percentage } (\%) = P(100) \qquad\qquad (2.4)$$

Thus, the percentage of Indian respondents who identified themselves as "Native American" is

.14(100) = 14%

The percentage who identified themselves as "Native American of multiple ancestry" is

.04(100) = 4%

and the percentage who identified themselves as "Native American of Indian descent" is

.82(100) = 82%

Proportion   A relative frequency obtained by dividing the frequency in each category by the total number of cases.

Percentage   A relative frequency obtained by dividing the frequency in each category by the total number of cases and multiplying by 100.

Learning Check.   *Calculate the proportion of males and females in your statistics class. What percentage is female?*

■ ■ ■ ■   **Percentage Distributions**

Percentages are usually displayed as *percentage distributions*. A **percentage distribution** is a table showing the percentage of observations falling into each category of the variable. For example, Table 2.2 presents the frequency distribution of categories of Native American identity (Table 2.1) along with the corresponding percentage distribution. Percentage distributions (or proportions) should always show the base (N) on which they were computed. Thus, in Table 2.2 the base on which the percentages were computed is $N = 6{,}754{,}800$.

Percentage distribution   A table showing the percentage of observations falling into each category of the variable.

Table 2.2 **Frequency and Percentage Distributions for Categories of Native American Identity**

Identity	Frequency (f)	Percentage (%)
Native American	947,500	14
Native American of multiple ancestry	269,700	4
Native American of Indian descent	5,537,600	82
Total (N)	6,754,800	100

Source: Adapted from C. Matthew Snipp, *American Indians: The First of This Land* (New York: Russell Sage Foundation, 1989), p. 51. Based on 1980 census data.

■ ■ ■ ■ **Comparisons**

In Table 2.2 we illustrated that there are three different categories of Native American identity: Native American, Native American of multiple ancestry, and Native American of Indian descent. These distinctions among the Native American population raise important questions about our understanding of who is considered Native American. For instance, population estimates vary considerably from fewer than 1 million Native Americans, if we restrict our definition to only persons who consistently identify their race and ethnicity as Native American, to almost 7 million persons if all three identity categories are pooled.

The decision to consider these groups separately or to pool them in considering the Native American population depends to a large extent on our research question. For instance, we know that 274,775 persons who identify their race and ethnicity as Native American were below the poverty line in 1980, and 288,988 were not in the labor force during the same period. Are these figures high or low? What do they tell us about the socioeconomic characteristics of Native Americans?

To answer these questions and determine whether the three categories of Native American identity have markedly different social characteristics we need to *compare* them. How do the numbers describing the poverty levels and unemployment rates of self-identified Native Americans compare with those numbers for Native Americans of multiple ancestry, Native Americans of Indian descent, or the population at large?

As students, as social scientists, and even as consumers, we are frequently faced with problems that call for some way to make a clear and valid comparison. For example, in 1993, 30.3 percent of elderly Americans lived alone.[4] Is this figure high or low? In 1997, 34.5 percent of those

[4]U.S. Bureau of the Census, "65+ in America," April 1996, P23-190.

between 25 and 34 years of age had never been married.[5] Does this reflect a change in the American family? In each of these cases comparative information is required to answer the question and reach a conclusion.

These examples also illustrate several ways in which comparisons can be made. What we compare depends largely on the question we pose. Without a clearly formulated research question, it is difficult to decide which type of comparison to make.

Several types of comparison are quite common in the social sciences. One type is the comparison between groups that have different characteristics—for example, comparisons between old and young, between white and Hispanic, or as in our chapter example, between different categories of Native American identity. Sometimes we may be interested in looking at regional differences among groups or in comparing groups from different segments of society. You may have read news stories about contrasts in voting patterns between the North and the South, or the percentage of homeowners in central cities and suburbs. Also, we may be interested in comparing changes in the same group over time, such as the percentage change in foreign-born residents in the United States over the past decade or how the population has shifted from the cities toward the suburbs.

■ ■ ■ ■ **Statistics in Practice:**
Labor Force Participation of Native Americans

Very often we are interested in comparing two or more groups that differ in size. Percentages are especially useful for making such comparisons. For example, we know that differences in socioeconomic status mark divisions between populations, indicating differential access to economic opportunities. Labor participation is an important indicator of access to economic opportunities and is strongly associated with socioeconomic status. Table 2.3 shows the raw frequency distributions for the variable *labor force participation* for all three categories of Native American identity.

Which group has the highest relative number of persons who are not in the labor force? Because of the differences in the population sizes of the three groups, this is a difficult question to answer based on only the raw frequencies. To make a valid comparison we have to compare the percentage distributions for all three groups. These are presented in Table 2.4. Notice that the percentage distributions make it easier to identify differences between the groups. Compared with Native Americans of multiple ancestry or Native Americans of Indian descent, Native Americans have the highest percentage not in the labor force (30.5% versus 21.6% and 19.9%). Conversely, among the three groups, Native Americans of Indian descent have the highest percentage (73.4% versus 68.1% and 61.6%) of persons who are employed.

[5]U.S. Bureau of the Census, "Marital Status and Living Arrangements," March 1996, P20-496.

Table 2.3   **Labor Force Participation Among Householders of Native American Background (raw frequencies)**

Labor Force Participation	Native American	Native American of Multiple Ancestry	Native American of Indian Descent
Employed	583,660	183,666	4,064,598
Unemployed	63,482	21,846	282,418
Not in labor force	288,988	58,255	1,101,982
Military and other	11,370	5,933	88,602
Total (N)	947,500	269,700	5,537,600

Source: Adapted from C. Matthew Snipp, *American Indians: The First of This Land* (New York: Russell Sage Foundation, 1989), p. 55. Based on 1980 census data.

Learning Check.   *Examine Table 2.4 and answer the following questions: What is the percentage of Native Americans who are employed? What is the base (N) for this percentage? What is the percentage of Native Americans of multiple ancestry who are not in the labor force? What is the base (N) for this percentage?*

Whenever one group is compared with another, the most meaningful conclusions can usually be drawn based on comparison of the relative frequency distributions. In fact, we are seldom interested in a single distribution. Most interesting questions in the social sciences are about differences

Table 2.4   **Labor Force Participation Among Householders of Native American Background (in percentages)**

Labor Force Participation	Native American	Native American of Multiple Ancestry	Native American of Indian Descent
Employed	61.6%	68.1%	73.4%
Unemployed	6.7%	8.1%	5.1%
Not in labor force	30.5%	21.6%	19.9%
Military and other	1.2%	2.2%	1.6%
Total	100.0%	100.0%	100.0%
(N)	947,500	269,700	5,537,600

Source: Adapted from C. Matthew Snipp, *American Indians: The First of This Land* (New York: Russell Sage Foundation, 1989), p. 55. Based on 1980 census data.

between two or more groups.[6] The finding that the labor force participation patterns of Native Americans vary depending on their race and ethnic identity raises a serious doubt about whether American Indians can be legitimately regarded as a single, relatively homogeneous ethnic group. Further analyses could examine *why* differences in Native American identity are associated with differences in labor force participation patterns. Other variables that explain these differences could be identified. These kinds of questions can be answered using more complex multivariate statistical techniques that involve more than two variables. The comparison of percentage distributions is an important foundation for those more complex techniques.

Before we continue, keep in mind that although we encourage you to begin thinking analytically about complex data, the basic procedures we'll review in the first ten chapters only allow you to draw some tentative conclusions about differences between groups. To make valid comparisons you will need to consider the more complex techniques of sampling and statistical inference, which are discussed in the last five chapters. As you proceed through this book and master all the statistical concepts necessary for valid inference, you will be able to provide more complex interpretations.

■ ■ ■ ■ **The Construction of Frequency Distributions**

Up to now you have been introduced to the general concept of a frequency distribution. We saw that data can be expressed as raw frequencies, proportions, or percentages. We also saw how to use percentages to compare distributions in different groups.

In this section you will learn how to construct frequency distributions. While most often this can be done by your computer, it is important to go through the process to understand how frequency distributions are actually put together.

For nominal and ordinal variables, constructing a frequency distribution is quite simple. Count and report the number of cases that fall into each category of the variable along with the total number of cases (N). For the purpose of illustration, let's take a small subsample of forty cases from our 1996 GSS sample and record their scores on the following variables: *gender,* a nominal level variable; *happiness,* an ordinal level variable; *age* and *number of children,* both interval-ratio level variables.

The gender of the respondents was recorded by the interviewer at the beginning of the interview. To measure happiness, respondents were asked, "Taken all together, how would you say things are these days—would you say that you are very happy, pretty happy, or not too happy?" Three categories (very happy, pretty happy, and not too happy) were offered for response. The first category represented the highest level of happiness. Respondent age was calculated based on the respondent's birth year. The

---

[6]David Knoke and George W. Bohrnstedt, *Basic Social Statistics* (New York: Peacock Publishers, 1991), p. 25.

number of children was determined by the question "How many children have you ever had?" The answers given by a subsample of forty respondents are displayed in Table 2.5. Note that each row in the table represents a respondent, whereas each column represents a variable. This format is conventional in the social sciences.

Table 2.5 **A GSS Subsample of Forty Respondents**

Sex of Respondent	General Happiness	Number of Children	Age of Respondent
Male	Very happy	2	34
Male	Pretty happy	7	63
Male	Pretty happy	1	33
Male	Not too happy	1	49
Male	Pretty happy	0	40
Male	Very happy	1	32
Male	Pretty happy	3	38
Male	Pretty happy	0	26
Male	Very happy	5	68
Male	Pretty happy	7	42
Male	Pretty happy	1	53
Male	Pretty happy	3	46
Male	Pretty happy	3	58
Male	Pretty happy	6	62
Male	Very happy	2	45
Male	Pretty happy	2	27
Male	Very happy	2	62
Male	Very happy	2	44
Male	Pretty happy	0	48
Male	Very happy	2	30
Male	Pretty happy	2	49
Male	Pretty happy	2	51
Male	Pretty happy	5	77
Male	Pretty happy	0	49
Male	Pretty happy	3	37
Male	Not too happy	1	20
Male	Pretty happy	2	53
Male	Very happy	2	67
Female	Not too happy	4	46
Female	Pretty happy	2	35
Female	Pretty happy	2	52
Female	Pretty happy	2	26
Female	Pretty happy	1	43
Female	Not too happy	5	74
Female	Not too happy	1	35
Female	Pretty happy	6	54
Female	Very happy	2	47
Female	Pretty happy	0	37
Female	Not too happy	1	25
Female	Pretty happy	1	38

You can see that it is going to be difficult to make sense of these data just by eyeballing Table 2.5. How many of these forty respondents are males? How many said they are very happy? How many were older than 50 years of age? To answer these questions, we will construct the frequency distributions for all four variables.

Frequency Distributions for Nominal Variables

Let's begin with the nominal variable, *gender*. First, we tally the number of males, then the number of females (the column of tallies has been included in Table 2.6 for the purpose of illustration). The tally results are then used to construct the frequency distribution presented in Table 2.6. The table has a title describing its content (Frequency Distribution of the Variable *Gender*: GSS Subsample). Its categories (male and female) and their associated frequencies are clearly listed; in addition, the total number of cases (*N*) is also reported. The Percentage column is the percentage distribution for this variable. To convert the Frequency column to percentages, simply divide each frequency by the total number of cases and multiply by 100. Percentage distributions are routinely added to almost any frequency table and are especially important if comparisons with other groups are to be considered. Immediately we can see that it is easier to read the information. There are 12 females and 28 males in this sample. Based on this frequency distribution, we can also conclude that the majority of sample respondents are male.

Learning Check. *Construct a frequency and percentage distribution for males and females in your statistics class.*

Table 2.6 **Frequency Distributions of the Variable *Gender*: GSS Subsample**

Gender	Tallies	Frequency (*f*)	Percentage (%)
Male	|||| |||| |||| |||| |||| |||	28	70
Female	|||| |||| ||	12	30
Total (*N*)		40	100

Frequency Distributions for Ordinal Variables

To construct a frequency distribution for ordinal level variables, follow the same procedures outlined for nominal level variables. Table 2.7 presents the

Table 2.7 **Frequency Distribution of the Variable *Happiness:* GSS Subsample**

Gender	Tallies	Frequency (*f*)	Percentage (%)
Very happy	ЖІ IIII	9	23
Pretty happy	ЖІ ЖІ ЖІ ЖІ ЖІ	25	63
Not too happy	ЖІ I	6	15
Total (*N*)		40	100

frequency distribution for the variable *happiness*. The table shows that 63 percent, a majority, felt that they were "pretty happy" with their current life.

The major difference between frequency distributions for nominal and ordinal variables is the order in which the categories are listed. The categories for nominal level variables do not have to be listed in any particular order. For example, we could list females first and males second without changing the nature of the distribution. Because the categories or values of ordinal variables are rank-ordered, however, they must be listed in a way that reflects their rank—from the lowest to the highest or from the highest to the lowest. Thus, the data on happiness in Table 2.7 are presented in declining order from "very happy" (the highest level of happiness) to "not too happy" (the least happy category).

Learning Check. *Figures 2.2, 2.3, and 2.4 illustrate the gender and happiness data in stages as presented in Tables 2.5, 2.6, and 2.7. To convince yourself that classifying the respondents by gender (Figure 2.3) and by happiness (Figure 2.4) makes the job of counting much easier, turn to Figure 2.2 and answer these questions: How many men are in the group? How many women? How many said they are very happy? Now turn to Figure 2.3: How many men are in the group? Women? Finally, examine Figure 2.4: How many said they were pretty happy? Very happy?*

Frequency Distributions for Interval-Ratio Variables

We hope you agree by now that constructing frequency distributions for nominal and ordinal level variables is rather straightforward. Simply list the categories and count the number of observations that fall into each category. Building a frequency distribution for interval-ratio variables with relatively few values is also easy. For example, when constructing a frequency distribution for *number of children*, simply list the number of children and report the corresponding frequency, as shown in Table 2.8.

Figure 2.2 **Forty Respondents from the GSS Subsample, Their Gender, and Their Level of Happiness (see Table 2.5)**

Figure 2.3   **Forty Respondents from the GSS Subsample, Classified by Gender (see Table 2.6)**

Figure 2.4   **Forty Respondents from the GSS Subsample, Classified by Gender and Level of Happiness (see Table 2.7)**

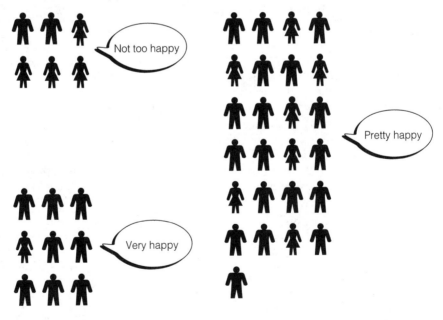

Table 2.8 **Frequency Distribution of the Variable *Number of Children*: GSS Subsample**

Number of Children	Frequency (*f*)	Percentage (%)
0	5	13
1	9	23
2	14	35
3	4	10
4	1	3
5	3	8
6	2	5
7	2	5
Total (*N*)	40	100

Very often interval-ratio variables have a wide range of values, which makes simple frequency distributions very difficult to read. For example, take a look at the frequency distribution for the variable *age* in Table 2.9. The distribution contains age values ranging from 20 to 77 years. For a more concise picture, the large number of different scores could be reduced into a smaller number of groups, each containing a range of scores. Table 2.10 displays such a grouped frequency distribution of the data in Table 2.9. Each group, known as a *class interval,* now contains ten possible scores instead of one. Thus, the ages of 20, 25, 26, and 27 all fall into a single class interval of 20–29. The second column of Table 2.10, Frequency, tells us the number of respondents that fall into each of the intervals—for example, that 5 respondents fall into the class interval of 20–29. Having grouped the scores, we can clearly see that the biggest single age group is between 40 and 49 years (12 out of 40, or 30% of sample). The percentage distribution we have added to Table 2.10 displays the relative frequency of each interval and emphasizes this pattern as well.

The decision as to how many groups to use and, therefore, how wide the intervals should be is usually up to the researcher and depends on what makes sense in terms of the purpose of the research. The rule of thumb is that an interval width should be large enough to avoid too many categories but not so large that significant differences between observations are concealed.[7] Obviously, the number of intervals depends on the width of each. For instance, if you are working with scores ranging from 10 to 60 and you establish an interval width of 10, you will have 5 intervals.

[7]Ibid., p. 41.

Table 2.9 **Frequency Distribution of the Variable *Age:* GSS Subsample**

Age of Respondent	Frequency	Age of Respondent	Frequency
20	1	46	2
25	1	47	1
26	2	48	1
27	1	59	3
30	1	51	1
32	1	52	1
33	1	53	2
34	1	54	1
35	2	58	1
37	2	62	2
38	2	63	1
40	1	67	1
42	1	68	1
43	1	74	1
44	1	77	1
45	1		

Table 2.10 **Grouped Frequency Distribution of the Variable *Age:* GSS Subsample**

Age	Frequency (*f*)	Percentage (%)
20–29	5	12.5
30–39	10	25
40–49	12	30
50–59	6	15
60–69	5	12.5
70–79	2	5
Total (N)	40	100

> ***Learning Check.***   *Can you verify that Table 2.10 was constructed correctly? Use Table 2.9 to determine the frequency of cases that fall into the categories of Table 2.10.*

> ***Learning Check.***   *If you are having trouble distinguishing between nominal, ordinal, and interval-ratio variables, go back to Chapter 1 and review the section on levels of measurement. The distinction between these three levels of measurement will be important throughout the book.*

■ ■ ■ ■ **Cumulative Distributions**

Sometimes we may be interested in locating the relative position of a given score in a distribution. For example, we may be interested in finding out how many or what percent of our sample was younger than 40 or older than 60. Frequency distributions can be presented in a cumulative fashion to answer such questions. A **cumulative frequency distribution** shows the frequencies at or below each category of the variable.

Cumulative frequencies are appropriate only for variables that are measured at an ordinal level or higher. They are obtained by adding to the frequency in each category the frequencies of all the categories below it.

Let's look at Table 2.11. It shows the cumulative frequencies based on the frequency distribution from Table 2.10. The cumulative frequency column, denoted by Cf, shows the number of persons at or below each interval. For example, you can see that 15 of the 40 respondents were 39 years old or younger, and 33 of the 40 were 59 or younger.

To construct a cumulative frequency distribution, start with the frequency in the lowest class interval (or with the lowest score, if the data are ungrouped) and add to it the frequencies in the next highest class interval. Continue adding the frequencies until you reach the last class interval. The cumulative frequency in the last class interval will be equal to the total number of cases (N). In Table 2.11 the frequency associated with the first class interval (20–29) is 5. The cumulative frequency associated with this interval is also 5 since there are no cases below this class interval. The frequency for the second class interval is 10. The cumulative frequency for this interval is 5 + 10 = 15. To obtain the cumulative frequency of 27 for the third interval we add its frequency (12) to the cumulative frequency associated with the second class interval (15). Continue this process until you reach the last class interval. Therefore, the cumulative frequency for the last interval is equal to 40, the total number of cases (N).

We can also construct a cumulative percentage distribution ($C\%$), which has wider applications than the cumulative frequency distribution (Cf).

Table 2.11   **Frequency Distribution and Cumulative Frequency for the Variable *Age:* GSS Subsample**

Age	Frequency (f)	Cf
20–29	5	5
30–39	10	15
40–49	12	27
50–59	6	33
60 years or older	7	40
Total (N)	40	

Box 2.1 Real Limits, Stated Limits, and Midpoints of Class Intervals

The class intervals presented in Table 2.10 constitute the categories of the variable *age* that we used to classify the survey's respondents. In Chapter 1 we noted that our variables need to be both exhaustive and mutually exclusive. These principles apply to the class intervals here as well. This means that each of the forty respondents can be classified into one and only one category. In addition, we should be able to classify all the possible scores.

In our example, these requirements are met: Each observation score fits into only one class interval, and there is an appropriate category to classify each individual score as recorded in Table 2.10. However, if you looked closely at Table 2.10 you may have noticed that there is actually a gap of one year between adjacent intervals. A gap could create a problem with scores that have fractional values. For example, let's suppose for a moment that respondent's age had been reported with more precision. Where would you classify a woman who was 49.25 years old? Notice that her age would actually fall between the intervals 40–49 and 50–59! To avoid this potential problem, use the real limits shown in the following table rather than the stated limits listed in Table 2.10.

Real limits extend the upper and lower limits of the intervals by .5. For instance, the real limits for the interval 40–49 are 39.5–49.5; the real limits for the interval 50–59 are 49.5–59.5; and so on. (Scores that fall exactly at the upper real limit or the lower real limit of the interval [for example, 59.5 or 49.5] are usually rounded to the closest even number. The number 59.5 would be rounded to 60 and would thus be included in the interval 59.5–69.5.) In the following table, we include both the stated limits and real limits for the grouped frequency distribution of respondent's age. So where would you classify a respondent who was 49.25 years old? (Answer: in the interval 39.5–49.5) How about 19.9? (in the interval 19.5–29.5)

The *midpoint* is a single number that represents the entire interval. A midpoint is calculated by adding the lower and upper real limits of the interval and dividing by 2. The midpoint of the interval 19.5–29.5, for instance, is (19.5 + 29.5) ÷ 2 = 24.5. The midpoint for all the intervals of the table are displayed in the third column.

Even though grouped frequency distributions are very helpful in summarizing information, remember that they are only a summary and therefore involve a considerable loss of detail. Since most researchers and students have access to computers, grouped frequencies are used only when the raw data are not available. Most of the statistical procedures described in later chapters are based on the raw scores.

Respondent's Age Stated Limits	Respondent's Age Real Limits	Midpoint	Frequency (f)
20–29	19.5–29.5	24.5	5
30–39	29.5–39.5	34.5	10
40–49	39.5–49.5	44.5	12
50–59	49.5–59.5	54.5	6
60–69	59.5–69.5	64.5	5
70–79	69.5–79.5	74.5	2
Total (N)			40

Table 2.12 **Frequency Distribution and Cumulative Percentages for the Variable**
Age: GSS Subsample

Age	Frequency (f)	Percentage (%)	C%
20–29	5	12.5	12.5
30–39	10	25	37.5
40–49	12	30	67.5
50–59	6	15	82.5
60 years or older	7	17.5	100
Total (N)	40	100	

A **cumulative percentage distribution** shows the percentage at or below each category (class interval or score) of the variable. A cumulative percentage distribution is constructed using the same procedure as for a cumulative frequency distribution except that the percentages—rather than the raw frequencies—for each category are added to the total percentages for all the previous categories.

In Table 2.12 we have added the cumulative percentage distribution to the frequency and percentage distributions shown in Table 2.10. The cumulative percentage distribution shows, for example, that 37.5% of the sample was younger than 40 years of age—that is, 39 years or younger.

Like the percentage distributions described earlier, cumulative percentage distributions are especially useful when you want to compare differences between groups. For an example of how cumulative percentages are used in a comparison, we have used the 1996 GSS data to contrast the opinions of black women and white men about their family income compared with other American families. Respondents were asked the following question: "Compared with American families in general, would you say your family income is far below average, below average, average, above average, or far above average?"

The percentage distribution and the cumulative percentage distribution for black women and white men is shown in Table 2.13. The cumulative percentage distributions suggest that relatively more black women than white men consider their family income to be average or lower. Whereas only 73 percent of the white males consider their family income average or lower (where lower includes the categories "below average" and "far below average"), 93 percent of the black women ranked their income as average or lower. What might explain these differences? Both gender and racial discrimination might play a role in the income gap between black women and white men. Black women experience a double disadvantage in income: the lower income associated with race and the lower wages of women. These data prompt many other questions about the role that both race and gender play in income inequality. For instance, are these differences primarily be-

Table 2.13 **Relative Family Income: White Men Versus Black Women**

Relative Family Income	White Men		Black Women	
	%	C%	%	C%
Far below average	7	7	8	8
Below average	20	27	37	45
Average	46	73	48	93
Above average	23	96	6	99
Far above average	4	100	1	100
Total	100		100	
(N)	521		128	

Source: GSS1996.

cause of gender or because of race? Racial and gender inequalities in income follow similar trends in other related areas such as occupation and education. Are there similar differences in the occupational and educational distributions of men and women? Blacks and whites?

Cumulative frequency distribution A distribution showing the frequency at or below each category (class interval or score) of the variable.

Cumulative percentage distribution A distribution showing the percentage at or below each category (class interval or score) of the variable.

■ ■ ■ ■ **Rates**

"More than a decade has passed since the 1970s' dramatic decline in marriage and rise in divorce revolutionized the American family. But echoes of those changes are still reverberating," states a report released by the Census Bureau in 1992. The report, "Marriage, Divorce, and Remarriage in the 1990s,"[8] finds that marriage rates, falling since World War II, continue to drop and divorce rates are twice the 1950s' level.

Terms such as *birth rate, unemployment rate,* and *marriage rate* are often used by social scientists and demographers and then quoted in the popular

[8]U.S. Bureau of the Census, "Marriage, Divorce, and Remarriage in the 1990s," *Current Population Reports* (Washington DC: GPO, 1992), pp. 23–180.

media to describe population trends. But what exactly are rates, and how are they constructed? A **rate** is a number obtained by dividing the number of actual occurrences in a given time period by the number of possible occurrences. For example, to determine the rate of marriage for 1990, the U.S. Census Bureau took the number of marriages performed in 1990 (actual occurrence) and divided it by the total population in 1990 (possible occurrences). The marriage rate for 1990 can be expressed as:

$$\text{Marriage rate, } 1990 = \frac{\text{Number of marriages in 1990}}{\text{Total population in 1990}}$$

Since 2,448,000 marriages were performed in 1990 and the number for the total population was 250,000,000, the marriage rate for 1990 can be expressed as:

$$\text{Marriage rate, } 1990 = \frac{2,448,000 \text{ marriages}}{250,000,000 \text{ Americans}} = .0098$$

This means that for every person in the United States, .0098 marriages occurred during 1990.

Rates are often expressed as rates per thousand or hundred thousand to eliminate decimal points and make the number easier to interpret. For example, to express the marriage rate per thousand we multiply it by 1,000:

$$\text{Marriage rate, } 1990 = \frac{2,448,000 \text{ marriages}}{250,000,000 \text{ Americans}} \times 1,000$$

This means that for every 1,000 people, 9.8 marriages occurred during 1990.

Similarly, the divorce rate for the U.S. population during 1990 can be obtained as follows:

$$\text{Divorce rate, } 1990 = \frac{1,175,000 \text{ divorces}}{250,000,000 \text{ Americans}} \times 1,000$$

$$= .0047 \times 1,000 = 4.7$$

For every 1,000 people, 4.7 divorces occurred in 1990.

The preceding marriage and divorce rates are referred to as *crude rates* because they are based on the total population. Rates can be calculated on the general population or on a more narrowly defined select group. For instance, marriage rates are often given for the number of people who are 15 years or older—people who are considered of "marriageable age." Rates can also be calculated separately for men and women. For example, the 1990 marriage rate for men 15 and older is obtained as follows:

$$\text{Marriage rate men 15 and older, } 1990 = \frac{2,448,000}{94,339,000} \times 1,000 = 25.9$$

and for women in the same age group:

$$\text{Marriage rate women 15 and older, 1990} = \frac{2,448,000}{101,457,000} \times 1,000 = 24.1$$

Rate A number obtained by dividing the number of actual occurrences in a given time period by the number of possible occurrences.

Learning Check. *Law enforcement agencies routinely record crime rates (the number of crimes committed relative to the size of a population), arrest rates (the number of arrests made relative to the number of crimes reported), and conviction rates (the number of convictions relative to the number of cases tried). Can you think of some variables that could be expressed as rates?*

■ ■ ■ ■ ■ **Statistics in Practice: Marriage and Divorce Rates over Time**

How can we examine the shifting marriage and divorce rates? Have marriage rates really declined? Have divorce rates risen? Like percentages, rates are useful in making comparisons between different groups and over time. The crude marriage rate of 9.8 for 1990 might be difficult to interpret by itself and will not answer our question of whether or not marriage rates have really changed. To illustrate how rates have changed over time, let's look at Table 2.14, which reports the marriage rates and divorce rates since 1970 for women 15 years and older. The table shows that over the past two decades marriage has declined rather steadily, whereas the divorce rate in 1990 was roughly the same as it was in 1975 but much higher than it was in 1970.

Table 2.14 Marriage and Divorce Rates per 1,000 Women 15 Years Old or Older

Year	Marriage Rate	Divorce Rate
1970	28.4	14.9
1975	25.6	20.3
1980	26.1	22.6
1985	24.8	21.7
1990	24.1	20.1
1995	(n.a.)	19.8

Source: Data from U.S. Bureau of the Census, *Statistical Abstract of the United States, 1997,* Table 145.

> **Learning Check.** *Make sure you understand how to read tables. Can you explain how we reached the preceding conclusions based on the information in Table 2.14?*

■ ■ ■ ■ Reading the Research Literature[9]: Statistical Tables

Statistical tables that display frequency distributions or other kinds of statistical information are found in virtually every book, article, or newspaper report that makes any use of statistics. However, the inclusion of statistical tables in a report or an article doesn't necessarily mean that the research is more scientific or convincing. You will always have to ask what the tables are saying and judge whether the information is relevant or accurately presented and analyzed. Most statistical tables presented in the social science literature are a good deal more complex than those we describe in this chapter. The same information can sometimes be organized in many different ways, and because of space limitations the researcher may present the information with minimum detail.

In this section we present some guidelines for how to read and interpret statistical tables displaying frequency distributions. The purpose is to help you see that some of the techniques described in this chapter are actually used in a meaningful way. Remember that it takes time and practice to develop the skill of reading tables. Even experienced researchers sometimes make mistakes when interpreting tables. So take the time to study the tables presented here; do the chapter exercises; and you will find that reading, interpreting, and understanding tables will become easier in time.

Basic Principles

The first step in reading any statistical table is to understand what the researcher is trying to tell you. There must be a reason for including the information, and usually the researcher tells you what it is. Begin your inspection of the table by reading its title. It usually describes the central contents of the table. Check for any source notes to the table. These tell the source of the data or the table and any additional information the author considers important. Next, examine the column and row headings and subheadings. These identify the variables, their categories, and the kind of statistics presented, such as raw frequencies or percentages. The main body of the table includes the appropriate statistics (frequencies, percentages, rates, and so on) for each variable and/or group as defined by each heading and subheading.

Table 2.15 was taken from an article written by Professor Marie Withers Osmond et al. about AIDS risks among women. In their study, Professor

[9]The idea of "Reading the Research Literature" sections that appear in most chapters was inspired by Joseph F. Healey, *Statistics: A Tool for Social Research*, 5th ed. (Belmont, CA: Wadsworth, 1999).

Table 2.15 **Frequency and Percentage Distribution for Race, Education, Income, Frequency of Condom Use, and the Decision to Use a Condom for Main Partner and Client Subsamples**

	GROUP			
	Main Partner		**Client**	
VARIABLES	*f*	%	*f*	%
Frequency of condom use:				
More than half the time	69	26	76	70
Less than half the time	199	74	33	30
Total (*N*)	268	100	109	100
Decision to use condom:				
Self	55	20	61	56
Both	123	46	20	18
Partner	17	6	9	8
Never discuss	73	27	19	17
Total (*N*)	268	99	109	99
Race/Ethnicity:				
European American	92	34	32	29
African American	127	47	59	54
Hispanic	35	13	18	17
Haitian	14	5	0	0
Total (*N*)	268	99	109	100
Education:				
Some college	47	17	19	17
High school	106	40	38	35
Less than high school	115	43	52	48
Total (*N*)	268	100	109	100
Income:				
Job	99	37	19	17
Welfare, government, or unemployment	26	10	7	6
Family, friends	63	23	16	15
Tricks, illegal	80	30	67	62
Total (*N*)	268	100	109	100

Source: Adapted from Marie Withers Osmond et al., "The Multiple Jeopardy of Race, Class, and Gender for AIDS Risk Among Women," *Gender and Society* 7, no. 1 (March 1993): 105. Used by permission of Sage Publications Inc.

Osmond and her co-authors examined the ways that high-risk sexual behaviors are related to decisions about using condoms by low-income, culturally diverse women in South Florida. The study focused on two subsamples of women who were randomly selected from various agencies including county jails, public and community health services, and drug and alcohol treatment centers. The first subsample ($N = 268$), which became known as the "main partner" group, included women who stated that they had a main sexual partner with whom they were sexually active. The second subsample ($N = 109$), which was known as the "client" group, included women who stated that they had sex in exchange for money or drugs.

In Table 2.15, the researchers display the frequencies and percentages for the major variables included in the study. Although the table is quite simple, it is important to examine it carefully, including its title and headings, to make sure you understand what the information means.

Learning Check.   *Inspect Table 2.15 and answer the following questions:*

- *What is the source of this table?*
- *How many variables are presented? What are their names?*
- *What is represented by the numbers presented in the first column? In the second column?*

What do the authors tell us about the table? The researchers use it to describe the social class and the racial/ethnic composition of the main partner and client subsamples. By analyzing the percentage columns, we can make a number of observations about the characteristics of the respondents included in the study and the differences between the two groups. First, note that the largest single group among both the main partner and client subsamples is African Americans (47% and 54%, respectively). Second, about two-fifths (40 percent) of the women in the main partner subsample graduated from high school and another 17 percent gained some additional trade or college training. Among the client subsample the corresponding figures are 35 percent and 17 percent. Third, with regard to income, 37 percent of the main partner subsample reported a job or business as their primary source, and "tricks" or other illegal activities (30%) were more common than welfare, government, or unemployment (10%). Among the client subsample 17 percent report jobs as their primary source of income and 62 percent say that their income came primarily from tricks or other illegal activities. What do these numbers tell us about the background differences between the two groups of women? The main partner group is slightly more educated than the client group. Moreover, for relatively more women in the client group, the primary sources of income are tricks and other illegal activities.

The two central variables in this study are *frequency of condom use* and *decision to use condom.* The researchers use the first as a measure of AIDS risk and the second to assess what power these women could assert in their relationships. Women who reported that they (self) made the decision to use a condom during sexual intercourse were seen as having more power than women whose response was either "both," "partner," or "never discuss."

A close examination of the percentages in Table 2.15 reveals some differences between the client and main partner subsamples on these two variables. Among the main partner subsample 26 percent use condoms more than half the time, but among women in the client subsample 70 percent report such frequency of use. Regarding the decision to use condoms, 20 percent of the women in the main partner subsample reported that they made this decision themselves, whereas 46 percent said the decision was made by both themselves and their partner. Among the client subsample these percentages are reversed: 56 percent of the women reported "self," whereas 18 percent said that "both" made the decision. The researchers conclude that power relations are different with clients than with main partners: women find it especially difficult to negotiate condom use with intimate sexual partners (that is, main partners). Thus, the women are more assertive and the men more compliant in the client relationship.

Finally, Table 2.15 provides some preliminary evidence that the two groups are different in terms of their power relations with their sexual partner and the extent to which they engage in high-risk sexual behavior. For a more detailed analysis of the relationships between these variables, you need to consider some of the more complex techniques of bivariate analysis and statistical inference. We consider these more advanced techniques beginning in Chapter 6.

Tables with a Different Format

Tables can sometimes present data for only a subset of the sample. For example, Table 2.16, based on 1996 census data for white Americans, shows the percentages of a number of variables. However, only partial information on each of the variables is included, and therefore the percentages do not add up to 100 percent. For instance, 17.2 percent of white Americans have had less than twelve years of school. The remaining 82.8 percent who have had twelve or more years of school are omitted from the table. Similarly, 69 percent owned homes in 1996; the 31 percent who are not homeowners are omitted from the table. In addition, Table 2.16 also presents the homicide rate (per 100,000) for white males in 1994.

Although the data displayed in Table 2.16 provide useful information, we are usually interested in answering questions that go beyond a simple description of how the variables are distributed. Most research usually goes on to make comparisons between groups or to compare one group at different times. For instance, to put the information on white Americans presented in

Table 2.16   **Selected Economic and Social Indicators for White Americans, 1996**

Indicators	Percentages (%)
Less than 12 years of school	17.2
Unemployed	3.1
Female-headed households	14.1
Own their own homes	69.0
Families below the poverty level	8.5
Homicide rate (1994 for males, per 100,000)	8.5

Source: Data from U.S. Bureau of the Census, *Statistical Abstract of the United States,* 1997, Tables 49 and 140.

Table 2.16 into a more meaningful context we may want to compare it with that of other groups of Americans. Such a comparison allows us to answer questions such as how "high" is a 69 percent home ownership rate, and is the 14 percent female-headed households figure "high" or "low"?

Take a look at Table 2.17. It includes the information from Table 2.16, plus corresponding information on black Americans. Information in the table indicates that living conditions for black Americans in the 1990s remained substantially below those of white Americans. Note the difference in the proportion of black and white families (26.4% compared with 8.5%) living below the poverty line, the substantially higher rates of unemployment for blacks (6.7% versus 3.1% for whites), and the notably higher rates of homicide (more than seven times as high).

Table 2.17   **Selected Economic and Social Indicators for White and Black Americans, 1996**

Indicators	Percentage of Whites (%)	Percentage of Blacks (%)
Less than 12 years of school	17.2	25.8
Unemployed	3.1	6.7
Female-headed households	14.1	46.8
Own their own homes	69.0	43.9
Families below the poverty level	8.5	26.4
Homicide rate (1994 for males, per 100,000)	8.5	65.1

Source: Data from U.S. Bureau of the Census, *Statistical Abstract of the United States,* 1997, Tables 49 and 140.

■ ■ ■ ■ Conclusion

In the introduction to this chapter we told you that constructing a frequency distribution is usually the first step in the statistical analysis of data; we hope that by now you agree that constructing a basic frequency or percentage distribution is a fairly straightforward task. As you have seen in the examples in this chapter, distribution tables help researchers to organize, summarize, display, and describe data. Trends within groups and differences or similarities between groups can be identified using a simple distribution table.

In the chapters that follow, you will find that frequency distribution tables provide the basic information for graphically displaying data and calculating measures of central tendency and variability. In other words, you will see frequency and percentage distributions again and again, so make sure you have confidence in your ability to construct and read distribution tables before you proceed to the next chapters.

MAIN POINTS

- The most basic method for organizing data is to classify the observations into a frequency distribution—a table that reports the number of observations that fall into each category of the variable we are analyzing.

- Constructing a frequency distribution is usually the first step in the statistical analysis of data.

- To obtain a frequency distribution for nominal and ordinal variables, count and report the number of cases that fall into each category of the variable along with the total number of cases (N).

- To construct a frequency distribution for interval-ratio variables that have a wide range of values, first combine the scores into a smaller number of groups—known as class intervals—each containing a number of scores.

- Proportions and percentages are relative frequencies. To construct a proportion, divide the frequency (f) in each category by the total number of cases (N). To obtain a percentage, divide the frequency (f) in each category by the total number of cases (N) and multiply by 100.

- Percentage distributions are tables that show the percentage of observations that fall into each category of the variable. Percentage distributions are routinely added to almost any frequency table and are especially important if comparisons between groups are to be considered.

- Cumulative frequency distributions allow us to locate the relative position of a given score in a distribution. They are obtained by adding to the frequency in each category the frequencies of all the categories below it.

- Cumulative percentage distributions have wider applications than cumulative frequency distributions. A cumulative percentage distribution is constructed by adding to the percentages in each category the percentages of all the categories below it.

- One other method of expressing raw frequencies in relative terms is known as a rate. Rates are defined as the number of actual occurrences in a given time period divided by the number of possible occurrences. Rates are often multiplied by some power of 10 to eliminate decimal points and make the number easier to interpret.

KEY TERMS

cumulative frequency distribution *percentage distribution*

cumulative percentage distribution *proportion*

frequency distribution *rate*

percentage

SPSS DEMONSTRATIONS

Demonstration 1 : Producing Frequency Distributions
[GSS/Module A]

In SPSS you can review the frequency distribution for a single variable or for several variables at once. The frequency procedure is found in the *Descriptive Statistics* menu under *Analyze.*

In the Frequencies dialog box (Figure 2.5), click on the variable name(s) in the left column and transfer the name(s) to the Variable(s) box. Remember, more than one variable can be selected at one time.

Figure 2.5

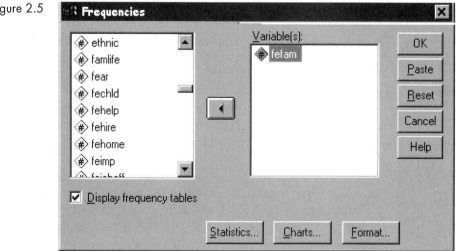

For our demonstration, let's select the variable FEFAM ("It is much better for everyone involved if the man is the achiever outside of the home and the woman takes care of the home and family"). Click on *OK* to process the frequency.

SPSS will produce two tables in a separate Output window, a statistics table and a frequency table. Use the Window scroll keys to move up and down the window to find the statistics and frequency tables for FEFAM. What level of measurement is this variable? (Refer to Chapter 1 to review definitions.)

In the first table, Statistics, SPSS identifies all the valid and missing responses to this question. Responses are coded missing if no answer was given, if the question was not applicable, or if the individual responded "don't know" to the question.

In the frequency table (see Figure 2.6), the variable is listed in the first line, along with the variable label. The first column lists the Value Label for each category of FEFAM. These correspond to numeric values listed in the column, so that 1=strongly agree, 2=agree, 3=disagree, 4=strongly disagree, 0=NAP (not applicable), 8=DK (don't know), and 9=NA (no answer). Review your output to locate all the columns and values we've just identified.

Figure 2.6

BETTER FOR MAN TO WORK, WOMAN TEND HOME

		Frequency	Percent	Valid Percent	Cumulative Percent
Valid	STRONGLY AGREE	79	5.5	6.8	6.8
	AGREE	353	24.7	30.3	37.0
	DISAGREE	528	37.0	45.3	82.3
	STRONGLY DISAGREE	206	14.4	17.7	100.0
	Total	1166	81.7	100.0	
Missing	NAP	229	16.0		
	DK	28	2.0		
	NA	4	.3		
	Total	261	18.3		
Total		1427	100.0		

The next four columns contain important frequency information about the variable. The Frequency column shows the number of respondents who gave a particular response. Thus, we can see that 79 people strongly agree with the statement, while 206 said that they strongly disagree with the statement. Is this consistent with what you expected? How many respondents said that they agree with the statement? We can also see that there are a total of 1,166 valid responses, while 261 are missing.

The Percent column calculates what percentage of the whole sample (1,427 cases) each of the responses represents. Thus, 5.5 percent of the sample said that they strongly agree with the statement that it is better for the man to work while the woman stays at home. Normally, data from the third column, Valid Percent, is more useful. This column removes all the cases defined as missing and recalculates percentages based only on the valid responses. Recalculated based only on valid cases (1,166), the percentage of those who strongly agree is 6.8.

The last column, Cumulative Percent, calculates cumulative percentages beginning with the first response. You can see that more than a third of the sample (37.0 %) either strongly agree or agree with the statement.

Demonstration 2: Recoding Variables

Some variables need to be collapsed or reduced into a smaller number of categories or intervals in order to better present and understand the data. We could, for example, collapse FEFAM into a dichotomous variable, agree or disagree. We would combine 1 (strongly agree) and 2 (agree) into one category, and 3 (disagree) and 4 (strongly disagree) into another. To accomplish this we could use the SPSS commands *Transform–Recode into Different Variable.*

For more detailed instruction on recoding variables, please refer to Appendix E: Recoding Variables. Appendix E explains the recode of the variable EDUC (respondent's years of education).

After reviewing Appendix E, recode FEFAM into a dichotomous variable called RFEFAM. Frequencies for RFEFAM, should look like Figure 2.7.

Figure 2.7

Recoded FEFAM

		Frequency	Percent	Valid Percent	Cumulative Percent
Valid	agree or strongly agree	432	30.3	37.0	37.0
	disagree or strongly disagree	734	51.4	63.0	100.0
	Total	1166	81.7	100.0	
Missing	System	261	18.3		
Total		1427	100.0		

SPSS PROBLEMS

1. Use the SPSS *Frequencies* command to produce a frequency table for the variable MARITAL from the GSS96. How would you describe the marital status of contemporary Americans?

    a. What percentage of the sample is married?
    b. What percentage is divorced?
    c. What percentage of the sample has ever been married?
    d. What percentage of the sample is currently unmarried? (Include all relevant categories in your total percentage.)

2. The GSS included a series of questions on women's employment. Variables included FEHELP (wife should help her husband's career), FEWORK (woman should work), and FECHLD (mother working doesn't hurt the child).
    a. Run frequencies for all three variables.
    b. Prepare a general statement summarizing your results from the three frequency tables. Identify the level of measurement for each variable. How would you describe the level of support for women's employment?

3. Produce the frequency tables for the variables AGE (respondent's age) and EDUC (respondent's highest year of school completed).
    a. What level of measurement is each variable?
    b. Reduce each variable so that it can be measured by four or five ordered categories. If you wanted to reduce the variables into categories, how would you redefine the variables?
    c. Recode AGE and EDUC. You may, based on Demonstration 2, recode the variables in SPSS. Report the frequencies for your revised variables of age and educational attainment. (You can also recalculate the categories with pencil/calculator.)

CHAPTER EXERCISES

1. Suppose you have surveyed thirty people and asked them whether they are white (W) or nonwhite (N), and how many traumas (serious accidents, rapes, or crimes) they have experienced in the past year. You also asked them to tell you whether they perceive themselves as being in the upper, middle, working, or lower class. Your survey resulted in the following raw data:

Race	Class	Traumas	Race	Class	Traumas
W	L	1	W	W	0
W	M	0	W	M	2
W	M	1	W	W	1
N	M	1	W	W	1
N	L	2	N	W	0
W	W	0	N	M	2
N	W	0	W	M	1
W	M	0	W	M	0

(continued on next page)

Race	Class	Traumas	Race	Class	Traumas
W	M	1	N	W	1
N	W	1	W	W	0
N	W	2	W	W	0
N	M	0	N	M	0
N	L	0	N	W	0
W	U	0	N	W	1
W	W	1	W	W	0

(Data based on General Social Survey files for 1987 to 1991)

    a. What level of measurement is the variable *race*? *class*?
    b. Construct raw frequency tables for race and class.
    c. What proportion of the thirty individuals are nonwhite? What percentage are white?
    d. What proportion identified themselves as middle class?

2. Using the data and your raw frequency tables from Exercise 1, construct a frequency distribution for class.
    a. Which is the smallest perceived class?
    b. Which two classes include the largest percentages of people?

3. Using the data from Exercise 1, construct a frequency distribution for trauma.
    a. What level of measurement is used for the trauma variable?
    b. Are people more likely to have experienced no traumas or only one trauma in the past year?
    c. What proportion have experienced one or more traumas in the past year?

4. Suppose you are using 1996 General Social Survey (GSS) data for a research project on education in the United States. The GSS includes a question that asks for the number of years of education. Based on 2,895 individuals, GSS reports the following frequency distribution for years of education:

Years of Education	Frequency
0	4
1	0
2	0
3	8
4	6
5	13
6	14
7	23

(continued on next page)

Years of Education	Frequency
8	80
9	69
10	121
11	155
12	849
13	296
14	339
15	153
16	392
17	120
18	125
19	42
20+	86

a. What is the level of measurement of years of education?

b. Construct a frequency table, with cumulative percentages, for years of education.

c. How many respondents have eight or less years of education? What percentage of the sample does this value represent?

d. Assume that you are really more interested in the general level of education than in the raw number of years of education, so you would like to group the data into four categories that better reflect your interests. Assume that anyone with twelve years of education is a high school graduate, and that anyone with sixteen years of education is a college graduate. Construct a cumulative frequency table for education in four categories based on these assumptions. What percentage of the sample has graduated from college? What percentage of the sample has *not* graduated from high school?

5. Suppose that an organization in your state is lobbying to make pornography illegal because they believe that it leads to a breakdown in morals. The leadership of this organization presumes that more women members than men support their stance toward pornography. The following frequency distribution shows how gender is related to belief that pornography leads to a breakdown in morals from a sample of members.

Does pornography lead to a breakdown in morals?

	No	Yes
Males	115	120
Females	120	110

a. Compute relative frequencies for men and women from the table. Use them to describe the relationship between attitude toward pornography and gender.

b. Do these data support the belief of the organization's leadership about their members?

6. The United States continues to receive large numbers of immigrants. Often these immigrants prefer to speak their own language, at least at home. In South River, New Jersey, 3,637 of the 12,788 residents above age 4 speak a language other than English at home, as reported by the U.S. Census Bureau. The following frequency distribution shows the languages:

Chinese	34
French	55
German	65
Hungarian	198
Indic	34
Italian	95
Polish	831
Portuguese	1,369
Russian	191
Spanish	473
Slavic	279
Other	13

a. Compute the relative frequency distribution for these data.

b. Why would a cumulative frequency distribution be less useful for this table?

c. What are the four most common languages among residents who don't speak English at home?

d. Which two languages are the least common?

7. Have you ever wondered whether television gives an accurate picture of life in the United States? *USA Today* (July 6, 1993) published the following data about characters appearing on television programs based on a week of watching network television.

Sex:	
Male	63%
Female	37%
Age:	
Children (0–12)	4%
Teens (13–17)	6%
Young adults (18–35)	42%
Middle age (36–59)	40%
Seniors (60+)	8%

Race/Ethnicity:
White	84%
Black	13%
Asian	1%
Hispanic	2%
Native American	0.4%

Employment:
Professional/executive	60%
Labor/service/clerical	21%
Law enforcement	19%

Miscellaneous:
Are handicapped	2%
Are overweight	10%
Wear glasses	14%

Compare these data with the following data describing the actual U.S. population.

Sex:
Male	49%
Female	51%

Age:
Children (0–12)	19%
Teens (13–17)	7%
Young adults (18–35)	30%
Middle age (36–59)	27%
Seniors (60+)	17%

Race/Ethnicity:
White	76%
Black	12%
Asian	3%
Hispanic	9%
Native American	1%

Employment:
Professional/executive	26%
Labor/service/clerical	72%
Law enforcement	2%

Miscellaneous:
Are handicapped	17%
Are overweight	68%
Wear glasses	38%

We can see from these data that some groups of people are underrepresented on television, whereas others are overrepresented. Use the percentages from the preceding tables to support your answers to the following questions:

a. Which gender is underrepresented?

b. Which two age groups are greatly overrepresented?

c. Which racial/ethnic group is the most underrepresented?

d. Does television provide an accurate depiction of the distribution of employment status among Americans?

e. Are blacks overrepresented or underrepresented on television?

f. Are handicapped people overrepresented or underrepresented?

8. In Table 2.13 in the chapter, you saw cumulative frequency tables for how white men and black women believe their family income compares with that for all Americans. The following two tables present relative frequency distributions for this 1996 GSS question, first for whites and blacks, then for males and females. Use the tables to answer the questions.

| | Race of Respondent | | | |
| | White | | Black | |
Opinion of Family Income	f	%	f	%
Far below average	74	6.4	16	8.0
Below average	252	21.9	61	30.7
Average	550	47.7	102	51.3
Above average	238	20.7	17	8.5
Far above average	38	3.3	3	1.5
Total (N)	1,152	100.0	199	100.0

| | Sex of Respondent | | | |
| | Male | | Female | |
Opinion of Family Income	f	%	f	%
Far below average	44	7.0	54	6.7
Below average	124	19.8	209	25.9
Average	296	47.2	390	48.4
Above average	139	22.2	136	16.9
Far above average	24	3.8	17	2.1
Total (N)	627	100.0	806	100.0

a. What is the cumulative percentage of females who think their family income is average or below? What is the cumulative percentage of males who believe their family income is below average?

b. Is the cumulative percentage of blacks who believe their family income is average or below greater or less than that for females who hold the same belief?

c. What is the cumulative percentage of whites who believe their family income is average or above?

d. What percentage of males believes their family income is below or far below average? What percentage of males believes their family income is above or far above average? Why might these two values be so close?

9. The tables below present the frequency distributions for education by gender and race. Use them to answer the following questions.

	Gender	
Education	Male	Female
Some high school	133	224
High school graduate	155	271
Some college	133	227
College graduate	132	125

	Race	
Education	White	Black
Some high school	174	183
High school graduate	264	162
Some college	195	165
College graduate	186	71

a. Construct frequency tables based on percentages and cumulative percentages of educational attainment for gender and race.

b. What percentage of males have gone beyond a high school education? What is the comparable percentage for females?

c. What percentage of whites have completed high school or less? What is the comparable percentage for blacks?

d. Are the cumulative percentages more similar for men and women or for blacks and whites? (In other words, where is there more inequality?) Explain.

10. First marriage rates per 1,000 for Canadian men and women in 1995, as reported by *Statistics Canada* (1997), are shown in the following table.

Age Category	Men	Women
15–19	1.4	6.0
20–24	26.8	43.2
25–29	43.0	39.8
30–34	20.5	14.2
35–39	7.5	4.9
40–44	2.8	1.7
45–49	1.1	0.7
Age not specified	0.3	0.2

    a. What level of measurement is age category?

    b. What can be said about the differences in first marriage rates between Canadian men and women?

11. A CNN/USA Today/Gallup Poll (January 20, 1998) compared Americans' attitudes toward abortion in 1975, 1992, and 1998. Poll results are shown in the following table.

Abortion should be:	1998	1992	1975
Always legal	23%	34%	21%
Legal in some circumstances	58%	48%	54%
Always illegal	17%	15%	22%

    a. How would you describe 1998 attitudes?

    b. Since 1975, have Americans changed their attitudes toward abortion? What is the basis for your answer?

12. Should the U.S. government pass legislation that "officially apologizes to American blacks for the fact that slavery was practiced before the Civil War in this country"? This question was asked by the Gallup Poll in 1997 (June 23–24 and 26–29) of 2,043 adult Americans. The following responses were collected.

	Should	Should Not	No Opinion
Total ($N = 2,043$)	633	1267	143
Whites ($N = 1,668$)	434	1117	117
Blacks ($N = 210$)	136	59	15

Transform the frequencies to percentages. How would you describe the results of this Gallup Poll?

13. Percentages and frequencies can also be examined in terms of their change over time. For example, have Americans grown more tolerant of gays? The Gallup Poll Archives (December 1997) reported on data from 1977 to 1996. Men and women were asked if they thought homosexuals should be hired for a select group of occupations. The percentage of respondents who said *yes* is reported for the set of five occupations.

	1977	1982	1985	1987	1989	1992	1996
Salesperson	68	70	71	72	79	82	90
Doctor	44	50	52	49	56	53	69
Armed forces	51	52	55	55	60	57	65
Elementary school teachers	27	32	36	33	42	41	55
Clergy	36	38	41	42	44	43	53

a. Overall, is there one occupational category where Americans are more accepting of homosexuals' being employed? What do you base your answer on?
b. Overall, is there one occupational category where Americans are least tolerant of homosexual employment? What evidence do you base your answer on?
c. Since 1977, have Americans become more tolerant of homosexuals in particular work settings? Explain your answer.

14. Researchers assessed the differences between black and white caregivers of the elderly in a midwestern state. Based on the characteristics listed below, describe what differences exist between black and white caregivers in this study.

Variables	Black (N=136) %	White (N=255) %
Female	86	83
Income		
Less than $10,000	21.3	9.2
$10,000–19,999	25.2	17.5
$20,000–29,999	18.1	21.5
$30,000–39,999	14.2	22.4
$40,000 or more	21.3	29.4
Marital Status		
Currently married	35.3	63.6
Never married	23.5	13.3
Previously married	41.2	23.1
Religion		
Catholic	5.8	59.1
Protestant	78.5	30.3
Jewish	—	6.9
Other	15.7	3.7

Source: Reprinted with the permission of the Gerontological Society of America, 1030 15th Street NW, Suite 250, Washington DC. "Religiosity and Perceived Rewards of Black and White Caregivers," S. Picot et al., *The Gerontologist*, Vol. 37, No. 1. Reproduced by permission of the publisher via Copyright Clearance Center.

15. What is the profile of the "typical" American woman that stays at home? *American Demographic* (September 1997) reported the following data on the "New Face of Homemakers."

American Women: Full-Time Homemakers

Variables	1976	1996
Total number	23 million	20 million
Age		
25–34	29.4%	26.3%
35–44	21.0	25.3
45–54	23.7	20.4
55–64	26.0	28.0
Race/Ethnicity		
Non-Hispanic white	84.3	69.1
Non-Hispanic black	8.8	12.9
Hispanic	5.3	12.9
Other	1.6	5.1
Marital Status		
Never married	3.1	11.7
Married	83.6	70.2
Separated/Divorced	6.7	12.4
Widowed	6.7	5.8
Education		
Less than high school	35.4	26.7
High school graduate	42.9	36.9
Some college	13.0	21.1
Bachelor's degree	6.4	11.9
Graduate/Professional degree	2.3	3.4
Presence of related children		
Any child under 18	56.0	45.3
Children under 6 only	8.5	9.6
Children aged 6–17 only	32.2	22.5
Children under 6 and 6–17	15.3	13.2

Source: Reprinted from *American Demographics* magazine, with permission, copyright 1998, PRIMEDIA Intertec, Stamford, Connecticut.

a. Based on these data, how would you describe differences between the profiles of 1976 and 1996 women?
b. For each of the variables, identify the level of measurement.
c. Review the list of variables presented in the table. What other variables would be useful in assessing the difference between women homemakers in these two periods?

3 Graphic Presentation

■ ■ ■ ■ **Introduction**

You have probably heard that "a picture is worth a thousand words." The same can be said about statistical graphs because they summarize hundreds or thousands of numbers. Many people are intimidated by statistical information presented in frequency distributions or in other tabular forms, but they find the same information to be readable and understandable when presented graphically. Graphs tell a story in "pictures" rather than in words or numbers. They are supposed to make us think about the substance rather than the technical detail of the presentation.

In this chapter you will learn about some of the most commonly used graphical techniques. We concentrate less on the technical details of how to create graphs and more on how to choose the appropriate graphs to make statistical information coherent. We also focus on how to interpret information presented graphically and how to recognize when a graph distorts what the numbers have to say. A graph is a device used to create a visual impression, and that visual impression can sometimes be misleading.

As we introduce the various graphical techniques, we also show you how to use graphs to tell a "story." The particular story we tell in this chapter is that of senior citizens in the United States. The different types of graphs introduced in this chapter demonstrate the many facets of the aging of American society over the next four decades. People have tended to talk about seniors as if they composed a homogeneous group, but the different graphical techniques we illustrate here dramatize the wide variations in economic characteristics, living arrangements, and family status among people aged 65 and older. Most of the statistical information presented in this chapter is based on the *1996 Census of Population and Housing* and numerous reports prepared by statisticians from the Census Bureau and other government agencies that gather information about senior citizens in the United States.

Numerous graphing techniques are available to you, but here we focus on just a few of the most widely used in the social sciences. The first two, the pie and bar charts, are appropriate for nominal and ordinal variables. The next two, histograms and frequency polygons, are used with interval-ratio variables. We also discuss statistical maps and time series charts. The statistical map is most often used with interval-ratio data. Finally, time series charts are used to show how some variables change over time.

■ ■ ■ ■ **The Pie Chart: The Race and Ethnicity of the Elderly**

The elderly population of the United States is racially and ethnically heterogeneous. As the data in Table 3.1 show, of the total elderly population (defined as persons 65 years and older) in 1994, about 29.8 million were white,[1] about 2.7 million black, and 752,052 Native American, Asian, and other.

[1]The census data group most Hispanic Americans as whites.

Table 3.1   **U.S. Population 65 Years and Over by Race and Ethnic Origin, 1994**

Race/Ethnicity	Frequency (f)	Percentages (%)
White	29,772,103	89.7
Black	2,677,912	8.1
Native American, Asian, and other	752,052	2.2
Total (N)	33,202,067	100.0

Source: U.S. Bureau of the Census, "65+ in America," *Current Population Reports,* 1996, Special Studies, P23-190.

A **pie chart** shows the differences in frequencies or percentages among categories of a nominal or an ordinal variable. The categories are displayed as segments of a circle whose pieces add up to 100 percent of the total frequencies. The pie chart shown in Figure 3.1 displays the same information that Table 3.1 presents. Although you can inspect these data in Table 3.1, you can interpret the information more easily by seeing it presented in the pie chart in Figure 3.1. It shows that the elderly population is predominantly white (89.7%), followed by black (8.1%).

> **Learning Check.**   *Notice that the pie chart contains all of the information presented in the frequency distribution. Like the frequency distribution, charts have an identifying number, a title that describes the content of the figure, and a reference to a source. The frequency or percentage is represented both visually and in numbers.*

Figure 3.1   **U.S. Population 65 Years and Over by Race and Ethnic Origin, 1994**

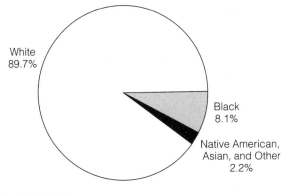

N = 33,202,067

Source: U.S. Bureau of the Census, *Current Population Reports,* 1996, P23-190.

Figure 3.2 **U.S. Population 65 Years and Over by Race and Ethinic Origin, 1990 and 2050**

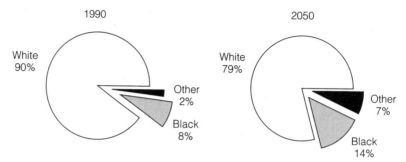

Source: U.S. Bureau of the Census, *Current Population Reports*, 1996, P23-178.

> *Pie chart* A graph showing the differences in frequencies or percentages among categories of a nominal or an ordinal variable. The categories are displayed as segments of a circle whose pieces add up to 100 percent of the total frequencies.

To compare two distributions we can use two pie charts. For example, Figure 3.2 shows two pie charts that display the U.S. population 65 years and older by race and ethnic origin for the years 1990 and 2050. This figure tells us at a glance that between 1990 and 2050 the percentage of blacks is expected to increase from 8 to 14 percent of the total elderly population. Similarly, the percentage of other ethnic groups (such as Asian, Pacific Islander, Native American, Eskimo, and Aleut) is expected to increase from 2 to 7 percent of the total elderly population.[2] We can highlight this growing racial diversity of the elderly population in the coming decades[3] by "exploding" the pie chart, moving the segments representing these groups slightly outward to draw them to the viewer's attention.

> *Learning Check.* Note that we could have "exploded" the segment of the pies representing the white population if we had wanted to highlight the shrinking proportion of whites.

---

[2]U.S. Bureau of the Census, "65+ in America," *Current Population Reports,* 1992, Special Studies, P23-178, p. 2-11.

[3]The 2050 figures are projected by the Census Bureau.

■ ■ ■ ■ **The Bar Graph: The Living Arrangements and Labor Force Participation of the Elderly**

The **bar graph** provides an alternative way to present nominal or ordinal data graphically. It shows the differences in frequencies or percentages among categories of a nominal or an ordinal variable. The categories are displayed as rectangles of equal width with their height proportional to the frequency or percentage of the category.

Let's illustrate the bar graph with an overview of the living arrangements of the elderly. Living arrangements change considerably with advancing age—an increasing number of the elderly live alone or with other relatives. Figure 3.3 is a bar graph displaying the percentage distribution of the elderly by living arrangements in 1993. This chart is interpreted similarly to a pie chart except that the categories of the variable are arrayed along the horizontal axis (sometimes referred to as the *x*-axis) and the percentages along the vertical axis (sometimes referred to as the *y*-axis). This bar graph is easily interpreted: It shows that in 1993, 54.7 percent of the elderly lived with a spouse, 30.3 percent lived alone, and the remaining 15.0 percent lived with other relatives or with nonrelatives.

Construct a bar graph by first labeling the categories of the variables along the horizontal axis. For these categories, construct rectangles of equal width, with the height of each proportional to the frequency or percentage of the category. Note that a space separates each of the categories to make clear that they are nominal categories.

Figure 3.3 **Living Arrangements of the Elderly (65 and Older) in the United States, 1993**

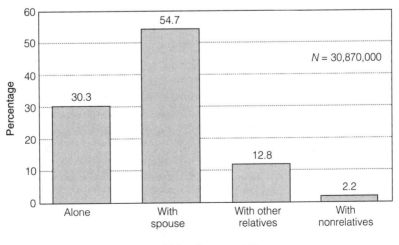

Source: U.S. Bureau of the Census, *Current Population Reports,* 1996, P23-190.

Figure 3.4   **Living Arrangement of U.S. Elderly (65 and Older) by Gender, 1993**

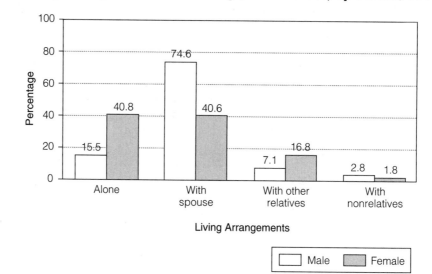

Source: U.S. Bureau of the Census, *Current Population Reports*, 1996, P23-190.

Bar graph A graph showing the differences in frequencies or percentages among categories of a nominal or an ordinal variable. The categories are displayed as rectangles of equal width with their height proportional to the frequency or percentage of the category.

Bar graphs are often used to compare one or more categories of a variable among different groups. For example, there is an increasing likelihood that women will live alone as they age. The longevity of women is the major factor in the gender differences in living arrangements.[4] In addition, elderly widowed men are more likely to remarry than elderly widowed women. Also it has been noted that the current generation of elderly women have developed more protective social networks and interests.[5]

Suppose we want to show how the patterns in living arrangements differ between men and women. Figure 3.4 compares the percentage of women and men 65 years and older who lived with others or alone in 1993. It clearly shows that elderly women are more likely than elderly men to live alone.

We can also construct bar graphs horizontally, with the categories of the variable arrayed along the vertical axis and the percentages or frequencies

[4]U.S. Bureau of the Census, "Marital Status and Living Arrangements: March 1996," *Current Population Reports*, 1998, P20-496, p. 5.

[5]U.S. Bureau of the Census, "65+ in America," *Current Population Reports*, 1996, Special Studies P23-190, pp. 6-1, 6-8.

Figure 3.5 **Income in 1992 of U.S. Elderly Householders Living Alone by Age and Sex (in percentages)**

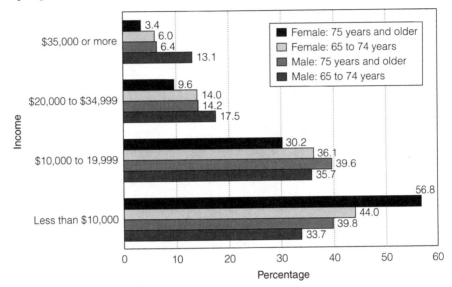

Source: U.S. Bureau of the Census, *Current Population Reports,* 1996, P23-190.

displayed on the horizontal axis. This format is illustrated in Figure 3.5, which compares the 1992 income of elderly men and women, by age groups.

From Figure 3.5, we see that the majority of all elderly persons had 1992 income below $20,000. Women living alone were more likely to have low income than men living alone, and people 75 years and over were poorer than people younger than 75 years. In the 65-to-74-year age group, 44.0 percent of the women, compared with 33.7 percent of the men, had income below $10,000; among people 75 years and older, the figures are 56.8 percent for women and 39.8 percent for men.

■ ■ ■ ■ **The Statistical Map: The Geographic Distribution of the Elderly**

Since the 1960s, the elderly have been relocating to the South and the West of the United States. It is projected that by 2020 these regions will increase their elderly population by as much as 80 percent. We can display these dramatic geographical changes in American society by using a statistical map. Maps are especially useful for describing geographical variations in variables, such as population distribution, voting patterns, crime rates, or labor force composition.

Figure 3.6   **Percentage Change in Population 65 Years and Over, 1993 to 2020**

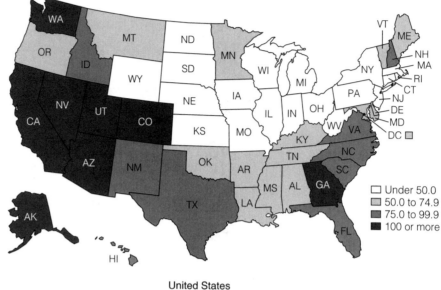

United States
62.7

Source: U.S. Bureau of the Census, 1993 from 1994 Press Release, *Updated National/State Population Estimates,* CB94-43; 2020 from "Population Projections for States, by Age, Sex, Race, and Hispanic Origin: 1993 to 2020," *Current Population Reports,* P25-111, U.S. Government Printing Office, Washington, DC, 1994.

Let's look at Figure 3.6. It presents a statistical map, by state, of the projected percentage increase from 1993 to 2020 of the population 65 years and older. The variable *percentage increase in the elderly population from 1993 to 2020* has four categories: under 50 percent, 50–74.9 percent, 75–99.9 percent, and 100 percent or more. Each category is represented by a different shading (or color code), and the states are shaded depending on their classification into the different categories. To make it easier to read a map that you construct and to identify its patterns, keep the number of categories relatively small—say, not more than five.

Figure 3.6 emphasizes that the greatest percentage of increase in the elderly population is predicted to occur mainly in the western states and the southern coastal states.

Figures 3.7 and 3.8 are two additional examples of statistical maps. They present a geographical breakdown by state of the distribution of elderly blacks and Hispanics. These maps show that in 1991 the southern states contained the largest concentration of black elderly, whereas the Hispanic elderly were concentrated mainly in the southwestern and western states.

Figure 3.7 **Percentage Black of Total State Population 65 Years and Over, 1991**

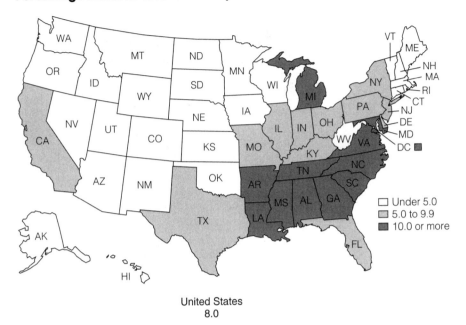

United States
8.0

Source: U.S. Bureau of the Census, "1991 Estimates of the Population of States by Age, Sex, Race, and Hispanic Origin," PE-16.

Figure 3.8 **Percentage Hispanic of Total State Population 65 Years and Over, 1991**

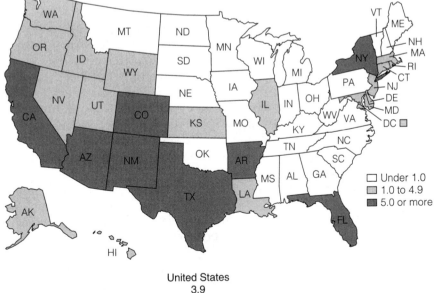

United States
3.9

Source: U.S. Bureau of the Census, "1991 Estimates of the Population of States by Age, Sex, Race, and Hispanic Origin," PE-16.

> **Learning Check.** *Because states vary so much in population, it might be informative to construct maps showing the percentage of the total population in each state that is black elderly (or Hispanic). How might such maps differ from Figures 3.7 and 3.8?*

The examples you have been exploring in this section are all limited to data on the state level. However, maps can also display geographical variations on the level of cities, counties, city blocks, census tracts, and other units. Your choice of whether to display variations on the state level or for smaller units will depend on the research question you wish to explore.

> **Learning Check.** *Can you think of a few other examples of data that could be described using a statistical map?*

■ ■ ■ ■ The Histogram

The **histogram** is used to show the differences in frequencies or percentages among categories of an interval-ratio variable. The categories are displayed as contiguous bars, with width proportional to the width of the category and height proportional to the frequency or percentage of that category. A histogram looks very similar to a bar chart except that the bars are contiguous to each other (touching) and may not be of equal width. In a bar chart, the spaces between the bars visually indicate that the categories are separate. Examples of variables with separate categories are *marital status* (married, single), *gender* (male, female), and *employment status* (employed, unemployed). In a histogram, the touching bars indicate that the categories or intervals are ordered from low to high in a meaningful way. For example, the categories of the variables *hours spent studying, age,* and *years of school completed* are contiguous, ordered intervals.

Figure 3.9 is a histogram displaying the percentage distribution of the population 65 years and over by age. The data on which the histogram is based are presented in Table 3.2. To construct the histogram of Figure 3.9, arrange the age intervals along the horizontal axis and the percentages (or frequencies) along the vertical axis. For each age category, construct a bar with the height corresponding to the percentage of the elderly in the population in that age category. The width of each bar corresponds to the number of years that the age interval represents. The area that each bar occupies tells us the proportion of the population that falls into a given age interval. The histogram is drawn with the bars touching each other to indicate that the categories are contiguous.

Figure 3.9   **Relative Frequency of U.S. Population 65 Years and Over, 1994**

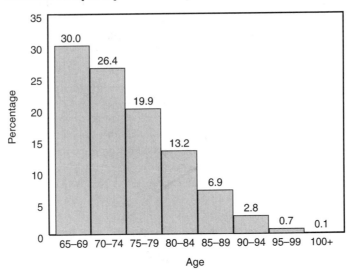

Source: U.S. Bureau of the Census, "65+ in America," *Current Population Reports,* 1996, Special Studies, P23-190.

Table 3.2   **Percentage Distribution of U.S. Population 65 Years and Over by Age, 1994**

Age	Percentage (%)
65–69	30.0
70–74	26.4
75–79	19.9
80–84	13.2
85–89	6.9
90–94	2.8
95–99	0.7
100+	0.1
Total	100.0
(N)	33,202,067

Source: U.S. Bureau of the Census, "65+ in America," *Current Population Reports,* 1996, Special Studies, P23-190.

> *Histogram* A graph showing the differences in frequencies or percentages among categories of an interval-ratio variable. The categories are displayed as contiguous bars, with width proportional to the width of the category and height proportional to the frequency or percentage of that category.

> **Learning Check.** *When bar charts or histograms are used to display the frequencies of the categories of a single variable, the categories are shown on the x-axis and the frequencies on the y-axis. In a horizontal bar chart or histogram this is reversed.*

■ ■ ■ ■ Statistics in Practice: The "Graying" of America

We can also use the histogram to depict more complex trends, as, for instance, the "graying" of America. Let's consider for a moment some of these trends: The elderly population today is ten times larger than it was in 1900, and it will more than double by the year 2030. Indeed, as a journalist has pointed out,

> if the automobile had existed in Colonial times, half the residents of the New Land . . . couldn't have taken a spin: One of every two people were under age 16. Most didn't live long enough to reach old age. Today, the population too young to drive has dropped to one in four while adults 65 and over account for one in eight.[6]

The histogram can give us a visual impression of these demographic trends. For an illustration, let's look at Figures 3.10 and 3.11. Both are applications of the histogram and are used to examine, by gender, actual and projected patterns of the age distribution in America in 1955 and 2010. Notice that in both figures, age groups are arranged along the vertical axis, whereas the frequencies (in millions of people) are along the horizontal axis. Each age group is classified by males on the left and females on the right. Because this type of histogram reflects age distribution by gender, it is also called an age-sex pyramid.

Visually compare the different pieces of data presented in these graphs. By observing where age groups are concentrated, you can discern major patterns in age distribution over time. Note the different shapes of Figure 3.10 and Figure 3.11. Whereas in 1955 the largest group in the population was 0 to 9 years old, in 2010 the largest group will be 45 to 54 years old. These dramatic changes reflect the "graying" of the baby boom (born 1946 to 1965) generation. Almost 84 million babies were born in the United States from 1946 to 1965—60 percent more than were born during the preceding

---

[6]*USA Today*, 10 November 1992.

Figure 3.10 **U.S. Population by Gender and Age, 1955 (in millions)**

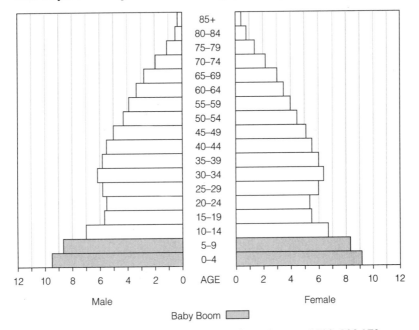

Source: U.S. Bureau of the Census, *Current Population Reports*, 1992, P23-178.

Figure 3.11 **U.S. Population by Gender and Age, 2010 (in millions)**

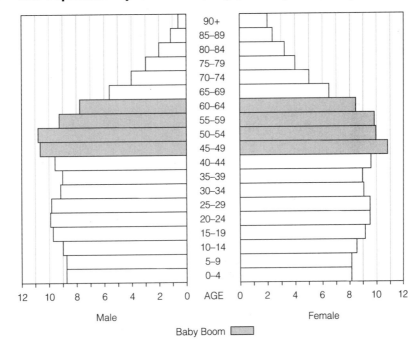

Source: U.S. Bureau of the Census, *Current Population Reports*, 1992, P23-178.

two decades. By 2010, as the baby boom generation reaches 45 to 64, the number of middle-aged and elderly Americans will increase dramatically.

Observe the differences in the number of men and women as age increases. These differences are especially noticeable in Figure 3.11. For example, between ages 70 and 74, women outnumber men 5 to 4; for those 85 years and over, women outnumber men almost 2 to 1. These differences reflect the fact that at every age male mortality exceeds female mortality.

> **Learning Check.** *Notice that when we want to use the histogram to compare groups, we must show a histogram for each group (see Figures 3.10 and 3.11). When we compare groups on the bar chart, we are able to compare two or more groups on the same bar chart (see Figure 3.4).*

■ ■ ■ ■ The Frequency Polygon

Numerical growth of the elderly population is worldwide, occurring in both developed and developing countries. In 1994 thirty nations had elderly populations of at least 2 million. Demographic projections indicate that there will be fifty-five such nations by 2020. Japan is one of the nations experiencing dramatic growth of its elderly population. Figure 3.12 is a frequency polygon displaying the elderly population of Japan by age.

The **frequency polygon** is another way to display interval-ratio distributions; it shows the differences in frequencies or percentages among cat-

Figure 3.12 **Population of Japan, Age 55 and Over, 1991**

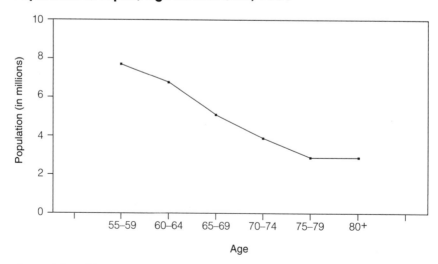

Source: Adapted from U.S. Bureau of the Census, Kevin Kinse, Center for International Research, International Data Base.

Figure 3.13 **Population of Japan, Age 55 and Over, 1991, 2000, and 2020**

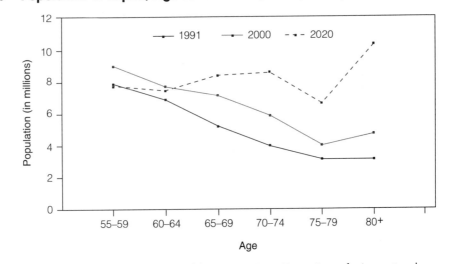

Source: Adapted from U.S. Bureau of the Census, Kevin Kinse, Center for International Research, International Data Base.

egories of an interval-ratio variable. Points representing the frequencies of each category are placed above the midpoint of the category and are joined by a straight line. Notice that in Figure 3.12 the age intervals are arranged on the horizontal axis and the frequencies along the vertical axis. Instead of using bars to represent the frequencies, however, points representing the frequencies of each interval are placed above the midpoint of the intervals. Adjacent points are then joined by straight lines.

Both the histogram and the frequency polygon can be used to depict distributions and trends of interval-ratio variables. How do you choose which one to use? To some extent, the choice is a matter of individual preference, but, in general, polygons are better suited for comparing how a variable is distributed across two or more groups or across two or more time periods. For example, Figure 3.13 compares the elderly population in Japan for 1991 with the projected elderly population for the years 2000 and 2020.

Let's examine this frequency polygon. It shows that Japan's population age 65 and over is expected to grow dramatically in the coming decades. According to projections, the percentage of Japan's population that is elderly could grow from 15.3 million (8.1 percent of the total population) to 21.4 million (16.7 percent) in 2000 and to 33.4 million (26.2 percent) by 2020.[7] Japan's oldest-old population is also projected to grow rapidly, from about 3.1 million (less than 3 percent of the total population) to 10.3 million

[7] U.S. Bureau of the Census, "65+ in America," *Current Population Reports*, 1992, Special Studies, P23-178, p. 2-19.

(8 percent) by 2020. This projected rise has already led to a reduction in retirement benefits and other adjustments to prepare for the economic and social impact of a rapidly aging society.[8]

Frequency polygon A graph showing the differences in frequencies or percentages among categories of an interval-ratio variable. Points representing the frequencies of each category are placed above the midpoint of the category and are joined by a straight line.

Learning Check. *Look closely at the frequency polygons shown in Figure 3.13. Verify the frequencies just described for the oldest-old population (80 years and over). Can you recalculate the percentages for these data?*

■ ■ ■ ■ **Time Series Charts**

We are often interested in examining how some variables change over time. For example, we may be interested in showing changes in the labor force participation of Hispanic women over the last decade, changes in the public's attitude toward abortion rights, or changes in divorce and marriage rates. A **time series chart** displays changes in a variable at different points in time. It involves two variables: *time,* which is labeled across the horizontal axis, and another variable of interest whose values (frequencies, percentages, or rates) are labeled along the vertical axis. To construct a time series chart, use a series of dots to mark the value of the variable at each time interval, and then join the dots by a series of straight lines.

Figure 3.14 shows a time series from 1900 to 2050 of the percentage of the total population that is 65 years or older (the figures for the years 2000 through 2050 are projections made by the Social Security Administration, as reported by the U.S. Census). This time series lets us clearly see the dramatic increase in the elderly population. The number of elderly increased from a little less than 5 percent in 1900 to about 12 percent in 1990. The rate is expected to nearly double by 2050, when almost 25 percent of the total population will be 65 years or older. This dramatic increase in the elderly population, especially beginning in the year 2010, is associated with the "graying" of the baby boom generation. This group, which was 0 to 9 years old in 1955 (see the age pyramid in Figure 3.10), will be 55 to 64 years old in the year 2010.

The implications of these demographic changes are enormous. To cite just a few, there will be more pressure on the health-care system and on pri-

[8]Ibid., p. 2-20.

Figure 3.14 **Percentage of Total U.S. Population 65 Years and Over, 1900 to 2050**

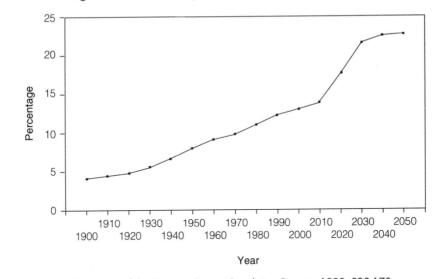

Source: U.S. Bureau of the Census, *Current Population Reports,* 1992, P23-178.

vate and public pension systems. In addition, because the voting patterns of the elderly differ from those of younger people, the "graying" of America will have major political effects.

Often we are interested in comparing changes over time for two or more groups. Let's examine Figure 3.15, which charts the trends in the percentage of divorced elderly from 1960 to 2040 for men and women. This time series graph shows that the percentage of divorced elderly men and elderly women was about the same until 1990. For both groups the percentage increased from less than 2 percent in 1960 to about 5 percent in 1990.[9] According to projections, however, there will be significant increases in the percentage of men and especially women who are divorced: from 5 percent of all the elderly in 1990 to 8 percent of all elderly men and 14 percent of all elderly women by the year 2020. This sharp upturn and the gender divergence are clearly emphasized in Figure 3.15.

> *Time series chart* A graph displaying changes in a variable at different points in time. It shows time (measured in units such as years or months) on the horizontal axis and the frequencies (percentages or rates) of another variable on the vertical axis.

[9]Ibid., p. 6-1

Figure 3.15 **Percentage of U.S. Population 65 Years and Over Currently Divorced, by Gender, 1960 to 2040**

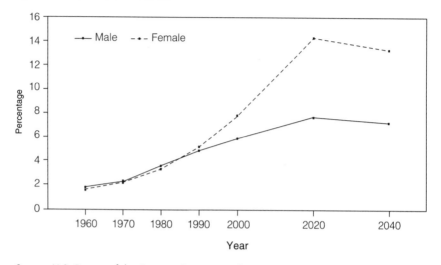

Source: U.S. Bureau of the Census, *Current Population Reports,* 1992, P23-178.

> **Learning Check.** *How does the time series chart differ from a frequency polygon? The difference is that frequency polygons display frequency distributions of a single variable, whereas time series charts display two variables. In addition, time is always one of the variables displayed in a time series chart.*

■ ■ ■ ■ Distortions in Graphs

In this chapter we have seen that statistical graphs can give us a quick sense of the main patterns in the data. However, graphs not only quickly inform us; they can also quickly deceive us. Because we are often more interested in general impressions than in detailed analyses of the numbers, we are more vulnerable to being swayed by distorted graphs. But what are graphical distortions? How can we recognize them? In this section we illustrate some of the most common methods of graphical deception so you will be able to critically evaluate information that is presented graphically. To help you learn more about graphical "integrity," we highly recommend *The Visual Display of Quantitative Information* (1983), by Edward Tufte. This book not only demonstrates the many advantages of working with graphs, but it also contains a detailed discussion of some of the pitfalls in the application and interpretation of graphics.

Figure 3.16   **Cost per Child Enrolled in Head Start, U.S.**

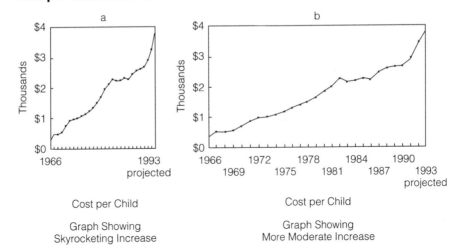

Cost per Child

Graph Showing
Skyrocketing Increase

Cost per Child

Graph Showing
More Moderate Increase

Source: Adapted from *USA TODAY,* March 15, 1993. Copyright 1993 USA TODAY.
Reprinted with permission.

Shrinking and Stretching the Axes: Visual Confusion

Probably the most common distortions in graphical representations occur
when the distance along the vertical or horizontal axis is altered in relation
to the other axis.[10] Axes can be stretched or shrunk to create any desired result. Let's look at the example presented in Figure 3.16a. It is taken from a
1993 issue of *USA Today,* showing changes in cost per child enrolled in
Head Start. The impression the graph gives is that from 1966 to 1993, cost
per child skyrocketed! However, although the cost has indeed gone up from
$271 to $3,849, these figures are not adjusted for inflation. Suppose that we
want to make the increase in cost look more moderate without adjusting for
inflation. We can stretch the horizontal axis to enlarge the distance between
the years, as shown in Figure 3.16b. Because of this stretching, the trend appears less steep and the increase in cost appears smaller. We have changed
the impression considerably without altering the data in any way. Another
way to decrease the steepness of the slope is to shrink the vertical axis so
that the dollar amounts are represented by smaller heights than they are in
Figure 3.16a or 3.16b.

The opposite effect can be obtained by shrinking the horizontal axis and
narrowing the distance between the points on the scale. That technique
makes the slope look steeper.[11] Consider the graphs in Figures 3.17a and

---

[10]R. Lyman Ott et al., *Statistics: A Tool for Social Sciences* (Boston: PWS-Kent Publishing Co.,
1992), pp. 92–95.
[11]Ibid.

Figure 3.17 **Women in U.S. Legislatures, 1973 to 1993**

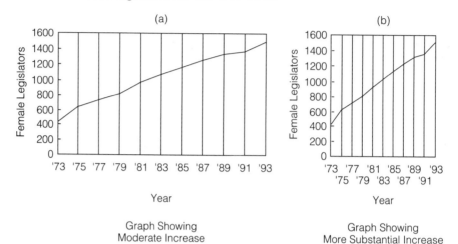

Source: Adapted from Marty Baumann, *USA TODAY*, February 12, 1993. Copyright 1993 USA TODAY. Reprinted with permission.

3.17b, which depict the increase in the number of women elected to state legislatures between 1973 and 1993. In Figure 3.17a the increase of more than 300 percent (from 424 to 1,516), although discernible, does not appear to be very great. This increase can be made to appear more dramatic by shrinking the horizontal axis so the years are moved closer together. This was done in Figure 3.17b, which represents the same data. The steeper slope, created by moving the years closer together, gives the impression of a more substantial increase.

> **Learning Check.** *When you are using a computer software program to draw a graph, the program will automatically adjust the size of the axes to avoid distortion. You will have to be creative to produce a distorted graph on your computer, if you can do it at all.*

Distortions with Picture Graphs

Another way to distort data with graphs is to use pictures to represent quantitative information. The problem with picture graphs is that the visual impression received is created by the picture's total area rather than by its height (the graphs we have discussed so far rely on height).

Take a look at Figure 3.18. It shows the estimated number of HIV-infected people in some of the hardest-hit areas around the world in 1992. Note that sub-Saharan Africa, where the virus may have originated, is the

Figure 3.18 **Estimated Number of HIV Infections in 1992**

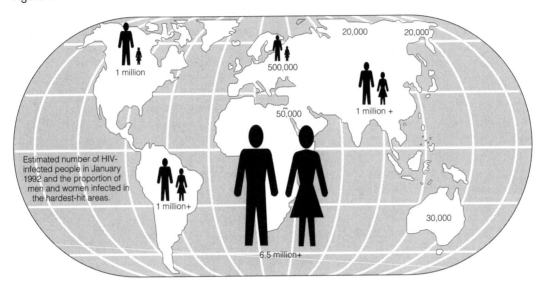

Source: Adapted from *The New York Times*, June 28, 1992. Copyright © 1992 The New York Times Co. Reprinted with permission.

hardest hit, with 6.5 million infected men and women. This number is more than six times the number of HIV infections in South and Central America, where the number of infections is about 1 million. Yet the human figures representing the number of infections for Africa are about twenty times larger in total area than the size of the human figures for South and Central America. The reason for this magnified effect is that although the data are one-dimensional (1 million compared with 6.5 million infected people in sub-Saharan Africa), the human figures representing these numbers are two-dimensional. Therefore, it is not only height that is represented but width as well, creating a false impression of the difference in the number of HIV infections.

These examples illustrate some potential pitfalls in interpreting graphs, emphasizing the point that a graph is a device used to create a visual impression, and that visual impressions can sometimes be misleading. Always interpret a graph in the context of the numerical information the graph represents.

■ ■ ■ ■ **Statistics in Practice: Diversity at a Glance**

In this chapter you are learning how to present statistical information using various graphical techniques. Graphs can tell a story in pictures rather than in words or numbers. Because this chapter tells the story of the elderly in

America, the different types of graphs you are being introduced to illustrate that the elderly in America are increasing not just in numbers but in diversity as well.

We now illustrate some additional ways in which graphics can be used to visually highlight diversity. In particular, we show how graphs can help us to (1) explore the differences and similarities between the many social groups coexisting within American society and (2) emphasize the rapidly changing composition of the U.S. population. Indeed, because of the heterogeneity of American society, the most basic question to ask when you look at data is "compared to what?" This question not only is at the heart of quantitative thinking[12] but underlies inclusive thinking as well.

Three types of graphs—the bar chart, the frequency polygon, and the time series chart—are particularly suitable for making comparisons among groups. Let's begin with the bar chart displayed in Figure 3.19. It compares elderly males and females that live alone, by age, gender, and race or Hispanic origin. Figure 3.19 shows that the percentage of elderly who live alone varies not only by age but also by both race and gender. For instance, we see that in every age category, elderly females are more likely than elderly males to live alone. This trend holds true regardless of race or Hispanic origin. Why the difference? Women who are divorced or widowed late in life have a lower rate of remarriage than men. Therefore, they are more likely to live alone. Figure 3.19 illustrates that by looking at age, race, Hispanic origin, and gender simultaneously we are able to see that elderly people have different experiences depending on these variables.

Learning Check.   *Examine Figure 3.19 again. Notice that regardless of their gender, elderly Americans of Hispanic origin are less likely to live alone than black or white elderly. Can you think of possible explanations for these differences?*

The frequency polygon provides another way of looking at differences based on gender and/or race/ethnicity, or on other attributes such as class, age, or sexual orientation. For example, Figure 3.20 compares years of school completed by black Americans ages 25 to 64 and 65 years and older with that of all Americans in the same age groups.

The data illustrate that in the United States the percentage of Americans who have completed only 8 years of education has declined dramatically, from about 24 percent among Americans 65 years and older to less than 6 percent for those 25 to 64 years old. The decline for black Americans is even more dramatic, from 46 percent of the black elderly to about 6 percent for those 25 to 64 years old. The corresponding trend illustrated in Figure 3.20 is the increase in the percentage of Americans (all races as well as black

---

[12]Edward R. Tufte, *The Visual Display of Quantitative Information* (Cheshire, CT: Graphics Press, 1983), p. 53.

Figure 3.19 **Percentage of U.S. Population 65 Years and Over Living Alone by Age, Gender, and Race or Hispanic Origin, March 1990**

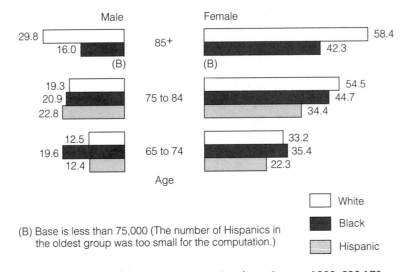

(B) Base is less than 75,000 (The number of Hispanics in the oldest group was too small for the computation.)

Source: U.S. Bureau of the Census, *Current Population Reports,* 1992, P23-178.

Figure 3.20 **Years of School Completed in the United States by Race and Age, 1993**

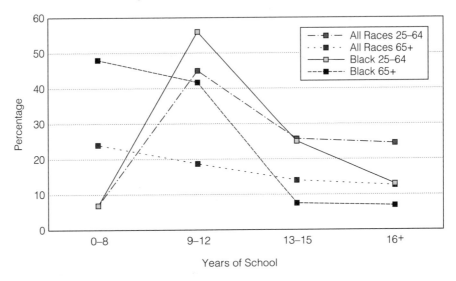

Source: U.S. Bureau of the Census, *Current Population Reports,* 1996, P23-190.

Americans) who have completed 9 to 12 years of schooling, 13 to 15 years, or 16 years or more. For example, about 41 percent of black Americans 65 years or older completed 9 to 12 years of schooling, compared with almost 56 percent of those 25 to 64 years.

The trends shown in Figure 3.20 reflect the development of mass education in the United States during the past fifty years. The percentage of Americans who have completed four years of high school or more has risen from about 40 percent in 1940 to almost 82 percent in 1997. Similarly in 1940 only about 5 percent of Americans completed four or more years of college, compared with about 24 percent in 1997.[13]

> **Learning Check.** *Figure 3.20 illustrates that overall, younger Americans (25 to 64 years old) are better educated than elderly Americans. However, despite these overall trends there are differences between the number of years of schooling completed by "Blacks" and "All races." Examine Figure 3.20 and find these differences. What do they tell you about schooling in America?*

Finally, Figure 3.21 is a time series chart showing changes over time in the percentage of divorced white, black, and Hispanic women. It shows that between 1975 and 1985, the percentage of divorce among white and black women increased steadily. However, between 1985 and 1990 there was a dramatic decline in the percentage of divorce among black women, whereas the percentage of divorce among white women changed very little. In part, the apparent decline in the divorce rate among black women is because they are more likely to separate without divorcing. For whatever reason, however, the result is a convergence in the percentage of divorce among white and black women. In contrast, the percentage of divorce among Hispanic women decreased slightly between 1980 and 1985 and remained almost unchanged between 1985 and 1990.

To conclude, the three examples of graphs in this section as well as other examples throughout this chapter have illustrated how graphical techniques can portray the complexities of the social world by emphasizing the distinct characteristics of age, gender, and ethnic groups. By depicting similarities and differences, graphs help us better grasp the richness and complexities of the social world.

MAIN POINTS

■ A pie chart shows the differences in frequencies or percentages among categories of nominal or ordinal variables. The categories of the variable are segments of a circle whose pieces add up to 100 percent of the total frequencies.

[13]U.S. Bureau of the Census, "Educational Attainment in the United States: 1997," *Current Population Reports,* P20-505, Table A.

Figure 3.21 **Percentage of Divorced U.S. Women (after first marriage) by Race and Hispanic Origin, 1975, 1980, 1985, and 1990**

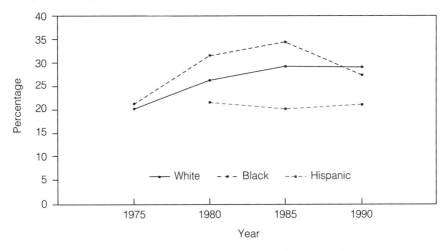

Note: Data not available for Hispanic women, 1975.
Source: U.S. Bureau of the Census, *Current Population Reports,* 1992, P23-178.

■ A bar graph shows the differences in frequencies or percentages among categories of a nominal or an ordinal variable. The categories are displayed as rectangles of equal width with their height proportional to the frequency or percentage of the category.

■ Histograms display the differences in frequencies or percentages among categories of interval-ratio variables. The categories are displayed as contiguous bars with their width proportional to the width of the category and height proportional to the frequency or percentage of that category.

■ A frequency polygon shows the differences in frequencies or percentages among categories of an interval-ratio variable. Points representing the frequencies of each category are placed above the midpoint of the category (interval). Adjacent points are then joined by a straight line.

■ A time series chart displays changes in a variable at different points in time. It displays two variables: time, which is labeled across the horizontal axis, and another variable of interest whose values (for example, frequencies, percentages, or rates) are labeled along the vertical axis.

KEY TERMS

bar graph   *pie chart*
frequency polygon   *time series chart*
histogram

SPSS DEMONSTRATIONS

Demonstration 1: Producing a Bar Chart
[GSS96/Module B]

SPSS for Windows greatly simplifies and improves the production of graphics. The program offers a separate choice from the main menu bar, *Graphs,* that lists fifteen separate types of graphs that SPSS can create. We will use GSS96 for this demonstration.

The first option under the *Graphs* menu is *Bar,* which will produce various types of bar charts. We will use bar charts to display the distribution of the nominal variable RACE (race of respondent). After clicking on *Graphs* and then *Bar,* you will be presented with the initial dialog box shown in Figure 3.22.

Figure 3.22

Almost all graphics procedures in SPSS begin with a dialog box that allows you to choose exactly which type of chart you want to construct. Many graph types can display more than one variable (the Clustered or Stacked choices). We will keep things simple here, so click on *Simple,* then on *Define.* When you do so, the main dialog box for simple bar charts opens (Figure 3.23).

The variable RACE should be placed in the box labeled "Category Axis." In the "Bars Represent" box, click on the "% of cases" radio button. This choice changes the default statistic from the number of cases to percentages, which are normally more useful for comparison purposes.

There is one more thing to do before telling SPSS to create the bar chart. Unlike the way SPSS works for the statistical procedures, SPSS automatically *includes* missing values in many graphs rather than deleting them. You

Figure 3.23

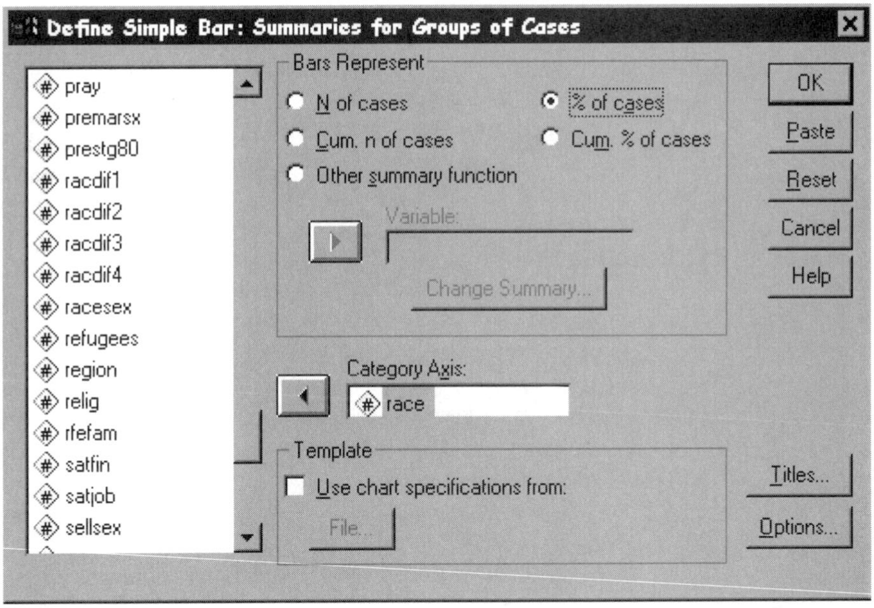

can, and should, change this by clicking on *Options*. You will see the dialog box shown in Figure 3.24. Click in the box labeled "Display groups defined by missing values" to turn off this choice. Then click on *Continue*, then on *OK* to submit your request to SPSS.

Figure 3.24

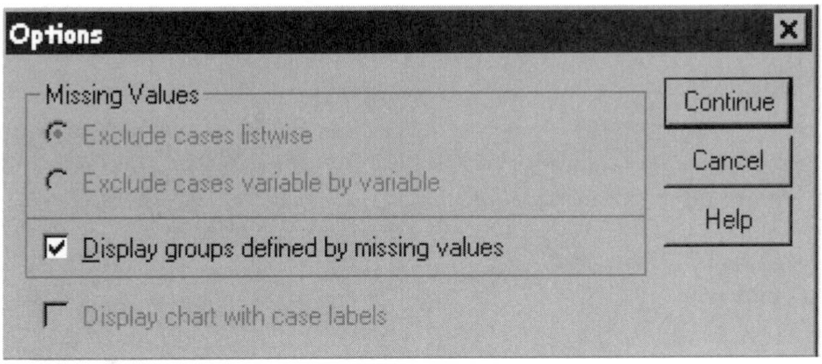

The bar chart for RACE is presented in an Output Window labeled SPSS Viewer. You can see in Figure 3.25 that the bar chart for RACE has only three bars because the only valid responses to this question are "white," "black," or "other."

SPSS graphs can be edited by clicking on the button labeled *Edit*, then *SPSS Chart Object*, which moves the graph from the Chart Carousel window to its own window and displays various editing tools and choices.

Figure 3.25

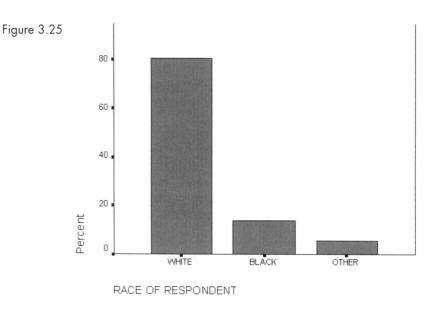

Demonstration 2: Producing a Histogram

Histograms are used to display interval or ratio variables. We'll use the variable AGE from the 1996 GSS file. Under the *Graphs* menu in SPSS is a *Histogram* option. Click on these choices and you will see the dialog box shown in Figure 3.26.

Figure 3.26

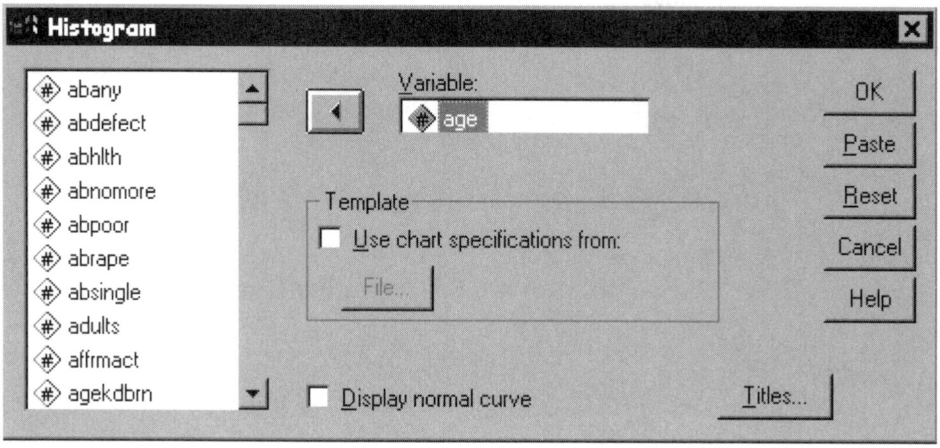

Histograms are created for one variable at a time (that's why there was no opening dialog box as for bar charts). You simply put the variable you want to display in the Variable box. You don't need to worry about missing values in histograms. SPSS automatically deletes them from the display, unlike the bar chart default action. So just click on the *OK* button to process this request. The resulting histogram is shown in Figure 3.27.

Figure 3.27

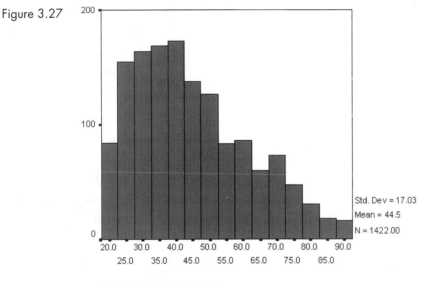

AGE OF RESPONDENT

You can see that there are more people at ages below 40 than above. SPSS automatically decided that the appropriate width for each interval was five years, based on the range of the variable AGE and the optimal number of bars to be displayed on a screen. The number displayed under each bar is the midpoint of that interval, so, for example, the bar for 50 years of age includes everyone from 47.5 to 52.5 (which in practice implies that it includes all the respondents who are 48, 49, 50, 51, and 52 years old).

Study this histogram and understand why it has the shape it does. Notice that SPSS also calculates three statistics—the standard deviation, mean, and *N* (the number of cases)—displayed in the bottom right corner of the histogram. The number of valid cases is 1,422, meaning that five people did not answer this question.

SPSS PROBLEMS

1. You have been assigned data for a presentation on perceived differences between whites and blacks. Using data from the GSS 1996, Module B,

you examine how Americans attribute the differences between whites and blacks. Is it due to discrimination, inborn ability, lack of educational opportunities, or lack of motivation? Variables appropriate for your report include RACDIF1, RACDIF2, RACDIF3, and RACDIF4 (although there are certainly others you could use).

a. Using SPSS, construct pie charts for the variables you have chosen. Be sure to include appropriate labels and to remove any missing responses from the chart. (*Hint:* With the SPSS menus, choose *Graphs–Pie* to reach the appropriate dialog box.)

b. Now construct bar charts for the same variables. (*Hint: From* the SPSS menus, choose *Graphs–Bar.*)

c. Briefly describe the attitude of Americans regarding the basis for social differences between whites and blacks, using either the pie or bar charts.

2. Is there a difference in responses between blacks and whites? Based on Exercise 1, reconstruct the bar charts for the same variables, separating the responses between blacks and whites.

The most straightforward way to produce this output is to first split the file into groups and have SPSS analyze each separately. To do this choose *Data–Split File* from the menu, then choose "Repeat analysis for each group" in the dialog box and place RACE in the Groups Based On box (and click on *OK*). All procedures run after this will be done separately for whites, blacks, and other categories.

Your output should present separate bar charts for blacks and whites for each of the variables you select. Compare the bar charts. What difference exists between the attitudes of blacks and whites?

3. Is there a difference in responses between blacks and whites in whether they believe that conditions for blacks have improved in the United States (BLKSIMP)? Based on the level of measurement, determine the appropriate graphic display for this variable. Produce separate graphics for blacks and whites. What differences, if any, can you report?

4. Determine how best to graphically represent the following variables:

REGION—region of residence

CLOSEBLK—How close to blacks does the respondent feel?

DISCAFF—Are whites hurt by affirmative action policies?

POLVIEWS—Political views of respondent

CHILDS—number of children in household

MELTPOT1—Do you agree that it is better to maintain distinct cultures?

MARITAL—marital status of respondent

Note: Before constructing the histogram or pie chart, you may want to review the variable first using the *Frequencies* or *Utilities–Variables* command.

Figure 3.28

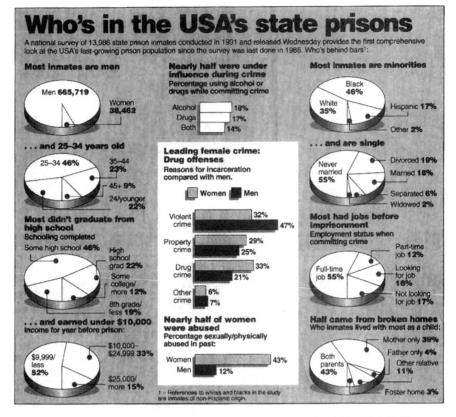

Who's in the USA's state prisons

A national survey of 13,986 state prison inmates conducted in 1991 and released Wednesday provides the first comprehensive look at the USA's fast-growing prison population since the survey was last done in 1986. Who's behind bars[1]:

Most inmates are men

Men **665,719**
Women **38,462**

... and 25–34 years old

25–34 **46%**
35–44 **23%**
45+ **9%**
24/younger **22%**

Most didn't graduate from high school
Schooling completed

Some high school **46%**
High school grad **22%**
Some college/more **12%**
8th grade/less **19%**

... and earned under $10,000
Income for year before prison:

$9,999/less **52%**
$10,000–$24,999 **33%**
$25,000/more **15%**

Nearly half were under influence during crime
Percentage using alcohol or drugs while committing crime

Alcohol **18%**
Drugs **17%**
Both **14%**

Leading female crime: Drug offenses
Reasons for incarceration compared with men.

□ Women ■ Men

Violent crime **32%** / **47%**
Property crime **29%** / **25%**
Drug crime **33%** / **21%**
Other crime **6%** / **7%**

Nearly half of women were abused
Percentage sexually/physically abused in past:

Women **43%**
Men **12%**

1 – References to whites and blacks in the study are inmates of non-Hispanic origin.

Most inmates are minorities

Black **46%**
White **35%**
Hispanic **17%**
Other **2%**

... and are single

Never married **55%**
Divorced **19%**
Married **18%**
Separated **6%**
Widowed **2%**

Most had jobs before imprisonment
Employment status when committing crime

Full-time job **55%**
Part-time job **12%**
Looking for job **16%**
Not looking for job **17%**

Half came from broken homes
Who inmates lived with most as a child:

Both parents **43%**
Mother only **39%**
Father only **4%**
Other relative **11%**
Foster home **3%**

Source: J. L. Albert, *USA TODAY*, May 20, 1993, p. 8A. Copyright 1993 USA TODAY. Reprinted with permission.

CHAPTER EXERCISES

1. Consider the series of pie charts shown in Figure 3.28, which describe characteristics of state prison inmates.
   a. What proportion of inmates are men?
   b. What percentage of inmates had a full-time job before being imprisoned?
   c. What percentage of inmates did not graduate from high school?
   d. What percentage of inmates are people of color?

2. In Chapter 2, Exercise 1, you surveyed thirty people and asked them whether they were white or nonwhite and how many traumas they had experienced in the past year. You also asked them to tell you whether they perceived themselves as being in the upper, middle, working, or lower class. The survey resulted in the following raw data:

Race	Class	Trauma	Race	Class	Trauma
W	L	1	W	W	0
W	M	0	W	M	2
W	M	1	W	W	1
N	M	1	W	W	1
N	L	2	N	W	0
W	W	0	N	M	2
N	W	0	W	M	1
W	M	0	W	M	0
W	M	1	N	W	1
N	W	1	W	W	0
N	W	2	W	W	0
N	M	0	N	M	0
N	L	0	N	W	0
W	U	0	N	W	1
W	W	1	W	W	0

(Data based on General Social Survey files for 1987 to 1991)

    a. Construct a pie chart depicting the percentage distribution of race. (*Hint:* Remember to include a title, percentages, and appropriate labels.)

    b. Construct a pie chart showing the percentage distribution of class.

    c. Construct a graph with two pie charts comparing the percentage distribution of the number of traumas experienced in the past year by race.

3. Based on the bar charts in Figure 3.28:

    a. What percentage of inmates were using both alcohol and drugs during their criminal act?

    b. What percentage were using either alcohol or drugs, or both?

    c. What percentage of women were incarcerated for drug crimes? What percentage of men?

    d. How many women were incarcerated for drug crimes?

4. Using the data from Exercise 2, construct bar graphs showing percentage distributions for race and class. Remember to include appropriate titles, percentages, and labels.

5. Suppose you want to compare the number of traumas experienced in the past year for blacks and whites.

    a. Using the data from Exercise 2, construct a grouped bar graph (similar to Figure 3.4) to show the percentage distribution of the number of traumas experienced in the past year by race.

    b. Which race is most likely to have experienced two traumas in the past year?

c. Why shouldn't you construct a grouped bar chart showing the frequencies rather than the percentages?

6. Imagine that you work for the Food and Drug Administration (FDA) and your current task is to write a report about pesticides found in fruits and vegetables. You've been given the following data to use in the report:

	Percentage of Produce with Pesticides (%)	Number of Types of Pesticides Detected
Tomatoes	47	42
Celery	74	16
Strawberries	73	38
Oranges	71	20
Apples	64	34
Cantaloupes	53	33
Pears	57	26
Cherries	64	23
Peaches	76	28
Spinach	56	24

(Data from the Environmental Working Group, based on FDA data, 1990 to 1992)

a. Construct a bar chart showing the distribution of pesticides in various types of produce.
b. Construct a bar chart showing the number of types of pesticides detected in each food.
c. Can you think of a way to combine both of these charts into a single graph? Construct such a graph.

7. Based on data from Chapter 2, Exercise 12, decide how best to graphically represent those data in an easily understood format.
a. Would you choose to use bar charts or pie charts? Explain the reason for your answer.
b. Construct bar or pie charts (depending on your answer) to represent all of the data. Remember to include appropriate titles, percentages, and labels.

8. The statistical map shown in Figure 3.29 represents the percentage of each state's population that was 65 years or older in 1990, based on U.S. census data.
a. Write a 200-word report for your local newspaper describing the variation in elderly population across the United States.
b. Think about what causes some states to have more elderly in their population. Then locate the states with 15.0 percent or more elderly and explain why the reasons for their relatively high proportion of elderly might be the same (or different).

Figure 3.29

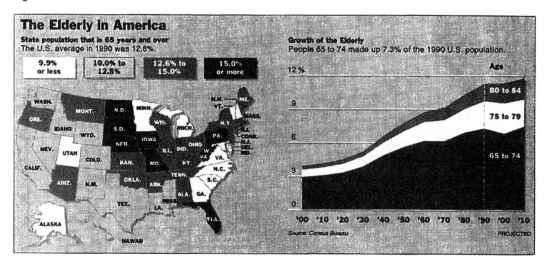

Source: New York Times, November 10, 1992. Copyright © 1992 by The New York Times Co. Reprinted by permission.

9. You are writing a research paper about teen smoking and have the following data. To make calculations simpler, assume that there were 100 students at each grade level in each year.

Year	Grade	Percentage Who Smoke Daily
1992	8	7
1992	10	12
1992	12	17
1993	8	8
1993	10	14
1993	12	19

(Data from the University of Michigan)

a. Suppose you want to argue that older teens are more likely to smoke daily. Construct a bar chart that supports this argument. (*Hint:* Group by year.)

b. Now suppose you want to argue that more teens were smoking daily in 1993 than in 1992. Construct a bar chart that supports this argument. (*Hint:* Group by grade.)

c. Explain why the graphs in (a) and (b) are appropriate for each situation.

10. Use the data on educational level in Chapter 2, Exercise 4, for this problem.
    a. What level of measurement is "years of education"? Why can you use a histogram to graph the distribution of education, in addition to a bar chart?
    b. Construct a histogram for years of education, using equal-spaced intervals of four years. Don't use percentages in this chart.

11. The 1996 GSS data on educational level can be further broken down by race, as follows:

Years of Education	Whites	Blacks
0	1	1
1	0	0
2	0	0
3	4	0
4	4	0
5	9	1
6	11	2
7	14	3
8	70	1
9	49	2
10	88	4
11	109	10
12	692	38
13	237	12
14	278	18
15	128	8
16	332	22
17	105	10
18	107	6
19	32	4
20	74	8

    a. Construct two histograms for education, one for blacks and one for whites.
    b. Now use the two graphs to describe the differences in educational attainment by race.

12. Use the data on abortion attitudes from Chapter 2, Exercise 11, for this problem.
    a. Construct a time series bar graph showing the change in abortion attitudes over the period from 1975 to 1998.
    b. Next, construct a time series polygon (similar to a frequency polygon) showing the same data. Which of these two graphs would you prefer to include in a report? Why?

13. Examine the time series chart concerning marriage and divorce rates shown in Figure 3.30.
    a. Has there been more fluctuation in the divorce rate or the first marriage rate? Which rate has exhibited a greater *percentage* change over time?
    b. There are two peaks in the remarriage rate. Can you suggest what might have caused these peaks?

Figure 3.30

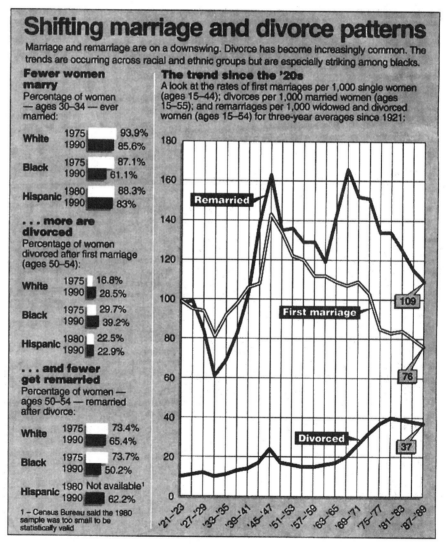

Source: J. L. Albert, *USA TODAY,* December 12, 1992, p. 12A. Copyright 1992 USA TODAY. Reprinted with permission.

14. As time has gone on, segregation in large American school systems has not decreased, as many expected after the civil rights reforms of the 1950s and 1960s. Evidence for this is provided by the following data, taken from the U.S. Department of Education, for several years. These numbers represent the typical percentage of whites in school districts attended by blacks (so the higher the number, the *more integrated* the district).

Year	Boston	Chicago	Los Angeles
1968	27.4%	5.4%	7.5%
1974	32.8%	3.2%	10.2%
1986	22.3%	6.0%	11.7%
1992	18.2%	4.6%	9.6%

a. Construct one time series chart to represent these numbers. Then describe the variation in segregation as measured by these data. Which city has had the most segregated school districts?

b. Let's say you wanted to exaggerate the changes in percentage of whites over time. How would you modify the time series chart to do this?

15. The U.S. Department of Justice analyzed the age patterns of victims of serious crime during the period 1992–1994 and presented its findings in a bar chart (Figure 3.31). Based on the bar chart, how would you describe the relationship between age and serious violent crimes in the United States?

Figure 3.31

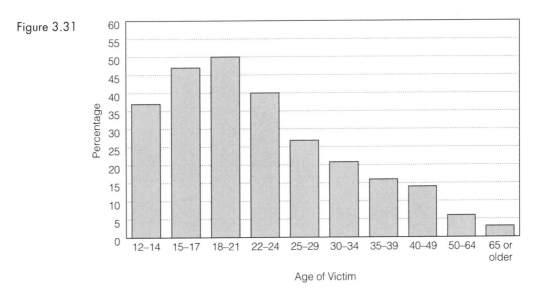

Age of Victim

Source: U.S. Department of Justice, "Age Patterns of Victims of Serious Violent Crime," *Bureau of Justice Statistics Special Report*, July 1997, NCJ-162031, p. 2.

16. In 1998 *American Demographics* reported on ethnic differences in the use of time. Based on America's Use of Time Survey by the University of Maryland, the following table is adapted from a list of the number of hours per week that individuals spent on selected free-time activities. What would be the best way to present this data graphically?

Free-Time Activity	White	Black	Asian	Hispanic
Education	2.7	2.4	10.3	5.3
Religion	1.0	2.7	0.6	0.8
Organization	0.9	1.0	0.3	1.2
Hobbies	2.9	1.6	2.8	2.5
TV	20.0	24.5	16.1	20.8

Source: Adapted from *American Demographics* 20(6), June 1998, p. 20. Reprinted with permission, copyright 1998, PRIMEDIA Intertec, Stamford, Connecticut.

4 **Measures of Central Tendency**

Skewed Distribution

Box 4.4 Statistics in Practice: Median Annual Earnings Among Subgroups

Symmetrical Distribution

MAIN POINTS

KEY TERMS

SPSS DEMONSTRATIONS

SPSS PROBLEMS

CHAPTER EXERCISES

■ ■ ■ ■ **Introduction**

In Chapters 2 and 3 we learned that frequency distributions and graphical techniques are useful tools for presenting information. The main advantage of using frequency distributions or graphs is to summarize quantitative information in ways that can be easily understood even by a lay audience. Often, however, we need to describe a large set of data involving many variables for which graphs and tables may not be the most efficient tools. For instance, let's say we want to present information on the income, education, and political party affiliation of both men and women. Presenting this information might require up to six frequency distributions or graphs. The more variables we add, the more complex the presentation becomes.

Another way of describing a distribution is by selecting a single number that describes or summarizes the distribution more concisely. Such numbers describe what is typical about the distribution—for example, the average income among Hispanics who are college graduates or the most common party identification among the rural poor. Numbers that describe what is average or typical of the distribution are called **measures of central tendency**.

Measures of central tendency Numbers that describe what is average or typical of the distribution.

In this chapter we will learn about three measures of central tendency: the *mode,* the *median,* and the *mean.* You are probably somewhat familiar with these measures—the terms *median* income and *average* income, for example, are used quite a bit even in the popular media. Each describes what is most typical, central, or representative of the distribution. In this chapter we will also learn about how these measures differ from one another. We will see that the choice of an appropriate measure of central tendency for representing a distribution depends on three factors: (1) the way the variables are measured (their level of measurement), (2) the shape of the distribution, and (3) the purpose of the research.

■ ■ ■ ■ ■ **The Mode**

The **mode** is the category or score with the largest frequency or percentage in the distribution. Of all the averages discussed in this chapter, the mode is the easiest one to identify. Simply locate the category represented by the highest frequency.

> *Mode* The category or score with the largest frequency (or percentage) in the distribution.

We can use the mode to determine, for example, the most common foreign language spoken in the United States today. English is clearly the language of choice in public communication in the United States, but you may be surprised by the Census Bureau finding that one out of every seven people living in the United States speaks one of 329 different languages other than English at home. Record immigration from many countries since 1980 has contributed to a sharp increase in the number of people who speak a foreign language.[1]

What is the most common foreign language spoken in the United States today? To answer this question look at Table 4.1, which lists the ten most commonly spoken foreign languages in the United States and the number of people who speak each language. The table shows that Spanish is the most common; more than 17 million people speak Spanish. In this example we refer to "Spanish" as the mode—the category with the largest frequency in the distribution.

The mode is always a category or score, *not* a frequency. Do not confuse the two. That is, the mode in the previous example is "Spanish," not its frequency of 17,339,000.

The mode is not necessarily the category with the majority (that is, more than 50%) of cases, as it is in Table 4.1; it is simply the category in which the largest number (or proportion) of cases fall. For example, Figure 4.1 is a pie chart showing the answers of 1996 GSS respondents to the following question: "Please indicate whether it is the government's responsibility to provide for the elderly." Note that the highest percentage (48%) of respondents is associated with the answer "probably should be." The answer "probably should be" is therefore the mode.

The mode is used to describe nominal variables. Recall that with nominal variables—such as foreign languages spoken in the United States, race/

[1]*USA Today*, April 28, 1993.

Table 4.1   **Ten Most Common Foreign Languages Spoken in the United States**

Language	Number of Speakers
Spanish	17,339,000
French	1,702,000
German	1,547,000
Italian	1,309,000
Chinese	1,249,000
Tagalog	843,000
Polish	723,000
Korean	626,000
Vietnamese	507,000
Portuguese	430,000

Source: U.S. Bureau of the Census, *Statistical Abstract of the United States: 1997,* 117th edition (Washington, DC: GPO, 1997).

ethnicity, or religious affiliation—we are only able to classify respondents based on a qualitative and not a quantitative property. By describing the most commonly occurring category of a nominal variable (such as Spanish in our example), the mode thus reflects the most important element of the distribution of a variable measured at the nominal level. The mode is the only measure of central tendency that can be used with nominal level variables. It can also be used to describe the most commonly occurring category in any distribution. For example, the variable *opinion of government responsibility* presented in Figure 4.1 is an ordinal variable.

In some distributions there are two scores or categories with the highest frequency. Such distributions have two modes and are said to be *bimodal.* For instance, Figure 4.2 is a bar graph showing the level of agreement of 1996 GSS respondents to the following statement: "Generally speaking, America is a better country than most other countries." The same percentage of respondents (40%) "strongly agree" and "agree" with the statement. Both response categories have the highest frequency, and therefore both are the modes. We can describe this distribution as bimodal. When two scores or categories with the highest frequencies are quite close (but not identical) in frequency, the distribution is still "essentially" bimodal. In these situations you should not rely on merely reporting the (true) mode, but instead report the two highest frequency categories.

Figure 4.1 **Is It the Government's Responsibility to Provide for the Elderly?**

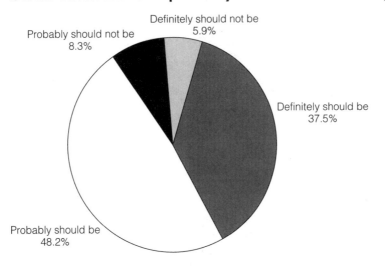

Source: General Social Survey, 1996.

Figure 4.2 **American Is a Better Country Than Most Other Countries**

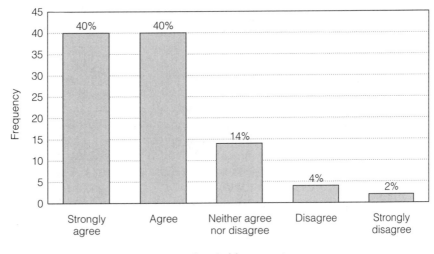

Source: General Social Survey, 1996.

Learning Check.   *Listed below are the political party affiliations of fifteen individuals. Find the mode.*

Democrat	*Republican*	*Democrat*	*Republican*
Republican	*Independent*	*Democrat*	*Democrat*
Democrat	*Republican*	*Independent*	*Democrat*
Independent	*Republican*	*Democrat*	

Why is the mode the only measure of central tendency you can use to describe this distribution?

■ ■ ■ ■   **The Median**

The **median** is a measure of central tendency that can be calculated for variables that are at least at an ordinal level of measurement. The median represents the exact middle of a distribution; it is the score that divides the distribution into two equal parts so that half the cases are above it and half below it. For example, the median household income in 1994 was $38,782. This means that half the households in the United States earned more than $38,782 and half earned less than $38,782. Since many variables used in social research are ordinal, the median is an important measure of central tendency in social science research.

Median   The score that divides the distribution into two equal parts so that half the cases are above it and half below it.

For instance, what are the opinions of Americans about the economy in the United States? How can we describe their rating of current economic conditions? To answer these questions, a Gallup Poll conducted in 1997 asked people whether they rated the economy as "excellent," "good," "only fair," or "poor."[2] Rating of economic conditions is an ordered (ordinal) variable. Thus, to estimate the average rating we need to use a measure of central tendency appropriate for ordinal variables. The median is a suitable measure for those variables whose categories or scores can be arranged in order of magnitude from lowest to highest. Therefore, the median can be used with ordinal or interval-ratio variables, for which scores can be at least rank-ordered, but cannot be calculated for variables measured at the nominal level.

[2]*Gallup Poll Archives*, November 1997.

Finding the Median in Sorted Data

It is very easy to find the median. In most cases it can be done by simple inspection of the sorted data. The location of the median score will differ somewhat, depending on whether the number of observations is odd or even. Let's first consider two examples with an odd number of cases.

An Odd Number of Cases Suppose we are looking at the responses of five people to the question "Thinking about the economy, how would you rate economic conditions in this country today?" Following are the responses of these five hypothetical persons:

Poor	Jim
Good	Sue
Only fair	Bob
Poor	Jorge
Excellent	Karen
Total (N)	5

To locate the median, first arrange the responses in order from lowest to highest (or highest to lowest):

Poor	Jim
Poor	Jorge
Only fair	*Bob*
Good	Sue
Excellent	Karen
Total (N)	5

The median is the response associated with the middle case. Find the middle case when N is odd by adding 1 to N and dividing by 2: $(N + 1) \div 2$. Since N is 5, you calculate $(5 + 1) \div 2 = 3$. The middle case is thus the third case (Bob).

The response associated with the third case (Bob) is the median. Notice that the median divides the distribution exactly in half so that there are two respondents who are more satisfied and two respondents who are less satisfied.

Now let's look at another example (see Figure 4.3). The following is a list of the number of hate crimes reported in the nine largest U.S. states for 1995.[3]

---

[3]*Statistical Abstract of the United States*, 1997. Table 323.

Number	State
1751	California
164	Florida
405	Michigan
768	New Jersey
845	New York
267	Ohio
282	Pennsylvania
326	Texas
266	Washington
Total (N)	9

To locate the median, first arrange the number of hate crimes in order from lowest to highest:

Number	State
164	Florida
266	Washington
267	Ohio
282	Pennsylvania
326	Texas
405	Michigan
768	New Jersey
845	New York
1751	California
Total (N)	9

The middle case is $(9 + 1) \div 2 = 5$, the fifth state, Texas. The median is 326, the number of hate crimes associated with Texas. It divides the distribution exactly in half so that there are four states with fewer hate crimes and four with more.

An Even Number of Cases Now let's delete the last score to make the number of states even. The scores have already been arranged in increasing order:

164, 266, 267, 282, 326, 405, 768, 845

When N is even, we no longer have a single middle case. The median is therefore located halfway between the two middle cases. Find the two middle cases by using the previous formula: $(N + 1) \div 2$, or $(8 + 1) \div 2 = 4.5$. For our example, this means that you average the scores for the fourth and fifth states, Pennsylvania and Texas. The numbers of hate crimes

Figure 4.3   **Finding the Median Number of Hate Crimes for Nine States**

Finding the Median for Nine States

1. Order the cases from lowest to highest:

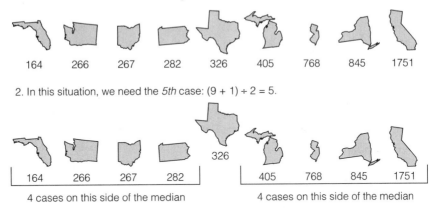

164   266   267   282   326   405   768   845   1751

2. In this situation, we need the *5th* case: (9 + 1) ÷ 2 = 5.

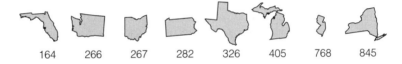

326

164   266   267   282   405   768   845   1751

4 cases on this side of the median   4 cases on this side of the median

Finding the Median for Eight States

1. Order the cases from lowest to highest:

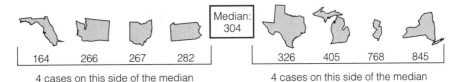

164   266   267   282   326   405   768   845

2. In this situation, we need "imaginary" case 4.5: (8 + 1) ÷ 2 = 4.5.

3. To find the value of this case, take the average of the two cases surrounding it:

Median:
304

164   266   267   282   326   405   768   845

4 cases on this side of the median   4 cases on this side of the median

associated with these states are 282 and 326. To find the median for an interval-ratio variable, simply average these two middle numbers:

$$\text{Median} = \frac{282 + 326}{2} = 304$$

The median is therefore 304.

When data are ordinal, averaging the middle two scores is no longer appropriate. The median simply falls between two middle values.

> **Learning Check.** *Find the median of the following distribution of an interval-ratio variable:* 22, 15, 18, 33, 17, 5, 11, 28, 40, 19, 8, 20.

Finding the Median in Frequency Distributions

Often our data are arranged in frequency distributions. The procedure for locating the median in a frequency distribution is a bit more involved than the procedure we just described. Take, for instance, the frequency distribution displayed in Table 4.2. It shows the political views of GSS respondents in 1996. We use a subsample of the GSS for demonstration.

To identify the median we have to find the category associated with the observation located at the middle of the distribution. To help locate this observation we construct a cumulative percentage distribution, as shown in the last column of Table 4.2. In this example, the percentages are cumulated from "extremely liberal" to "extremely conservative." We could also cumulate the other way, from "extremely conservative" to "extremely liberal." The observation located at the middle of the distribution is the one that has a cumulative percentage value equal to 50 percent. The median is the value of the category associated with this observation.[4] This middle observation falls within the category "moderate." The median for this distribution is therefore "moderate." If you are not sure why the middle of the distribution—the 50 percent point—is associated with the category "moderate," look again at the cumulative percentage column (*C*%). Notice that 27.29 percent of the observations are accumulated below the category "moderate" and that 65.7 percent are accumulated up to and including the category "moderate." We know, then, that 50 percent is located somewhere within the "moderate" category.

Alternatively, the median can be computed based on the cumulative frequency distribution (*Cf*) shown in column 3 of Table 4.2. In that case, you

Table 4.2 **Political Views of GSS Respondents, 1996**

Political View	Frequency (*f*)	*Cf*	Percentage (%)	*C*%
Extremely liberal	15	15	2.69	2.69
Liberal	65	80	11.67	14.36
Slightly liberal	72	152	12.93	27.29
Moderate	214	366	38.42	65.71
Slightly conservative	76	442	13.64	79.35
Conservative	92	534	16.52	95.87
Extremely conservative	23	557	4.13	100.00
Total (*N*)	557		100.0	

first identify the observation located in the middle of the distribution and then find the category associated with that observation. Using the formula $(N + 1) \div 2$, we find that the middle observation is $(557 + 1) \div 2 = 279$. The cumulative frequency (Cf) of 279 falls in the category "moderate." Therefore, the median is "moderate."

When interval-ratio data are arranged in grouped distributions, the procedure for calculating the median is a bit more complicated. In most situations it is unlikely that you will need to calculate the median from grouped distributions by hand. We have presented an example involving a grouped distribution in Box 4.1.

> **Learning Check.** *If you are confused about cumulative distributions, review Chapter 2, pp. 46–49.*

Statistics in Practice: Changes in Age at First Marriage

We can use the median to compare groups. Consider the significant changes that have taken place during the past two decades in marriage patterns in the United States. They have profoundly influenced our lives both socially and economically. Delayed first marriage is associated with increased education and work experience for both men and women.[5]

Figure 4.4 compares the median age at first marriage for men and women in 1970 and in 1990. Because the median is a single number summarizing central tendency in the distribution, we can use it to note differences between subgroups of the population or changes over time. In this example, the increase in median age at first marriage for both men and women from 1970 to 1990 clearly shows a movement away from first marriage at an early age.

> **Learning Check.** *Examine Figure 4.4 and contrast median ages at first marriage of women and men over the two decades. What can you learn about gender and age at first marriage?*

Locating Percentiles in a Frequency Distribution

The median is a special case of a more general set of measures of location called *percentiles*. A **percentile** is a score at or below which a specific percentage of the distribution falls. The *n*th percentile is a score below which *n*

[4]This rule was adapted from David Knoke and George W. Bohrnstedt, *Basic Statistics* (New York: Peacock Publishers, 1991), pp. 56–57.

[5]U.S. Bureau of the Census, "Population Profile of the United States, 1991," *Current Population Reports*, 1991, P23-173.

Box 4.1 Finding the Median in Grouped Data

When interval-ratio variables are arranged in grouped frequency distributions, we use the following formula to locate the median:

$$\text{Median} = L + \left[\frac{N(.5) - Cf}{f}\right] w$$

where

L = the lower limit of the interval containing the median
Cf = the cumulative sum of the frequencies below the interval containing the median
f = the frequency of the interval containing the median
w = the width of the interval containing the median
N = the total number of cases

Grouped Frequency Distribution of Hours Worked: A GSS Subgroup

Hours Worked Stated Limits	Hours Worked Real Limits	Frequency (f)	Cf
00–09	–0.5– 9.5	2	2
10–19	9.5–19.5	3	5
20–29	19.5–29.5	3	8
30–39	29.5–39.5	4	12
40–49	39.5–49.5	19	31
50–59	49.5–59.5	3	34
		Total (N)	34

To illustrate the computation of the median, consider the grouped distribution of the variable "hours worked," which we first presented in Chapter 2 (Box 2.1). Here we have added a cumulated frequencies column (Cf) to the original table.

To locate the median, we first identify the middle case in the distribution by dividing the total number of cases (N) by 2. Since N = 34, the middle case is 34 ÷ 2, or 17. We need to locate the seventeenth case and identify the score corresponding to it. The cumulative frequency column shows that there are twelve observations preceding the interval 39.5–49.5. This interval contains nineteen more observations. Hence, the seventeenth observation is located within that interval and, therefore, we know that the median is located somewhere within that interval. To find the exact value of the median, we apply the formula for medians in grouped data:

$$\text{Median} = 39.5 + \left[\frac{34(.5) - 12}{19}\right] 10 = 42.1$$

The median number of hours worked is 42.1.

Figure 4.4 **Median Age at First Marriage for Men and Women, 1970 and 1990**

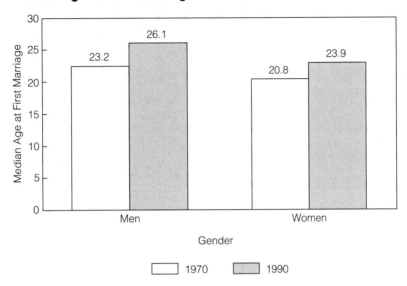

percent of the distribution falls. For example, the 75th percentile is a score that divides the distribution so that 75 percent of the cases are below it. The median is the 50th percentile. It is a score that divides the distribution so that 50 percent of the cases fall below it. Like the median, percentiles require that data be ordinal or higher in level of measurement. Percentiles are easy to identify when the data are arranged in frequency distributions.

To help illustrate how to locate percentiles in a frequency distribution, we display in Table 4.3 the frequency distribution, the percentage distribution, and the cumulative percentage distribution for the variable *number of children per family* for the 1996 GSS sample. The 50th percentile (the median) is 2 children, meaning that 50 percent of the respondents have 2 children or less (as you can see from the cumulative percentage column, 50% falls somewhere in the third category, associated with 2 children). Similarly, the 80th percentile is 3 children because 80 percent of the respondents have 3 children or less. The procedure for calculating percentiles from grouped distributions is similar to the one suggested for calculating medians. This procedure is illustrated in Box 4.2.

> *Percentile* A score below which a specific percentage of the distribution falls.

Percentiles are widely used to evaluate relative performance on standardized achievement tests, such as the SAT or ACT. Let's suppose that your ACT score was 29. To evaluate your performance for the college

Table 4.3 **Frequency Distribution for Number of Children: GSS 1996 Sample**

Number of Children	Frequency (f)	Percentage (%)	C%
0	404	28	28
1	214	15	43
2	399	28	70
3	219	15	86
4	115	8	94
5	41	3	96
6	26	2	98
7	16	1	99
8 or more	11	1	100
Total (N)	1445	100	

Box 4.2 Finding Percentiles in Grouped Data

The formula for medians shown in Box 4.1 can also be used to calculate percentiles from grouped data. The only adjustment required in the formula is to multiply the total number of cases by the desired percentile and determine the interval where the appropriate percentile is located. For instance, the following formula was adapted to find the 25th percentile for the data shown in the table in Box 4.1.

$$25\text{th percentile } (Q_1) = L + \left[\frac{N(.25) - Cf}{f}\right] w$$

$$= 29.5 + \left[\frac{34(.25) - 8}{4}\right] 10 = 30.75$$

In other words, 25 percent of the respondents worked less than 30.75 hours.

---

admissions officer, the testing service translated your score into a percentile rank. Your percentile rank was determined by comparing your score with the scores of all other seniors who took the test at the same time. Suppose for a moment that 90 percent of all students received a lower ACT score than you (and 10 percent scored above you). Your percentile rank would have been 90. If, however, there were more students who scored better than you—let's say that 15 percent scored above you and 85 percent scored lower than you—your percentile rank would have been only 85.

Another widely used measure of location is the *quartile*. The lower quartile is equal to the 25th percentile and the upper quartile is equal to the 75th percentile. (Can you locate the upper quartile in Table 4.3?) A college admissions office interested in accepting the top 25 percent of its applicants based on their SAT scores could calculate the upper quartile (the 75th percentile) and accept everyone whose score is equivalent to the 75th percentile or higher. (Note that they would be calculating percentiles based on the scores of their applicants, not of all students in the nation who took the SAT.)

> **Learning Check.** *Can you provide the formula for calculating the 20th percentile? The 20th percentile is sometimes referred to as the lower quintile. Can you determine the percentile that marks the upper quintile?*

■ ■ ■ ■ **The Mean**

The arithmetic **mean** is by far the best-known and most widely used average. The mean is what most people call the "average." The mean is typically used to describe central tendency in interval-ratio variables such as income, age, and education. You are probably already familiar with how to calculate the mean. Simply add up all the scores and divide by the total number of scores.

> *Mean*   The arithmetic average obtained by adding up all the scores and dividing by the total number of scores.

Crime statistics, for example, are often analyzed using the mean. Each year about 25 percent of U.S. households are victims of some form of crime. Although violent crimes in the United States are the least common types of crimes, they are nonetheless, the highest of any industrialized nation. For instance, murder rates in the United States are approximately five times as high as those in Europe.

Table 4.4 shows the 1995 murder rates (per 100,000 population) for the fifteen largest cities in the United States. We want to summarize the information presented in this table by calculating some measure of central tendency. Because the variable "murder rate" is an interval-ratio variable, we will select the arithmetic mean as our measure of central tendency.

To find the mean murder rate for the data presented in Table 4.4, add up the murder rates for all the cities and divide the sum by the number of cities:

$$\text{Mean} = \frac{315.7}{15} = 21.10$$

Table 4.4 **1995 Murder Rate (per 100,000 population) for the Fifteen Largest Cities in the United States**

City	Murder Rate per 100,00
New York	16.1
Los Angeles	24.5
Chicago	30.0
Houston	18.2
Philadelphia	28.2
San Diego	7.9
Phoenix	19.7
Dallas	26.5
San Antonio	14.2
Detroit	47.6
Honolulu	4.3
San Jose	4.6
Las Vegas	14.9
San Francisco	13.4
Baltimore	45.6

Source: Statistical Abstract of the United States, 1997.

The mean murder rate for the fifteen largest cities in the United States is 21.[6]

Using a Formula to Calculate the Mean

Another way to calculate the arithmetic mean is to use a formula. Beginning with this section, we introduce a number of formulas that will help you calculate some of the statistical concepts we are going to present. A *formula* is a shorthand way to explain what operations we need to follow to obtain a certain result. So instead of saying "add all the scores together and then divide by the number of scores," we can define the mean by the following formula:

$$\overline{Y} = \frac{\sum Y}{N}$$

4.1

Let's take a moment to consider these new symbols because we continue to use them in later chapters. We use Y to represent the raw scores in the distribution of the variable y; $\overline{Y}$ is pronounced "Y-bar" and is the mean of

[6]The rates presented in Table 4.4 are computed for aggregate units (cities) of different sizes. The mean of 21 is therefore called an *unweighted* mean. It is not the same as the murder rate for the population in the combined cities.

the variable y. The symbol represented by the Greek letter Σ is pronounced "sigma," and it is used often from now on. It is a summation sign (just like the + sign) and directs us to sum whatever comes after it. Therefore, ΣY means "add up all the raw y scores." Finally, the letter N, as you know by now, represents the number of cases (or observations) in the distribution.

Let's summarize:

Y = the raw scores of the variable y
$\overline{Y}$ = the mean of y
ΣY = the sum of all the y scores
N = the number of observations or cases

Now that we know what the symbols mean, let's work through another example. The following are the ages of the ten students in a graduate research methods class:

21, 32, 23, 41, 20, 30, 36, 22, 25, 27

What is the mean age of the students?

For these data the ages included in this group are represented by Y; $N = 10$, the number of students in the class; and ΣY is the sum of all the ages:

$$\Sigma Y = 21 + 32 + 23 + 41 + 20 + 30 + 36 + 22 + 25 + 27 = 277$$

Thus, the mean age is

$$\overline{Y} = \frac{\Sigma Y}{N} = \frac{277}{10} = 27.7$$

The mean can also be calculated when the data are arranged in a grouped distribution. We have presented an example involving a grouped distribution in Box 4.3.

Learning Check. *The following distribution is the same as the one you used to calculate the median in an earlier Learning Check: 22, 15, 18, 33, 17, 5, 11, 28, 40, 19, 8, 20. Can you calculate the mean? Is it the same as the median, or is it different?*

Understanding Some Important Properties of the Arithmetic Mean

The following three mathematical properties make the mean the most important measure of central tendency. It is, in fact, a concept that is basic to numerous and more complex statistical operations.

Interval-Ratio Level of Measurement   Because it requires the mathematical operations of addition and division, the mean can only be calculated for variables measured at the interval-ratio level. This is the only level of measurement that provides numbers that can be added and divided.

Box 4.3   Finding the Mean in a Frequency Distribution

When data are arranged in a frequency distribution, we must give each score its proper weight by multiplying it by its frequency. We can use the following modified formula to calculate the mean:

$$\bar{Y} = \frac{\Sigma fY}{N}$$

where

$\bar{Y}$ = the mean
fY = a score multiplied by its frequency
ΣfY = the sum of all the fY's
N = the total number of cases in the distribution

We now illustrate how to calculate the mean from a frequency distribution using the preceding formula. In the 1988 General Social Survey respondents were asked what they think is the ideal number of children for a family. Their responses are presented in the following table.

Ideal Number of Children: GSS 1988

Number of Children (Y)	Frequency (f)	Frequency × Y (fY)
0	12	0
1	25	25
2	733	1466
3	333	999
4	183	732
5	26	130
6	15	90
7	12	84
Total	$N = 1{,}339$	$\Sigma fY = 3{,}526$

Notice that to calculate the value of ΣfY (column 3), each score (column 1) is multiplied by its frequency (column 2), and the products are then added together. When we apply the formula

$$\bar{Y} = \frac{\Sigma fY}{N}$$

$$\bar{Y} = \frac{3{,}526}{1{,}339} = 2.6$$

we find that the mean for the ideal number of children is 2.6.

Learning Check. If you are having difficulty understanding how to find the mean in a frequency distribution examine this table. It explains the process without using any notation.

Finding the Mean in a Frequency Distribution

	Number of people per house	Number of houses like this	Number of people such houses contribute
	1	3	3
	2	5	10
	3	1	3
	4	1	4

Total number of people: 20 Total number of houses: 10
Mean number of people per house: 20 ÷ 10 = 2

Here is another example that requires finding the average education of working-class and middle-class African American women.* We will calculate the mean from the following frequency distributions. The tables are frequency distributions of years of education of middle-class and working-class African American women. Try to apply the formula for means in a frequency distribution to the two tables to find the mean education for the two groups.

Years of Education of Working-Class African American Women: GSS 1988 to 1990

Education (Y)	Frequency (f)	Frequency × Y (fY)
4	1	4
6	3	18
7	2	14
8	2	16
9	5	45
10	3	30
11	9	99

(continued on next page)

*Social class rank is based on the respondents' self-identification.

Box 4.3 *Finding the Mean in a Frequency Distribution (continued)*

Years of Education of Working-Class African American Women: GSS 1988 to 1990 *(continued)*

Education (Y)	Frequency (f)	Frequency × Y (fY)
12	16	192
13	4	52
14	8	112
15	1	15
16	4	64
17	1	17
18	2	36
20	1	20
Total	$N = 62$	$\Sigma fY = 734$

$$\overline{Y} = 11.8$$

Years of Education of Middle-Class African American Women: GSS 1988 to 1990

Education (Y)	Frequency (f)	Frequency × Y (fY)
4	1	4
6	3	18
7	1	7
8	2	16
9	2	18
10	3	30
11	7	77
12	12	144
13	6	78
14	10	140
15	2	30
16	6	96
17	0	0
18	6	108
20	2	40
Total	$N = 63$	$\Sigma fY = 806$

$$\overline{Y} = 12.8$$

Examine the tables showing years of education for working-class and middle-class African American women. Note how similar working-class and middle-class African American women are on education. This similarity is striking given that education is a major component of social class. You may want to explore this issue further by comparing the level of education of working-class and middle-class European American women or of working-class and middle-class men of both races. SPSS Problem 5 at the end of this chapter provides specific instructions that will help you explore this question further.

Center of Gravity   Because the mean (unlike the mode and the median) incorporates *all* the scores in the distribution, we can think of it as the *center of gravity* of the distribution. That is, the mean is the point that perfectly balances all the scores in the distribution. If we subtract the mean from each score and add up all the differences, the sum will always be zero!

Learning Check.   *Why is the mean considered the center of gravity of the distribution? Think of the last time you were in a park on a seesaw (it may have been a long time ago) with a friend who was much heavier than you. You were left hanging in the air until your friend moved closer to the center. In short, to balance the seesaw a light person far away from the center (the mean) can balance a heavier person who is closer to the center. Can you illustrate this principle with a simple income distribution?*

Illustrating the Seesaw Principle

a. Three men, weights 60, 120, and 180, all stand on a seesaw. The fulcrum is placed at 120. The mean is (60 +120 + 180)/3. The seesaw balances.

060 070 080 090 100 110 120 130 140 150 160 170 180 190 200 210 220 230 240

b. The 180-pound man is replaced by a 240-pound man, but we do not move the fulcrum. The seesaw slowly falls to the right.

060 070 080 090 100 110 120 130 140 150 160 170 180 190 200 210 220 230 240

(continued on next page)

c. We move the fulcrum to 140. The new mean is (60 + 120 + 240)/3. The seesaw balances again.

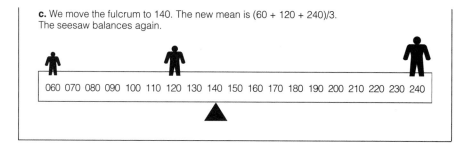

060 070 080 090 100 110 120 130 140 150 160 170 180 190 200 210 220 230 240

Sensitivity to Extremes   The examples we have used to show how to compute the mean demonstrate that, unlike with the mode or the median, every score enters into the calculation of the mean. This property makes the mean sensitive to extreme scores in the distribution. The mean is pulled in the direction of either very high or very low values. A glance at Figure 4.5 should convince you of that. Figures 4.5a and 4.5b each show the incomes of ten individuals. In Figure 4.5b, the income of one individual has shifted from $5,000 to $35,000. Notice the effect it has on the mean; it shifts from $3,000 to $6,000! The mean is disproportionately affected by the relatively high income of $35,000 and is misleading as a measure of central tendency for this distribution. Notice that the median's value is not affected by this extreme score; it remains at $3,000. Thus, the median gives us better information on the typical income for this group. In the next section we will see that because of the sensitivity of the mean, it is not suitable as a measure of central tendency in distributions that have a few very extreme values on one side of the distribution. (A few extreme values are no problem if they are not mostly on one side of the distribution.)

> **Learning Check.**   *When asked to choose the appropriate measure of central tendency for a distribution, remember that the level of measurement is not the only consideration. When variables are measured at the interval-ratio level, the mean is usually the measure of choice, but remember that extreme scores in one direction make the mean unrepresentative and the median or mode may be the better choice.*

■   ■   ■   ■   **The Shape of the Distribution:
The Experience of Traumatic Events**

In this chapter we have looked at the way in which the mode, median, and mean reflect central tendencies in the distribution. Distributions (this discussion is limited to distributions of interval-ratio variables) can also be described by their general shape, which can be easily represented visually. A distribution can be either *symmetrical* or *skewed*, depending on whether there are a few extreme values at one end of the distribution.

Figure 4.5 **The Value of the Mean Is Affected by Extreme Scores**

a. No extreme scores: the mean is $3,000

Income (Y)	Frequency (f)	fY
1,000	1	1,000
2,000	2	4,000
3,000	4	12,000
4,000	2	8,000
5,000	1	5,000
	$N = 10$	$\Sigma fY = 30,000$

$$\text{Mean} = \frac{\Sigma fY}{N} = \frac{30,000}{10} = \$3,000$$

Median = $3,000

b. One extreme score: the mean is $6,000

Income (Y)	Frequency (f)	fY
1,000	1	1,000
2,000	2	4,000
3,000	4	12,000
4,000	2	8,000
⋮	⋮	⋮
35,000	1	35,000
	$N = 10$	$\Sigma fY = 60,000$

$$\text{Mean} = \frac{\Sigma fY}{N} = \frac{60,000}{10} = \$6,000$$

Median = $3,000

A distribution is **symmetrical** (Figure 4.6a) if the frequencies at the right and left tails of the distribution are identical, so that if it is divided into two halves, each will be the mirror image of the other. In a *unimodal* symmetrical distribution the mean, median, and mode are identical.

In **skewed** distributions, there are a few extreme values on one side of the distribution. Distributions that have a few extremely low values are referred to as **negatively skewed** (Figure 4.6b), and those with a few extremely high values are said to be **positively skewed** (Figure 4.6c). In a negatively skewed distribution, the mean will be pulled in the direction of

Figure 4.6   **Types of Frequency Distributions**

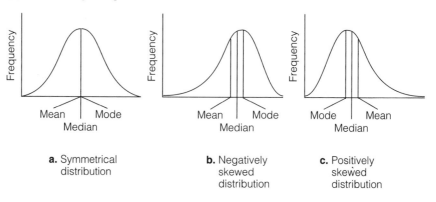

a. Symmetrical distribution

b. Negatively skewed distribution

c. Positively skewed distribution

the lower scores; in a positively skewed distribution, it will be pulled toward the higher scores.

We can illustrate the differences among symmetrical and positively and negatively skewed distributions by comparing how different groups in our society experience traumatic events. Severe illness, death in the family, divorce, and unemployment are traumatic events that are a part of life. But are some of us more prone to trauma than others? Are some groups likely to suffer more traumatic events than others?

Every year the GSS asks a sample of respondents to report on the number of different traumatic events (deaths, divorces, unemployment, and hospitalizations or disabilities) that happened to them during the previous five years. Here we look at the responses to this question by the following three groups: (1) working-class black males, (2) middle-class black males, and (3) working-class black females. The frequency distributions for the number of traumas for these three groups are presented in Tables 4.5 through 4.7, and the corresponding bar graphs are depicted in Figures 4.7 through 4.9.

Symmetrical distribution   The frequencies at the right and left tails of the distribution are identical; each half of the distribution is the mirror image of the other.

Skewed distribution   A distribution with a few extreme values on one side of the distribution.

Negatively skewed distribution   A distribution with a few extremely low values.

Positively skewed distribution   A distribution with a few extremely high values.

The Symmetrical Distribution

First, let's examine Table 4.5 and Figure 4.7, displaying the distribution of trauma reported by working-class black males. Notice that the largest number (31) experienced 1 trauma during the last five years (mode = 1), and about an equal number (23 and 20, respectively) reported either 0 or 2 traumas. As shown in Figure 4.7, the mode, median, and mean are almost identical, and they coincide at about the middle of the distribution.

Table 4.5 **Number of Traumas During the Last Five Years: Working-Class Black Males**

Number of Traumas (Y)	Frequency (f)	fY	Percentages (%)	C%
0	23	0	29.9	29.9
1	31	31	40.2	70.1
2	20	40	26.0	96.1
3+*	3	9	3.9	100.0
Total	77	80	100.0	

$$\bar{Y} = \frac{\sum fY}{N} = \frac{80}{77} = 1.04$$

Median = 1.00
Mode = 1.00

*The category 3+ is assumed to be 3 for the purpose of calculating the mean.

Figure 4.7 **Number of Traumas During the Last Five Years: Working-Class Black Males**

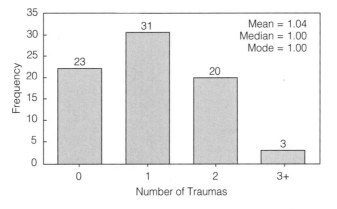

Source: General Social Survey, 1988 to 1990.

The distribution of traumas for working-class black males as depicted in Table 4.5 and Figure 4.7 is a nearly symmetrical distribution. The mean, median, and mode are almost identical, and the distribution below the center (where the mean, median, and mode are located) is almost a mirror image of that above the center.

The Positively Skewed Distribution

Now let's examine Table 4.6 and Figure 4.8, displaying the distribution of traumas reported by middle-class black males. Note that the largest num-

Table 4.6 **Number of Traumas During the Last Five Years: Middle-Class Black Males**

Number of Traumas (Y)	Frequency (f)	fY	Percentages (%)	$C\%$
0	22	0	38.6	38.6
1	15	15	26.3	64.9
2	11	22	19.3	84.2
3+	9	27	15.8	100.0
Total	57	64	100.0	

$$\bar{Y} = \frac{\sum fY}{N} = \frac{64}{57} = 1.12$$

Median = 1.00
Mode = 0.00

Figure 4.8 **Number of Traumas During the Last Five Years: Middle-Class Black Males**

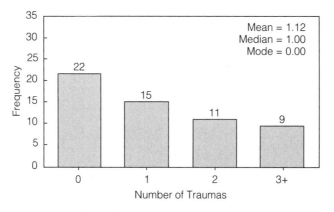

Mean = 1.12
Median = 1.00
Mode = 0.00

Source: General Social Survey, 1988 to 1990.

ber of respondents (22) is concentrated at the low end of the scale (0 traumas), with few people reporting that they experienced a high number of traumas (3+). Notice also that in this distribution the mean, median, and mode have different values, with the mean having the highest value (mean = 1.12), the median the second highest value (median = 1.00), and the mode the lowest value (mode = 0.00). The distribution of traumas for middle-class black males as depicted in Table 4.6 and Figure 4.8 is positively skewed. As a general rule, for skewed distributions the mean, median, and mode do not coincide. The mean, which is always pulled in the direction of extreme scores, falls closest to the tail of the distribution where a small number of extreme scores are located.

The Negatively Skewed Distribution

Now turn to Table 4.7 (below) and Figure 4.9 (page 136) for working-class black females. Here you can see the opposite pattern. The distribution of the number of traumas reported by working-class black women is a negatively skewed distribution. First, note that the largest numbers of women are concentrated at the high end of the scale (3+ traumas) and that there are fewer women at the low end. The mean, median, and mode also differ in values as they did in the previous example. However, here the mode has the highest value (mode = 3.00+), the median has the second highest (median = 2.00), and the mean has the lowest value ($\overline{Y} = 1.76$).

Table 4.7 **Number of Traumas During the Last Five Years: Working-Class Black Females**

Number of Traumas (Y)	Frequency (f)	fY	Percentages (%)	C%
0	13	0	25.5	25.5
1	6	6	11.8	37.3
2	12	24	23.5	60.8
3+	20	60	39.2	100.0
Total	51	90	100.0	

$$\overline{Y} = \frac{\Sigma fY}{N} = \frac{90}{51} = 1.76$$

Median = 2.00

Mode = 3.00

Figure 4.9   **Number of Traumas During the Last Five Years: Working-Class Black Females**

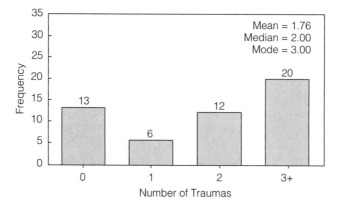

Source: General Social Survey, 1988 to 1990.

Guidelines for Identifying the Shape of a Distribution

Following are some useful guidelines for identifying the shape of a distribution.

1. In unimodal distributions, when the median, mode, and mean coincide or are almost identical, the distribution is symmetrical.
2. When the mean is higher than the median (or is positioned to the right of the median), the distribution is positively skewed.
3. When the mean is lower than the median (or is positioned to the left of the median), the distribution is negatively skewed.

> ***Learning Check.***   *To identify positively and negatively skewed distributions, look at the tail on the chart. If the tail points to the right (the positive end of the x-axis), the distribution is positively skewed. If the tail points to the left (the negative, or potentially negative, end of the x-axis), the distribution is negatively skewed.*

■ ■ ■ ■   **Considerations for Choosing a Measure of Central Tendency**

So far we have considered three basic kinds of averages: the mode, the median, and the mean. Each can represent a central tendency of a distribution. But which one should we use? the mode? the median? the mean? Or, perhaps, all of them? There is no simple answer to this question. However, in general, we tend to use only one of the three measures of central tendency, and the choice of the appropriate one involves a number of considerations. These considerations and how they affect our choice of the appropriate measure are presented in the form of a decision tree in Figure 4.10.

Figure 4.10 **How to Choose a Measure of Central Tendency**

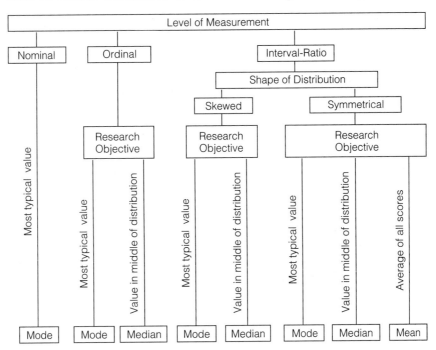

Level of Measurement

One of the most basic considerations in choosing a measure of central tendency is the variable's level of measurement. Valid use of any of the three measures requires that the data be measured at the level appropriate for that measure or higher. Thus, as shown in Figure 4.10, with nominal variables our choice is restricted to the mode as a measure of central tendency.

However, with ordinal data we have two choices: the mode or the median (or sometimes both). Our choice will depend on what we want to know about the distribution. If we are interested in showing what is the most common or typical value in the distribution, then our choice is the mode. If, however, we want to show which value is located exactly in the middle of the distribution, then the median is our measure of choice.

When the data are measured on an interval-ratio level, the choice between the appropriate measures is a bit more complex and is restricted by the shape of the distribution.

Skewed Distribution

When the distribution is skewed, the mean may give misleading information on the central tendency because its value is affected by extreme scores in the distribution. The median (see, for example, Box 4.4) or the mode can

Box 4.4   Statistics in Practice:
Median Annual Earnings Among Subgroups

Personal income is frequently positively skewed because there are fewer people with high income; therefore, studies on earnings often report median income. The mean tends to overestimate both the earnings of the most typical earner (the mode) and the earnings represented by the 50th percentile (the median). In the following example the median is used to compare annual earnings of white, black, and Hispanic men and women.

Median Annual Earnings by Race and Gender,
Full-Time Year-Round Workers, 1997

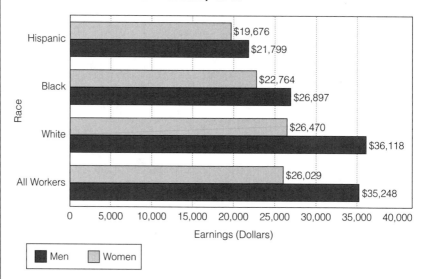

Source: U.S. Bureau of Census, *Current Population Report,* 1998, P60-200, Table 7.

The graph compares the 1997 median annual earnings of full-time male and female workers in the entire population and among white, black, and Hispanic workers. Since the earnings of white males are the highest in comparison with all other groups, it is useful to look at each group's median earnings relative to the earnings of white males. For example, white women were paid just 73 cents for every $1 paid to white men ($26,470/36,118 = .73). For women of color, the gap is greater. In 1997 black women were paid approximately 63 cents ($22,764/36,118 = .63) and Hispanic women 54 cents ($19,676/36,118 = .54) for every $1 paid to white men.

be chosen as the preferred measure of central tendency because neither is influenced by extreme scores. For instance, for both middle-class black males and working-class black females (Figures 4.8 and 4.9), the mean does not provide as accurate a representation of the "typical" number of traumas an individual has experienced during the last five years as the median and the mode. Thus, either one could be used as an "average," depending on the research objective.

Symmetrical Distribution

When the distribution we want to analyze is symmetrical, we can use any of the three averages. Again, our choice depends on the research objective and what we want to know about the distribution. In general, however, the mean will be our best choice because it contains the greatest amount of information and is easier to use in more advanced statistical analyses.

MAIN POINTS

- The mode, the median, and the mean are measures of central tendency—numbers that describe what is average or typical about the distribution.

- The mode is the category or score with the largest frequency (or percentage) in the distribution. It is often used to describe the most commonly occurring category of a nominal level variable.

- The median is a measure of central tendency that represents the exact middle of the distribution. It is calculated for variables measured on at least an ordinal level of measurement.

- The mean is typically used to describe central tendency in interval-ratio variables, such as income, age, or education. We obtain the mean by summing all the scores and dividing by the total (N) number of scores.

- In a symmetrical distribution the frequencies at the right and left tails of the distribution are identical. In skewed distributions there are either a few extremely high (positive skew) or a few extremely low (negative skew) values.

KEY TERMS

mean
measures of central tendency
median
mode
negatively skewed distribution

percentile
positively skewed distribution
skewed distribution
symmetrical distribution

SPSS DEMONSTRATIONS

Demonstration 1: Producing Measures of Central Tendency
with Frequencies [GSS1996/Module A]

The Frequencies command, which we demonstrated in Chapter 2, also has
the ability to produce the three measures of central tendency discussed in
this chapter. We will use Frequencies to calculate measures of central tendency for AGE and FEHIRE ("Employers should make special efforts to
hire and promote women").

Click under *Analyze, Descriptive Statistics,* then *Frequencies.* Place AGE
and FEHIRE in the Variable(s) box. Then click on the *Statistics* button. You
will see that the Central Tendency box (Figure 4.11) lists four choices, but
we will only click on the first three. Then click on *Continue,* then on *OK* to
process this request.

Figure 4.11

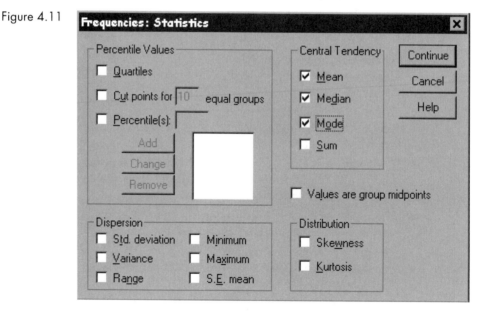

The frequency table for AGE is quite lengthy, so only the Statistics box is
displayed here (Figure 4.12). AGE is an interval-ratio variable, which means
that the mode, median, and mean are all appropriate measures of central
tendency. The mean of AGE is about 44.54, but remember that the GSS file
includes only adults, so the mean value (and the other measures of central
tendency) should not be taken as representative of the American population
as a whole.

The median is 42, which is close to the mean value. Roughly half of the
respondents are above the age of 42 and half are below. However, since the

Figure 4.12 ▶ **Frequencies**

Statistics

		AGE OF RESPON DENT	should hire and promote women
N	Valid	1422	710
	Missing	5	717
Mean		44.54	2.65
Median		42.00	2.00
Mode		33ᵃ	2

a. Multiple modes exist. The smallest value is shown

median is lower than the mean, it implies that the distribution of age is positively skewed. The mode is 33, so more people are 33 years old than any other age (43 people, to be exact). The output indicates in the summary box that there are multiple modes. Can you identify the other mode(s)?

In the same Statistics box, measures of central tendency are also displayed for FEHIRE. SPSS has no idea that this variable is measured on an ordinal scale. In other words, SPSS produces exactly the output we asked for, without regard for whether the output is correct for this type of variable. It's up to you to select the proper measure of central tendency.

Since FEHIRE is an ordinal variable, we can use the median and mode to summarize its distribution. The median is 2.0, which corresponds to the response "agree." We can confirm from the cumulative percentages that about 55 percent either "strongly agree" or "agree" with this statement, while the remaining 45 percent are either neutral or "disagree" or "strongly disagree." The mode is 2, which means that the most frequent response is to "agree" that there should be efforts to hire and promote women in the workplace.

Demonstration 2: Producing Measures of Central Tendency with Descriptives

When you want to calculate the mean of interval-ratio variables but you don't need to view the actual frequency table listing the responses in each category, the Descriptives procedure is often the best choice. Descriptives can be found in the *Statistics–Summarize* menu.

The Descriptives dialog box (Figure 4.13) is uncomplicated and only requires that you place the variables of interest in the Variable(s) box. By default, Descriptives will calculate the mean, standard deviation (to be discussed in Chapter 5), minimum, maximum, and the number of cases with a valid response.

Figure 4.13

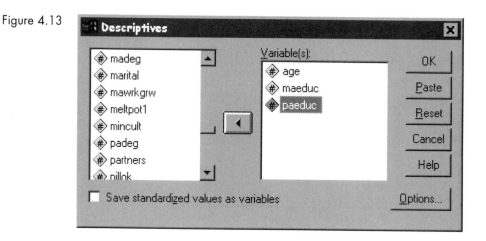

Figure 4.14 displays these descriptive statistics for AGE, MAEDUC, and PAEDUC (respondent's age and the highest year of school completed by respondent's mother and father).

The output from Descriptives automatically lists the variables in the order we specified in the dialog box. The mean of AGE is the same as SPSS calculated in Frequencies. For MAEDUC, mother's highest year of schooling, we can see that the mean value of 11.52 is slightly higher than the mean reported for father's educational attainment (PAEDUC), 11.49.

Figure 4.14 ▸ **Descriptives**

Descriptive Statistics

	N	Minimum	Maximum	Mean	Std. Deviation
AGE OF RESPONDENT	1422	18	89	44.54	17.03
HIGHEST YEAR SCHOOL COMPLETED, MOTHER	1214	0	20	11.52	3.39
HIGHEST YEAR SCHOOL COMPLETED, FATHER	1008	0	20	11.49	4.08
Valid N (listwise)	940				

SPSS PROBLEMS

1. In Chapter 2, we reviewed the frequencies for FEFAM. [Module A]
   a. Which measure of central tendency, mean or median, is most appropriate to summarize the distribution of FEFAM? Explain why.

b. Suppose we are interested in whether males and females have the same attitudes about women working outside the home? Run frequencies and selected statistics for FEFAM, this time separating results for men and women. (Use the *Data–Split File* command. Remember to reset the *Split File* to include all cases when you are done comparing men and women.) Are there any differences in their measures of central tendency? Explain.

c. Repeat b, this time using FEIMP (the importance of women's issues to the respondent) as your variable.

2. We are interested in investigating whether males and females have equal levels of education. Use the variable EDUC with the Frequencies procedure to produce frequency tables and the mean, median, and mode separately for males and females (as described in 1b). [Module A]

a. On average, do men and women have equal levels of education? Use all the available information to answer this question.

b. When we use statistics to describe the social world, we should always go beyond merely using statistics to describe the condition of various social groups. Just as important is our interpretation of the statistics and some judgment as to whether any differences we find between groups seem of practical importance; that is, do they make a practical difference in the world? Do you think any differences you discovered between male and female educational levels are important enough to have an effect on such things as the ability to get a job or the salary that someone makes? Defend your answer.

3. Some people believe that minorities have more children than whites. Use SPSS to investigate this question with either GSS data file. (The variable CHILDS records the respondent's number of children.) To get the necessary information, have SPSS split the file by RACE, then run Frequencies for CHILDS. [Module A]

a. What is the best measure of central tendency to represent the number of children in a household? Why?

b. Which race has more children per respondent?

c. Rerun your analysis, this time with the variable CHLDIDEL (ideal number of children). Is there a difference among the racial categories? Explain.

4. Picking an appropriate statistic to describe the central tendency of a distribution is a critical skill. Based on the GSS [Module A], determine the appropriate measure(s) of central tendency for the following variables:

a. How often do respondents attend church? [ATTEND]

b. Respondents' political views. [POLVIEWS]

c. The marital status of those sampled. [MARITAL]

d. Income of respondents. [INCOME]

e. How successful respondents are in their family life. [FAMLIFE]

f. How much housework does the respondent's spouse do? [SPHMEWRK]

5. The educational attainment of working-class and middle-class African American women was compared in Box 4.3. You can use SPSS to do the same for European American women and for men of both races. There is more than one method to get the frequency distribution of education for these groups, but the easiest might be to use the *Split File* menu choice. Use the 1996 GSS [Module A] for this exercise.

   a. In the Split File dialog box (found under the *Data* menu), click on the "Repeat analysis for each group" then place the variables RACESEX and CLASS (in that order) in the Groups Based On box. RACESEX has four categories: 1 = white males, 2 = white females, 3 = black males, and 4 = black females. CLASS has five valid values, with working-class = 2 and middle-class = 3. After you run the Split File procedure, SPSS will create a separate set of output for each group defined by the combination of the values of RACESEX and CLASS.

   b. Now run Frequencies on the variable EDUC. SPSS will create a great deal of output; all you need to do is find the appropriate frequency tables and calculate the mean from the table. For example, to find the frequency table for middle-class black males, look for the section with values of RACESEX = 3 and CLASS = 3.

   c. Is the gap in education between African American working-class and middle-class men greater than that between European American working-class and middle-class men? For females?

CHAPTER EXERCISES

1. In Chapter 2, Exercise 1, you surveyed thirty people about their race (white or nonwhite) and how many traumas they had experienced in the past year. You also asked each about his or her self-perceived social class. Find the mode for all three variables.

2. You are interested in understanding public attitudes toward women working outside of the home. The General Social Survey 1996 contains a question about whether the respondent agrees or disagrees that it is better for a man to work and a woman to stay at home. Here is a random sample of 490 responses to this question.

Attitude	Frequency
Strongly agree	39
Agree	138
Disagree	223
Strongly disagree	90

   a. At what level is this variable measured? What is the mode for attitude toward women working outside of the home?

   b. Calculate the median for this variable. (*Hint:* The number of cases is even.) In general, how would you characterize the public's attitude?

3. In Chapter 2, Exercise 4, we looked at the level of education of the American population, based on 1996 GSS data.
   a. What is the level of measurement for "years of education"? What is the mode for education? What is the median for education?
   b. Construct quartiles for education. What is the 25th percentile? the 50th percentile? the 75th percentile? Why don't you need to calculate the 50th percentile to answer this question?

4. Using a sample from the 1996 GSS, you find the following grouped distribution for respondent's age:

Age Category	Frequency
18–29	114
30–39	130
40–49	131
50–59	84
60-69	63
70–89	44

   a. Calculate the median for age, using the formula for a grouped distribution. What is its value?
   b. Also calculate the 20th and 80th percentile values for age.

5. Religion has been and continues to be important to many Americans. Nevertheless, there are demographic differences in religious behavior, including age. The following table, taken from the GSS, depicts how often people pray within various age groups (not all ages are displayed).

	Age Group			
Prayer Frequency	18–29	30–39	40–49	50–59
Several times a day	3	12	13	8
Once a day	12	12	14	9
Several times a week	5	2	5	6
Once a week	8	2	3	1
Less than once a week	9	9	9	8

   a. Calculate the median and mode for each age group.
   b. Use this information to characterize how prayer behavior varies by age. Does the median or mode provide a better description of the data? Do the statistics support the idea that there is a prayer "generation gap," such that some age groups engage in more prayer than others?

6. AIDS is a serious health problem for this country (and many others). Data from the National Centers for Disease Control for 1996 and 1997 show the number of AIDS cases in various Metropolitan Statistical Areas (the top ten MSAs are listed).

Metropolitan Statistical Area	1996	1997
New York	10,357	9,897
Miami	2,049	1,672
Newark	1,424	1,356
Washington, DC	2,155	1,789
Baltimore	1,522	1,277
Chicago	1,838	1,568
Los Angeles	3,654	2,629
Boston	1,097	766
Fort Lauderdale	1,200	1,015
Philadelphia	1,669	1,492

Source: Data adapted from Centers for Disease Control and Prevention, *HIV/AIDS Surveillance Report*, Vol. 9, no. 2 (1997), Table 2, pp. 8–9.

a. Calculate the mean number of AIDS cases in these urban areas for both 1996 and 1997. How would you characterize the difference in the number of AIDS cases between 1996 and 1997? Does the mean adequately represent the central tendency of the distribution of AIDS cases in each year? Why or why not?

b. Recalculate the mean for each year after removing the New York MSA. Is the mean now a better representation of central tendency for the remaining nine MSAs? Explain.

7. U.S. households have become smaller over the years. The following table from the 1996 GSS lists the number of people aged 18 years or older living in a household.

Number of People	Frequency
1	190
2	316
3	54
4	17
5	2
6	2

Calculate the mean number of people living in a U.S. household in 1996.

8. In Exercise 6 you calculated the mean number of AIDS cases. We now want to test whether the distribution of AIDS cases is symmetrical or skewed.

a. Calculate the median and mode for each year, using all MSAs. Based on these results and the means, how would you characterize the distribution of AIDS cases for each year?

b. What value best represents the central tendency of each distribution?

c. If you found the distributions to be skewed, what is the statistical cause?

9. In Exercise 7 you examined U.S. household size in 1996. Using these data, construct a bar chart to represent the distribution of household size.
   a. From the appearance of the bar chart, would you say the distribution is positively or negatively skewed? Why?
   b. Now calculate the median and mode for the distribution. Do these numbers provide further evidence to support your decision about how the distribution is skewed? Why do you think the distribution of household size is asymmetrical?

10. Exercise 3 used GSS data on the educational level of U.S. adults.
   a. Calculate the mean for years of education.
   b. Compare the value of the mean with those you have already calculated for the median and mode. Without constructing a bar chart, describe whether and how the distribution of years of education is skewed.

11. You listen to a debate between two politicians discussing the economic health of the United States. One politician says that the average income of U.S. adults is $25,000; the other says that the average American makes only $20,000, so Americans are not as well off as the first politician claims. Is it possible for both these politicians to be correct? If so, explain how.

12. You have examined the educational level of Americans in several previous exercises. Discuss the advantages and disadvantages of all three measures of central tendency when they are used to summarize with one number the years of education of American adults. Are you confident that one of these three is the best measure of central tendency for education? If so, why?

13. Do murder rates in cities vary with city size? Investigate this question using the following data for large and small cities. Calculate the mean and median for each group of cities. Where is the murder rate highest? Do the mean and median have the same pattern for the two groups?

Murder Rate per 100,000 in 1995			
Large Cities	Murder Rate	Small Cities	Murder Rate
New York	16.1	Norfolk, VA	21.7
Los Angeles	24.5	Riverside, CA	12.4
Chicago	30.0	St. Petersburg, FL	12.4
Houston	18.2	Raleigh, NC	7.5
Philadelphia	28.2	Lexington-Fayette, KY	5.8
San Diego	7.9	Rochester, NY	23.0
Phoenix	19.7	Baton Rouge, LA	28.4
Dallas	26.5	Jersey City, NJ	11.0
San Antonio	14.2	Stockton, CA	18.8
Detroit	47.6	Akron, OH	8.1

14. The Gallup Poll (May 29, 1998) asked white and nonwhite adults about their opinion on U.S. race relations. Responses to the question "How would you rate the state of race relations in the United States these days?" for each group are presented in the following table. (Those who answered no opinion have been omitted.)

State of Race Relations	White	Nonwhites
Very good	41	5
Somewhat good	281	40
Neither good nor bad	182	32
Somewhat bad	248	59
Very bad	58	29
Total	810	165

Source: Data from David Moore, The Gallup Poll, May 29, 1998. Used by permission of The Gallup Poll.

Calculate the appropriate measure of central tendency for whites and nonwhites. How would you describe their position on the state of race relations in the United States?

15. The following table shows the number of marriages and divorces (including reported annulments) in December 1997 for ten states.

State	Number of Marriages	Number of Divorces
Florida	11,716	6,505
New York	11,303	4,849
Ohio	8,263	4,432
Illinois	6,642	4,024
Nevada	6,161	2,866
New Jersey	5,699	2,206
Virginia	4,823	3,104
North Carolina	4,197	2,778
Georgia	4,016	2,353
Michigan	3,994	3,366

Source: Data from the National Center for Health Statistics, *Monthly Vital Statistics Report,* Vol. 46, no. 12 (July 28, 1998), Table 3.

a. Calculate the mean number of marriages and divorces for the ten states.

b. Recalculate the mean number of marriages and divorces without Florida figures. Explain how your mean numbers have changed.

5   Measures of Variability

KEY TERMS
SPSS DEMONSTRATIONS
SPSS PROBLEMS
CHAPTER EXERCISES

■ ■ ■ ■ **Introduction**

In the last chapter we looked at measures of central tendency: the mean, the median, and the mode. With these measures, we can use a single number to describe what is average for or typical of a distribution. Although measures of central tendency can be very helpful, they tell only part of the story. In fact, when used alone they may mislead rather than inform. Another way of summarizing a distribution of data is by selecting a single number that describes how much variation and diversity there is in the distribution. Numbers that describe diversity or variation are called **measures of variability**. Researchers often use measures of central tendency along with measures of variability to describe their data.

> *Measures of variability* Numbers that describe diversity or variability in the distribution.

In this chapter we discuss five measures of variability: the index of qualitative variation, the range, the interquartile range, the standard deviation, and the variance. Before we discuss these measures, let's explore why they are important.

■ ■ ■ ■ **The Importance of Measuring Variability**

The importance of looking at variation and diversity can be illustrated by thinking about the differences in the experiences of U.S. women. Are women united by their similarities or divided by their differences? The answer is *both*. To address the similarities without dealing with differences is "to misunderstand and distort that which separates as well as that which binds women together."[1] Even when we focus on one particular group of women, it is important to look at the differences as well as the commonalities. Take, for example, Asian American women. As a group they share a number of characteristics.

> Their participation in the workforce is higher than that of women in any other ethnic group. Many . . . live life supporting others, often

[1]Johnneta B. Cole, "Commonalities and Differences," in *Race, Class, and Gender*, ed. Margaret L. Andersen and Patricia Hill Collins (Belmont, CA: Wadsworth, 1992), pp. 128–129.

allowing their lives to be subsumed by the needs of the extended family. . . . However, there are many circumstances when these shared experiences are not sufficient to accurately describe the condition of a particular Asian American woman. Among Asian American women there are those who were born in the United States . . . and . . . those who recently arrived in the United States. Asian American women are diverse in their heritage or country of origin: China, Japan, the Philippines, Korea . . . and . . . India. . . . Although the majority of Asian American women are working class—contrary to the stereotype of the "ever successful" Asians—there are poor, "middle-class," and even affluent Asian American women.[2]

As this example illustrates, one basis of stereotyping is treating a group as if it is totally represented by its central value, ignoring the diversity within the group. Sociologists often contribute to this type of stereotyping when their empirical generalizations, based on a statistical difference between averages, are interpreted in an overly simplistic way. All this argues for the importance of using measures of variability as well as central tendency whenever we want to characterize or compare groups. Whereas the similarities and commonalties in the experiences of Asian American women are depicted by a measure of central tendency, the diversity of their experiences can be described only by using measures of variation.

The concept of variability has implications not only for describing the diversity of social groups such as Asian American women, but also for issues that are important in your everyday life. One of the most important issues facing the academic community is how to reconstruct the curriculum to make it more responsive to students' needs. Let's consider the issue of statistics instruction on the college level.

Statistics is perhaps the most anxiety-provoking course in any social science curriculum. Statistics courses are often the last "roadblock" preventing students from completing their major requirements. One factor, identified in numerous studies as a handicap for many students, is the "math anxiety syndrome." This anxiety often leads to a less than optimum learning environment, with students often trying to memorize every detail of a statistical procedure rather than understand the general concept involved.

Let's suppose that a university committee is examining the issue of how to better respond to the needs of students. In its attempt to evaluate statistics courses offered in different departments, the committee compares the grading policy in two courses. The first, offered in the sociology department, is taught by Professor Brown; the second, offered through the school of social work, is taught by Professor Yamato. The committee finds that over the years the average grade for Professor Brown's class has been C+. The average grade in Professor Yamato's class is also C+. We could easily be misled by these statistics into thinking that the grading policy of both instructors is about the same. However, we need to look more closely into

Figure 5.1 **Distribution of Grades for Professors Brown and Yamato's Statistics Classes**

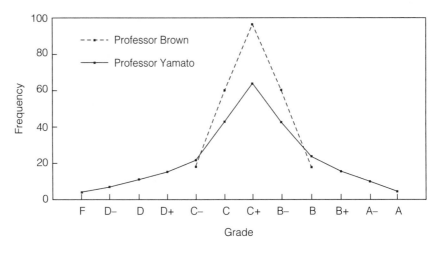

how the grades are distributed in each of the classes. The differences in the distribution of grades are illustrated in Figure 5.1, which displays the frequency polygon for the two classes.

Compare the shapes of these two distributions. Notice that while both distributions have the same mean, they are shaped very differently. The grades in Professor Yamato's class are more spread out, ranging from A to F, while the grades for Professor Brown's class are clustered around the mean and range only from B to C–. Although the means for both distributions are identical, the grades in Professor Yamato's class vary considerably more than the grades given by Professor Brown. The comparison between the two classes is more complex than we first thought it would be.

As this example demonstrates, information on how scores are spread from the center of a distribution is as important as information about the central tendency in a distribution. This type of information is obtained by measures of variability.

> **Learning Check.** *Look closely at Figure 5.1. Whose class would you choose to take? If you were worried that you might fail statistics, your best bet would be Professor Brown's class where no one fails. However, if you want to keep up your GPA and are willing to work, Professor Yamato's class is the better choice. If you had to choose one of these classes based solely on the average grades, your choice would not be well informed.*

■ ■ ■ ■ **The Index of Qualitative Variation (IQV)**

The United States is undergoing a demographic shift from a predominantly European population to one characterized by increased racial, ethnic, and cultural diversity. These changes challenge us to rethink every conceptualization of society based solely on the experiences of European populations and force us to ask questions that focus on the experiences of different racial/ethnic groups. For instance, we may want to compare the racial/ethnic diversity in different cities, regions, or states or to find out if a group has become more racially and ethnically diverse over time.

The **index of qualitative variation (IQV)** is a measure of variability for nominal variables like race and ethnicity. The index can vary from 0.00 to 1.00. When all the cases in the distribution are in one category, there is no variation (or diversity) and IQV is 0.00. In contrast, when the cases in the distribution are distributed evenly across the categories, there is maximum variation (or diversity) and IQV is 1.00.

> *Index of qualitative variation (IQV)* A measure of variability for nominal variables. It is based on the ratio of the total number of differences in the distribution to the maximum number of possible differences within the same distribution.

Suppose that you attend "Northeastern College," a small college where the majority of students are white and a small minority is either black or Hispanic. In contrast, your best friend is going to "Southwestern College," a small liberal arts college, where the number of white, black, and Hispanic students is about equal. The frequency and percentage distributions for the two colleges are presented in Table 5.1. Which college is more diverse? Clearly, Southwestern College, where whites, blacks, and Hispanics are more or less equally represented, is more diverse than Northeastern College,

Table 5.1 **Racial/Ethnic Groups at Two Colleges**

Race/Ethnicity	Northeastern College		Southwestern College	
	Frequency (f)	Percentages* (%)	Frequency (f)	Percentages* (%)
White	480	90	224	34
Black	31	6	200	30
Hispanic	23	4	236	36
Total (N)	534	100	660	100

*Rounded

Figure 5.2 **Racial/Ethnic Groups at Northeastern and Southwestern Colleges**

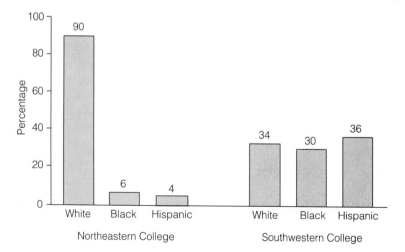

where blacks and Hispanics are but a small minority. You can also get a visual feel for the relative diversity in the two colleges by examining the two bar charts presented in Figure 5.2.

Steps for Calculating the IQV

To substantiate these observations let's compute and compare the IQV for the two colleges. The IQV is based on the ratio of the total number of differences in the distribution to the maximum number of possible differences within the same distribution.

Calculating the Total Number of Differences[3]   Suppose the student council at Northeastern College included four students divided into three racial groups. For the purpose of illustration, we have assigned names to each of the students. Our hypothetical data are presented in Table 5.2.

Whereas Ruth and Justin are both white, Ruth differs racially from Gabriel, who is black, and from Rachel, who is Hispanic. Similarly, Justin differs racially from both Gabriel and Rachel. Finally, Gabriel differs racially from Rachel. Computing the index of qualitative variation (IQV) involves counting these differences. To help us count the total number of racial differences, let's list all the pairs that differ racially:

1. Ruth, Gabriel
2. Ruth, Rachel

[3]The following discussion is based on John H. Mueller et al., *Statistical Reasoning in Sociology* (New York: Houghton Mifflin Company, 1970), pp. 174–178.

Table 5.2   **The Student Council at Northeastern College**

Race	Student	Frequency
White	Ruth, Justin	2
Black	Gabriel	1
Hispanic	Rachel	1
Total (N)		4

3. Justin, Gabriel
4. Justin, Rachel
5. Gabriel, Rachel

Counting each pair as one difference, we come up with a total of five differences.

Learning Check. *Figure 5.3 illustrates how to count the total number of racial differences in the student council of Northeastern College. Examine this figure and determine the total number of gender differences for the same individuals.*

A simpler method of finding the total number of differences in the distribution is to multiply the frequency in each category by the frequency in every other category in the distribution and sum the products. Thus, if we use the frequencies listed in Table 5.2 to count the total number of differences we have: $(2 \times 1) + (2 \times 1) + (1 \times 1) = 2 + 2 + 1 = 5$

We can express the procedure for finding the total number of differences in a distribution with this formula:

$$\text{Total observed differences} = \sum f_i f_j$$

where

f_i = the frequency of category i
f_j = the frequency of category j

Applying this formula, let's count the total number of racial differences at Northeastern College:

$$\text{Total observed differences} = (480 \times 31) + (480 \times 23) + (31 \times 23) = 26,633$$

And now let's count the differences at Southwestern College:

$$\text{Total observed differences} = (224 \times 200) + (224 \times 236) + (200 \times 236)$$
$$= 144,864$$

Figure 5.3   **Counting the Total Number of Racial Differences:
The Student Council at Northeastern College**

Ruth   Gabriel   Rachel   Justin

All Possible Pairs		Racially Same or Different?

Ruth   Justin   same

Ruth   Gabriel   different

Ruth   Rachel   different

Justin   Gabriel   different

Justin   Rachel   different

Gabriel   Rachel   different

Total number of racial differences: 5

Calculating the Maximum Possible Differences The total number of differences at the colleges is not only a function of the degree of racial/ethnic diversity, but also of the total number of students in each college. (It is also a function of the number of racial/ethnic groups in each college; in this example, the number of racial and ethnic categories is identical in both colleges.) Therefore, the number of differences in each college can only be interpreted relative to the maximum diversity possible in each college. Maximum diversity would be attained if there were an equal number of students in each racial/ethnic group. For example, Southwestern would reach maximum diversity if its 660 students were equally divided among the three racial groups (220 in each racial group).

To calculate the maximum possible differences follow this formula:

$$\text{Maximum possible differences} = \frac{K(K-1)}{2}\left(\frac{N}{K}\right)^2$$

where

K = the number of categories in the distribution
N = the total number of cases in the distribution

For Northeastern the maximum possible differences are

$$\frac{3(3-1)}{2}\left(\frac{534}{3}\right)^2 = 95{,}052$$

and for Southwestern

$$\frac{3(3-1)}{2}\left(\frac{660}{3}\right)^2 = 145{,}200$$

Computing the Ratio The IQV is the ratio between the total observed differences and the maximum possible differences:

$$IQV = \frac{\text{Total observed differences}}{\text{Maximum possible differences}}$$

$$= \frac{\sum f_i f_j}{\frac{K(K-1)}{2}\left(\frac{N}{K}\right)^2} \tag{5.1}$$

The IQV for Northeastern is

$$IQV = \frac{(480 \times 31) + (480 \times 23) + (31 \times 23)}{\frac{3(3-1)}{2}\left(\frac{534}{3}\right)^2} = 0.28$$

and for Southwestern it is

$$IQV = \frac{(224 \times 200) + (224 \times 236) + (200 \times 236)}{\frac{3(3-1)}{2}\left(\frac{660}{3}\right)^2} = 0.998$$

Notice that the values of the IQV for the two colleges support our earlier observation: At Southwestern College, where IQV = 0.998, there is considerably more racial/ethnic variation than at Northeastern College, where IQV = 0.28.

To summarize, these are the steps we follow to calculate the IQV:

1. To find the total number of observed differences, multiply the frequency in each category by the frequency in every other category in the distribution and sum the products:

Total observed differences = $\sum f_i f_j$

2. Find the maximum possible differences:

Maximum possible differences = $\dfrac{K(K-1)}{2}\left(\dfrac{N}{K}\right)^2$

3. Find the IQV:

$$IQV = \frac{\text{Total observed differences}}{\text{Maximum possible differences}}$$

Expressing the IQV as a Percentage The IQV can also be expressed as a percentage, rather than a proportion: simply multiply the IQV by 100. Expressed as a percentage, the IQV would reflect the percentage of racial/ethnic differences relative to the maximum possible differences in each distribution. Thus, an IQV of 0.28 indicates that the number of racial/ethnic differences in Northeastern College is 28 percent (0.28 × 100) of the maximum possible differences. Similarly, for Southwestern College, an IQV of 0.998 means that the number of racial/ethnic differences is 99.8 percent (0.998 × 100) of the maximum possible differences.

Calculating the IQV from Percentage or Proportion Distributions The IQV can also be calculated from percentage or proportion distributions by substituting these percentages or proportions for f's into the formula and 100 or 1.00 for N. For instance, we can calculate the IQV for Northeastern College using the percentage distribution (from Table 5.1) instead:

$$IQV = \frac{(90 \times 6) + (90 \times 4) + (6 \times 4)}{\frac{3(3-1)}{2}\left(\frac{100}{3}\right)^2} = 0.28$$

As you can see, the IQV is the same whether we use the frequency or the percentage distribution.

Box 5.1 The IQV Formula: What's Going On Here?

At first, the denominator of the IQV formula looks mean and nasty. However, it is actually quite easy to understand what it's doing. Let's say we have a frequency distribution that looks like this:

Whites	8
Blacks	4
Asians	2 Observed Differences = 84
Hispanics	2
Total	16

It's pretty easy to tell that the situation with the maximum possible differences would be this:

Whites	4
Blacks	4 Maximum Differences = 96
Asians	4
Hispanics	4

Now let's look at the formula's denominator:

$$\frac{K(K-1)}{2}\left(\frac{N}{K}\right)^2$$

Let's look at what's in parentheses first. $N \div K$ in this case is $16 \div 4$, or 4, so $N \div K$ is simply the frequency in each cell when we have maximum differences.

Now let's tackle that exponent. Why square 4? Well, in the frequency distribution with maximum differences, instead of multiplying 8×4, 8×2, and so on, we will be multiplying 4×4, or 4 squared to count differences.

But how many times do we multiply 4 by 4? If we look at the figure, we can easily see that it's 6. That's exactly what the left side of the formula is telling us: $K(K-1) \div 2$ in this case is $4(4-1) \div 2$, or 6. This tells us the number of 4×4's we need:

So don't let formulas scare you. They're only systematic ways of doing things you are already doing intuitively.

Learning Check.   *Calculate the maximum possible differences for the data shown in Table 5.2 (racial makeup of the student council at Northeastern College). Calculate the IQV for the student council (we have already calculated total differences).*

Box 5.2   Statistics in Practice: Diversity at Berkeley Through the Years*

"BERKELEY, Calif.—The photograph in Sproul Hall of the 10 Cal 'yell leaders' from the early 1960's, in their Bermuda shorts and letter sweaters, leaps out like an artifact from an ancient civilization. They are all fresh-faced, and in a way that is unimaginable now, they are all white."[†]

On the flagship campus of the University of California system, the center of the affirmative action debate in higher education today, the ducktails and bouffant hairdos of those 1960s cheerleaders seems indeed out of date. The University of California's Berkeley campus was among the first of the nation's leading universities to embrace elements of affirmative action in its admission policies and now boasts that it has one of the most diverse campuses in the United States.

The following pie charts show the racial and ethnic breakdown of undergraduates at U.C. Berkeley for 1984 and 1994. The IQVs were calculated using the percentage distribution (as shown in the pie charts) for race and ethnicity for each year. The IQVs illustrate the changes in Berkeley's student body from mostly white in 1984 to one of the most diverse campuses in the United States.

Racial/Ethnic Composition of Student Body at U.C. Berkeley, 1984 and 1994

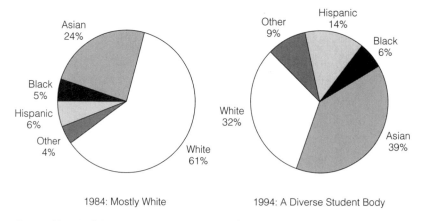

1984: Mostly White   1994: A Diverse Student Body

Source: New York Times, June 4, 1995. Copyright © 1995 by The New York Times Co. Reprinted by permission.

$$IQV_{1984} = \frac{\begin{array}{c}(61\times24)+(61\times5)+(61\times6)+(61\times4)+(24\times5)\\ +(24\times6)+(24\times4)+(5\times6)+(5\times4)+(6\times4)\end{array}}{\dfrac{5(5-1)}{2}\left(\dfrac{100}{5}\right)^2}$$

$$= \frac{2,813}{4,000} = 0.70$$

$$IQV_{1994} = \frac{\begin{array}{c}(32\times39)+(32\times6)+(32\times14)+(32\times9)+(39\times6)\\ +(39\times14)+(39\times9)+(6\times14)+(6\times9)+(14\times9)\end{array}}{\dfrac{5(5-1)}{2}\left(\dfrac{100}{5}\right)^2}$$

$$= \frac{3,571}{4,000} = 0.89$$

*Adapted from *The New York Times*, June 4, 1995. Copyright © 1995 by The New York Times Co. Used by permission.
†Ibid.

Learning Check. *Examine Box 5.2 and consider the impact that the number of categories of a variable has on the IQV. What would happen to the Berkeley case in 1994 if Asians were broken down into two categories with 20 percent in one and 19 percent in the other? (To answer this question you will need to recalculate the IQV for 1994 with these new data.)*

Statistics in Practice: Diversity in U.S. Society

According to demographers' projections, by the middle of the next century the United States will no longer be a predominantly white society. The combined population of the four largest minority groups—African Americans, Asian Americans, Hispanic Americans, and Native Americans—reached an estimated 7.1 million in 1996.[4] Population shifts during the 1990s indicate geographic concentration of minority groups in specific regions and metropolitan areas of the United States.[5] Demographers call it chain migration: "Migrants don't go randomly to various spots in the USA. They use friends, neighbors and relatives to locate their new residences."[6]

[4]*Statistical Abstract of the United States*, 1997, Table 19.
[5]William H. Frey, "The Diversity Myth," *American Demographics* 20, no. 6 (1998): 39–43.
[6]*USA Today*, June 9, 1993.

For example, Los Angeles is home to one-fifth of the Hispanic population, placing first in total growth and accounting for 18 percent of all Hispanic population gains between 1990 and 1996.[7]

How do you compare the amount of diversity in different cities or states? Diversity is a characteristic of a population many of us can sense intuitively. For example, the ethnic diversity of a large city is seen in the many members of various groups encountered when walking down its streets or traveling through its neighborhoods.[8]

We can use the IQV to measure the amount of diversity in different states. Table 5.3 displays population by race and ethnicity for all fifty states. Based on the data in Table 5.3 and using Formula 5.1 as in our earlier example, we have calculated the IQV for each state in Table 5.4.

The advantage of using a single number to express diversity is demonstrated in Figure 5.4, which depicts the regional variations in diversity as expressed by the IQVs from Table 5.4. Figure 5.4 shows the wide variation in racial/ethnic diversity that exists in the United States. Notice that California, with an IQV of .78, is the most diverse state. At the other extreme, Maine and Vermont, whose populations are overwhelmingly white, both have IQVs of 0.05 and are the most homogeneous states.

> **Learning Check.** *What regional variations in racial/ethnic diversity are depicted in Figure 5.4? Can you think of at least two explanations for these patterns?*

■ ■ ■ ■ **The Range**

The simplest and most straightforward measure of variation is the **range,** which measures variation in interval-ratio variables. It is the difference between the highest (maximum) and lowest (minimum) scores in the distribution:

Range = highest score − lowest score

In the 1996 GSS, the oldest person included in the study was 89 years old and the youngest was 18. Thus, the range was 89 − 18, or 71 years.

> *Range* A measure of variation in interval-ratio variables. It is the difference between the highest (maximum) and lowest (minimum) scores in the distribution.

---

[7]Frey, p. 39.

[8]Michael White, "Segregation and Diversity Measures in Population Distribution," *Population Index* 52, no. 2 (1986): 198–221.

Table 5.3 **Population (in thousands) by Race and Ethnicity for the Fifty States, 1994**

State	Hispanic	White	Black	American Indian, Eskimo, Aleut	Asian or Pacific Islander
Alabama	26	3,070	1,079	17	27
Alaska	18	445	25	94	24
Arizona	770	2,853	139	238	75
Arkansas	23	2,011	389	14	16
California	8,371	16,840	2,415	302	3,504
Colorado	462	2,925	158	34	77
Connecticut	211	2,699	295	8	63
Delaware	16	549	129	2	12
Florida	1,742	9,930	1,025	45	209
Georgia	121	4,848	1,966	15	105
Hawaii	40	354	29	7	749
Idaho	61	1,040	5	16	12
Illinois	984	8,604	1,795	25	344
Indiana	106	5,123	464	13	45
Iowa	39	2,697	54	8	31
Kansas	99	2,239	154	23	39
Kentucky	23	3,505	272	6	22
Louisiana	87	2,795	1,364	19	50
Maine	7	1,215	5	6	8
Maryland	131	3,358	1,320	14	178
Massachusetts	263	5,222	364	14	178
Michigan	201	7,748	1,364	58	126
Minnesota	61	4,229	122	55	100
Mississippi	15	1,672	958	9	15
Missouri	63	4,561	582	20	52
Montana	12	785	3	51	5
Nebraska	44	1,487	62	13	16
Nevada	160	1,113	103	26	55
New Hampshire	12	1,105	7	2	11
New Jersey	744	5,654	1,134	19	353
New Mexico	621	824	40	148	20
New York	1,894	12,192	3,179	69	839
North Carolina	80	5,263	1,568	87	70
North Dakota	4	598	4	27	5
Ohio	140	9,559	1,235	21	108
Oklahoma	84	2,619	257	263	41
Oregon	129	2,771	55	44	88
Pennsylvania	226	10,482	1,157	16	171
Rhode Island	45	879	47	4	21
South Carolina	29	2,496	1,102	9	27
South Dakota	5	657	4	54	4

(continued on next page)

Table 5.3 **Population (in thousands) by Race and Ethnicity for the Fifty States, 1994** *(continued)*

State	Hispanic	White	Black	American Indian, Eskimo, Aleut	Asian or Pacific Islander
Tennessee	37	4,247	839	11	41
Texas	4864	10,767	2,235	81	430
Utah	97	1,724	15	29	44
Vermont	4	568	2	2	4
Virginia	171	4,884	1,277	18	202
Washington	235	4,560	176	98	274
West Virginia	8	1,745	58	2	9
Wisconsin	100	4,594	276	44	69
Wyoming	25	433	4	10	3

Source: *Statistical Abstract of the United States*, 1997, Table 34.

Table 5.4 **Racial and Ethnic Diversity as Measured by the IQVs for the Fifty States, 1994**

State	IQV	State	IQV
California	0.78	Tennessee	0.38
New Mexico	0.75	Washington	0.33
Texas	0.71	Massachusetts	0.31
New York	0.63	Missouri	0.30
Hawaii	0.63	Ohio	0.30
Louisiana	0.60	Pennsylvania	0.29
Mississippi	0.60	Kansas	0.28
Maryland	0.60	Rhode Island	0.27
Arizona	0.59	Indiana	0.25
New Jersey	0.57	Oregon	0.24
Georgia	0.56	Utah	0.23
South Carolina	0.56	Wisconsin	0.22
Illinois	0.54	South Dakota	0.21
Alaska	0.54	Wyoming	0.21
Alabama	0.51	Nebraska	0.20
Virginia	0.51	Kentucky	0.20
Nevada	0.50	Idaho	0.19
North Carolina	0.49	Montana	0.19
Florida	0.48	Minnesota	0.18
Delaware	0.46	North Dakota	0.15
Oklahoma	0.43	Iowa	0.11
Colorado	0.43	West Virginia	0.10
Michigan	0.39	New Hampshire	0.07
Connecticut	0.39	Maine	0.05
Arkansas	0.38	Vermont	0.05

Figure 5.4 **Racial/Ethnic Diversity in the United States, 1994 IQV**

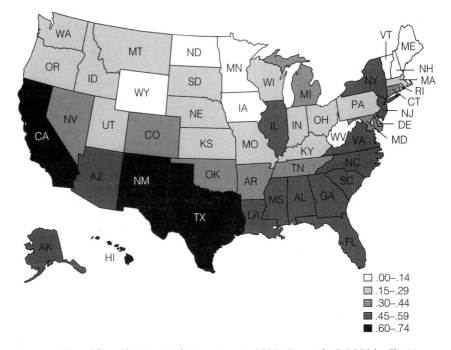

□ .00–.14
▨ .15–.29
▧ .30–.44
■ .45–.59
■ .60–.74

Source: Adapted from *The New York Times,* June 4, 1995. Copyright © 1995 by The New York Times Co. Used by permission.

The range can also be calculated on percentages. For example, since the 1980s, relatively large communities of the elderly have become noticeable, not just in the traditional retirement meccas of the Sun Belt, but also in the Ozarks of Arkansas and the mountains of Colorado and Montana. The number of elderly persons increased in every state during the 1980s, but by different amounts. Table 5.5 displays the *percentage* increase in the elderly population from the 1980s to the 1990s by region and by state.[9]

What is the range in the percentage increase in state elderly population for the United States? To find the ranges in a distribution, simply pick out the highest and lowest scores in the distribution and subtract. Nevada has the highest percentage increase, with 92.5 percent, and the District of Columbia has the lowest increase, with 3.8 percent. The range is 88.7 percentage points, or 92.5 percent – 3.8 percent.

Although the range is simple and quick to calculate, it is a rather crude measure because it is based on only the lowest and highest scores. These two scores might be extreme and rather atypical, which might make the

[9]The percentage increase in the population 65 years and over for each state and region was obtained by the following formula:

$$\text{Percentage increase} = \left(\frac{1990 \text{ population} - 1980 \text{ population}}{1980 \text{ population}} \right) 100$$

Table 5.5 Percentage Increase in the Population 65 Years and Over by Region, Division, and State, 1980–1990

Region, Division, and State	Percentages %	Region, Division, and State	Percentages %
United States	21.6	**South** (cont.)	
		South Atlantic (cont.)	
Northeast	14.4	Virginia	30.9
New England	15.9	West Virginia	12.6
Maine	15.5	North Carolina	32.7
Vermont	13.3	South Carolina	37.1
New Hampshire	20.9	Georgia	25.9
Massachusetts	12.2	Florida	39.6
Rhode Island	18.0		
Connecticut	21.6	East South Central	15.9
		Kentucky	13.5
Middle Atlantic	14.0	Tennessee	19.0
New York	8.3	Alabama	18.2
New Jersey	19.2	Mississippi	10.4
Pennsylvania	19.0		
		West South Central	19.6
Midwest	15.4	Arkansas	11.6
East North Central	17.5	Louisiana	15.4
Ohio	20.0	Oklahoma	12.4
Indiana	18.5	Texas	24.6
Illinois	13.3		
Michigan	21.0	**West**	33.5
Wisconsin	15.2	Mountain	42.9
		Montana	25.6
West North Central	11.2	Idaho	29.0
Minnesota	13.8	Wyoming	26.3
Iowa	9.6	Colorado	32.7
Missouri	10.4	New Mexico	39.7
North Dakota	13.0	Arizona	54.9
South Dakota	12.2	Utah	36.9
Nebraska	8.3	Nevada	92.5
Kansas	11.6		
		Pacific	30.4
South	25.7	Washington	32.7
South Atlantic	32.8	Oregon	28.5
Delaware	35.7	California	28.9
Maryland	30.0	Alaska	91.3
Washington, DC	3.8	Hawaii	62.5

Source: U.S. Bureau of the Census, "65+ in the United States," *Current Population Reports,* 1996, Special Studies P23-190, Table 5-1.

range a misleading indicator of the variation in the distribution. For instance, notice that among the fifty states and the District of Columbia listed in Table 5.5, no other has a percentage increase nearly as low as the District of Columbia's and only Alaska has a percentage increase nearly as high as Nevada's. The range of 88.7 percentage points does not give us information about the variation in states between the District of Columbia and Alaska.

> **Learning Check.** *Why can't we use the range to describe diversity in nominal variables? The range can be used to describe diversity in ordinal variables (for example, we can say that responses to a question ranged from "somewhat satisfied" to "very dissatisfied"), but it has no quantitative meaning. Why not?*

■ ■ ■ ■ The Interquartile Range: Increases in Elderly Populations

To remedy this limitation we can employ an alternative to the range—the *interquartile range*. The **interquartile range** (IQR), a measure of variation for interval-ratio variables, is the width of the middle 50 percent of the distribution. It is defined as the difference between the lower and upper quartiles ($Q1$ and $Q3$).

$$IQR = Q3 - Q1$$

Recall that the first quartile ($Q1$) is the 25th percentile, the point at which 25 percent of the cases fall below it and 75 percent above it. The third quartile ($Q3$) is the 75th percentile, the point at which 75 percent of the cases fall below it and 25 percent above it. The interquartile range, therefore, defines variation for the middle 50 percent of the cases.

Like the range, the interquartile range is based on only two scores. However, because it is based on intermediate scores, rather than on the extreme scores in the distribution, it avoids some of the instability associated with the range.

These are the steps for calculating the IQR:

1. To find $Q1$ and $Q3$, order the scores in the distribution from the highest to the lowest score, or vice versa. Table 5.6 presents the data of Table 5.5 arranged in order from Nevada (92.5%) to the District of Columbia (3.8%).

2. Next, we need to identify the first quartile, $Q1$ or the 25th percentile. We have to identify the percentage increase in the elderly population associated with the state that divides the distribution so that 25 percent of the states are below it and 75 percent of the states are above it. To find $Q1$ we multiply N by .25:

$$(N) (.25) = (51) (.25) = 12.75$$

The first quartile falls between the 12th and 13th states. Counting from the bottom, the 12th state is West Virginia, and the percentage increase associated with it is 12.6. The 13th state is North Dakota, with a percentage

Table 5.6 **Percentage Increase in the Population 65 Years and Over, 1980–1990, by State, Ordered from Highest to Lowest**

State	Percentage %	State	Percentage %	State	Percentage %
Nevada	92.5	Wyoming	26.3	Minnesota	13.8
Alaska	91.3	Georgia	25.9	Kentucky	13.5
Hawaii	62.5	Montana	25.6	Illinois	13.3
Arizona	54.9	Texas	24.6	Vermont	13.3
New Mexico	39.7	Connecticut	21.6	North Dakota	13.0
Florida	39.6	Michigan	21.0	West Virginia	12.6
South Carolina	37.1	New Hampshire	20.9	Oklahoma	12.4
Utah	36.9	Ohio	20.0	Massachusetts	12.2
Delaware	35.7	New Jersey	19.2	South Dakota	12.2
Colorado	32.7	Pennsylvania	19.0	Arkansas	11.6
North Carolina	32.7	Tennessee	19.0	Kansas	11.6
Washington	32.7	Indiana	18.5	Mississippi	10.4
Virginia	30.9	Alabama	18.2	Missouri	10.4
Maryland	30.0	Rhode Island	18.0	Iowa	9.6
Idaho	29.0	Maine	15.5	New York	8.3
California	28.6	Louisiana	15.4	Nebraska	8.3
Oregon	28.5	Wisconsin	15.2	Washington, DC	3.8

increase of 13.0. To find the first quartile we take the average of 12.6 and 13.0. Therefore, $(12.6 + 13.0) \div 2 = 12.8$ is the first quartile (Q1).

3. To find Q3, we have to identify the state that divides the distribution in such a way that 75 percent of the states are below it and 25 percent of the states are above it. We multiply N this time by .75:

$$(N)(.75) = (51)(.75) = 38.25$$

The third quartile falls between the 38th and 39th states. Counting from the bottom, the 38th state is Maryland, and the percentage increase associated with it is 30.0. The 39th state is Virginia, with a percentage increase of 30.9. To find the third quartile we take the average of 30.8 and 31.5. Therefore, $(30.0 + 30.9) \div 2 = 30.45$ is the third quartile (Q3).

4. We are now ready to find the interquartile range:

$$IQR = Q3 - Q1 = 30.45 - 12.80 = 17.65$$

The interquartile range of percentage increase in the elderly population is 17.65 percentage points.

Notice that the IQR gives us better information than the range. The range gave us an 88.7-point spread, from 92.5 percent to 3.8 percent, but the IQR tells us that half the states are clustered between 30.45 and 12.80—a much narrower spread. The extreme scores represented by Nevada and the District of Columbia have no effect on the IQR because they fall at the extreme ends of the distribution.

> *Interquartile range (IQR)* The width of the middle 50 percent of the distribution. It is defined as the difference between the lower and upper quartiles ($Q1$ and $Q3$).

> **Learning Check.** *Why is the IQR better than the range as a measure of variability, especially when there are extreme scores in the distribution? To answer this question you may want to examine Figure 5.5.*

Figure 5.5 **The Range Versus the Interquartile Range: Number of Children Among Two Groups of Women**

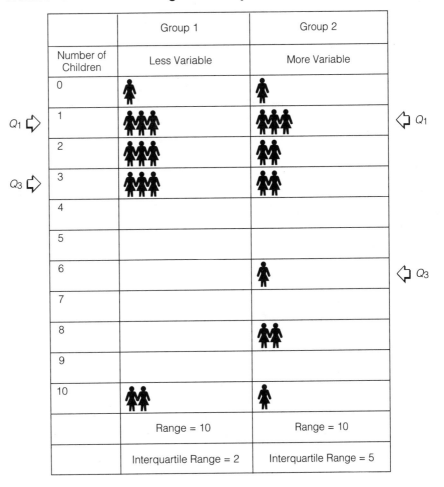

■ ■ ■ ■ **The Box Plot**

A graphic device called the *box plot* can visually present the range, the interquartile range, the median, the lowest (minimum) score, and the highest (maximum) score. The box plot provides us with a way to visually examine the center, the variation, and the shape of distributions of interval-ratio variables.

Figure 5.6 is a box plot of the distribution of the 1980–1990 percentage increase in elderly population displayed in Table 5.6. To construct the box plot in Figure 5.6 we used the lowest and highest values in the distribution, the upper and lower quartiles, and the median. We can easily draw a box plot by hand following these instructions:

1. Draw a box between the lower and upper quartiles.
2. Draw a solid line within the box to mark the median.
3. Draw vertical lines (called whiskers) outside the box, extending to the lowest and highest values.

What can we learn from creating a box plot? We can obtain a visual impression of the following properties: First, the center of the distribution is easily identified by the solid line inside the box. Second, since the box is drawn between the lower and upper quartiles, the interquartile range is reflected in the height of the box. Similarly, the length of the vertical lines drawn outside the box (on both ends) represents the range of the distribution. Both the interquartile range and the range give us a visual impression of the spread in the distribution. Finally, the relative position of the box and/or the position of the median within the box tell us whether the distri-

Figure 5.6   **Box Plot of the Distribution of the Percentage Increase in Elderly Population, 1980–1990**

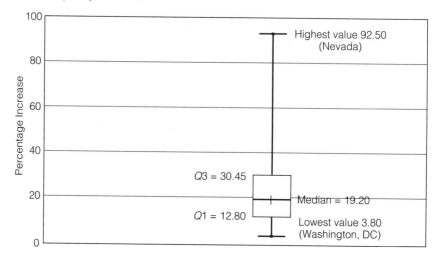

bution is symmetrical or skewed. A perfectly symmetrical distribution would have the box at the center of the range as well as the median in the center of the box. When the distribution departs from symmetry, the box and/or the median will not be centered; it will be closer to the lower quartile when there are more cases with lower scores or to the upper quartile when there are more cases with higher scores.

> **Learning Check.** *Is the distribution shown in the box plot in Figure 5.6 symmetrical or skewed?*

Box plots are particularly useful for comparing distributions. To demonstrate box plots that are shaped quite differently, in Figure 5.7 we have used the data on the percentage increase in the elderly population (Table 5.5) to compare the pattern of change occurring between 1980 and 1990 in the northeastern and western regions of the United States. As you can see, the box plots differ from each other considerably. What can you learn from comparing the box plots for the two regions? First, the positions of the medians highlight the dramatic increase in the elderly population in the western United States. While the Northeast (median = 18%) has experienced a steady rise in its elderly population, the West is showing a much higher percentage increase (median = 32.7%). Second, both the range (illustrated by the position of the whiskers in each box plot) and the interquartile range (illustrated by the height of the box) are much wider in the West (range = 66.9%; IQR = 29.75 %) than in the Northeast (range = 13.3%; IQR = 5.65%),

Figure 5.7 **Box Plots of the Percentage Increase in the Elderly Population, 1980–1990, for Northeast and West Regions**

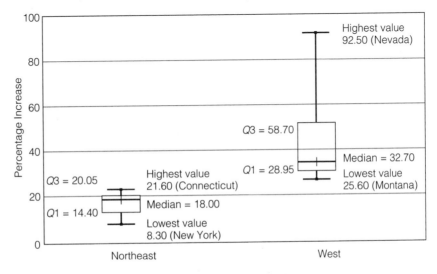

indicating that there is more variability among states in the West than among those in the Northeast. Finally, the relative positions of the boxes tell us something about the different shapes of these distributions. Because its box is at about the center of its range, the Northeast distribution is almost symmetrical. In contrast, with its box off center and closer to the lower end of the distribution, the distribution of percentage change in the elderly population for the western states is positively skewed. (In comparing these two distributions notice that although it is positively skewed, the lowest values in the western distribution are higher than the highest value in the northeastern distribution.)

■ ■ ■ ■ The Variance and the Standard Deviation: Changes in the Nursing Home Population

The elderly population in the United States today is ten times as large as in 1900 and is projected to more than double from 1990 to 2030. The pace and direction of these demographic changes will create compelling social, economic, and ethical choices for individuals, families, and governments, especially regarding the living arrangements of the elderly.[10]

Most of the elderly live in households rather than in nursing homes, but the likelihood of living in a nursing home increases with age. Table 5.7 presents the percentage change in the nursing home population for all regions of the United States. These percentage changes were calculated by the Census Bureau using the following formula:

$$\text{Percentage change} = \frac{1990 \text{ population} - 1980 \text{ population}}{1980 \text{ population}}(100)$$

For example, the nursing home population in the Pacific region was 156,404 in 1980. In 1990 the nursing home population increased to 180,977. Therefore, the percent change from 1980 to 1990 is

$$\text{Percentage change} = \frac{180,977 - 156,404}{156,404}(100) = 15.7$$

Table 5.7 shows that between 1980 and 1990 the size of the nursing home population in the United States increased by an average of 31.5 percent.[11] But this average increase does not inform us about the regional variation in the nursing home population. For example, do the New England states show a smaller-than-average increase because of the outmigration of the

---

[10]U.S. Bureau of the Census, 1992, "65+ in America," *Current Population Reports*, Special Studies, P2-19.

[11]Because the percentage changes in the nursing home population were computed for aggregate units (region), the mean increase of 31.5 percent in the nursing home population is the *unweighted* average of the regional increases.

Table 5.7 **Percentage Change in the Nursing Home Population by Region, 1980–1990**

Region	Percentage (%)
Pacific	15.7
West North Central	16.2
New England	17.6
East North Central	23.2
West South Central	24.3
Middle Atlantic	28.5
East South Central	38.0
Mountain	47.9
South Atlantic	71.7
Mean $(\bar{Y})$	31.46

Source: U.S. Bureau of the Census, "65+ in the United States," *Current Population Reports*, 1996, Special Studies P23-190, Table 6-5.

elderly population to the warmer climate of the Sun Belt states? Is the increase higher in the South because of the immigration of the elderly?

Although it is important to know the average percentage increase for the nation as a whole, you may also want to know whether regional increases differ from the national average. If the regional increases are close to the national average, the figures will cluster around the mean, but if the regional increases deviate much from the national average, they will be widely dispersed around the mean.

Table 5.7 suggests that there is considerable regional variation. The percentage change ranges from 71.7 percent in the South Atlantic states to 15.7 percent in the Pacific states, so the range is 56 percent (71.7 − 15.7 = 56%). Moreover, except for the Middle Atlantic region, most of the regions deviate considerably from the national average of 31.5 percent. How large are these deviations on the average? We want a measure that will give us information about the overall variations among all regions in the United States and, unlike the range or the interquartile range, will not be based on only two scores.

Such a measure will reflect how much, on the average, each score in the distribution deviates from some central point, such as the mean. We use the mean as the reference point rather than other kinds of averages (the mode or the median) because the mean is based on all the scores in the distribution. Therefore, it is more useful as a basis from which to calculate average deviation. The sensitivity of the mean to extreme values carries over to the calculation of the average deviation, which is based on the mean. Another reason for using the mean as a reference point is that more advanced measures of variation require the use of algebraic properties that can be assumed only by using the arithmetic mean.

The *variance* and the *standard deviation* are two closely related measures of variation that increase or decrease based on how closely the scores cluster around the mean. The **variance** is the average of the squared deviations from the center (mean) of the distribution, and the **standard deviation** is the square root of the variance. Both measure variability in interval-ratio variables.

> *Variance* A measure of variation for interval-ratio variables; it is the average of the squared deviations from the mean.
>
> *Standard deviation* A measure of variation for interval-ratio variables; it is equal to the square root of the variance.

Calculating the Deviation from the Mean

Consider again the distribution of the percentage change in the nursing home population for the nine regions of the United States. Because we want to calculate the average difference of all the regions from the national average (the mean), it makes sense to first look at the difference between each region and the mean. This difference, called a *deviation from the mean*, is symbolized as $(Y - \bar{Y})$. The sum of these deviations can be symbolized as $\Sigma(Y - \bar{Y})$.

The calculations of these deviations for each region are displayed in Table 5.8 and Figure 5.8. We have also summed these deviations. Note that

Table 5.8   Percentage Change in the Nursing Home Population, 1980–1990, by Region and Deviation from the Mean

Region	Percentages (%)	$Y - \bar{Y}$
Pacific	15.7	$15.7 - 31.5 = -15.8$
West North Central	16.2	$16.2 - 31.5 = -15.3$
New England	17.6	$17.6 - 31.5 = -13.9$
East North Central	23.2	$23.2 - 31.5 = -8.3$
West South Central	24.3	$24.3 - 31.5 = -7.2$
Middle Atlantic	28.5	$28.5 - 31.5 = -3.0$
East South Central	38.0	$38.0 - 31.5 = 6.5$
Mountain	47.9	$47.9 - 31.5 = 16.4$
South Atlantic	71.7	$71.7 - 31.5 = 40.2$
	$\Sigma Y = 283.10$	$\Sigma(Y - \bar{Y}) = 0$

$$\text{Mean} = \bar{Y} = \frac{\Sigma Y}{N} = \frac{283.10}{9} = 31.5$$

Figure 5.8 **Illustrating Deviations from the Mean**

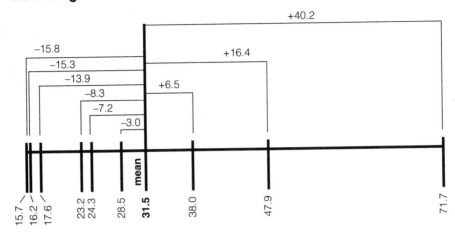

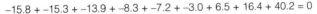

$$-15.8 + -15.3 + -13.9 + -8.3 + -7.2 + -3.0 + 6.5 + 16.4 + 40.2 = 0$$

each region has either a positive or a negative deviation score. The deviation is positive when the percentage increase in the nursing home population is above the mean. It is negative when the percentage increase is below the mean. Thus, for example, New England's deviation score of –13.9 means that its percentage change in the nursing home population was 13.9 below the mean.

You may wonder if we could calculate the average of these deviations by simply adding up the deviations and dividing them? Unfortunately we cannot, because the sum of the deviations of scores from the mean is always zero, or algebraically $\Sigma(Y - \overline{Y}) = 0$. In other words, if we were to subtract the mean from each score and then add up all the deviations as we did in Table 5.8, the sum would be zero, which in turn would cause the average deviation (that is, average difference) to compute to zero. This is always true because the mean is the center of gravity of the distribution.

Mathematically, we can overcome this problem either by ignoring the plus and minus signs, using instead the absolute values of the deviations, or by squaring the deviations—that is, multiplying each deviation by itself—to get rid of the negative sign. Since absolute values are difficult to work with mathematically, the latter method is used to compensate for the problem.

Table 5.9 presents the same information as Table 5.8, but here we have squared the actual deviations from the mean and added together the squares. The sum of the squared deviations is symbolized as $\Sigma(Y - \overline{Y})^2$. Note that by squaring the deviations we end up with a sum representing the deviation from the mean, which is positive. (Note that this sum will equal zero if all the cases have the same value as the mean case.) In our example, this sum is $\Sigma(Y - \overline{Y})^2 = 2733.92$.

Table 5.9 **Percentage Change in the Nursing Home Population,
1980–1990, by Region and Deviation from the Mean**

Region	Percentages (%)	$Y - \bar{Y}$	$(Y - \bar{Y})^2$
Pacific	15.7	$15.7 - 31.5 = -15.8$	249.64
West North Central	16.2	$16.2 - 31.5 = -15.3$	234.09
New England	17.6	$17.6 - 31.5 = -13.9$	193.21
East North Central	23.2	$23.2 - 31.5 = -8.3$	68.89
West South Central	24.3	$24.3 - 31.5 = -7.2$	51.84
Middle Atlantic	28.5	$28.5 - 31.5 = -3.0$	9.00
East South Central	38.0	$38.0 - 31.5 = 6.5$	42.25
Mountain	47.9	$47.9 - 31.5 = 16.4$	268.96
South Atlantic	71.7	$71.7 - 31.5 = 40.2$	1616.04

$$\sum Y = 283.10 \qquad \sum (Y - \bar{Y}) = 0 \qquad \sum (Y - \bar{Y})^2 = 2733.92$$

$$\text{Mean} = \bar{Y} = \frac{\sum Y}{N} = \frac{283.10}{9} = 31.5$$

Learning Check. *Examine Table 5.9 again and note the disproportionate contribution of the South Atlantic region to the sum of the squared
deviations from the mean (it actually accounts for almost 60% of the sum
of squares). Can you explain why? Hint: It has something to do with the
sensitivity of the mean to extreme values.*

Calculating the Variance and the Standard Deviation

The average of the squared deviations from the mean we just calculated is
known as the *variance*. The variance is symbolized as S_Y^2. Remember that
we are interested in the *average* of the squared deviations from the mean.
Therefore, we need to divide the sum of the squared deviations by the number of scores (N) in the distribution. However, unlike the calculation of the
mean, we will use $N - 1$ rather than N in the denominator.[12] The formula for
the variance can be stated as

$$S_Y^2 = \frac{\sum (Y - \bar{Y})^2}{N - 1} \tag{5.2}$$

Variance

where

[12]$N - 1$ is used in the formula for computing variance because usually we are computing
from a sample with the intention of generalizing to a larger population. $N - 1$ in the formula
gives a better estimate and is also the formula used in SPSS.

S_Y^2 = the variance

$Y - \overline{Y}$ = the deviations from the mean

$\Sigma(Y - \overline{Y})^2$ = the sum of the squared deviations from the mean

N = the number of scores

Notice that the formula incorporates all the symbols we defined earlier. This formula means: The variance is equal to the average of the squared deviations from the mean.

Follow these steps to calculate the variance:

1. Calculate the mean, $\overline{Y} = \Sigma Y \div N$.
2. Subtract the mean from each score to find the deviation, $(Y - \overline{Y})$.
3. Square each deviation, $(Y - \overline{Y})^2$.
4. Sum the squared deviations, $\Sigma(Y - \overline{Y})^2$.
5. Divide the sum by $N - 1$, $\Sigma(Y - \overline{Y})^2 \div (N - 1)$.
6. The answer is the variance.

To assure yourself that you understand how to calculate the variance, go back to Table 5.7 and follow this step-by-step procedure for calculating the variance. Now plug the required quantities into Formula 5.2. Your result should look like this:

$$S_Y^2 = \frac{\Sigma(Y - \overline{Y})^2}{N - 1} = \frac{2733.92}{8} = 341.74$$

One problem with the variance is that it is based on squared deviations and therefore is no longer expressed in the original units of measurement. For instance, it is difficult to interpret the variance of 341.74, which represents the distribution of the percentage change in the nursing home population, because this figure is expressed in squared percentages. Thus, we often take the square root of the variance and interpret it instead. This gives us the *standard deviation*, S_Y.

The standard deviation, symbolized as S_Y, is the square root of the variance, or

$S.D.$
$$S_Y = \sqrt{S_Y^2}$$

The standard deviation for our example is

$$S_Y = \sqrt{S_Y^2} = \sqrt{341.74} = 18.49$$

The formula for the standard deviation uses the same symbols as the formula for the variance:

$$S_Y = \sqrt{\frac{\Sigma(Y - \overline{Y})^2}{N - 1}} \tag{5.3}$$

As we interpret the formula, we can say that the standard deviation is equal to the square root of the average of the squared deviations from the mean.

The advantage of the standard deviation is that unlike the variance, it is measured in the same units as in the original data. For instance, the standard deviation for our example is 18.49. Because the original data were expressed in percentages, this number is expressed as a percentage as well. In other words, you could say, "The standard deviation is 18.49 percent." But what does this mean? The actual number tells us very little by itself, but it allows us to evaluate the dispersion of the scores around the mean.

In a distribution where all the scores are identical, the standard deviation is zero (0). Zero is the lowest possible value for the standard deviation. In an identical distribution, all of the points would be the same, with the same mean, mode, and median. There is no variation or dispersion in the scores.

The more the standard deviation departs from zero, the more variation there is in the distribution. There is no upper limit to the value of the standard deviation. In our example, we can conclude that a standard deviation of 18.49 percent means that the percentage change in the nursing home population for the nine regions of the United States is widely dispersed around the mean of 31.5 percent.

Another way to interpret the standard deviation is to compare it with another distribution. For instance, Table 5.10 displays the means and standard deviations of employee age for two samples drawn from a Fortune 100 corporation. Samples are divided into female clerical and female technicians. Note that the mean ages for both samples are about the same—approximately 39 years of age. However, the standard deviations suggest that the distribution of age is dissimilar between the two groups. Figure 5.9 illustrates this dissimilarity in the two distributions.

The relatively low standard deviation for female technicians indicates that this group is relatively homogeneous in age. That is to say, most of the women's ages, while not identical, are fairly similar. The average deviation from the mean age of 39.87 is 3.75 years. In contrast, the standard deviation for female clerical employees is about twice the standard deviation for female technicians. This suggests a wider dispersion or greater heterogeneity in the ages of clerical workers. We can say that the average deviation from the mean age of 39.46 is 7.80 years for clerical workers. The larger standard

Table 5.10 **Age Characteristics of Female Clerical and Technical Employees**

Characteristics	Female Clerical N = 22	Female Technical N = 39
Mean age	39.46	39.87
Standard deviation	7.80	3.75

Source: Adapted from Marjorie Armstrong-Stassen, "The Effect of Gender and Organizational Level on How Survivors Appraise and Cope with Organizational Downsizing," *Journal of Applied Behavioral Science* 34, no. 2 (June 1998): 125–142. Used by permission.

Figure 5.9 **Illustrating the Means and Standard Deviations for Age Characteristics**

Female clerical: mean = 39.46, standard deviation = 7.80

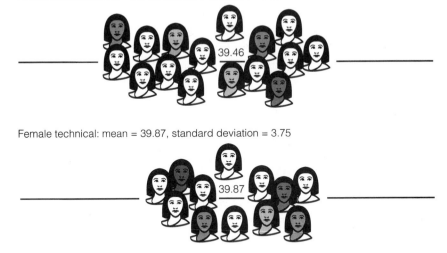

Female technical: mean = 39.87, standard deviation = 3.75

deviation indicates a wider dispersion of points below or above the mean. More clerical employees are either older or younger than the mean age of 39.46.

Learning Check. *Take time to understand the section on standard deviation and variance. You will see these statistics in more advanced procedures. Although your instructor may require you to memorize the formulas, it is more important for you to understand how to interpret standard deviation and variance and when they can be appropriately used. Many hand calculators and all statistical computer software programs will calculate these measures of diversity for you, but they won't tell you what they mean. Once you understand the meaning behind these measures, the formulas will be easier to remember.*

■ ■ ■ ■ **Considerations for Choosing a Measure of Variation**

So far we have considered five measures of variation: the IQV, the range, the interquartile range, the variance, and the standard deviation. Each measure can represent the degree of variability in a distribution. But which one should we use? There is no simple answer to this question. However, in general, we tend to use only one measure of variation, and the choice of the appropriate one involves a number of considerations. These considerations and how they affect our choice of the appropriate measure are presented in the form of a decision tree in Figure 5.10.

Box 5.3 Computational Formula for the Variance and the Standard Deviation

We have learned how to use the definitional formulas for the standard deviation and the variance. These formulas are easy to follow conceptually, but they are tedious to compute, especially when working with a large number of scores. The following computational formulas are easier and faster to use and give exactly the same result:

$$S_Y^2 = \left(\frac{\sum Y^2}{N-1}\right) - \left(\frac{N}{N-1}\right)\left(\frac{\sum Y}{N}\right)^2$$

$$S_Y = \sqrt{\left(\frac{\sum Y^2}{N-1}\right) - \left(\frac{N}{N-1}\right)\left(\frac{\sum Y}{N}\right)^2}$$

where

$\sum Y^2$ = the sum of the squared scores (find this by first squaring each score and adding up the squared scores)

N = the number of scores in the distribution

$\left(\dfrac{\sum Y}{N}\right)^2$ = the sum of the scores divided by N and then squared (find this quantity by first dividing the sum of the scores by N and then squaring the answer, which is equivalent to squaring the mean)

To illustrate how to calculate the variance and standard deviation using the computational formula, we will use data on the suicide rates per 100,000 population in the nine largest states in the United States as reported in the *Statistical Abstract of the United States, 1997* (Table 132). In the following table, we add an additional column to the original data to help us generate the following quantities required by the formula: $\sum Y^2$ and $\sum Y$.

Suicide Rates per 100,000 Population in the Nine Largest U.S. States for 1994

City	Suicide Rate (Y)	Y²
New Jersey	7.3	53.29
New York	8.2	67.24
Ohio	9.9	98.01
Michigan	10.9	118.81
Pennsylvania	11.0	121.00
California	11.8	139.24
Georgia	11.8	139.24
Texas	12.7	161.29
Florida	14.9	222.02
	$\sum Y = 98.5$	$\sum Y^2 = 1120.14$

$$\text{Mean} = \bar{Y} = \frac{\sum Y}{N} = \frac{98.5}{9} = 10.94$$

Now plug the results into the formula for the variance:

$$S_Y^2 = \left(\frac{\sum Y^2}{N-1}\right) - \left(\frac{N}{N-1}\right)\left(\frac{\sum Y}{N}\right)^2 = \frac{1120.14}{8} - \left(\frac{9}{8}\right)\left(\frac{98.5}{9}\right)^2$$

$$= 140.02 - 134.75 = 5.27$$

The standard deviation can be found by taking the square root of the variance. For our example, the standard deviation is

$$S_Y = \sqrt{5.27} = 2.30$$

Hence, for the nine largest states, the suicide rate has a mean* of 10.94 suicides per 100,000 population and a standard deviation of 2.30 suicides per 100,000 population.

*Unweighted mean

Figure 5.10 **How to Choose a Measure of Variation**

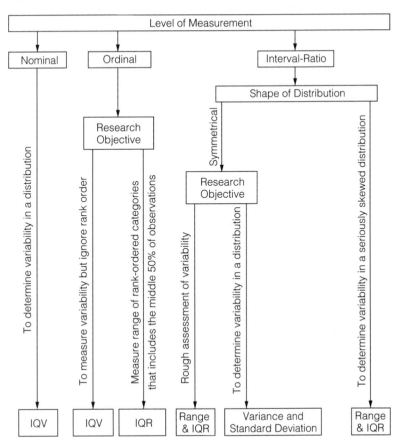

As in choosing a measure of central tendency, one of the most basic considerations in choosing a measure of variability is the variable's level of measurement. Valid use of any of the measures requires that the data are measured at the level appropriate for that measure or higher, as shown in Figure 5.10.

■ *Nominal level.* With nominal variables, your choice is restricted to the IQV as a measure of variability.

■ *Ordinal level.* The choice of measure of variation for ordinal variables is more problematic. The IQV can be used to reflect variability in distributions of ordinal variables, but because it is not sensitive to the rank ordering of values implied in ordinal variables, it loses some information. Another possibility is to use the interquartile range. However, the interquartile range relies on distance between two scores to express variation, information that cannot be obtained from ordinal measured scores. The compromise is to use the interquartile range alongside the median, interpreting the interquartile range as the range of rank-ordered values that includes the middle 50 percent of the observations.[13]

■ *Interval-ratio level.* For these variables, the variance (or standard deviation), range, or interquartile range can be chosen. Because the range, and to a lesser extent the interquartile range, is based on only two scores in the distribution (and therefore tends to be sensitive if either of the two points is extreme), the variance and/or standard deviation is usually preferred. However, if a distribution is extremely skewed so that the mean is no longer representative of the central tendency in the distribution, the range and the interquartile range can be used. The range and the interquartile range will also be useful when you are reading tables or quickly scanning data to get a rough idea of the extent of dispersion in the distribution.

■ ■ ■ ■ **Reading the Research Literature: Gender Differences in Caregiving**

In Chapter 2 we discussed how frequency distributions are presented in the professional literature. We noted that most statistical tables presented in the social science literature are considerably more complex than those we describe in this book. The same can be said about measures of central tendency and variation. Most research articles use measures of central tendency and variation in ways that go beyond describing the central tendency and variation of a single variable. In this section, we refer to both the mean and standard deviation because in most research reports the standard deviation is reported along with the mean.

---

[13]Herman J. Loether and Donald G. McTavish, *Descriptive and Inferential Statistics: An Introduction* (Boston: Allyn and Bacon, 1980), pp. 160–161.

Table 5.11   **Gender and Caregiving: Number and Hours Helped per Month***

	Wives		Husbands	
	Mean	**Standard Deviation**	**Mean**	**Standard Deviation**
I. INFORMAL CAREGIVING				
Number Helped:				
Kin[†]	5.25	2.86	4.02	2.59
Friends	3.44	2.54	2.26	2.17
Total people	8.71	5.34	6.29	3.58
Hours Helped:				
All kin[†]	42.77	30.82	15.06	14.02
Parents	11.52	19.10	3.76	3.74
Parents-in-law	6.20	12.23	3.95	7.47
Adult children	20.79	37.27	4.60	10.45
Friends	10.93	14.48	6.35	10.30
II. FORMAL CAREGIVING				
Number of groups	2.08	2.26	3.14	3.54
Volunteer hours	8.09	13.68	9.41	13.67
III. TOTAL CAREGIVING				
Total hours[‡]	61.79	60.49	30.82	27.76

*Measures computed for all respondents (N = 273), except Hours Helped Parents (includes only those with at least one living parent, N = 165), Parents-in-law (includes only those with at least one living parent-in-law, N = 162), and Adult Children (includes only those with at least one adult child, N = 126).

[†]Kin includes parents, parents-in-law, adult children, siblings, grandparents, aunts, uncles, and any other kin mentioned.

[‡]Total Hours = Informal (Hours for All Kin and Friends) + Formal (Volunteer Hours).

Source: Adapted from Naomi Gerstel and Sally Gallagher, "Caring for Kith and Kin: Gender, Employment, and the Privatization of Care," *Social Problems* 41, no. 4 (November 1994): 525.

Table 5.11, taken from an article by Professors Naomi Gerstel and Sally Gallagher,[14] illustrates a common research application of the mean and standard deviation. Professors Gerstel and Gallagher examined gender differences in caregiving to relatives and friends, as well as in volunteering to groups. Despite growing acceptance among Americans of governmental

---

[14]Naomi Gerstel and Sally Gallagher, "Caring for Kith and Kin: Gender, Employment, and Privatization of Care," *Social Problems* 41, no. 4 (November 1994): 519–537.

aid for the disabled, the majority of Americans continue to believe it is the responsibility of women to provide personal and household assistance to elderly parents and in-laws, as well as to aging siblings and adult children.

Gerstel and Gallagher's major hypothesis is "that wives will give care in far greater breadth . . . and depth than husbands. That is, wives are far more likely to give help to a larger number of people than husbands, including to more relatives, more friends, as well as more volunteer groups."[15] The researchers (1) assess the amount and types of care provided by wives compared with husbands and (2) look at the relevance of employment status to the amount and type of care provided by wives and husbands.

Data for this study come from household interviews conducted in 1990 with 273 married respondents—179 married women and 94 of their husbands. The sample was limited to whites (86 percent) and blacks (14 percent) over the age of 21 in Springfield, Massachusetts. Table 5.11 lists the most important variables used in the study. It presents the means and standard deviations for the breadth and depth of informal and formal caregiving.

To measure the breadth of informal caregiving, interviewers named a number of different categories of people—including mother and father, adult children, other relatives, and friends. After naming a category (for example, mother), the interviewer gave the respondent a list of tasks and asked if she or he had done each task for the person named within the past month. The total number of people given care and the number of people in each category provide a measure of the "breadth of informal caretaking." In addition, respondents were asked how many hours in the past month they provided care to each category of person. The total number of hours of care given to all kin and hours given to people in each category (parents, parents-in-law, adult children) is a measure of the "depth of informal caregiving." To measure the "breadth and depth of formal caregiving," respondents were asked to list the number of voluntary organizations they belonged to and in which they did charity and volunteer work, and how many hours they spent on that work. Finally, because gender is a central focus of this study, the means and standard deviations are reported separately for men (husbands) and women (wives).

What can you conclude from examining the standard deviations for these variables? The first thing to look for is variables with a great deal of variation, based on a large standard deviation score. Based on the summary in Table 5.11, you can see that this is the case with the variable "hours helped" in all categories, as well as with both aspects of formal caregiving. Notice that for both men and women (except for parent hours helped for men), the standard deviations are larger than the mean. Under the category of adult children for "hours helped," wives have a mean of 20.79 hours with 37.27 hours as the standard deviation. On the other hand, husbands reported an average of 4.60 helping hours (about a fifth of the time wives reported), with a standard deviation of 10.45 hours. This indicates that there is a great deal of variation in the hours of care among both men and women in the study.

[15]Ibid., p. 522.

In describing the data displayed in Table 5.11, the researchers focused on the differences between men and women for each of the variables:

> The table shows striking differences between wives and husbands in the breadth, depth, and distribution of caregiving. Compared to husbands, wives help . . . a larger number of people, both kin and friends. Moreover, wives give . . . more hours of care to friends. The differences in the amount of time wives, compared to husbands, spend providing for their own parents are even larger. . . . Mothers spend more than four times more hours than fathers helping their adult children. Overall, wives give help to more relatives and spend almost three times as much time doing so. Clearly, wives are the major caregivers.[16]

MAIN POINTS

- Measures of variability are numbers that describe how much variation and diversity there is in a distribution.

- The index of qualitative variation (IQV) is used to measure variation in nominal variables. It is based on the ratio of the total number of differences in the distribution to the maximum number of possible differences within the same distribution. IQV can vary from 0.00 to 1.00.

- The range measures variation in interval-ratio variables and is the difference between the highest (maximum) and lowest (minimum) scores in the distribution. To find the range, subtract the lowest from the highest score in a distribution.

- The interquartile range (IQR) measures the width of the middle 50 percent of the distribution. It is defined as the difference between the lower and upper quartiles (Q1 and Q3).

- The box plot is a graphical device that visually presents the range, the interquartile range, the median, the lowest (minimum) score, and the highest (maximum) score. The box plot provides us with a way to visually examine the center, the variation, and the shape of a distribution.

- The variance and the standard deviation are two closely related measures of variation for interval-ratio variables that increase or decrease based on how closely the scores cluster around the mean. The variance is the average of the squared deviations from the center (mean) of the distribution; the standard deviation is the square root of the variance.

KEY TERMS

index of qualitative variation (IQV) *range*

interquartile range (IQR) *standard deviation*

measures of variability *variance*

[16]Ibid., p. 525.

SPSS DEMONSTRATIONS

Demonstration 1: Producing Measures of Variability with Frequencies
[Module A]

Except for the IQV, the SPSS Frequencies procedure can produce all the measures of variability we've reviewed in this chapter. (SPSS can be programmed to calculate the IQV, but the programming procedures are beyond the scope of our book.)

We'll begin with Frequencies and calculate various statistics for AGE. If we click on *Frequencies,* then on the *Statistics* button, we can select the appropriate statistics.

The measures of variability available are listed in the Dispersion box on the bottom of the dialog box (see Figure 5.11). We've selected the standard deviation, variance, and range, plus the mean (in the Central Tendency box) for reference. In the Percentile Values box, we've selected Quartiles to tell SPSS to calculate the values for the 25th, 50th, and 75th percentiles. SPSS also allows us to specify exact percentiles in this section (such as the 34th percentile) by typing a number in the box after "Percentile(s)" and then clicking on the *Add* button.

Figure 5.11

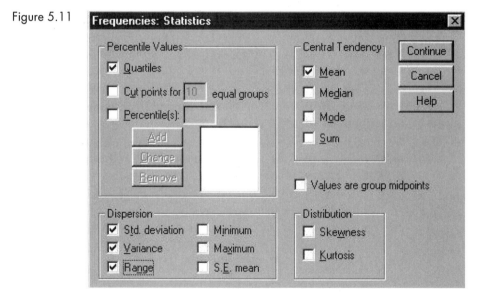

We have already seen the frequency table for the variable AGE, so after clicking on *Continue,* we click on *Format* to turn off the display table. This is done by clicking on the button for "Suppress tables with more than 10 categories" (see Figure 5.12). There are other formatting options here that you may explore later when using SPSS.

Figure 5.12

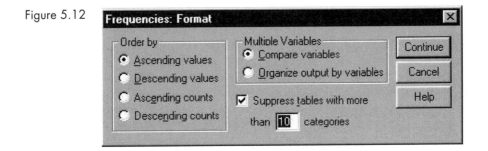

Click on *Continue,* then *OK* to run the procedure. SPSS produces the mean and the other statistics we requested (Figure 5.13). The range of age is 71 years (from 18 to 89). The standard deviation is 17.03, which indicates that there is a moderate amount of dispersion in the ages (this can also be seen from the histogram of AGE in Chapter 3). The variance, 290.01, is the square of the standard deviation.

Figure 5.13

Statistics

AGE OF RESPONDENT

N	Valid	1422
	Missing	5
Mean		44.54
Std. Deviation		17.03
Variance		290.01
Range		71
Percentiles	25	31.00
	34	35.00
	50	42.00
	75	56.00

The value of the 25th percentile is 31; the value of the 50th percentile (which is also the median) is 42; and the value of the 75th percentile is 56. Although Frequencies does not calculate the interquartile range, it can easily be calculated by subtracting the value of the 25th percentile from the 75th percentile, which yields a value of 25 years. Compare this value with the standard deviation.

Demonstration 2: Producing Variability Measures and Box Plots with Explore [Module A]

Another SPSS procedure that can produce the usual measures of variability is Explore, which also produces box plots. The Explore procedure is located in the *Descriptive Statistics* section of the *Analyze* menu. In its main dialog box (Figure 5.14), the variables for which you want statistics are placed in the Dependent List box. You have the option of putting one or more nominal variables in the Factor List box. Explore will display separate statistics for each category of the nominal variable(s) you've selected.

Figure 5.14

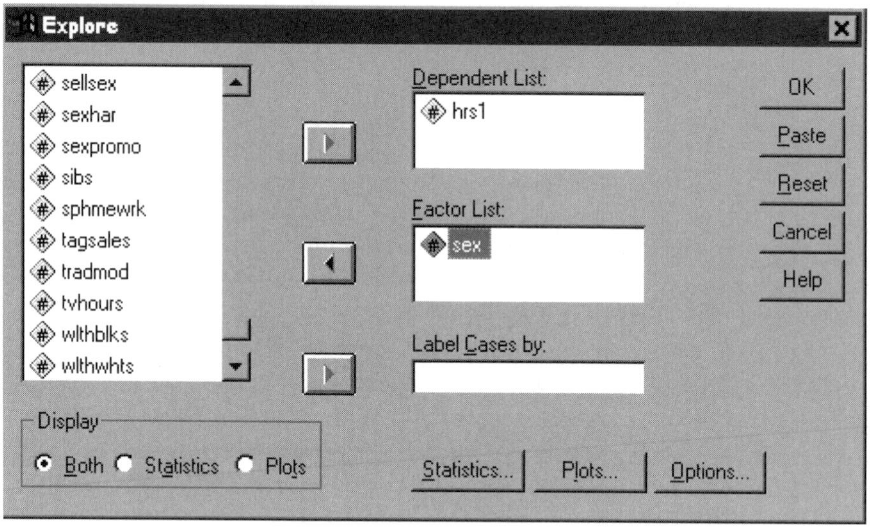

Place the variable HRS1 (hours worked last week) in the Dependent box and SEX in the Factor box, to provide separate output for males and females. By default, Explore will produce statistics and plots, so we don't need to make any other choices. Although our request will not produce percentiles or create a histogram, Explore has options to do both plus several other tasks.

Selected output for males is shown in Figure 5.15. Though not replicated here, you'll notice that the first table is the "Case Processing Summary Table." It indicates that 476 males were working last week and answered this question. The valid sample of females is also reported, 482. Based on the second table, "Descriptives," we know that for males, the mean number of hours worked was 44.79; the median was 41. The standard deviation is 14.16, the range is 87, and the IQR is 10 hours, which is quite narrow compared with the range or standard deviation. (A stem and leaf plot—another way to visually present and review data—is also displayed by default. However, we will not cover stem and leaf plots in this text. The option for the stem and leaf plot can be changed so that it will not be displayed.)

Although not displayed here, the mean of HRS1 for females is 38.84; the median is 40. The standard deviation is 13.18, the IQR is 8, and the range, 86—values close to those for males but slightly smaller. The variation in hours worked last week for females is slightly lower than for males.

Figure 5.15

Number of Hours Worked Last Week

RESPONDENTS SEX			Statistic	Std. Error
MALE	Mean		44.79	.65
	95% Confidence Interval for Mean	Lower Bound	43.51	
		Upper Bound	46.06	
	5% Trimmed Mean		44.79	
	Median		41.00	
	Variance		200.611	
	Std. Deviation		14.16	
	Minimum		2	
	Maximum		89	
	Range		87	
	Interquartile Range		10.00	
	Skewness		.083	.112
	Kurtosis		1.181	.223

Explore displays separate box plots for males and females in the same window for easy comparison (Figure 5.16). Although the SPSS box plot has some differences from those discussed in this chapter, some things are the same. The solid dark line is the value of the median. The width of the shaded box (in color on the screen) is the IQR (10 hours for males, 8 hours for females). Notice that the median is centered in the box for females but not for males. This is partially because so many males report working 40 hours.

Figure 5.16

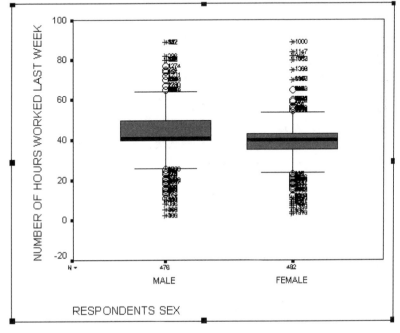

Unlike the box plots in this chapter—in which the "whiskers" extend out to the minimum and maximum values—SPSS extends whiskers from the box edges to 1½ times the box width (the IQR). If there are additional values beyond 1½ times the IQR, SPSS displays the individual cases. Those that are somewhat extreme (1½ to 3 box widths from the edge of the box) are marked with an open circle; those considered very extreme (more than 3 box widths from the box edge) are marked with an asterisk. The number of valid cases in each group is listed at the bottom of each box plot. There is no way to force SPSS to produce a box plot similar to the ones shown in this chapter.

The box plot shows us that variability in hours worked last week for males and females is similar, although the IQR for males runs from 40 to 50 hours, while the IQR for females runs from 35 to 44 hours. Both genders have outlying cases beyond the edge of the whiskers for people who worked many hours, or few hours, last week.

SPSS PROBLEMS

1. Use the 1996 GSS file [Module A] to investigate the variability of the respondent's current age (AGE) and age when the respondent's first child was born (AGKDBRN). You can use either the Frequencies or the Explore procedure.
   a. Which variable has more variability? Use more than one statistic to answer this question.
   b. Why should one variable have more variability than the other, from a societal perspective?

2. Using the Explore procedure, separate the statistics for AGKDBRN for men and women, selecting SEX as a factor variable. What difference exists in the age of men and women at the birth of their first child? Assess the difference between men and women based on measures of central tendency and variability.

3. Use the 1996 GSS file [Module A] to study the number of hours that blacks and whites work each week. The variable HRS1 measures the number of hours a respondent worked the week before the interview.
   a. Use the Explore procedure to study the variability of hours worked, comparing blacks and whites (RACE) in the GSS sample. Be sure to request a box plot by leaving the "Both" choice selected in the Display box.
   b. Is there much difference in the variability of work hours between the two groups?
   c. Write a short paragraph describing the box plot that SPSS created as if you were writing a report and had included the box plot as a chart to support your conclusions about the variability (and central tendency) of hours worked between blacks and whites.

4. Repeat the procedure in Exercise 3, investigating the dispersion in the variables EDUC (education) and PRESTG80 (occupational prestige

score). Select your own factor (nominal) variable to make the comparison (such as SEX, RACE, or some other factor). Write a brief statement summarizing your results.

5. Using Module B, investigate respondents' attitudes toward immigrants in the United States, based on two GSS variables, IMMAMECO (immigrants are generally good for America's economy) and IMMIDEAS (immigrants make America more open to new ideas and cultures).
   a. First, identify the level of measurement for each variable.
   b. Based on the level of measurement, what would be the appropriate set of measures of central tendency and variability?
   c. Use SPSS to calculate the different set of measures for blacks and whites. Is there a difference in attitudes among blacks and whites on these variables?

CHAPTER EXERCISES

1. Americans often think of themselves as quite diverse in their political opinions, within the continuum of liberal to conservative. Let's use the data from the 1996 GSS to investigate the diversity of political views. The following frequency table displays respondents' self-rating of their political position. (The statistics box is not displayed; cases with no response were removed for this example.)

THINK OF SELF AS LIBERAL OR CONSERVATIVE

		Frequency	Percent	Valid Percent	Cumulative Percent
Valid	EXTREMELY LIBERAL	15	2.6	2.7	2.7
	LIBERAL	65	11.2	11.7	14.4
	SLIGHTLY LIBERAL	72	12.4	12.9	27.3
	MODERATE	214	36.8	38.4	65.7
	SLGHTLY CONSERVATIVE	76	13.1	13.6	79.4
	CONSERVATIVE	92	15.8	16.5	95.9
	EXTRMLY CONSERVATIVE	23	4.0	4.1	100.0
	Total	557	95.9	100.0	
Missing	DK	24	4.1		
	Total	24	4.1		
Total		581	100.0		

a. What is the maximum possible number of differences, given this table?
b. What is the observed number of differences?
c. What is the IQV for this variable? Do you find it to be higher (closer to 1) or lower (closer to 0) than you might have expected for political views? Or to put it another way, did you expect that Americans would be diverse in their political views, or more narrowly concentrated in certain categories? Does this value of IQV support your expectation and what you observe from the table?

2. Traditionally, minorities and females have had few jobs in the construction trades. Even in 1990, most construction workers were still male and white. Perhaps the recent hiring of more females in construction jobs has also produced more racial diversity. That is, perhaps there are higher percentages of blacks and Hispanics among female construction workers than among male workers. U.S. census data from 1990 provide the following frequency distributions for male and female carpenters in a large midwestern city.

	Number of Carpenters	
	Male	Female
White	6,448	101
Black	1,458	35
Hispanic	2,156	26
Total	10,062	162

a. Calculate the total number of observed differences for males and females separately.
b. Calculate the maximum number of possible differences for each gender.
c. Use the values you calculated in (a) and (b) to calculate the IQV for males and females. Is there more diversity by race for male or female carpenters? Notice that there are more Hispanic than black male carpenters, but more black than Hispanic female carpenters.

3. The Census Bureau annually estimates the percentage of Americans below the poverty level for various geographic areas. Use the information in the following table to characterize poverty in the southern versus the western portion of the United States.

**Percentage of Americans Below the Poverty Level,
Southern and Western Regions, 1995**

	Percentage		Percentage
South		**West**	
Alabama	20.1	Alaska	7.1
Florida	16.2	Arizona	16.1
Georgia	12.1	California	16.7
Kentucky	14.7	Idaho	14.5
Louisiana	19.7	Montana	15.3
Mississippi	23.5	Nevada	11.1
North Carolina	12.6	New Mexico	25.3
South Carolina	19.9	Oregon	11.2
Tennessee	15.5	Utah	8.4
		Washington	12.5

Source: Statistical Abstract of the United States, 1997, Table 741.

   a. What is the range of poverty rates in the South? the West? Which is greater?

   b. What is the interquartile range (IQR) for the South? for the West? Which is larger?

   c. Using these calculations, compare the variability of the poverty rate of the states in the West with those in the South.

   d. Calculate the standard deviation for each region.

   e. Which region appears to have more variability as measured by the standard deviation? Are these results consistent with what you found using the range and the IQR?

4. Use the data from Exercise 3 again. This time your task is to create box plots to display the variation in poverty level by region.

   a. First, combine both regions and create a box plot for all the states. What is the 75th percentile for poverty rate? Are there any outlying cases (that is, outside the whiskers)?

   b. Now create a separate box plot for the West and one for the South. Do these box plots add to your discussion from Exercise 3? If so, how?

5. Use Table 5.5 for this exercise to continue comparisons by region. Use only the information for states in the West (Mountain) and Midwest (West North Central).

   a. Compare the Mountain states with those in the West North Central on the percentage increase in the elderly population by calculating the range. Which region had a greater range?

   b. Calculate the IQR for each region. Which is greater?

   c. Use the statistics to characterize the variability in population increase of the elderly in the two regions. Why do you think one region is more variable than another?

6. *Occupational prestige* is a statistic developed by sociologists to measure the status of one's occupation. It is measured on a scale from 1 to 100 (though no occupation has a score at either extreme). A lower score corresponds to lower occupational prestige. Occupational prestige is also a component of what sociologists call *socioeconomic status*, a composite measure of one's status in society. On average, blacks have lower occupational prestige than whites, but is it still possible that the variation among prestige scores for blacks is no greater than it is for whites? We investigate this question, using 1996 GSS data to generate the following SPSS output from the Explore procedure.

Descriptives

	RACE				Statistic	Std. Error
PRESTG80 RS OCCUPATIONAL PRESTIGE SCORE (1980)	1 WHITE	Mean			44.25	.64
		95% Confidence Interval for Mean	Lower Bound		42.99	
			Upper Bound		45.50	
		5% Trimmed Mean			43.81	
		Median			44.00	
		Variance			184.873	
		Std. Deviation			13.60	
		Minimum			17	
		Maximum			86	
		Range			69	
		Interquartile Range			17.00	
		Skewness			.513	.114
		Kurtosis			-.129	.228
	2 BLACK	Mean			38.08	1.46
		95% Confidence Interval for Mean	Lower Bound		35.16	
			Upper Bound		41.00	
		5% Trimmed Mean			37.44	
		Median			36.00	
		Variance			156.493	
		Std. Deviation			12.51	
		Minimum			17	
		Maximum			69	
		Range			52	
		Interquartile Range			15.50	
		Skewness			.701	.281
		Kurtosis			-.101	.555

(*Note:* The Explore procedure produces a variety of statistics, including some we have not discussed, including skewness, kurtosis, and the standard error ("Std. Error"). Don't worry about these in this exercise.)

a. Notice that SPSS supplies the interquartile range (IQR), the median, and the minimum and maximum values for each race. Looking at the values of the mean and median, do you think the distribution of prestige is skewed for blacks? for whites? Why or why not?

b. Explain why you think there is more variability of prestige for whites or for blacks, or why the variability of prestige is similar for the two groups.

7. A child psychologist is studying the behavior of children during play by unobtrusively observing them during recess periods at an elementary school. Her work is part of a long-term study on how cooperation develops among humans and how a person's characteristics are related to cooperative behavior. In this phase of the study, she has recorded the number of incidents of cooperative behavior among two groups of children over a 15-day period.

GROUP 1: 4 9 11 3 10 12 6 9 11 9 4 12 20 8 9
GROUP 2: 9 4 13 2 5 10 6 4 7 7 12 8 5 10 8

The psychologist's first task is to study the average level of cooperation and its dispersion in each group before she takes into account group differences.
a. Calculate the mean and standard deviation for the number of incidents of cooperative behavior in each group.
b. Use these values to discuss differences in cooperative behavior between the two groups of children, from a statistical standpoint.
c. What if you learn that there are twice as many children in Group 1 as in Group 2? Would that new information modify your answer to (b)? Why or why not?

8. A group of investigators has just finished a study that measured the amount of time each partner in a marriage spends doing housework. The investigators classified each couple as traditional or nontraditional, depending on the attitudes of both partners. (Traditional couples commonly grant more authority to the male; nontraditional couples share more in decision making.) The investigators provide you with the following data for males only.

Hours of Housework per Week

Traditional Family	Nontraditional Family
$\bar{Y} = 6.3$	$\bar{Y} = 12.4$
$\Sigma Y^2 = 1{,}104$	$\Sigma Y^2 = 2{,}889$
$\Sigma Y = 63$	$\Sigma Y = 186$
$N = 10$	$N = 15$

a. Calculate the variance and standard deviation from these statistics for each family type.
b. What can you say about the variability in the amount of time men spend doing housework in traditional versus nontraditional marriages? Why might there be a difference? Why might there be more variability for one type of family than another?

c. Was it necessary in this problem to provide you with the value of $\overline{Y}$ to calculate the variance and standard deviation?

9. You are interested in studying the variability of violent crime and incarceration in the eastern and midwestern United States. The U.S. Census Bureau collected the following statistics on these two variables for twenty-one states in the East and Midwest in 1995.

	Violent Crime Rate per 100,000 People	Total Federal and State Prisoners
Maine	131	1,396
New Hampshire	115	2,014
Vermont	118	1,279
Massachusetts	687	11,687
Rhode Island	368	2,902
Connecticut	406	14,801
New York	842	68,489
New Jersey	600	27,066
Pennsylvania	427	32,410
Ohio	483	44,663
Indiana	525	16,125
Illinois	996	37,658
Michigan	688	41,112
Wisconsin	281	11,199
Minnesota	356	4,846
Iowa	354	5,906
Missouri	664	19,134
North Dakota	87	608
South Dakota	208	1,841
Nebraska	382	3,074
Kansas	421	7,054

Source: *Statistical Abstract of the United States,* 1997, Tables 315 and 358.

a. Calculate the mean for each variable.
b. Calculate the standard deviation for each variable.
c. Calculate the interquartile range for each variable.
d. Compare the mean with the standard deviation and IQR for each variable. Does there appear to be more variability in the rate of violent crime or the rate of incarceration in these states? Which states contribute more to the greater variability for each variable?
e. Suggest why one variable has more variability than the other. In other words, what social forces would cause one variable to have a relatively larger standard deviation than its mean?

10. Construct a box plot for both variables in Exercise 9. Discuss how the box plot reinforces the conclusions you drew about the variability of the rates of violent crime and incarceration.

11. Use the data in Table 5.6 for this exercise.
   a. Calculate the standard deviation for the percentage increase in the elderly population from 1980 to 1990 by state.
   b. Compare this statistic with the IQR and the box plot shown in Figure 5.6. Which is larger, the IQR or the standard deviation?
   c. Would the standard deviation lead you to the same conclusion about the variability of the increase in the elderly population as the IQR and the box plot?

12. You decide to use GSS data from 1996 to investigate how Americans feel about spending federal government money on welfare and to improve the condition of blacks. You obtain the following output on these two variables, where "Too little" means that the federal government is spending too little, "About right" means that the level of government spending on this issue is about right, and "Too much" means the government is spending too much.

WELFARE

		Frequency	Percent	Valid Percent	Cumulative Percent
Valid	TOO LITTLE	48	8.3	17.7	17.7
	ABOUT RIGHT	76	13.1	28.0	45.8
	TOO MUCH	147	25.3	54.2	100.0
	Total	271	46.6	100.0	
Missing	NAP	300	51.6		
	DK	9	1.5		
	NA	1	.2		
	Total	310	53.4		
Total		581	100.0		

IMPROVING THE CONDITIONS OF BLACKS

		Frequency	Percent	Valid Percent	Cumulative Percent
Valid	TOO LITTLE	89	15.3	35.5	35.5
	ABOUT RIGHT	105	18.1	41.8	77.3
	TOO MUCH	57	9.8	22.7	100.0
	Total	251	43.2	100.0	
Missing	NAP	300	51.6		
	DK	28	4.8		
	NA	2	.3		
	Total	330	56.8		
Total		581	100.0		

(*Note:* The label of NAP stands for "Not Applicable," DK stands for "Don't Know," and NA for "No Answer.")

a. What would an appropriate measure of variability be for these variables? Why?

b. Calculate the appropriate measure of variability for each variable.

c. In 1996, was there more variability for attitudes toward spending on welfare or improving the conditions of blacks?

13. The percent of male and female arrestees who tested positive for marijuana use at the time of their 1995 arrests are reported below. Is there a difference between male and female arrestees? Are the percentages about the same for both? Calculate the appropriate measures of central tendency and variability.

City	Male	Female
Atlanta, GA	32	13
Birmingham, AL	36	12
Cleveland, OH	29	11
Dallas, TX	37	21
Denver, CO	33	21
Detroit, MI	42	18
Fort Lauderdale, FL	33	18
Houston, TX	29	18
Indianapolis, IN	38	24
Los Angeles, CA	23	14
Manhattan, NY	28	16
New Orleans, LA	32	16
Omaha, NE	42	24
Philadelphia, PA	34	20
Phoenix, AZ	29	19
Portland, OR	29	16
St. Louis, MO	39	18
San Antonio, TX	34	16
San Diego, CA	35	20
San Jose, CA	27	12
Washington, DC	32	18

Source: Statistical Abstract of the United States, 1997, Table 331.

14. The following table presents member characteristics of a secular self-help organization for alcohol abusers. Based on the means and standard deviations (in parentheses) provided in the table, prepare a brief statement on the characteristics of the group's members.

Characteristics	Men N = 116	Women N = 42
Age first used	20.5 (7.7)	21.7 (8.5)
Age first high	20.1 (7.7)	22.5 (8.9)
If abstinent, years of current abstinence	7.8 (6.8	7.6 (6.0)
Years of problem	6.8 (6.6)	10.2 (9.0)

Source: Data from Gerard Conners and Kurt Dermen, "Characteristics of Participants in Secular Organizations for Society," *American Journal of Drug and Alcohol Abuse* 22, no. 2 (1996): 281–295. Used by permission of Marcel Dekker Journals.

15. The following table presents U.S. Department of Justice data from selected states on the officer and civilian workforce, showing the actual number of male and female employees. What measure(s) of variability would be appropriate to summarize these data?

State	Officers		Civilians	
	Males	Females	Males	Females
California	57,189	7,049	9,976	20,314
Maine	1,919	99	311	340
Indiana	8,496	676	2,089	2,684
South Carolina	1,047	48	268	318
Ohio	18,428	1,998	3,912	3,063

Source: U.S. Department of Justice, "Crime in the United States 1996," *Uniform Crime Reports,* September 28, 1997.

16. The following table summarizes gender differences found in a study of university faculty pay in Canada. Based on the means and standard deviations (in parentheses), what conclusions can be drawn about the gender differences in faculty pay?

	Women	Men
Annual pay	$40,800 (9,700)	$49,300 (12,600)
Age at first full-time teaching job	32.5 (6.0)	31.5 (5.1)
Years at current rank	5.1 (4.2)	7.5 (5.7)

Source: Adapted from M. Ornstein and P. Stewart, "Gender and Faculty Pay in Canada," *Canadian Journal of Sociology* 21, no. 4 (1996): 461–481. Used by permission.

6 Relationships Between Two Variables: Cross-Tabulation

■ ■ ■ ■ **Introduction**

Bivariate analysis is a method designed to detect and describe the relationship between two variables. Before now, you may have had an intuitive sense of the terms *relationship* and *association*. We have seen in earlier chapters that by comparing the properties of different groups one can often think in terms of "relationships." In this and the following two chapters we look at the concept of relationships between variables in more depth.

> *Bivariate analysis* A statistical method designed to detect and describe the relationship between two variables.

You should be familiar with the idea of "relationship" simply because you are aware that in the world around you things (and people) "go together." For example, as children grow their weight increases; larger cities have more crime than smaller cities. In fact, many of the reports in our daily newspapers are statements about relationships. For example, a news story based on a census report on immigration to the United States documents the struggles many new immigrants have with the English language.[1] The article compares the English proficiency of native-born Americans with that of foreign-born Americans by using a set of simple bar charts (Figure 6.1). Notice that there is a pattern in Figure 6.1. More native-born than foreign-born Americans speak English well. This simple example illustrates a relationship or association between the variables *nativity* (native-born versus foreign-born) and *English proficiency* (speak English very well).

One of the main objectives of social science is to make sense out of human and social experience by uncovering regular patterns among events. Therefore, the language of relationships is at the heart of social science inquiry. Consider the following examples from articles and research reports:

Example 1 Students who had a history of earning good grades were less likely to miss class than students who did not.[2] (This example indicates a relationship between grade point average and absenteeism among college students.)

Example 2 Contrary to the stereotype, whites use government safety net programs more than blacks or Hispanics, and they are more likely than minorities to be lifted out of poverty by the taxpayer money they get.[3] (This example indicates a relationship between race and receipt of government aid.)

---

[1]*USA Today,* November 5, 1993.

[2]Gary Wyatt, "Skipping Class: An Analysis of Absenteeism Among First-Class College Students," *Teaching Sociology* 20 (July 1992): 201–207.

[3]*USA Today,* October 9, 1992.

Figure 6.1 **English Proficiency and Nativity**

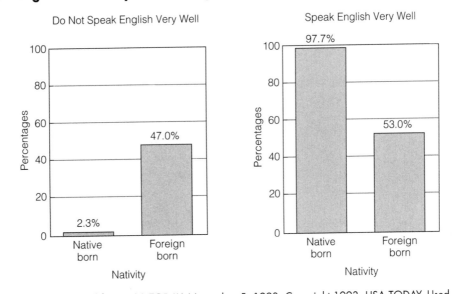

Source: Adapted from *USA TODAY*, November 5, 1993. Copyright 1993, USA TODAY. Used by permission.

Example 3 On average, blacks and Hispanics have lower access to health care than whites.[4] (This example indicates a relationship between race and access to health care.)

In each of these examples, a relationship means that certain values of one variable tend to "go together" with certain values of the other variable. In Example 1, lower grades "go together" with irregular class attendance; higher grades go with regular class attendance. In Example 2, being white "goes together" with frequent use of government aid; being black or Hispanic goes with less frequent use of government aid. Finally, in Example 3, white "goes together" with greater likelihood of access to health care; being black or Hispanic "goes together" with a lower likelihood of access.

In this chapter we introduce one of the most common techniques used in the analysis of relationships between two variables: *cross-tabulation*. **Cross-tabulation** is a technique for analyzing the relationship between two variables that have been organized in a table. We demonstrate not only how to detect whether two variables are associated, but also how to determine the strength of the association and, when appropriate, its direction. We will also see how these methods are applied in "real" research situations.[5]

[4]National Center for Health Statistics, *Health, United States, 1998.*

[5]Full consideration of the question of detecting the presence of a bivariate relationship requires the use of inferential statistics. Inferential statistics is discussed in Chapters 11 through 15.

> *Cross-tabulation* A technique for analyzing the relationship between two variables that have been organized in a table.

■ ■ ■ ■ ■ **Independent and Dependent Variables**

In the social sciences, an important aspect in research design and statistics is the distinction between the *independent variable* and the *dependent variable*. These terms, first introduced in Chapter 1, are used throughout this chapter as well as in the following chapters, and therefore it is important that you understand the distinction between them. Let's take our example about nativity and English proficiency, in which nativity—whether a person is native born or foreign born—has some influence on the level of English proficiency. Even though immigration status is not necessarily a direct cause of language ability, immigration status is assumed to be connected to English proficiency through a complex set of experiences—such as education, employment, and other socialization experiences—all of which do have an influence on the acquisition of language. If we hypothesize that English proficiency (which is the variable to be explained by the researcher) varies by whether a person is native born or foreign born (which is the variable assumed to influence English proficiency), then English proficiency is the dependent variable and nativity is the independent variable.

In each of the illustrations given, there are two variables, an independent and a dependent variable. In Example 1, the purpose of the research is to explain absenteeism. One of the variables hypothesized as being connected to absenteeism is grades. Therefore, absenteeism is the dependent variable and grades the independent variable. In Example 2, the object of the investigation is to examine the common stereotype that people of color use government aid more than white Americans. The investigator is trying to explain differences in utilization of government aid using race as an explanatory variable. Therefore, utilization of government aid is the dependent variable and race is the independent variable. Similarly, in Example 3, access to health care is the dependent variable because it is the variable to be explained, whereas race, the explanatory variable, is the independent variable.

The statistical techniques discussed in this and the following two chapters help the researcher decide the strength of the relationship between the independent and dependent variables.

> *Learning Check.* *For some variables, whether it is the independent or dependent variable depends on the research question. If you are still having trouble distinguishing between an independent and a dependent variable, go back to Chapter 1 (pp. 9–11) for a detailed discussion.*

■ ■ ■ ■ ■ **How to Construct a Bivariate Table: Race and Home Ownership**

A **bivariate table** displays the distribution of one variable across the categories of another variable. It is obtained by classifying cases based on their joint scores for two variables. It can be thought of as a series of frequency distributions joined to make one table. The data in Table 6.1 represent a sample of respondents by race and whether they own or rent their home.

To make sense out of these data we must first construct the table in which these individual scores will be classified. In Table 6.2, the fifteen respondents have been classified according to joint scores on race and home

Table 6.1 **Race and Home Ownership for Fifteen GSS Respondents**

Respondent	Race	Home Ownership
1	Black	Own
2	Black	Own
3	White	Rent
4	White	Rent
5	White	Own
6	White	Own
7	White	Own
8	Black	Rent
9	Black	Rent
10	Black	Rent
11	White	Own
12	White	Own
13	White	Rent
14	White	Rent
15	Black	Rent

Table 6.2 **Home Ownership by Race (absolute frequencies)**

HOME OWNERSHIP	RACE			
	Black	White		
Own	2	5	7	Row Marginals (Row total)
Rent	4	4	8	
	6	9	15	Total Cases (N)
	Column Marginals (Column total)			

ownership. The table has the following features typical of most bivariate tables:

1. The table's title is descriptive, identifying its content in terms of the two variables and level of measurement.

2. It has two dimensions, one for race and one for home ownership. The variable *home ownership* is represented in the rows of the table, with one row for owners and another for renters. The variable *race* makes up the columns of the table, with one column for each racial group included. A table may have more columns and more rows, depending on how many categories the variables represent. For example, had we included a group of Hispanics, there would have been three columns (not including the Row Total column). Usually, the independent variable is the **column variable** and the dependent variable is the **row variable**.

3. The intersection of a row and a column is called a **cell**. For example, the two individuals represented in the upper left cell are blacks who are also homeowners.

4. The column and row totals are the frequency distribution for each variable, respectively. The column total is the frequency distribution for *race*, the row total for *home ownership*. Row and column totals are sometimes called **marginals**. The total number of cases (*N*) is the number reported at the intersection of the row and column totals. (These elements are all labeled in the table.)

5. The table is a 2 × 2 table because it has two rows and two columns (not counting the marginals). We usually refer to this as an *r* × *c* table, in which *r* represents the number of rows and *c* the number of columns. Thus, a table in which the row variable has three categories and the column variable two categories would be designated as a 3 × 2 table.

6. The source of the data should also be clearly noted in a source note to the table. This is consistent with what we reviewed in Chapter 2, Organization of Information.

Bivariate table   A table that displays the distribution of one variable across the categories of another variable.

Column variable   A variable whose categories are the columns of a bivariate table.

Row variable   A variable whose categories are the rows of a bivariate table.

Cell   The intersection of a row and a column in a bivariate table.

Marginals   The row and column totals in a bivariate table.

> **Learning Check.** *Examine Table 6.2. Make sure you can identify all of the parts just described and that you understand how the numbers were obtained. Can you identify the independent and dependent variables in the table? You will need to know this to convert the frequencies to percentages.*

How to Compute Percentages in a Bivariate Table

To compare home ownership status for blacks and whites, we need to convert the raw frequencies to percentages because the column totals are not equal. Recall from Chapter 2 that percentages are especially useful for comparing two or more groups that differ in size. There are two basic rules for computing and analyzing percentages in a bivariate table:

1. Calculate percentages within each category of the independent variable.
2. Interpret the table by comparing the percentage point difference for different categories of the independent variable.

Calculating Percentages Within Each Category of the Independent Variable

The first rule means that we have to calculate percentages within each category of the variable that the investigator defines as the independent variable. When the independent variable is arrayed in the *columns,* we compute percentages within each column separately. The frequencies within each cell and the row marginals are divided by the total of the column in which they are located, and the column totals should sum to 100 percent. When the independent variable is arrayed in the *rows,* we compute percentages within each row separately. The frequencies within each cell and the column marginals are divided by the total of the row in which they are located, and the row totals should sum to 100 percent.

In our example, we are interested in *race* as the independent variable and in its relationship with *home ownership.* Therefore, we are going to calculate percentages by using the column total of each racial group as the base of the percentage. For example, the percentage of black respondents who own their homes is obtained by dividing the number of black homeowners by the total number of blacks in the sample:

$$(100)\frac{2}{6} = 33\%$$

Table 6.3 presents percentages based on the data in Table 6.2. Notice that the percentages in each column add up to 100 percent, including the total column percentages. Always show the *N*s that are used to compute the percentages—in this case, the column totals.

Table 6.3 **Home Ownership by Race (in percentages)**

HOME OWNERSHIP	RACE		
	Black	White	Total
Own	33%	56%	47%
Rent	67%	44%	53%
Total	100% (6)	100% (9)	100% (15)

Comparing the Percentages Across Different Categories of the Independent Variable

The second rule tells us to compare how home ownership varies between blacks and whites. Comparisons are made by examining differences between percentage points across different categories of the independent variable. Some researchers limit their comparisons to categories with at least a 10 percent difference. In our comparison, we can see that there is a 23 percent difference between the percentage of white homeowners (56%) versus black homeowners (33%). In other words, in this group[6] whites are more likely to be homeowners than blacks. Therefore, we can conclude that one's race appears to be associated with the likelihood of being a homeowner.

Notice that the same conclusion would be drawn had we compared the percentage of black and white renters. However, since the percentages of homeowners and renters within each racial group sum to 100 percent, we need to make only one comparison. In fact, for any 2 × 2 table only one comparison needs to be made to interpret the table. For a larger table, more than one comparison can be made and used in interpretation.

Learning Check. *Practice constructing a bivariate table. Use Table 6.1 to create a percentage bivariate table. Compare your table with Table 6.3. Did you remember all of the parts? Are your calculations correct? If not, go back and review this section. It might be helpful to examine Box 6.1. It illustrates the process of constructing and percentaging bivariate tables. Remember, you must correctly identify the independent variable so you know whether to percentage across the rows or down the columns.*

---

[6]Note that this group is but a small sample taken from the GSS national sample. The relationship between home ownership and race noted here may not necessarily hold true in other (larger) samples.

Box 6.1   Percentaging a Bivariate Table

1. Black and white homeowners and renters:

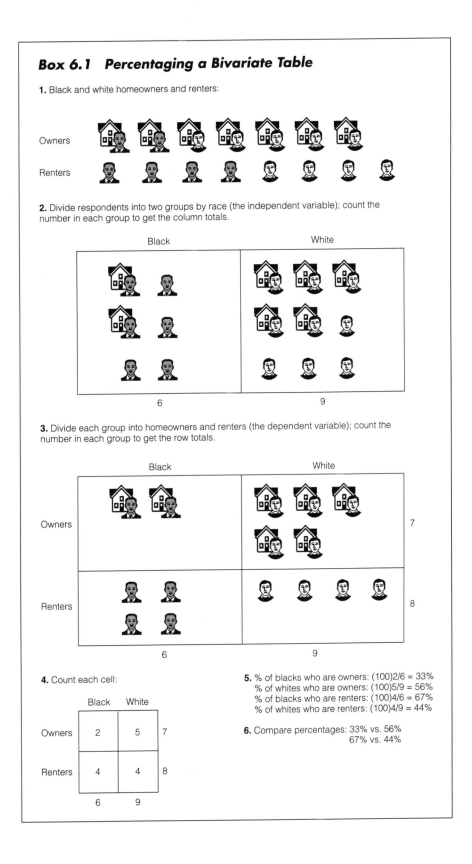

Owners

Renters

2. Divide respondents into two groups by race (the independent variable); count the number in each group to get the column totals.

Black   White

6   9

3. Divide each group into homeowners and renters (the dependent variable); count the number in each group to get the row totals.

Black   White

Owners   7

Renters   8

6   9

4. Count each cell:

	Black	White	
Owners	2	5	7
Renters	4	4	8
	6	9	

5. % of blacks who are owners: (100)2/6 = 33%
% of whites who are owners: (100)5/9 = 56%
% of blacks who are renters: (100)4/6 = 67%
% of whites who are renters: (100)4/9 = 44%

6. Compare percentages: 33% vs. 56%
67% vs. 44%

■ ■ ■ ■ **How to Deal with Ambiguous Relationships Between Variables**

Sometimes it isn't apparent which variable is independent or dependent; sometimes the data can be viewed either way. In this case, you might compute both row and column percentages. For example, Table 6.4 presents three sets of figures for the variables *attitude toward abortion* and *job security* for a sample of 109 GSS 1996 respondents: (a) the absolute frequencies, (b) the column percentages, and (c) the row percentages. *Job security* (labeled "Job Find") is measured with the survey question "About how easy would it be for you to find a job with another employer with approximately the same income and fringe benefits you now have?" The variable *attitude toward abortion* is measured in terms of the respondent's approval or disapproval of three reasons for obtaining an abortion: (1) the woman does not want the baby because the family has a very low income and cannot afford

Table 6.4 **The Different Ways Percentages Can Be Computed: Support for Abortion by Job Security**

a. Absolute Frequencies

Abortion	Job Find Easy	Job Find Not Easy	Row Total
Yes	36	24	60
No	26	23	49
Column Total	62	47	109

b. Column Percentages (column totals as base)

Abortion	Job Find Easy	Job Find Not Easy	Row Total
Yes	58%	51%	55%
No	42%	49%	45%
Column Total	100%	100%	100%
	(62)	(47)	(109)

c. Row Percentages (row totals as base)

Abortion	Job Find Easy	Job Find Not Easy	Row Total
Yes	60%	40%	100%
			(60)
No	53%	47%	100%
			(49)
Column Total	57%	43%	100%
			(109)

Source: General Social Survey, 1996.

Figure 6.2 **Bar Chart Comparing Column and Row Percentages Shown in Tables 6.4b and 6.4c**

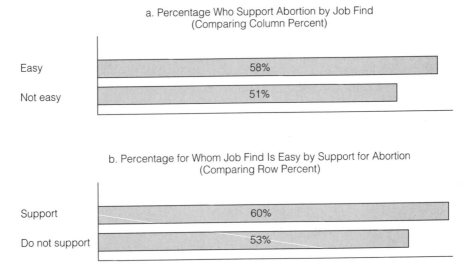

a. Percentage Who Support Abortion by Job Find
(Comparing Column Percent)

b. Percentage for Whom Job Find Is Easy by Support for Abortion
(Comparing Row Percent)

Source: General Social Survey, 1996.

more children; (2) the woman is not married and does not want to marry the father; and (3) the woman does not want to have more children. Table 6.4b shows that respondents who feel more secure economically are slightly more likely to support the right to abortion than those who feel less secure economically (58% compared with 51%). Table 6.4c shows that individuals who support abortion are more likely to feel economically secure than those who are against abortion (60% compared with 53%).

Thus, percentaging within each *column* (Table 6.4b) allows us to examine the hypothesis that job security (the independent variable) is associated with support for abortion (the dependent variable). When we percentage within each *row* (Table 6.4c), the hypothesis is that attitudes toward abortion (the independent variable) may be related to one's sense of job security (the dependent variable).[7] Figures 6.2a and 6.2b are simple bar charts illustrating the two methods of calculating and comparing percentages as depicted in Tables 6.4b and 6.4c.

Finally, it is important to understand that ultimately what guides the construction and interpretation of bivariate tables is the theoretical question posed by the researcher. Although the particular example in Table 6.4

---

[7]One other way in which percentages are sometimes expressed is with the total number of cases (*N*) used as the base. These overall percentages express the proportion of the sample who share two properties. For example, 36 of 109 respondents (33%) are for abortion and have job security. Overall percentages do not have as much research utility as row and column percentages and are used less frequently.

makes sense if interpreted using row or column percentages, not all data can be interpreted this way. For example, a table comparing women's and men's attitudes toward the Equal Rights Amendment could provide a sensible explanation in only one direction. Gender might influence a person's attitude toward the amendment, but a person's attitude toward the amendment certainly couldn't influence her or his gender. Therefore, either row or column percentages are appropriate, depending on the way the variables are arrayed, but not both.

> **Learning Check.** *Figures 6.2a and 6.2b each show only one set of bars. Figure 6.2a shows only the percentage who support abortion in each group. What are the percentages of those who do not support abortion? Figure 6.2b shows only the percentages with job security (easy). What are the percentages of those who do not have job security (not easy)? Hint: The percentages add to 100 percent within each category of the variable treated as independent.*

■ ■ ■ ■   **Reading the Research Literature: Medicaid Use Among the Elderly**

The guidelines for constructing and interpreting bivariate tables discussed in this chapter are not always strictly followed. Most bivariate tables presented in the professional literature are a good deal more complex than those we have just been describing. Let's conclude this section with a typical example of how bivariate tables are presented in social science literature. The following example is drawn from a study by Madonna Harrington Meyer on Medicaid use among the elderly.[8]

Access to health care for all Americans is at the top of the U.S. domestic agenda today. The rise of long-term care as a politically salient topic is fueled by the increase in the elderly population, which is ten times larger than it was in 1900 and will more than double by the year 2030. Financing of long-term care is a problem for older persons because Medicare, the universal health care program for the aged, excludes most long-term care. Only Medicaid, the poverty-based health care program, includes long-term care coverage. Therefore, only the poor elderly receive assistance from the state for long-term care.[9]

In this study, Professor Meyer explores the distribution of Medicaid benefits to the frail elderly. She examines the hypothesis that gender and race are important determinants of Medicaid use "because the U.S. long-term care system stratifies by gender and race by perpetuating, rather than alleviating, inequality created by social and market forces."[10]

---

[8]Madonna Harrington Meyer, "Gender, Race, and the Distribution of Social Assistance: Medicaid Use Among the Frail Elderly," *Gender & Society* 8, no. 1 (1994): 8–28.

[9]Ibid., p. 9.

[10]Ibid., p. 12.

The study examines differences in Medicaid use in 1984 by age, education, marital status, gender, and race. Meyer analyzed data from the National Long Term Care survey conducted by the Department of Health and Human Services in 1982 and 1984. The data set is based on a national random sample of 6,000 functionally impaired older persons who resided in the community in 1982. By 1984, respondents were either still living in the community, had entered a nursing home, or had died.

Table 6.5 shows the results of the survey. Follow these steps in examining it:

1. Identify the dependent variable and the type of unit of analysis it describes (such as individual, city, or child). Here the dependent variable is *received Medicaid in 1984*. The categories for this variable are "yes" and "no." The type of unit used in this table is individual.

Table 6.5 **Percentage Medicaid Use in 1984, by Age, Education, Marital Status, Gender, and Race of Functionally Impaired Older Persons**

| | RECEIVED MEDICAID IN 1984 | | |
	Yes	No	N
Age			
65–74	19.5	80.5	1,561
75–84	23.8	76.2	1,943
85+	25.4	74.6	1,007
Education			
8th grade or less	29.5	70.5	2,326
9th–12th grade	16.5	83.5	1,523
Some college	8.1	91.5	530
Marital Status			
Married	13.7	86.3	1,947
Widowed	28.6	71.4	2,079
Divorced, separated, never married	34.6	65.4	437
Gender			
Men	17.1	82.9	1,488
Women	25.4	74.6	3,024
Race			
White	19.1	80.9	3,942
Black and Hispanic	47.7	52.3	570

Source: Adapted from Madonna Harrington Meyer, "Gender, Race, and the Distribution of Social Assistance: Medicaid Use Among the Frail Elderly," *Gender & Society* 8, no. 1 (1994): 8–28. Used by permission of Sage Publications, Inc.

2. Identify the independent variables included in the table and the categories of each. There are five independent variables: age, education, marital status, gender, and race. Age consists of three categories: 65–74, 75–84, and 85+. Education consists of the categories "8th grade or less," "9th–12th grade," and "some college." "Married," "widowed," and "divorced, separated, never married" are the categories for marital status. Gender consists of "men" and "women"; and finally, "white" and "black and Hispanic" are the categories for race.

3. Clarify the structure of the table. Note that the independent variables are arrayed in the rows of the table, while the dependent variable, *received Medicaid in 1984,* is arrayed in the columns. The table is divided into five panels, one for each independent variable. There are actually five bivariate tables here—one for each independent variable.

Since the independent variables are arrayed in the rows, percentages are calculated within each row separately, with the row totals serving as the base for the percentages. For example, there were 1,561 respondents who were 65 to 74 years old. Of these, 19.5 percent received Medicaid in 1984 and 80.5 percent did not. Similarly, of the 1,943 respondents who were 75 to 84, 23.8 percent received Medicaid in 1984 and 76.2 percent did not. Although not shown in the table, the percentages within each row add to 100 percent.

4. Using Table 6.5, we can make a number of comparisons, depending on which independent variable we are examining. For example, to determine the relationship between age and the propensity to use Medicaid, compare the percentages of respondents of the different age groups who received Medicaid in 1984 (19.5%, 23.8%, and 25.4%). Alternatively, you can compare the percentages of respondents in the three age groups who did not receive Medicaid (80.5%, 76.2%, and 74.6%). Based on these percentage comparisons, we can conclude that among the frail elderly age is associated with Medicaid use: the oldest-old are more likely than the youngest-old to receive Medicaid.

Next look at the relationship between education and the propensity to use Medicaid. You can compare percentages of respondents with 8th-grade education or less who received Medicaid (29.5%) with those of respondents with 9th- to 12th-grade education (16.5%) and those with some college (8.1%) who received Medicaid. You can make similar comparisons to determine the association between marital status, gender, or race and the receipt of Medicaid.

The bivariate relations between age, education, marital status, gender, and receipt of Medicaid presented in Table 6.4 are also illustrated in Figure 6.3. Notice that only the percentages of elderly who have received Medicaid in 1984 ("yes") are shown. Since the percentage of elderly who responded "yes" and the percentage that responded "no" sum to 100, there is no need to show both sets of figures.

5. Finally, what conclusion can you draw about variations in the propensity to use Medicaid? The author offers this interpretation of the findings presented in the table:

Figure 6.3 **Percentage Who Received Medicaid in 1984 by Age, Education, Marital Status, Gender, and Race**

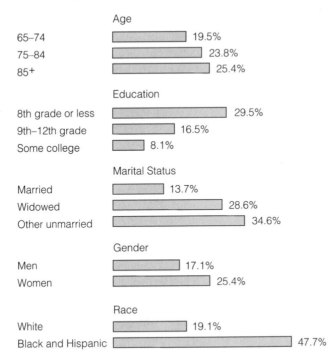

Advancing age is a . . . determinant of Medicaid use, because with age, the need for chronic care increases while available resources decrease. The magnitude of the relationship is small, however; Medicaid use is only slightly higher for the oldest old than for the youngest old. Level of education is also . . . related to Medicaid use, in part because of the link between education and income. Those with an 8th grade level of education or less are nearly 4 times as likely as those with some college education to receive Medicaid. Marital status is . . . [also] related to Medicaid use. Widowed persons, for example, are more than twice as likely as married persons to rely on Medicaid. Older persons who are divorced, separated, or never married are most likely to receive Medicaid. . . . Finally, gender and race are also . . . predictors of Medicaid use. Women are somewhat more likely than men to rely on Medicaid, and other races are considerably more likely than whites to rely on Medicaid.[11]

[11]Ibid., pp. 14–15.

> **Learning Check.** *Use Table 6.5 to verify each of the following conclusions drawn by the researcher about Medicaid use among the elderly: (1) Advancing age is a determinant of Medicaid use. (2) Level of education is related to Medicaid use. (3) Marital status is related to Medicaid use. (4) Gender and race are also related to Medicaid use. Can you explain these patterns? What other questions do these patterns raise about Medicaid use among the elderly?*

■ ■ ■ ■ The Properties of a Bivariate Relationship

So far we have looked at the general principles of a bivariate relationship as well as the more specific "mechanics" involved in examining bivariate tables. In this next section we present some detailed observations we may want to make about the "properties" of a bivariate association. These properties can be expressed as three questions to ask when examining a bivariate relationship:[12]

1. Does there appear to be a relationship?
2. How strong is it?
3. What is the direction of the relationship?

The Existence of the Relationship

We have seen earlier in this chapter that calculating percentages and comparing them are the two operations necessary to analyze a bivariate table. Based on Table 6.6, we want to examine whether the frequency of traumatic events experienced by women during the preceding five-year period had an effect on their support for abortion. Support for abortion was measured with the following question: "Please tell me whether or not you think it should be possible for a pregnant woman to obtain a legal abortion if the woman wants it for any reason." The frequency of traumatic events was determined by asking respondents to indicate the number of traumas they had experienced during the preceding five years. Among the traumatic events considered were death in the family, unemployment, and hospitalizations.

Let's hypothesize that women who have suffered more traumatic events are more likely to be pro-choice. We are not suggesting that suffering trauma necessarily "causes" pro-choice attitudes, but that perhaps there is an indirect connection between the two. For example, perhaps women who suffered more traumas feel a loss of control over their lives and are thus more likely to want to control their own bodies through the right to choose

---

[12]The same three properties are also discussed by Joseph F. Healey in *Statistics: A Tool for Social Research*, 5th ed. (Belmont, CA: Wadsworth, 1999), pp. 314–320.

Table 6.6 **Support for Abortion by Trauma (women)**

	NUMBER OF TRAUMAS			
ABORTION	0	1	2+	Total
Yes	19%	41%	76%	46%
No	81%	59%	24%	54%
Total	100%	100%	100%	100%
(N)	(27)	(44)	(33)	(104)

Source: General Social Survey, 1982 to 1990.

an abortion. (Indirect associations often can be elaborated further by looking at other variables. We discuss elaboration in more detail later in this chapter.)

In this formulation trauma is said to "influence" attitudes toward abortion, so it is the independent variable; therefore, percentages are calculated within each category of trauma (trauma is the column variable). We now want to establish whether a relationship exists between the two variables.

A relationship is said to exist between two variables in a bivariate table if the percentage distributions *vary across* the different categories of the independent variable. In our example, we would expect the percentages of those who support abortion and those who oppose it to differ across the three categories of trauma. We can easily see that the percentage that supports abortion changes across the different levels of trauma. Of women experiencing 0 traumas, 19 percent are pro-choice; of women experiencing 1 trauma, 41 percent are pro-choice; and of those experiencing 2 or more traumas, 76 percent are pro-choice.

Table 6.6 indicates that trauma and support for abortion are associated, as hypothesized.

If the number of traumas were unrelated to attitudes toward abortion among women, then we would expect to find equal percentages of women who are pro-choice (or anti-choice), regardless of the number of traumas experienced. Table 6.7 is a fictional representation of a strictly hypothetical pattern of no association between abortion attitudes and traumas. The percentage of women who are pro-choice in each category of trauma is equal to the overall percentage of women in the sample who are pro-choice (46%).

The Strength of the Relationship

In the preceding section we saw how to establish whether an association exists in a bivariate table. We now need to establish how to determine the strength of an association between the two variables. A quick method is to

Table 6.7   **Support for Abortion by Trauma (a hypothetical illustration of no relationship)**

	NUMBER OF TRAUMAS			
ABORTION	0	1	2+	Total
Yes	46%	46%	46%	46%
No	54%	54%	54%	54%
Total	100%	100%	100%	100%
(N)	(27)	(44)	(33)	(104)

examine the percentage difference across the different categories of the independent variable. The larger the percentage difference across the categories, the stronger the association.

In the hypothetical example of no relationship between trauma and attitude toward abortion (Table 6.7), there is a 0 percent difference between the columns. At the other extreme, if all women who suffered 1 or more traumas were pro-choice, and none of the women with 0 traumas was prochoice, a perfect relationship would be manifested in a 100 percent difference. Most relationships, however, will be somewhere in between these two extremes. In fact, we rarely see a situation with either a 0 percent or a 100 percent difference. Going back to the observed percentages in Table 6.6, we find the largest percentage difference between 0 and 2+ traumas (76% – 19% = 57%). The differences between 0 and 1 trauma (41% – 19% = 22%) and between 1 and 2+ traumas (76% – 41% = 35%), though not as large, are nonetheless substantial, indicating a strong relationship between number of traumas experienced by women and their attitudes toward abortion.

Percentage differences are a *rough* indicator of the strength of a relationship between two variables. In later chapters we discuss measures of association that provide a more standardized indicator of the strength of an association.

The Direction of the Relationship

When both the independent and dependent variables in a bivariate table are measured at the ordinal level or higher, we can talk about the relationship between the variables as being either positive or negative. A **positive** bivariate relationship exists when the variables vary in the same direction. Higher values of one variable "go together" with higher values of the other variable. In a **negative** bivariate relationship the variables vary in opposite directions: higher values of one variable "go together" with lower values of the other variable (and the lower values of one go together with the high values of the other).

> *Positive relationship* A bivariate relationship between two variables measured at the ordinal level or higher in which the variables vary in the same direction.
>
> *Negative relationship* A bivariate relationship between two variables measured at the ordinal level or higher in which the variables vary in opposite directions.

Table 6.8 from the GSS survey displays a positive relationship between health condition and social class. Examine each class category separately. For individuals in the lower social class (lowest score), a poor health condition is most typical (39%); for the middle-class group, fair health is most common (45%); and finally, the high social class (the highest score) exhibits the highest percentage (63%) of instances of good health. This is a positive relationship, with higher class positions associated with better health condition and lower class positions associated with poorer health.

Table 6.8 **Health Condition by Social Class: A Positive Relationship**

	CLASS		
HEALTH	Low	Middle	High
Poor	39%	12%	9%
Fair	36%	45%	28%
Good	25%	43%	63%
Total	100%	100%	100%
(N)	(39)	(254)	(202)

Source: General Social Survey, 1987 to 1992.

Table 6.9 from the GSS shows a negative association between the frequency of trauma and social class. For individuals in the lower social class, a frequency of 2+ is the most typical (47%); for the middle-class group the most common (42%) trauma level is 1; and finally, 0 trauma is most frequently associated (48%) with the upper social class. The relationship is a negative one because as class position increases the frequency of trauma decreases.[13]

[13]Note that the statement "as class position increases the frequency of trauma decreases" applies to the category 2+ only in a limited sense: the frequency (percentage) of trauma decreases as class position increases from low to middle class (from 47% to 17%) or from low to high class (from 47% to 32%), but not from middle to high class, where the percentages actually increase (from 17% to 32%).

Table 6.9 **Frequency of Trauma by Social Class: A Negative Relationship**

TRAUMA	CLASS		
	Low	Middle	High
0	31%	41%	48%
1	22%	42%	20%
2+	47%	17%	32%
Total	100%	100%	100%
(N)	(48)	(220)	(180)

Source: General Social Survey, 1987 to 1992.

In the next chapter we will see that measures of relationship for ordinal or interval-ratio variables take on a positive or a negative value, depending on the direction of the relationship.

> **Learning Check.** *Based on Table 6.9, collapse the trauma categories into two —0 versus 1 or more. Recalculate the bivariate table, estimating the percentages. Compare your results to Table 6.9. What can you say about the changes in the relationship between trauma and social class?*

■ ■ ■ ■ **Elaboration**

In the preceding sections we have looked at relationships between two variables—an independent and a dependent variable. The examination of a possible relationship between two variables is, however, only a first step in data analysis. Having established through bivariate analysis that the independent and dependent variables are associated, we seek to further interpret and understand the nature of this relationship. In this section we discuss a procedure called *elaboration*. **Elaboration** is a process designed to further explore a bivariate relationship, involving the introduction of additional variables, called control variables. Each potential control variable represents an alternative explanation for the bivariate relationship under consideration.

> ***Elaboration*** A process designed to further explore a bivariate relationship; it involves the introduction of additional variables, called control variables. Each potential control variable represents an alternative explanation for the bivariate relationship under consideration.

The introduction of additional, control variables into a bivariate relationship serves three primary goals in data analysis.

- Elaboration allows us to *test for nonspuriousness.* Establishing cause-and-effect relations requires not only showing that an independent and a dependent variable are associated, but also establishing the time order between them and providing theoretical and empirical evidence that the association is nonspurious—that is, that it cannot be "explained away" by other variables.

- Elaboration *clarifies the causal sequence* of bivariate relationships by introducing variables hypothesized to intervene between the independent and dependent variable.

- Elaboration *specifies the different conditions* under which the original bivariate relationship might hold.

In the preceding sections we learned how to establish that two variables are associated; in this section we explore the theoretical and statistical considerations involved in elaborating bivariate relationships. We illustrate the process of elaboration using three examples. The first is an example of testing for nonspuriousness; the second is a research example illustrating a causal sequence in which a third variable intervenes between the independent and dependent variables; and finally, the third research example illustrates how elaboration can uncover conditional relationships.

Testing for Nonspuriousness: Firefighters and Property Damage

Let's begin with a favorite example of a spurious relationship. Researchers have confirmed a strong bivariate relationship between *number of firefighters* (the independent variable) at a fire site and *amount of property damage* (the dependent variable). The more firefighters at the site, the greater the amount of damage. This association might lead you to the embarrassing conclusion (depicted in Figure 6.4) that firefighters cause property damage at fire sites.

Figure 6.4 depicts what might be a *direct causal relationship* between firefighters and the amount of damage. The relationship between two variables is said to be a **direct causal relationship** when it cannot be accounted for by other theoretically relevant variables. Clearly, in this case the relationship between the number of firefighters and amount of damage can be

Figure 6.4 **The Bivariate Relation Between Number of Firefighters and Property Damage**

Number of Firefighters ──────────────▶ Property Damage

| independent variable | | dependent variable |

accounted for by a third, causally prior variable—the size of the fire. When the fire is large, more firefighters are sent to the site and there is a great deal of property damage. Similarly, when the fire is small, fewer firefighters are at the site and there is probably very little damage.

This alternative explanation is shown in Figure 6.5. Note that according to the hypothesized causal order suggested in Figure 6.5, both the number of firefighters and the extent of property damage are related to the variable *size of fire,* but not to each other. The size of the fire is called a *control variable,* and the relation between the number of firefighters and property damage as depicted in Figure 6.5 is *spurious.* A **spurious relationship** is a relationship between two variables in which both the independent and dependent variables are influenced by a causally prior control variable and there is no causal link between them. The bivariate relationship between the independent and dependent variables can thus be "explained away" through the introduction of the control variable.

> *Direct causal relationship* A bivariate relationship that cannot be accounted for by other theoretically relevant variables.
>
> *Spurious relationship* A relationship in which both the independent and dependent variables are influenced by a causally prior control variable and there is no causal link between them. The relationship between the independent and dependent variables is said to be "explained away" by the control variable.

Figure 6.5 **Spurious Relationship**

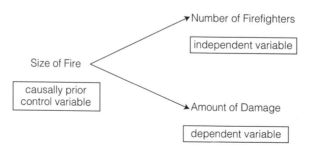

Size of Fire

| causally prior control variable |

Number of Firefighters

| independent variable |

Amount of Damage

| dependent variable |

Researchers have adopted the following rule of thumb for determining whether a relationship between two variables is either direct (causal) or spurious: If the bivariate relationship between the two variables remains about the same after controlling for the effect of one or more causally prior and theoretically relevant variables, then the original bivariate relationship is said to be a direct (causal relationship) association. On the other hand, if the original bivariate relationship decreases considerably (or vanishes), then the bivariate relationship is said to be spurious.

Let's see how we can apply this rule of thumb to the firefighter example. One way to control for the effect of the size of the fire on the relationship between the number of firefighters and the extent of damage is to divide the fire sites into large and small fires and then reexamine the bivariate association between the other two variables within each group of fire sites. If the original bivariate relationship vanishes (or diminishes considerably), then the explanation suggested by Figure 6.5 would seem more likely. If, however, the original relationship is maintained, then we may need to hold onto the original explanation suggested by Figure 6.4 or go back to the drawing board and think of other alternative explanations for the puzzling relationship between the number of firefighters and the extent of property damage.

Figure 6.6 illustrates the bivariate association between the number of firefighters and the extent of property damage (6.6a), and the process of controlling for the variable *size of fire* (6.6b). Note that the control for size of fire resulted in a substantial decrease (from 40% to 12% difference) in the size of the relationship between the number of firefighters and property damage. This result supports the notion, as depicted in Figure 6.5, that the size of the fire explains both the number of firefighters and the extent of property damage, and that the relationship between the number of firefighters and property damage is therefore spurious.

The introduction of the control variable *size of fire* into the original bivariate relationship between *number of firefighters* and amount of damage illustrates the process of elaboration. These are the steps:

1. Divide the observations into subgroups on the basis of the control variable. We have as many subgroups as there are categories in the control variable. (In our case there were two subgroups: small and large fires.)

2. Reexamine the relationship between the original two variables separately for each of the subgroups. The bivariate relationship in each of the separate tables is called a **partial relationship**. The tables are called **partial tables**.

3. Compare the partial relationships with the original bivariate relationship for the total group. In a direct causal pattern, the partial relationships will be very close to the original bivariate relationship. In a spurious pattern, the partial relationship will be much weaker than in the original bivariate relationship.

Figure 6.6   **Elaborating a Bivariate Relationship**

1. A bivariate relationship between the number of firefighters and the extent of the property damage at 20 fire sites.

(a)

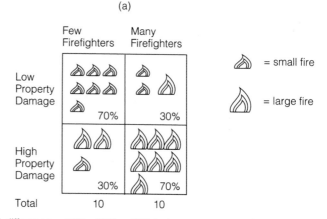

% difference = 70% – 30% = 40% (column percentages)

2. Control for size of fire: divide fire sites into small and large fires. In each group, recalculate the bivariate relationship between the number of firefighters and the extent of the property damage.

(b)

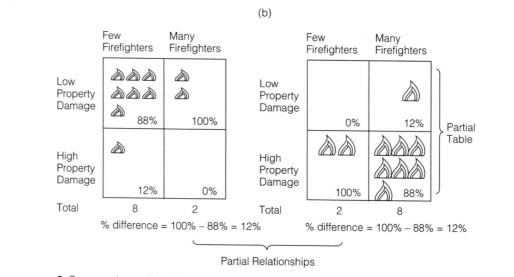

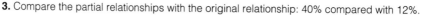

Partial Relationships

3. Compare the partial relationships with the original relationship: 40% compared with 12%.

> *Partial relationship* The relationship between the independent and dependent variables shown in a partial table.
>
> *Partial tables* Bivariate tables produced when controlling for a third variable.

We have employed the elaboration procedure to test for a spurious relationship between the number of firefighters and the amount of property damage. Now let's see how elaboration is employed to interpret the causal sequence of bivariate relationships by introducing a control variable hypothesized to *intervene* between the independent and dependent variables.

An Intervening Relationship: Religion and Attitude Toward Abortion

The research on the relationship between religious affiliation and attitudes toward abortion has shown a consistent pattern: religious affiliation is related to the level of support for abortion.[14] In particular, it has been shown that Catholics oppose abortion more than Protestants or Jews.[15]

To test the hypothesis that religion and abortion attitudes are related, we used data from the 1996 GSS sample. We limited the analysis to Catholics and Protestants and to male respondents at least 35 years of age. While the GSS measures attitudes toward abortion under several conditions and circumstances, we restrict our analysis to respondents' approval or disapproval of abortion if a married woman does not want any more children.

The findings are presented in Table 6.10 and illustrated in Figure 6.7. Since, according to the hypothesis, religious affiliation is the independent variable, we use column percentages for our analysis. The results provide some support for the hypothesis that religion is related to attitudes toward abortion. We see that 47 percent of Protestants, compared with 36 percent of Catholics, support a married woman's right to an abortion if she does not want any more children.

These results may suggest the existence of a causal relationship between religion and attitudes toward abortion. According to this interpretation of the relationship, being either Protestant or Catholic leads to a different abortion orientation regardless of other factors. Graphically, this hypothesized relationship is shown in Figure 6.8.

Another body of research findings dealing with religion challenges the conclusion that there is a direct causal link (as suggested by Figure 6.8) between religious affiliation and support for abortion. According to this

[14]For example, see Harris Mills, "Religion, Values and Attitudes Toward Abortion," *Journal for the Scientific Study of Religion* 24, no. 2 (1985): 119–236.

[15]Mario Renzi, "Ideal Family Size as an Intervening Variable Between Religion and Attitudes Towards Abortion," *Journal for the Scientific Study of Religion* 14, (1975): 23–27.

Table 6.10 **Religious Affiliation and Support for Abortion, Men Only**

| SUPPORT | RELIGIOUS AFFILIATION | | Total |
	Catholic	Protestant	
Yes	36% (12)	47% (36)	44%
No	64% (21)	53% (40)	56%
Total (N)	100% (33)	100% (76)	100% (109)

Source: General Social Survey, 1996.

Figure 6.7 **Percentage Who Support Abortion by Religious Affiliation**

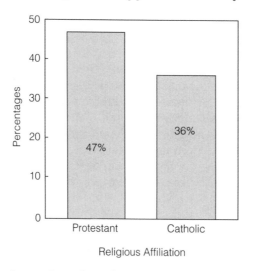

Source: General Social Survey, 1996.

Figure 6.8 **The Bivariate Relationship Between Religion and Support for Abortion**

Table 6.11 **Religious Affiliation and Preferred Family Size, Men Only**

PREFERRED FAMILY SIZE	RELIGIOUS AFFILIATION		Total
	Catholic	Protestant	
2 or fewer children	48% (16)	53% (40)	51%
More than 2 children	52% (17)	47% (36)	49%
Total (N)	100% (33)	100% (76)	100% (109)

Source: General Social Survey, 1996.

research literature, some of the difference between Catholics and Protestants can be explained by the variable *preferred family size.*[16] It is argued that religion is systematically related to desired family size: Catholics prefer larger numbers of children than non-Catholics. Similarly, if one conceptualizes abortion as an alternative device to control family size, then support for abortion may also be associated with preferred family size. Therefore, preferred family size operates as an intervening mechanism through which the relationship between religion and abortion attitudes occurs.

To check these ideas, we analyzed the bivariate associations between preferred family size and religion (Table 6.11) and between preferred family size and support for abortion (Table 6.12).[17] Notice that because the theory suggests that preferred family size operates as an intervening mechanism between religious affiliation and support for abortion, it is analyzed as the dependent variable in Table 6.11 and as the independent variable in Table 6.12.

The data in Tables 6.11 and 6.12 confirm the linkages between preferred family size and religion and preferred family size and support for abortion. First, slightly more Catholics (52%) than Protestants (47%) prefer larger families (Table 6.11). Second, more respondents who prefer smaller families support a woman's right to abortion (55%) compared with those who prefer larger families (45%) (Table 6.12). According to this interpretation of the relationship between religion and abortion attitudes, not only is preferred family size associated with both religious affiliation and support for abortion, but it also intervenes between religious affiliation and support for abortion. Thus, it is hypothesized that the relation between religion and

[16]Renzi, "Ideal Family Size," pp. 23–27.

[17]Preferred family size was measured by responses to a question about the ideal number of children for a family. Those respondents who said 2 or fewer children as ideal were classified as preferring small families; those who answered 3 or more were classified as preferring large families.

Table 6.12 **Preferred Family Size and Support for Abortion, Men Only**

| SUPPORT | PREFERRED FAMILY SIZE | | Total |
	2 or fewer children	More than 2 children	
Yes	55% (31)	32% (17)	48%
No	45% (25)	68% (36)	56%
Total (N)	100% (56)	100% (53)	100% (109)

Source: General Social Survey, 1996.

attitudes toward abortion is *indirect,* and *linked* via the control variable—preferred family size.

The hypothetical causal sequence suggested by this interpretation is shown in Figure 6.9. In this formulation the control variable (preferred family size) is called an *intervening variable.* An **intervening variable** is a control variable that follows an independent variable but precedes the dependent variable in a causal sequence. Because preferred family size follows the independent variable, religion, but precedes the dependent variable, abortion attitudes, it is considered an intervening variable. The relationship between religion and support for abortion shown in Figure 6.9 is called an *intervening relationship.* An **intervening relationship** is one between two variables in which a control variable intervenes between the independent and dependent variables.

> *Intervening variable* A control variable that follows an independent variable but precedes the dependent variable in a causal sequence.
>
> *Intervening relationship* A relationship in which the control variable intervenes between the independent and dependent variables.

Figure 6.9 **Intervening Relationship**

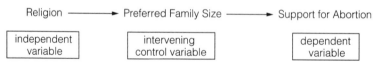

Religion ⟶ Preferred Family Size ⟶ Support for Abortion

| independent variable | intervening control variable | dependent variable |

We can test the model shown in Figure 6.9 by controlling for preferred family size and repeating the original bivariate analysis between religious affiliation and support for abortion. We control for family size by separating the respondents who indicated that they preferred larger families from those who prefer smaller families. If the causal sequence hypothesized by Figure 6.9 is correct, then the association between religion and abortion attitudes should disappear or diminish considerably once preferred family size has been controlled.

The results presented in Table 6.13 and Figure 6.10 support the notion, as depicted in Figure 6.9, that preferred family size intervenes between religion and abortion attitudes. Among those who prefer smaller families, there is an increase in the level of support among both religious groups: 44 percent of Catholics and 60 percent of Protestants reported support for a

Table 6.13 **Religious Affiliation and Support for Abortion After Controlling for Desired Family Size, Men Only**

Family Size: 2 or fewer children

| | RELIGIOUS AFFILIATION | | |
SUPPORT	Catholic	Protestant	Total
Yes	44% (7)	60% (24)	55%
No	56% (9)	40% (16)	45%
Total (N)	100% (16)	100% (40)	100% (56)

Family Size: More than 2 children

| | RELIGIOUS AFFILIATION | | |
SUPPORT	Catholic	Protestant	Total
Yes	29% (5)	33% (12)	32%
No	71% (12)	67% (24)	68%
Total (N)	100% (17)	100% (36)	100% (53)

Source: General Social Survey, 1996.

Figure 6.10 **Percentage Supporting Abortion by Religious Affiliation After Controlling for Preferred Family Size**

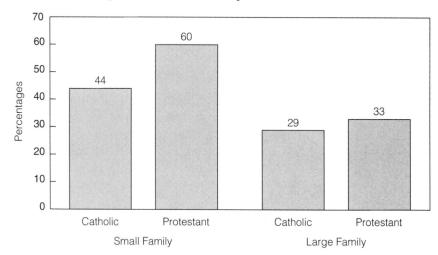

Source: General Social Survey, 1996.

married woman's right to an abortion if she does not want any more children. Among male respondents who prefer larger families, both religious groups report lower levels of support for abortion—29 percent of Catholics and 33 percent of Protestants. These findings increase our understanding of the original bivariate relationship between religious affiliation and attitudes toward abortion.

Learning Check. *You may have noticed that the tests for spuriousness and for an intervening relationship are identical: they both require that the partial associations disappear or diminish considerably! So how can you differentiate between the two? The differentiation is made on a theoretical rather than on an empirical ground. When a relationship is spurious, there is no causal link between the independent and dependent variables, and both are influenced by a causally prior control variable. In contrast, in an intervening relationship, there is an indirect causal link between the independent and dependent variables, and the control variable follows the independent variable but precedes the dependent variable in the causal sequence.*

Conditional Relationships: More on Abortion

What other variables may explain the relationship between religion and attitudes toward abortion? One possible variable is *religious participation*. In their research on abortion attitudes, Arney and Trescher[18] found that when religious participation is controlled for, there is little difference in abortion attitudes between Catholics and Protestants who attend church less than once a month. In contrast, among Catholics and Protestants who attend church more than once a month, Catholics were more likely than Protestants to oppose abortion.[19] Other researchers note that age and gender may also influence the relationship between religion and abortion attitudes.

What do these examples have in common? They all specify different conditions under which the relationship between religion and abortion attitudes are expected to hold. For example, Arney and Trescher indicate that the differences in abortion attitudes between Protestants and Catholics might only hold under one condition (attend church more than once a month) of the control variable *religious participation* but not under another (attend church less than once a month). Similarly, the relationship may differ for men and women, or for older and younger individuals. When a bivariate relationship differs for different conditions of the control variable, we say that it is a **conditional relationship**. Another way to describe a conditional relationship is to say that there is a *statistical interaction* between the control variable and the independent variable.

Because conditional relationships are very common, sociology offers many research examples illustrating this pattern of elaboration. One such example comes from a study by Jacqueline Scott on the relationship between stance on legal abortion and opinions about the morality of abortion. The study shows that although nearly all opponents of legal abortion view abortion as morally wrong, not all pro-choice supporters view abortion as morally right. Instead, many pro-choice supporters favor legal abortion despite personal moral reservations.[20] This bivariate relationship between abortion morality and stance on legal abortion is displayed in Table 6.14.

> *Conditional relationship* A relationship between the independent and dependent variables that differs for different conditions of the control variable.

Because stance on legal abortion is the independent variable, percentages are calculated in the columns. The results of this analysis support Scott's hypothesis. Among those who oppose abortion, there is almost unanimous agreement (98%) that abortion is morally wrong. Among those who favor legal abortion, however, the level of incongruence is relatively

[18]William R. Arney and William H. Trescher, "Trends in Attitudes Toward Abortion, 1972–1975," *Family Planning Perspective* 8 (1976): 117–124.

[19]William V. D'Antonio and Steven Stack, "Religion, Ideal Family Size, and Abortion: Extending Renzi's Hypothesis," *Journal for the Scientific Study of Religion* 19 (1980): 397–408.

[20]Jacqueline Scott, "Conflicting Belief About Abortion: Legal Approval and Moral Doubts," *Social Psychology Quarterly* 52, no. 4 (1989): 319–326.

Table 6.14 **Abortion Morality and Stance on Legal Abortion**

| | STANCE ON LEGAL ABORTION | | |
ABORTION MORALITY	Pro-Choice	Pro-Life	Total
Always wrong or depends	37%	98%	57%
Not wrong	63%	2%	43%
Total	100%	100%	100%
(N)	(337)	(162)	(499)

Source: Adapted from Jacqueline Scott, "Conflicting Belief About Abortion: Legal Approval and Moral Doubts," *Social Psychology Quarterly,* 52, no. 4 (1989): 319–326. Copyright © 1989 by the American Sociological Association. Reprinted by permission.

high: 37 percent support legal abortion despite viewing it as morally wrong or ambiguous.[21]

Although there is little difference between men's and women's attitudes toward the legality of abortion, some argue that women are far more likely to feel that abortion is morally wrong. For example, Carol Gilligan[22] argues that whereas men tend to be more concerned with rights and rules, women are more concerned with caring and relationships. Abortion, therefore, may pose a greater moral dilemma for women than for men. To examine the hypothesis that women are more likely than men to favor legal abortion despite moral reservations, Scott controlled for gender and compared the original relationship between stance on legal abortion and abortion morality among men and women. The cross-tabulation of abortion morality by stance on legal abortion, controlling for gender, is given in Table 6.15. The table shows a marked gender difference in the relationship between abortion morality and stance on legal abortion. Although we can still conclude from Table 6.15 that stance on legal abortion and abortion morality are associated, we need to qualify this conclusion by saying that this association is stronger for men (percentage difference is 96% − 29% = 67%) than for women (percentage difference is 100% − 46% = 54%).

Because the relationship between the independent and dependent variables is different in each of the partial tables, the relationship is said to be a conditional relationship; that is, the original bivariate relationship depends upon the control variable. In our example, the strength of the relationship between abortion morality and stance on legal abortion is conditioned on

[21]Ibid., p. 322.
[22]Carol Gilligan, *In a Different Voice* (Cambridge, MA: Harvard University Press, 1982).

Table 6.15 **Abortion Morality and Stance on Legal Abortion After Controlling for Gender**

ABORTION MORALITY	MEN STANCE ON LEGAL ABORTION			WOMEN STANCE ON LEGAL ABORTION		
	Pro-Choice	Pro-Life	Total	Pro-Choice	Pro-Life	Total
Always wrong or depends	29%	96%	50%	46%	100%	64%
Not wrong	71%	4%	50%	54%	0%	36%
Total	100%	100%	100%	100%	100%	100%
(N)	(172)	(78)	(250)	(165)	(84)	(249)

Source: Adapted from Jacqueline Scott, "Conflicting Belief About Abortion: Legal Approval and Moral Doubts," *Social Psychology Quarterly* 52, no. 4 (1989): 319–326. Copyright © 1989 by the American Sociological Association. Reprinted by permission.

Figure 6.11 **A Conditional Relationship**

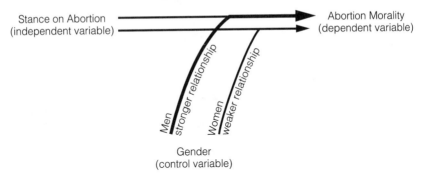

gender. The conditional relationship between stance on abortion and abortion morality is depicted in Figure 6.11.

The Limitations of Elaboration

The elaboration examples discussed in this section point to the complexity of the social world. We started this chapter by stating that most things around us "go together." It is more accurate to say that most things around us are "tangled," and one of the goals of social science is to "untangle" them. Elaboration is a procedure that helps us "untangle" bivariate relations.

In the illustrations presented in this section, we looked at bivariate relationships that were clarified and reinterpreted when a control variable was introduced. How do we know which variables to control for? In reality, theory provides significant guidance both as to the relationships we look for and the sorts of variables that should be introduced as controls. Without theory as a guide, elaboration can become a series of exercises that more closely resemble random shots in the dark than scientific analysis. Even with theory as our guide, the statistical analysis is often more complex than the presentation in this section may suggest. In our examples, when the control variable was introduced, the "real" nature of the relationship "jumped right out at us." It's not always that easy. In fact, most often there is a perilous gap between theory and analysis. This does not mean that you have to abandon your effort to "untangle" bivariate relationships, only that you should be aware of both the importance of theory as a guide to your analysis and the limitations of the statistical analysis.

> **Learning Check.**   *In this section you have been introduced to a number of important new terms. See if you can write out definitions for the following terms: elaboration, control variable, intervening variable, causally prior variable, spurious relationship, partial relationship, partial table, and conditional relationship. If you cannot provide a definition for each of these terms, you are not clear on the process of elaboration. Go back and review.*

■   ■   ■   ■   ■   **Statistics in Practice: Family Support for the Transition from High School**

In earlier chapters we saw that statistics helps us analyze how race, class, age, or gender shapes our experiences. However, we focused primarily on how these categories of experience operate separately (we compared men and women, young and old, working class and middle class, and so on). Now we need to think about how these systems interlock in shaping our experience as individuals in society.[23] Everyone has his or her own particular combination of race, social class, age, and gender. These factors act as lenses through which we experience the world. Through analysis of the intersecting effects of these factors, we can understand the ways that others experience the world from their different or similar perspectives.

When we start to see race, class, and gender as intersecting systems of experience, we see, for example, that while white women and women of

---

[23]Patricia Hill Collins, "Toward a New Vision: Race, Class, and Gender as Categories of Analysis and Connection," keynote address at Integrating Race and Gender into the College Curriculum, a workshop sponsored by the Center for Research on Women, Memphis State University, Memphis, TN, 1989.

color may share some common experience based on their gender, their racial experiences are distinct. Similarly, depending on their race and class, men experience gender differently. For example, we know that the removal of manufacturing jobs has increased black and Hispanic male job loss. At the same time, many women of color have found their work opportunities expanded in service and high-tech jobs.[24]

The methods of bivariate analysis and, in particular, the statistical techniques of elaboration are especially suitable for the examination of how race, class, and gender are linked with social behavior. In Table 6.16 we present the findings of a study that examined the kind of family support women who are now in professional and managerial positions received when they made their transition from high school to college. This example illustrates how to analyze the simultaneous operation of race and class using the method of elaboration. The example also demonstrates the drastic differences in conclusions that would have been drawn had either or both of these factors been ignored.[25]

Table 6.16 includes five types of family support, representing five dependent variables: (1) information on entrance examinations and colleges, (2) information on admission requirements, (3) financial support in paying tuition and fees, (4) emotional support, and (5) encouragement for career. Only one category is given for each dependent variable. This category represents the percentage of women who received family support in each of the specified areas. For example, 42 percent of black women who were raised in middle-class families ($N = 50$) reported that family members had helped them in procuring information on entrance examinations and colleges. The remaining 58 percent of this group (not shown) did not receive family help in this area.

Race and class origin are the two independent variables in this analysis. To estimate the effect of class on family support we compare middle-class-raised women with working-class-raised women among black and white women. The data show that there are large class differences in all types of family support provided to these women by their families. For example, whereas 90 percent of black middle-class families paid tuition for their daughters, only 56 percent of black working-class families did so. A similar pattern is observed among the white women. Similarly, more middle-class families, both black (86%) and white (70%), provided emotional support to their daughters during the transition from high school to college.

The second step involved in looking at a table like this one is to examine whether there are racial differences in family support. To estimate the effect of race we compare black and white women who were raised in working-

[24]D. Stanley Eitzen and Maxine Baca Zinn, "Structural Transformation and Systems of Inequality," in *Race, Class, and Gender,* ed. Margaret L. Andersen and Patricia Hill Collins (Belmont, CA: Wadsworth, 1998), pp. 233–237.

[25]Lynn Weber Cannon, Elizabeth Higginbotham, and Marianne L. A. Leung, "Race and Class Bias in Research on Women: A Methodological Note," Research paper 5, presented at the Center for Research on Women, Memphis, TN, 1987.

Table 6.16   **Race and Class Origin Differences in Percentage of Women Reporting Family Support for the Transition from High School to College**

	BLACK		WHITE	
	Working Class	Middle Class	Working Class	Middle Class
TYPE OF FAMILY SUPPORT	(N = 50)	(N = 50)	(N = 50)	(N = 50)
Information				
Entrance exams and colleges	22%	42%	24%	40%
Admission requirements	20%	34%	20%	30%
Financial				
Paid tuition and fees	56%	90%	62%	88%
Emotional				
Emotional support	64%	86%	56%	70%
Encouragement for career	56%	60%	40%	52%

Source: Adapted from Lynn Weber, Elizabeth Higginbotham, and Marianne L. A. Leung, "Race and Class Bias in Research on Women: A Methodological Note." Research paper 5, presented at the Center for Research on Women, University of Memphis, 1987.

class and in middle-class families. This comparison reveals virtually no relationship between race and either procurement of information or financial support. For instance, 20 percent of working-class blacks and 20 percent of working-class whites received help with information on college admissions requirements; similarly, 34 percent of middle-class blacks and 30 percent of middle-class whites received support in this category. However, examination of the emotional support category reveals fairly substantial race differences: 86 percent of middle-class-raised black respondents report that their families provided emotional support, compared with only 70 percent of the middle-class-raised whites. Similarly, more working-class black families (64% vs. 56%) provided emotional support to their daughters in the transition from high school to college.

The group that differs most from the others on both emotional support and encouragement for career are black middle-class women, who received the highest degree of family support in each of these categories.

In conclusion, the data reveal a strong relationship between class origin and both information and financial support provided by the family. In addition, the data show relationships between both race and class and emotional encouragement and support. Had the study failed to address both the race and class background of these professional and managerial women, we would have drawn very different conclusions about the role of families in supporting women as they moved from high school to college.[26]

This is another example of the pattern of elaboration examined earlier. In this case, class origin is used as a control variable to elaborate on the relationship between race and family support. In American society, race is associated with class (blacks are more likely to be raised in a working-class family), and class is associated with family support (working-class families are less likely to provide family support). Had we not analyzed the effect of the class background of the women as well as their race, we would have concluded that black women receive far less support in all areas than white women. Such a conclusion could have reinforced a stereotype—that black families are less supportive of their children's education. Such a conclusion would represent a distortion of the real process since it is working-class women, both black and white, who receive less family support.[27]

Finally, this example demonstrates the importance of looking at the simultaneous effects of race and class on the lives of women. This is only one among many ways in which race, class, and gender comparisons can be incorporated in a statistical analysis. Moreover, examining the linkages between race, class, and gender cannot be limited to women. While integrating these variables into our analysis introduces complexity to our research, it also suggests new possibilities for thinking that will enrich us all.

MAIN POINTS

- Bivariate analysis is a statistical technique designed to detect and describe the relationship between two variables. A relationship is said to exist when certain values of one variable tend to "go together" with certain values of the other variable.

- A bivariate table displays the distribution of one variable across the categories of another variable. It is obtained by classifying cases based on their joint scores for two variables.

- Percentaging bivariate tables is a method used to examine the relationship between two variables that have been organized in a bivariate table. The percentages are always calculated within each category of the independent variable.

- Bivariate tables are interpreted by comparing percentages across different categories of the independent variable. A relationship is said to exist if the percentage distributions vary across the categories of the independent variable.

[26]Ibid.
[27]Ibid.

- Variables measured at the ordinal or interval-ratio levels may be positively or negatively associated. With a positive association, higher values of one variable correspond to higher values of the other variable. When there is a negative association between variables, higher values of one variable correspond to lower values of the other variable.

- Elaboration is a technique designed to clarify bivariate associations. It involves the introduction of control variables to interpret the links between the independent and dependent variables.

- In a spurious relationship, both the independent and dependent variables are influenced by a causally prior control variable, and there is no causal link between them.

- In an intervening relationship, the control variable follows the independent variable but precedes the dependent variable in the causal sequence.

- In a conditional relationship, the bivariate relationship between the independent and dependent variables is different in each of the partial tables.

KEY TERMS

bivariate analysis

bivariate table

cell

column variable

conditional relationship

cross-tabulation

direct causal relationship

elaboration

intervening relationship

intervening variable

marginals

negative relationship

partial relationship

partial table

positive relationship

row variable

spurious relationship

SPSS DEMONSTRATIONS

Demonstration 1: Producing Bivariate Tables

SPSS has a separate procedure designed specifically to produce cross-tabulation tables. It is called the Crosstabs procedure and can be found under *Descriptive Statistics* in the *Analyze* menu. The dialog box for Crosstabs (Figure 6.12) requires us to specify both a variable that will define the rows and one that defines the columns of a table. We will investigate the relationship between support for a legal abortion for a woman who is married but does not want any more children and religious affiliation.

By default, SPSS displays the count in each cell of the table. Normally, then, you should click on the *Cells* button to request percentages. (See Figure 6.13.) As usual, we percentage the table based on the independent or

Figure 6.12

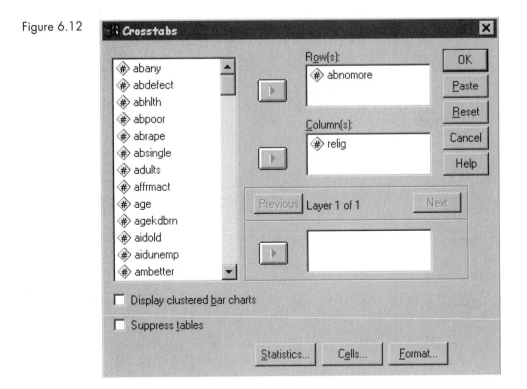

Figure 6.13

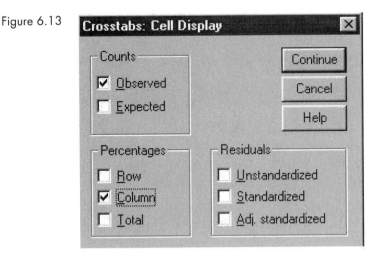

predictor variable, which is religious affiliation. The independent variable is placed in the columns, while the dependent variable is placed in rows. We click on the checkbox for "Column" to percentage the table by religious affiliation. (Note that "Observed" is already checked by default in the Counts section.)

Click on *Continue*, then *OK*, to obtain the table shown in Figure 6.14. SPSS displays both the count and the column percentage in each cell. In the upper left corner of the table, the labels "Count" and "Col Pct" are displayed as a reminder of what SPSS has placed in each cell. Row totals and column totals are supplied automatically, as is the overall total (891 respondents gave valid responses to both questions). The number of missing responses on one or both variables is also displayed.

Figure 6.14

ABNOMORE MARRIED--WANTS NO MORE CHILDREN * RELIG RS RELIGIOUS PREFERENCE Crosstabulation

| | | | RELIG RS RELIGIOUS PREFERENCE | | | | | |
			1 PROTESTANT	2 CATHOLIC	3 JEWISH	4 NONE	5 OTHER	Total
ABNOMORE MARRIED--WANTS NO MORE CHILDREN	1 YES	Count	226	80	18	63	20	407
		% within RELIG RS RELIGIOUS PREFERENCE	42.8%	39.2%	100.0%	58.3%	60.6%	45.7%
	2 NO	Count	302	124		45	13	484
		% within RELIG RS RELIGIOUS PREFERENCE	57.2%	60.8%		41.7%	39.4%	54.3%
Total		Count	528	204	18	108	33	891
		% within RELIG RS RELIGIOUS PREFERENCE	100.0%	100.0%	100.0%	100.0%	100.0%	100.0%

The table shows great differences in support across religious categories for a legal abortion for a married woman who does not want any more children. A majority of Protestants and Catholics oppose abortion in this instance, but a majority of Jews, the nonreligious, and others (Hindus, Buddhists, Muslims, and others) support abortion. Do you find these differences surprising, or are they consistent with your understanding of the social world?

Demonstration 2: Producing Tables with a Control Variable

As we've seen in this chapter, the analysis of data is enhanced when a third variable—a control variable—is added to a bivariate table. In the Crosstabs procedure, the third variable is added in the Layer section of the main dialog box (see Figure 6.15). (This box is labeled "Layer 1 of 1" because it is possible to have additional levels of control, which are accessed by clicking on the *Next* button.) We will keep the same dependent variable, ABNOMORE, but make RACE the column variable and SEX the control variable. There is no need to change the numbers displayed in the cells: the count and column percentages are still correct choices. Figure 6.16 shows the bivariate tables of support for legal abortions for married women by race, separately for males and females (males are listed first). SPSS labels each table with the value of the control variable for easy reference.

Figure 6.15

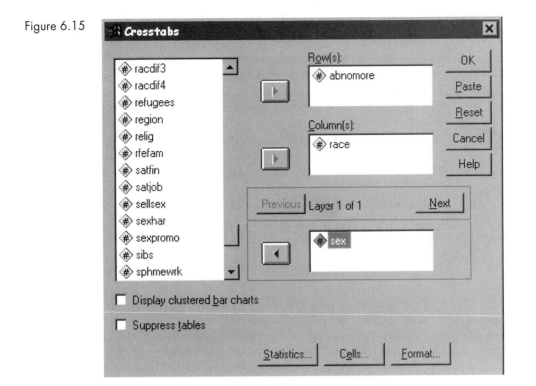

Figure 6.16

ABNOMORE MARRIED--WANTS NO MORE CHILDREN * RACE RACE OF RESPONDENT * SEX RESPONDENTS SEX Crosstabulation

SEX RESPONDENTS SEX				RACE RACE OF RESPONDENT			Total
				1 WHITE	2 BLACK	3 OTHER	
1 MALE	ABNOMORE MARRIED--WANTS NO MORE CHILDREN	1 YES	Count	140	20	8	168
			% within RACE RACE OF RESPONDENT	42.6%	55.6%	29.6%	42.9%
		2 NO	Count	189	16	19	224
			% within RACE RACE OF RESPONDENT	57.4%	44.4%	70.4%	57.1%
	Total		Count	329	36	27	392
			% within RACE RACE OF RESPONDENT	100.0%	100.0%	100.0%	100.0%
2 FEMALE	ABNOMORE MARRIED WANTS NO MORE CHILDREN	1 YES	Count	188	42	10	240
			% within RACE RACE OF RESPONDENT	47.8%	48.8%	47.6%	48.0%
		2 NO	Count	205	44	11	260
			% within RACE RACE OF RESPONDENT	52.2%	51.2%	52.4%	52.0%
	Total		Count	393	86	21	500
			% within RACE RACE OF RESPONDENT	100.0%	100.0%	100.0%	100.0%

For males, we see that a majority of blacks support abortion for married women, but a larger majority of whites do not, nor do those of other races (although there are only 10 respondents). In the second table (for females), a majority of all races oppose abortion, with blacks more strongly opposed than whites. Finding differences like these are one reason why researchers use control variables in analyses. These tables display a conditional relationship between race and support for abortion for married women when sex is introduced as a control variable.

In later chapters we will use the *Statistics* button in the Crosstabs dialog box to request additional output to further interpret and evaluate bivariate tables.

SPSS PROBLEMS

1. The 1996 GSS data contain responses to questions about the respondent's general happiness (HAPPY) and his or her subjective class identification (CLASS). Analyze the relationship between responses to these two questions with the SPSS Crosstabs procedure, requesting counts and appropriate cell percentages. [Module B]
    a. What percentage of working-class people responded that they were "very happy"?
    b. What percentage of the upper class were "very happy"?
    c. What percentage of those who were "pretty happy" were also middle class?
    d. Most of the people who said they were "very happy" were from which two classes?
    e. Is there a relationship between perceived class and perceived happiness? If there is a relationship, describe it. Is it strong or weak?
    f. Rerun your analysis, this time adding RACE as a control variable. Is there a difference in the relationship between perceived class and happiness for whites and blacks in the sample?

2. Suppose an editorial in your local newspaper reports that people with more education oppose busing as a means of reducing racial segregation in schools. You decide to use GSS1996 data to verify this reported relationship between attitude toward busing (BUSING) and the level of education (use the variable DEGREE, which groups educational attainment in five categories). [Module B]
    a. Create the appropriate table using SPSS and the Crosstabs procedure.
    b. Create a bar chart grouped by DEGREE to graphically display this same table. (*Hint:* In SPSS, use a clustered bar chart.)
    c. Is there a relationship between the variables? What is its direction?
    d. Do the data support the newspaper's assertion?
    e. Use the Crosstabs procedure with RACE as a control variable to create separate tables of DEGREE by BUSING for blacks and whites.

f. Is there a difference in support for busing between blacks with different levels of education?

g. Is there a difference in attitude between whites with different levels of education?

h. Is the relationship between attitude toward busing and level of education a spurious one?

3. Is there a relationship between liberal views about the role of women and attitudes about abortion rights for women? [Module A]

a. Use SPSS to construct a table showing the relationship between attitudes about women working (FEWORK) and attitudes about women having access to abortion for any reason (ABANY). (*Hint:* Use ABANY as the dependent variable.) Next, use SPSS to construct tables showing the same relationship controlling for sex. Then answer questions b–e.

b. Overall, are women or men more likely to agree that a woman should be able to have an abortion if she wants to for any reason?

c. Do men who approve of women working have different opinions on this question than men who disapprove of women working?

d. Is there a difference in support for abortion between women who approve of women working and those who disapprove? If so, describe the relationship.

e. Finally, use SPSS to construct tables showing the relationship between attitudes toward abortion and toward women working, controlling for race. Does attitude toward women working have a different effect on beliefs about abortion for whites and blacks?

4. Is there a difference in attitudes about abortion, depending on the circumstance of the woman's pregnancy or her reason for an abortion? Separately assess the relationship between SEX and two abortion items, ABPOOR (should a woman have an abortion if she can't afford any more children) and ABHLTH (should a woman have an abortion if her health is seriously endangered). What do you conclude? [Module A]

CHAPTER EXERCISES

1. Use the following data on fear, race, and home ownership for this exercise. Variables measure respondent's race, whether the respondent fears walking alone at night, and his/her home ownership.

a. Construct a bivariate table of frequencies for race and fear of walking alone at night. Which is the independent variable?

b. Calculate percentages for the table based on the independent variable. Describe the relationship between race and fear of walking alone using the table. What social factors might account for the relationship?

c. Use the data to construct a bivariate table to compare fear of walking alone at night between people who own their homes and those who rent. Use percentages to show whether there is a difference between homeowners and renters in fear of walking alone.

Respondent	Race	Fear of Walking Alone	Rent/Own
1	B	N	O
2	B	Y	O
3	W	N	R
4	W	N	O
5	W	Y	R
6	B	N	R
7	W	N	O
8	B	N	R
9	W	Y	O
10	B	Y	R
11	W	N	O
12	B	Y	O
13	B	Y	R
14	W	Y	O
15	W	N	R
16	B	N	R
17	W	N	O
18	B	Y	R
19	W	N	O
20	B	Y	R
21	W	Y	O

Race: B = black, W = white; Fear: Y = yes, N = no; R = rent, O = own.
Source: Data based on the 1987 to 1991 GSS.

2. The following bivariate table shows the relationship between support for school busing for racial purposes (reducing segregation at school) and the race of the respondent, based on a 1996 GSS subsample.

BUSING ATTITUDE TOWARD RACIAL BUSING * RACE RACE OF RESPONDENT Crosstabulation

| | | | RACE RACE OF RESPONDENT | | |
			1 WHITE	2 BLACK	Total
BUSING ATTITUDE TOWARD RACIAL BUSING	1 FAVOR	Count	57	17	74
		% within RACE RACE OF RESPONDENT	40.1%	63.0%	43.8%
	2 OPPOSE	Count	85	10	95
		% within RACE RACE OF RESPONDENT	59.9%	37.0%	56.2%
Total		Count	142	27	169
		% within RACE RACE OF RESPONDENT	100.0%	100.0%	100.0%

   a. Which is the independent variable?

   b. Is there a difference in attitude between whites and blacks? What is the percentage difference?

   c. What might account for the differences you see in the table? Suggest at least two reasons.

3. Advocates of gay rights often argue that homosexuality is not a "preference" or a choice, but rather an "orientation" that cannot be changed. Suppose one of your classmates has a close friend or family member who is gay or lesbian, and he thinks people who don't have similar relationships have different beliefs about the origins of homosexuality than people who do have a gay or lesbian friend or family member. Use the following table to answer the questions.

	Has gay or lesbian close friend or family member?		
Homosexuality is	Yes	No	Total
Choice	81	427	508
Can't change	144	352	496
Don't know	29	121	150
Total	254	900	1,154

Source: Data based on 1992 *New York Times*/CBS News poll.

   a. What is the dependent variable? the independent variable?

   b. What proportion of those polled have a close friend or family member who is gay or lesbian?

   c. Are people who have a close relationship with a gay or a lesbian more likely to believe that homosexuality is a choice or that it is something that cannot be changed? (*Hint:* Calculate percentages.)

4. Refer to Exercise 3. Suppose another of your classmates says that it can't be very significant to know whether a person believes homosexuality is a choice or not, since people are equally split (44% and 43%, respectively) between the two positions. Your professor presents the following table based on the same survey.

Should gays and lesbians be allowed in the military?	**Is homosexuality a choice or is it an orientation that cannot change?**			
	Choice	Can't change	Don't know	Total
Yes	162	268	66	496
No	276	160	62	498
Don't know	70	68	22	160
Total	508	496	150	1,154

Source: Data based on 1992 *New York Times*/CBS News poll.

      a. Which is the dependent variable in this table? Which is the independent variable? Discuss why assigning the variables to these categories is problematic.

      b. Is there a relationship between believing that homosexuality is a choice and attitude toward allowing gays and lesbians in the military? (Use percentages to support your answer.)

5. Americans are very concerned about their health-care system. Everyone would like to say that they are in good health, but some may have more difficulty in dealing with their health-care problems. A neighborhood clinic wants to develop a health promotion program aimed at people in their community. The clinic used data from the GSS1996 to measure the self-perceived health of people by gender, summarized in the following table.

**HEALTH CONDITION OF HEALTH * SEX RESPONDENTS SEX
Crosstabulation**

Count

		SEX RESPONDENTS SEX		
		1 MALE	2 FEMALE	Total
HEALTH CONDITION OF HEALTH	1 EXCELLENT	162	195	357
	2 GOOD	248	347	595
	3 FAIR	93	95	188
	4 POOR	12	36	48
Total		515	673	1188

      a. What proportion of respondents feel that they are in good health? in excellent health? What proportion feel that their health is only fair or poor?

      b. Is there a difference in self-perceived health of women versus men? What is the difference?

      c. Should the clinic focus its health promotion program on men or women? Why? (Use percentages to support your answer.)

6. The educational level of Americans has been increasing throughout the twentieth century. The following U.S. census data show the relationship between time and the level of education attained by American adults over the age of 25 for two periods.

	Education Level		
Year	Less than 9th grade	High school graduate or more	Bachelor's degree or more
1980	18.3%	66.5%	16.2%
1990	10.4%	75.2%	20.3%

a. What is the direction of this relationship?
b. Use percentage differences to describe the relationship. Why don't the percentages add to 100 percent by year? Is this a problem in analyzing the table?
c. Do these data support the idea that Americans were getting more education in 1990 than ten years before?

7. A census of Los Angeles County shows the following data for the proportion of people living in neighborhoods where they are of the same race/ethnicity as most of their neighbors. Thus, for example, 35 percent of blacks in 1980 lived in neighborhoods that were predominantly black.

Percentage of Each Race Living in Neighborhoods Where Most People Are of the Same Ethnicity

	Race	
Year	Black	Hispanic
1980	35%	17%
1990	13%	27%

Source: Based on U.S. census data as reported in the *Los Angeles Times*, May 6, 1991.

a. Did residential segregation patterns change over the decade for blacks? What is the direction of the change? Is the same true for Hispanics? How did their residential segregation pattern change?
b. Is there a relationship between ethnicity and the direction of the changes in segregation?

8. In Exercise 1 you found that blacks are more likely than whites to fear walking alone in their neighborhoods. You now wonder if this difference exists because whites are more likely to own their own homes and so live in safer neighborhoods. In other words, you want to try some elaboration.
a. Use the data from Exercise 1 to construct tables showing the relationship between fear of walking alone and race, controlling for whether the individual rents or owns his or her dwelling.
b. Does renting versus owning one's dwelling "explain" the difference in fear between whites and blacks? (Use percentage differences to support your answer).
c. Has introducing home ownership shown that the relationship between race and fear is spurious, or is home ownership an intervening variable? Explain.

9. Black students in the United States typically attend public schools in which they are a majority (and therefore white students a minority). In the past three decades, in many large cities such as New York, black

students have been attending schools with a decreasing proportion of white students.

a. Based on the following data for Los Angeles, New York, Dallas, and Chicago, is the direction of the trend the same or different in these four cities?

b. Is the trend stronger in New York or Dallas?

c. If the trends are different, use the column percentages to describe the differences in each city.

A Typical Black Student's Classmates Are	Los Angeles				New York			
	1968	1974	1986	1992	1968	1974	1986	1992
White Students	7.5%	10.2%	11.7%	9.6%	23.2%	16.5%	10.4%	8.4%
Students of Color	92.5%	89.8%	88.3%	90.4%	76.8%	83.5%	89.6%	91.6%

A Typical Black Student's Classmates Are	Dallas				Chicago			
	1968	1974	1986	1992	1968	1974	1986	1992
White Students	5.6%	14.4%	11.8%	9.3%	5.4%	3.2%	6.0%	4.6%
Students of Color	94.4%	86.6%	88.2%	90.7%	94.6%	98.8%	94.0%	95.4%

Source: Data from a *USA Today* article from 12 May 1994, based on 1991/92 U.S. Department of Education and National School Boards Association data.

10. An organization in your state is lobbying to make pornography illegal because its members believe that pornography leads to a breakdown in morals. You believe that people with conservative views about women are more likely to hold such beliefs about pornography and that people with liberal views about women are more likely to disagree with this view. The GSS has a question about whether people believe that pornographic materials lead to a breakdown in morals, and a question about whether people approve or disapprove of women working (the liberal position is to approve).

a. Do the GSS data in the following table support your beliefs or not? Why?

		Should women work outside the home?		
		Approve	Disapprove	Total
Does pornography lead to a breakdown in morals?	Yes	251	71	322
	No	175	28	203
	Total	426	99	525

Source: Data from the 1987 to 1991 GSS.

b. Your friend argues that there are gender differences in the effect that attitude about women working outside the home has on views about pornography. Do the GSS survey data in the following table support her belief? Why or why not?

Should women work outside the home?

Does		Males		Females	
pornography lead to a		Approve	Disapprove	Approve	Disapprove
breakdown	Yes	86	29	165	42
in morals?	No	90	15	85	13

Source: Data from the 1987 to 1991 GSS.

c. What can you conclude about the relationship between views about women working and attitudes about pornography? Is this an example of a conditional relationship?

11. In the previous exercise you examined the relationship between liberal attitudes about women and the belief that pornographic materials lead to a breakdown in morals. The relationship was not different for men and women. Do liberal attitudes about women working have a conditional effect for whites and blacks?

a. Use the following table to answer this question and describe the relationship between race, attitude toward women working, and attitude toward the effect of pornography.

Should women work outside the home?

Does		White		Black	
pornography lead to a		Approve	Disapprove	Approve	Disapprove
breakdown	Yes	157	36	94	35
in morals?	No	102	14	73	14

Source: Data from the 1987 to 1991 GSS.

b. Does race show that the relationship between attitude toward women working and attitude toward the effect of pornography is spurious? Why or why not?

12. Is there a relationship between recidivism and the age of a juvenile offender at his/her first offense? The following data highlight the relationship between age of juvenile offender at his/her first offense and percentage of second and third referrals to the juvenile court. Based on the tables, describe the relationship between age and repeat referrals, separately for the second and third referrals.

	Age at first referral		
	12 years or younger	13–15 years	16 years or older
Second Referral (N = 157)			
Percent referred	40.91	36.47	20.00
Percent not referred	59.09	63.53	80.00
Third referral (N = 76)			
Percent referred	63.89	49.48	20.83
Percent not referred	36.11	50.52	79.17

Source: Data from Kevin Minor, David Hartmann, and Sue Terry, "Predictors of Juvenile Court Actions and Recidivism," *Crime and Delinquency* 43, no. 3 (1997): 328–344. Used by permission.

13. Suicide among older males has drawn increasing public attention. White males 65 years or older have the highest suicide rate compared to women, younger males, and other ethnic groups. In 1996 Mark Kaplan, Margaret Adamek, and Olga Geling reported the results of their study based on 14,887 suicide death records for elderly white males. Based on the following bivariate table, what can you conclude about the relationship between age and method of suicide?

Age	65–74	75–84	85+
Suicide by firearms	5,687	4,570	1,167
Suicide by other methods	1,647	1,303	513

Source: Reprinted with permission of the Gerontological Society of America, 1030 15th Street, NW, Suite 250, Washington, DC 20005. "Sociodemographic Predictors of Firearm Suicide Among Older White Males," M. Kaplan, M. Adamek, and O. Geling, *The Gerontologist* 36(4): 530–533. Reproduced by permission of the publisher via Copyright Clearance Center, Inc.

14. How does work affect student academic performance? Based on data from the National Center for Education Statistics, Laura Horn and Andrew Malizio report that students' work schedules do affect some students' performance. The following table presents average number of hours worked and reported effects on overall academic performance. How would you describe the relationship of the variables in the bivariate table? (Note: The table percentages are calculated using average hours worked as the independent variable.)

Effect on Academic Performance	Average Hours Worked While Enrolled			
	1–15	16–20	21–34	35 or more
Positive effect	22	14	12	10
Negative effect	17	34	46	55
No effect	61	52	43	35

Source: Adapted from U.S. Department of Education, "Undergraduates Who Work," July 1998, NCES 98-137.

7   Measures of Association for Nominal and Ordinal Variables

Reading the Research Literature:
Worldview and Abortion Beliefs

Examining the Data

Interpreting the Data

MAIN POINTS

KEY TERMS

SPSS DEMONSTRATION

SPSS PROBLEMS

CHAPTER EXERCISES

■ ■ ■ ■ **Introduction**

In the previous chapter we focused on one bivariate technique—the method of cross-tabulation—in which the pattern of relationship between two variables was analyzed by making a number of percentage comparisons. In Chapters 7 and 8, we review special **measures of association** for nominal, ordinal, and interval-ratio variables. These measures enable us to use a single summarizing measure or number for analyzing the pattern of relationship between two variables. Such measures of association reflect the strength of the relationship and, at times, its direction (whether it is positive or negative). They also indicate the usefulness of predicting the dependent variable from the independent variable.

In this chapter we discuss three measures of association: lambda (a measure of association for nominal variables) and gamma and Somers' d (both suitable for measuring associations between ordinal variables). In Chapter 8 we introduce Pearson's correlation coefficient, measuring bivariate association between interval-ratio variables.

> *Measures of association* A single summarizing number that reflects the strength of the relationship, indicates the usefulness of predicting the dependent variable from the independent variable, and often shows the direction of the relationship.

To introduce the logic of the measures of association, imagine for a moment that we are asked to identify or predict support for affirmative action of each of 20 entering freshmen. Let's suppose that we know nothing about these students individually, but we find out that about 60 percent of all students on our campus tend to be in favor of affirmative action. One simple way would be to identify all 20 freshmen as affirmative action supporters (Prediction 1 in Figure 7.1). Chances are that we will make about 12 correct

Figure 7.1 **Reducing Prediction Error**

Prediction 1: All students are pro–affirmative action.

pro pro pro anti pro anti pro anti anti pro

anti anti pro pro pro anti pro pro pro anti

Using this prediction, we would make 8 errors.

Prediction 2: All Republicans are anti–affirmative action; all Democrats are pro–affirmative action.

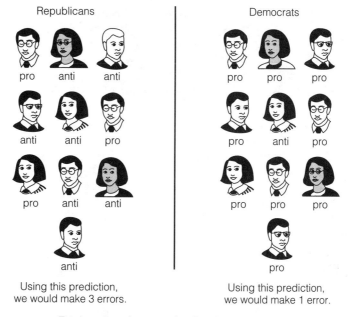

Republicans

pro anti anti

anti anti pro

pro anti anti

anti

Using this prediction,
we would make 3 errors.

Democrats

pro pro pro

pro anti pro

pro pro pro

pro

Using this prediction,
we would make 1 error.

Total number of errors using Prediction 2: 4 errors

guesses or predictions (60% of 20 = 12), but we will make about 8 errors. Suppose instead that we are told that half the students (10) identify themselves as Republicans and half (10) as Democrats. We could try to improve our predictions by identifying all Democrats as supporters and all Republicans as nonsupporters. Although not all Democrats support affirmative action and not all Republicans oppose it, chances are that we will reduce the number of errors by using political party affiliation to identify who is a supporter of affirmative action policies. Let's assume that 3 students who identify themselves as Republicans indeed support affirmative action, while 1 of those who self-identifies as a Democrat opposes it. Although we will still end up with 4 errors (Prediction 2 in Figure 7.1), this second method is nevertheless an improvement over our earlier method of guessing. We will have reduced the number of errors by 4 (or 8 − 4), a 50 percent improvement [(4 ÷ 8)100]! This 50 percent improvement represents the *proportional reduction of error* when political party affiliation is used to identify students who are supporters of affirmative action.

■ ■ ■ ■ **Proportional Reduction of Error and Degree of Association**

All the measures of association discussed in Chapters 7 and 8 are based on the concept of the **proportional reduction of error**, which is often abbreviated as **PRE**. According to the concept of PRE, two variables are associated when information about one can help us improve our prediction of the other. The variables *political party affiliation* and *position on affirmative action* are associated because political party affiliation helped improve our prediction of the students' positions on affirmative action by 50 percent.

Table 7.1 may help us grasp intuitively the general concept of PRE. Table 7.1 shows a strong relationship between the independent variable *number of children* and the dependent variable *support for abortion for any reason*. GSS1996 data are presented only for women aged 45 years or younger. The table shows that 67 percent of the women who had two or more children were anti-abortion, compared to only 39 percent of women who had one child or no children.

Table 7.1 **Support for Abortion by Number of Children (women 45 years or younger)**

Support Abortion for Any Reason	Number of Children		Total
	None or 1 child	2 or more children	
Yes	36	15	51
	(61%)	(33%)	(49%)
No	23	31	54
	(39%)	(67%)	(51%)
Total	59	46	105
	100%	100%	100%

Source: General Social Survey, 1996.

The conceptual formula for all[1] PRE measures of association is

$$PRE = \frac{E1 - E2}{E1} \qquad (7.1)$$

where

$E1$ = errors of prediction made when the independent variable is ignored (Prediction 1)

$E2$ = errors of prediction made when the prediction is based on the independent variable (Prediction 2)

All PRE measures are based on comparing predictive error levels that result from each of the two methods of prediction. Let's say that we want to predict a woman's position on abortion but we do not know anything about the number of children she has. Based on the row totals in Table 7.1 we could predict that every woman in the sample is anti-abortion because this is the modal category of the variable *abortion position*. With this prediction we would make 51 errors, because in fact 54 women in this group are anti-abortion but 51 women are pro-choice. (See the row marginals in Table 7.1.) Thus,

$E1 = 51$

How can we improve this prediction by using the information we have on each woman's number of children? For our new prediction, we will use the following rule: If a woman has two or more children, we predict that she will be anti-abortion; if a woman has one child or none, we predict that she is pro-choice. It makes sense to use this rule because we know, based on Table 7.1, that women with larger families are more likely to be anti-abortion, while women who have one child or none are more likely to be pro-choice. Using this prediction rule, we will make 38 errors (instead of 51) because 23 of the women who have one child or none are actually anti-abortion, whereas 15 of the women who have two children or more are pro-choice (23 + 15 = 38). Thus,

$E2 = 38$

Our first prediction method, ignoring the independent variable (number of children), resulted in 51 errors. Our second prediction method, using information we have about the independent variable (number of children), resulted in 38 errors. If the variables are *associated*, the second method will result in fewer errors of prediction than the first method. The stronger the relationship is between the variables, the larger will be the reduction in the number of errors of prediction.

[1]Although this general formula provides a framework for all PRE measures of association, only lambda is illustrated with this formula. Gamma and Somers' *d*, which are discussed in the next section, are calculated with a different formula. Yet they are still interpreted as PRE measures.

Let's calculate the proportional reduction of error for Table 7.1 using Formula 7.1. The proportional reduction of error resulting from using number of children to predict position on abortion is

$$PRE = \frac{51-38}{51} = .25$$

PRE measures of association can range from 0.0 to ±1.0. A PRE of zero indicates that the two variables are not associated; information about the independent variable will not improve predictions about the dependent variable. A PRE of ±1.0 indicates a perfect positive or negative association between the variables; we can predict the dependent variable without error using information about the independent variable. Intermediate values of PRE will reflect the strength of the association between the two variables and therefore the utility of using one to predict the other. The more the measure of association departs from 0.00 in either direction, the stronger the association. PRE measures of association can be multiplied by 100 to indicate the percentage improvement in prediction.

A PRE of .25 indicates that there is a weak relationship between women's number of children and their position on abortion. (Refer to Box 7.1 for a discussion of the strength of a relationship.) A PRE of .25 means that we have improved our prediction of women's position on abortion by 25 percent (.25 × 100 = 25%) by using information on their number of children.

> *Proportional reduction of error (PRE)* The concept that underlies the definition and interpretation of several measures of association. PRE measures are derived by comparing the errors made in predicting the dependent variable while ignoring the independent variable with errors made when making predictions that use information about the independent variable.

■ ■ ■ ■ Lambda: A Measure of Association for Nominal Variables

For most Americans, owning a home is a source of security both psychologically and financially. Psychologically, ownership provides stability and privacy. Financially, ownership not only is a symbol of wealth, but also represents the primary means of accumulating wealth in this society.

If home ownership is a primary source of psychological and financial security, is it associated with a sense of financial satisfaction? To examine this question, let's look at Table 7.2, which shows 321 female respondents of the GSS1996 classified by their home ownership status (whether they own or rent) and satisfaction with their financial situation. In this example, we will consider home ownership status to be the independent variable and level of

Box 7.1 What Is Strong? What Is Weak? A Guide to Interpretation

The more you work with various measures of association, the better feel you will have for what particular values mean. Until you develop this instinct, though, here are some guidelines regarding what is generally considered a strong relationship and what is considered a weak relationship.

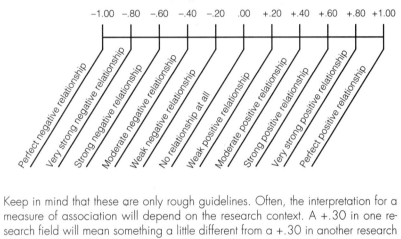

Keep in mind that these are only rough guidelines. Often, the interpretation for a measure of association will depend on the research context. A +.30 in one research field will mean something a little different from a +.30 in another research field. Zero, however, always means the same thing: no relationship.

Table 7.2 **Financial Satisfaction by Home Ownership, Women Only**

Financial Satisfaction	Home Ownership		
	Own	Rent	Row Total
Satisfied	104	18	122
More or Less	85	46	131
Not Satisfied	50	18	68
Column Total	239	82	321

Source: General Social Survey, 1996.

financial satisfaction—which may be explained or predicted by the independent variable—to be the dependent variable.

Because home ownership is a nominal variable, we need to apply a measure of association suitable for calculating relationships between nominal variables. Such a measure will help us determine how strongly associated home ownership is with financial satisfaction. **Lambda** is such a measure; it is also a PRE measure that follows the basic formula

$$PRE = \frac{E1 - E2}{E1}$$

A Method for Calculating Lambda

Take a look at Table 7.2 and examine the row totals, which show the distribution of the variable *financial satisfaction.* If we had to predict the level of financial satisfaction of everyone presented here, our best bet would be to guess the mode, which is that everyone is "more or less satisfied." This prediction will result in the smallest possible error. The number of wrong predictions we make using this method is actually 190, since 131 out of 321 were more or less satisfied (321 − 131 = 190).

Now take another look at Table 7.2, but this time let's consider home ownership status when we predict financial satisfaction. Again, we can use the mode of financial satisfaction, but this time we apply it separately to homeowners and renters. The mode for female homeowners is "satisfied" (104 homeowners); therefore, we can predict that all homeowners are "satisfied." With this method of prediction we make 135 errors, since 104 out of 239 homeowners were in fact satisfied (239 − 104 = 135). Next we look at female renters. The mode for this group is "more or less satisfied"; this will be our prediction for the entire group of renters. This method of prediction results in 36 errors (82 − 46 = 36). The total number of errors is thus 135 + 36, or 171.

Let's now put it all together and state the procedure for calculating lambda in more general terms.

1. Find *E*1, the errors of prediction made when the independent variable is ignored. To find *E*1, find the mode of the dependent variable and subtract its frequency from *N*. For Table 7.2,

   *E*1 = *N* − *Modal Frequency*

   *E*1 = 321 − 131 = 190

2. Find *E*2, the errors made when the prediction is based on the independent variable. To find *E*2, find the modal frequency for each category of the independent variable, subtract it from the category total to find the number of errors, then add up all the errors. For Table 7.2,

   Own errors = 239 − 104 = 135

   Rent errors = 82 − 46 = 36

   *E*2 = 135 + 36 = 171

3. Calculate lambda using Formula 7.1:

   $$lambda = \frac{E1 - E2}{E1} = \frac{190 - 171}{190} = 0.10$$

By comparing *E*1 with *E*2, we can determine whether using the independent variable to predict the dependent variable results in fewer errors. In our example, there is an advantage in using home ownership status to predict financial satisfaction because we have reduced the number of errors of prediction by 19, from 190 to 171.

The proportional reduction of error indicated by lambda, when multiplied by 100, can be interpreted as follows: By using information on respondents' home ownership status to predict their level of financial satisfaction, we have reduced our error in predicting financial satisfaction by 10 percent (.10 × 100 = 10%).

Lambda may range in value from 0.0 to 1.0. Zero indicates that there is nothing to be gained by using the independent variable to predict the dependent variable. A lambda of 1.0 indicates that by using the independent variable as a predictor we are able to predict the dependent variable without any error. In our case, a lambda of .10 is less than one quarter of the distance between 0.0 and 1.0, indicating that for this subsample of female respondents, home ownership status and financial satisfaction are only slightly associated.

> *Lambda* An asymmetrical measure of association, lambda is suitable for use with nominal variables and may range from 0.0 to 1.0. It provides us with an indication of the strength of an association between the independent and dependent variables.

> **Learning Check.** *Explain why lambda would not assume negative values.*

Statistics in Practice: Home Ownership, Financial Satisfaction, and Race

Historically, home ownership has been as important in the black community as in society as a whole. The home has served as the center of black family life in a society in which racial segregation was practiced in public places. Researchers have found that home ownership has a positive effect on the social participation of black residents.[2]

Despite the value that blacks place on home ownership, evidence exists that they do not receive the economic returns from home ownership that are experienced by whites. Therefore, the association between home ownership and financial satisfaction may differ for black and white respondents. To examine this, we selected a sample of women from our GSS data file and divided it into two groups by race. Let's examine the association between

[2]Hayward Derrick Horton, "Race and Wealth: A Demographic Analysis of Black Homeownership," *Sociological Inquiry* 62, no. 4 (November 1992): 480–489.

Table 7.3   **Financial Satisfaction by Home Ownership,
Black Women Only**

Financial Satisfaction	Home Ownership		
	Own	Rent	Row Total
Satisfied	5	2	7
More or Less	5	11	16
Not Satisfied	9	8	17
Column Total	19	21	40

Source: General Social Survey, 1996.

Table 7.4   **Financial Satisfaction by Home Ownership,
White Women Only,**

Financial Satisfaction	Home Ownership		
	Own	Rent	Row Total
Satisfied	94	14	108
More or Less	76	31	107
Not Satisfied	40	10	50
Column Total	210	55	265

Source: General Social Survey, 1996.

home ownership and financial satisfaction, as measured with lambda, in the two groups. Tables 7.3 and 7.4 present the bivariate tables for black and white women.

Using Formula 7.1 and the information in Table 7.3, we find that for the black female sample, $E1 = 23$ and $E2 = 20$. Plugging these figures into the formula for lambda, we find:

$$\frac{23 - 20}{23} = .13$$

For our sample of white females, presented in Table 7.4, we find that $E1 = 157$ and $E2 = 140$, and lambda is

$$\frac{157 - 140}{157} = .11$$

> **Learning Check.**   *Calculate E1 and E2 for each table and use Formula 7.1 to find lambda. Confirm that our answers are correct.*

A lambda of 0.13 for black females means that home ownership status has a weak association with financial satisfaction. Knowledge of black respondents' home ownership status improves the prediction of their level of financial satisfaction by 13 percent. A weak relationship also exists between home ownership and financial satisfaction among white females. A lambda of .11 reflects a proportional reduction of error of 11 percent.

Learning Check. *Understanding PRE measures is important because the concept of PRE is used quite often in statistics (it will come up again in Chapter 8). To get a better grasp of what PRE means, investigate the following cross-tabulations and accompanying pie charts.*

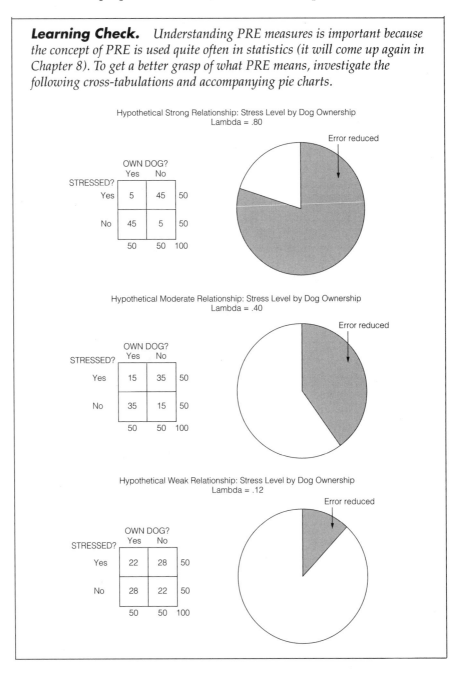

Some Guidelines for Calculating Lambda

Lambda is an **asymmetrical measure of association**. This means that lambda will vary depending on which variable is considered the independent variable and which the dependent variable. In our Table 7.2 example, we predicted financial satisfaction based on home ownership, and not vice versa. Had we instead considered financial satisfaction the independent variable and home ownership the dependent variable, the value of lambda for blacks would have been slightly larger at .21.

Asymmetrical measure of association A measure whose value may vary depending on which variable is considered the independent variable and which the dependent variable.

The method of calculation follows the same guidelines even when the variables are switched. However, exercise caution in calculating lambda, especially when the independent variable is arrayed in the rows rather than in the columns. To avoid confusion it is safer to switch the variables and follow the convention of arraying the independent variable in the columns; then follow the exact guidelines suggested for calculating lambda. Remember, however, that although lambda can be calculated either way, ultimately what guides the decision of which variables to consider as independent or dependent is the theoretical question posed by the researcher.

Lambda is always zero in situations in which the modes for each category of the independent variable fall into the same category of the dependent variable. A problem with interpreting lambda arises in situations in which lambda is zero while other measures of association indicate that the variables are associated. To avoid this potential problem examine the percentage differences in the table whenever lambda is exactly equal to zero. If the percentage differences are very small (usually 5% or less) lambda is an appropriate measure of association for the table. However, if the percentage differences are larger, indicating that the two variables may be associated, lambda will be a poor choice as a measure of association. In such a case, we may want to discuss the association in terms of the percentage differences or select an alternative measure of association.

Learning Check. Based on Table 7.4, calculate lambda using financial satisfaction as the independent variable and home ownership as the dependent variable. Is your answer .00?

Now calculate the percentage difference for both Tables 7.3 and 7.4. (Don't forget to percentage the tables first.) Is it safe to use lambda as a measure of association?

■ ■ ■ ■ Gamma and Somers' *d*: Ordinal Measures of Association

In the previous section we looked at a method for measuring the strength of association between *nominal* variables. The method was based on calculating the proportionate reduction in error that occurs when the independent variable is used to predict the dependent variable. In this section, we discuss how to measure and interpret an association between two *ordinal* variables. The two measures we discuss—gamma and Somers' *d*—are both PRE measures. This means that if there is an association between the two variables, knowledge of one variable will enable us to make better predictions of the other variable.

Before we illustrate how measures of association between ordinal variables are calculated and interpreted, let's review for a moment the definition of ordinal level variables. Whenever we assign numbers to rank-ordered categories ranging from low to high, we have an ordinal level variable. *Social class* is an example of an ordinal variable. We might classify individuals with respect to their social class status as "upper class," "middle class," or "working class." We can say that a person in the category "upper class" has a higher class position than a person in a "middle class" category (or that a "middle class" position is higher than a "working class" position).

Ordinal variables are very common in social science research. The General Social Survey contains many questions that ask people to indicate their response on an ordinal scale—for example, "very often," "fairly often," "occasionally," or "almost never."

Analyzing the Association Between Ordinal Variables: Job Security and Job Satisfaction

Let's look at a research example in which the association between two ordinal variables is considered. We want to examine the hypothesis that the higher a person's perceived job security, the higher his or her job satisfaction. To examine this hypothesis we selected two questions from the General Social Survey. The following question is a measure of *job satisfaction*:

> On the whole, how satisfied are you with the work you do—would you say you are very satisfied, moderately satisfied, a little dissatisfied, or very dissatisfied?[3]

To measure *job security* they asked:

> Thinking about the next 12 months, how likely do you think it is that you will lose your job or be laid off—very likely, fairly likely, not too likely, or not at all likely?[4]

---

[3]We have recoded the original response categories for this question into the following new categories: very satisfied = high, moderately satisfied = moderate, a little dissatisfied or very dissatisfied = low.

[4]We have recoded responses to this question into three levels of job security: not at all likely = high, not too likely = medium, and fairly likely and very likely = low job security.

Table 7.5 **Job Satisfaction by Job Security***

JOB SATISFACTION (Y)	JOB SECURITY (X)			
	High	Medium	Low	Total
High	a 36% (16)	b 22% (8)	c 11% (14)	18% (38)
Moderate	d 43% (19)	e 47% (17)	f 46% (60)	46% (96)
Low	g 20% (9)	h 31% (11)	i 43% (56)	36% (76)
Total (N)	100% (44)	100% (36)	100% (130)	100% (210)

*The cells in this table have been labeled from a through i.

Table 7.5 displays the cross-tabulation of these two variables, with *job security* as the independent variable and *job satisfaction* as the dependent variable. We find that 36 percent of those who indicate high job security, but only 11 percent of those who indicate low job security, are "satisfied" with their job. The percentage difference (36% − 11% = 25%) indicates that the variables are related. Several other percentage differences that can be computed on these data (for example, 36% − 22% = 14%; 22% − 11% = 11%) yield smaller percentage differences but lead to the same conclusion—that job satisfaction and job security are associated.

Comparison of Pairs Let's explore this relationship a bit further. To understand the logic of the ordinal measures of association discussed in this section, we have to restate the relationship not in terms of individual observations but in terms of **paired observations** and their relative position (or rank order) on the two variables. To make the discussion a bit more concrete, let's suppose that we narrow each variable into only two categories, high and low job security and high and low job satisfaction, and that there are only four people involved, one in each cell. For the purpose of this illustration we have assigned names to each individual. Our hypothetical data are presented in Figure 7.2. We now pair four people (six combinations can be created) and describe their rank order on the question of job security and job satisfaction. These results are presented in Figure 7.3.

Figure 7.2 **Job Satisfaction by Job Security of Four People (hypothetical)**

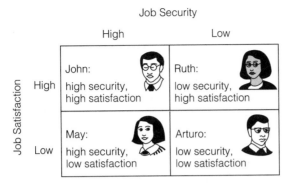

Figure 7.3 **Rank Order of Four People on Job Security and Job Satisfaction**

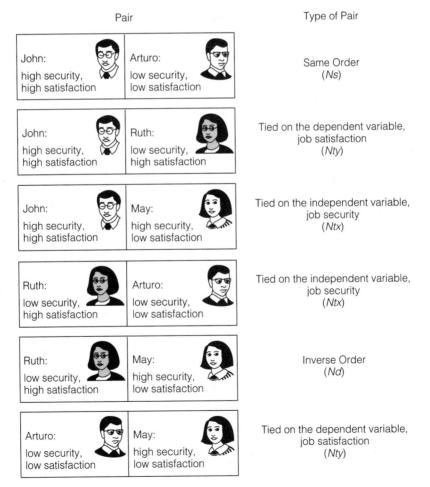

> *Paired observations* Observations compared in terms of their relative rankings on the independent and dependent variables.

Let's consider the first pair, John and Arturo. Notice that John, who is high on job security is also high on job satisfaction, and that Arturo, who has low job security, is also low on job satisfaction. When we consider the pair, we can say that the person who has higher job security (John) is also the more satisfied of the two. This pair would lead us to conclude that the higher one's job security the higher one's job satisfaction, or that job satisfaction increases with job security.

Next consider Ruth and May. Ruth, who has low job security, is satisfied with her job. On the other hand, May, who has high job security, is not satisfied with her job. Regarding the pair, we could say that the person who has the lower job security is the more satisfied. This pair would lead us to conclude that the lower one's job security the higher one's job satisfaction, or that job satisfaction decreases with job security.

Types of Pairs Because ordinal variables have direction—that is, their categories can range from low to high—the relationship between ordinal variables also has direction. The direction of the relationship can be *positive* or *negative*. With a positive relationship, if one person is ranked above another on one variable, he or she would rank above the other person on the second variable. Such a relative ranking of two observations is called a **same order pair**. We label the count of these types of pairs as *Ns*. Same order pairs show a positive association. John and Arturo are a same order pair, displaying a positive association because John, who is higher than Arturo on job security, is also more satisfied than Arturo. For John and Arturo, job satisfaction increases with job security.

> *Same order pair (Ns)* Paired observations that show a positive association; the member of the pair ranked higher on the independent variable is also ranked higher on the dependent variable.

With a negative relationship, if one person is ranked above another on one variable, he or she ranks below the other person on the second variable. Such relative ranking of a pair of observations is called an **inverse order pair,** and their count is labeled as *Nd*. Inverse order pairs show a negative association. Ruth and May are an inverse order pair. They display a negative association because Ruth, who is lower than May on job security, is higher (more satisfied) than May on job satisfaction. For Ruth and May, job satisfaction decreases with job security.

> *Inverse order pair (Nd)*   Paired observations that show a negative association; the member of the pair ranked higher on the independent variable is ranked lower on the dependent variable.

Note that there are four other pairs in Figure 7.3. These pairs all have the same value on either job satisfaction or job security. For example, whereas Ruth and Arturo have different levels of job satisfaction (Ruth's level of job satisfaction is high while Arturo's is low), they each have low job security.

Pairs that have the same value on a variable are called **tied pairs**. If the pairs share the same value on the independent variable (*X*), they are called pairs *tied on the independent variable* (their count is designated as **Ntx**); if the pairs are tied on the dependent variable (*Y*), they are called pairs *tied on the dependent variable* (**Nty**). Pairs can also share the same value on both the independent and the dependent variables; such pairs are designated as **Ntxy**.

> *Tied pairs (Ntx, Nty, or Ntxy)*   Paired observations that share the same value on the independent variable (*Ntx*), the dependent variable (*Nty*), or both variables (*Ntxy*).

Uses for Information About Pairs   We now discuss two ordinal measures of association that use the information about these types of pairs: gamma and Somers' *d*. Gamma, the most frequently used measure of association for ordinal variables, considers only untied pairs—that is, *same order* and *inverse order* pairs. Somers' *d* includes pairs that are tied on the dependent variable. Therefore, our discussion will be limited to same order pairs, inverse order pairs, and pairs tied on the dependent variable.

To calculate gamma you must first count the number of same order (*Ns*) pairs and inverse order (*Nd*) pairs that can be obtained from a bivariate table. To calculate Somers' *d* you must also count the number of pairs that are tied on the dependent variable (*Nty*). Once we find *Ns*, *Nd*, and *Nty*, the calculation of gamma and Somers' *d* is straightforward.

Because our illustration is based on only four cases (Figure 7.2), the number of pairs involved is very small and can easily be identified (as in Figure 7.3) and counted. However, because the number of cases in most bivariate tables is considerably larger (as it is in our original table, Table 7.5), the number of pairs that can be generated becomes very large, and the process of identifying the types of pairs becomes a bit more complicated.

Counting Pairs

To illustrate the process of identifying and counting all the same order pairs (*Ns*), inverse order pairs (*Nd*), and pairs tied on the dependent variable

(*Nty*) that can be generated from a bivariate table,[5] let's go back to Table 7.5. Note that the table is constructed so that the cell in the upper left corner represents the highest category on both X and Y (*high* job security and *high* job satisfaction), with levels of X decreasing from left to right, and levels of Y decreasing from top to bottom. Always make sure the table you are analyzing is arranged in this way before following the procedure outlined here to find *Ns* and *Nd*. If your table is not arranged in this way, you can always rearrange it to follow this format.

Same Order Pairs (Ns)　To find the number of same order (*Ns*) pairs, multiply the frequency in each cell in the table by the sum of the frequencies of all the cells that are lower on *both* variables—that is, both below *and* to the right of that cell. Repeat this process for each cell that has cells below it and to its right, and then sum the products. The total of these products is *Ns*.

We begin with the upper left cell, *a*, in Table 7.5. The frequency (16) is multiplied by the sum of the frequencies in cells *e*, *f*, *h*, and *i* (17 + 60 + 11 + 56). These are the cells that lie below and to the right of cell *a*. The product equals 2,304 pairs.

This computation is illustrated in Figure 7.4, in which cell *a* and the cells that lie below it and to its right (*e*, *f*, *h*, *i*) are shaded. Move to the next cell, cell *d*. Its frequency, 19, is multiplied by (11 + 56), the sum of the frequencies in cells *h* and *i*, and the result equals 1,273 pairs. (The direction of the multiplication is illustrated in Figure 7.4.) This procedure is repeated for cells *b* and *e*. Because cells *g* and *h* have no cells below them and cells *c*, *f*, and *i* have no cells to the right of them, they are not included in the computations of *Ns*. Now let's sum the total number of same order (*Ns*) pairs:

$$Ns = 2{,}304 + 1{,}273 + 928 + 952 = 5{,}457$$

Each one of these 5,457 pairs meets the definition of same order (*Ns*) pairs presented earlier. For instance, take a pair formed from cell *a* and cell *h*. The member of the pair from cell *a* is high on job security and high on job satisfaction, whereas the member of the pair from cell *h* is medium on job security and low on job satisfaction. Therefore, we can say that the member of the pair that is higher on job security (high versus medium) is also higher on job satisfaction (high versus low).

> **Learning Check.** *To convince yourself that each of the 5,457 Ns pairs we counted is indeed a same order pair, try for yourself any of the combinations as depicted in Figure 7.4.*

Inverse Order Pairs (Nd)　To calculate the number of inverse order pairs (*Nd*) we proceed in exactly the same way, except for one difference. Because we

[5]This process of counting pairs also appears in Chava Frankfort-Nachmias and David Nachmias, *Research Methods in the Social Sciences* (New York: St. Martin's Press, 1996), pp. 408–412.

Figure 7.4   **Counting All Same Order Pairs (Ns) from Table 7.5**

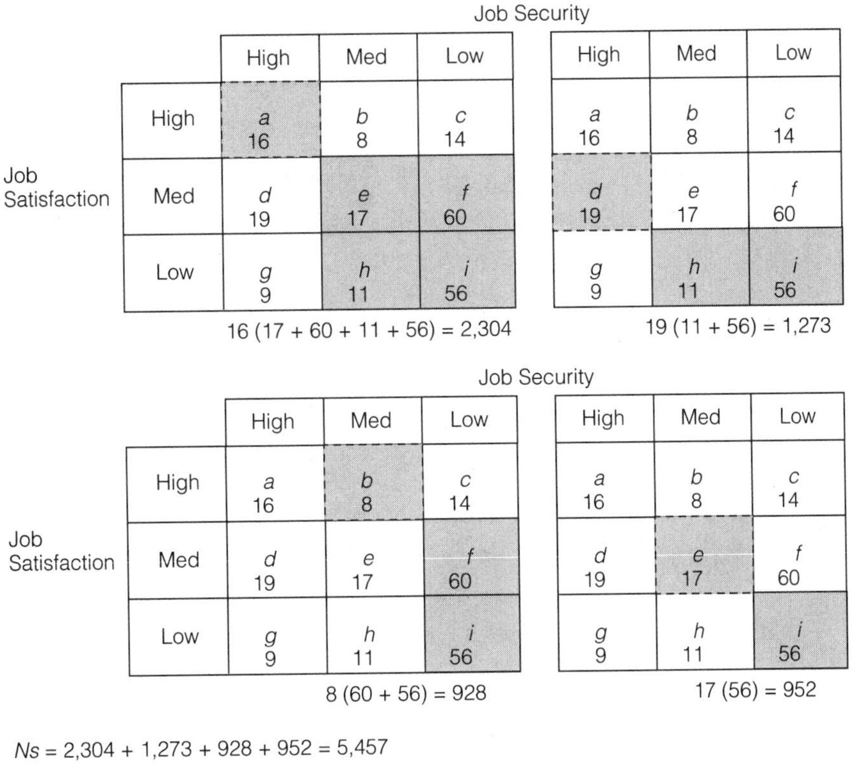

Job Security

		High	Med	Low
	High	a 16	b 8	c 14
Job Satisfaction	Med	d 19	e 17	f 60
	Low	g 9	h 11	i 56

16 (17 + 60 + 11 + 56) = 2,304

Job Security

		High	Med	Low
	High	a 16	b 8	c 14
Job Satisfaction	Med	d 19	e 17	f 60
	Low	g 9	h 11	i 56

19 (11 + 56) = 1,273

Job Security

		High	Med	Low
	High	a 16	b 8	c 14
Job Satisfaction	Med	d 19	e 17	f 60
	Low	g 9	h 11	i 56

8 (60 + 56) = 928

		High	Med	Low
	High	a 16	b 8	c 14
Job Satisfaction	Med	d 19	e 17	f 60
	Low	g 9	h 11	i 56

17 (56) = 952

Ns = 2,304 + 1,273 + 928 + 952 = 5,457

are interested in pairs in which the relative ranking of the variables is reversed, we reverse our process and begin with the upper right cell (cell c). We then multiply its frequency by the cells below it and to its left. Repeat this process for each cell that has cells below it and to its left, and then sum the products. The total of these products is Nd. These computations are illustrated in Figure 7.5. Now let's sum to compute Nd:

$$Nd = 784 + 1,200 + 224 + 153 = 2,361$$

All these pairs follow the definition of inverse order. Take a pair formed from cells c and g. The member of the pair from cell c is low on job security and high on job satisfaction. Conversely, the member of the pair from cell g is high on job security and low on job satisfaction. Therefore, we can say that the member of the pair that is lower on job security is higher on job satisfaction.

Pairs Tied on the Dependent Variable (Nty)   In Table 7.5 the dependent variable is the row variable, so all pairs from the same row will be tied on the dependent variable. For example, if we select two cases from Table 7.5, one from cell a and one from cell b, each has a high level of job satisfaction.

Figure 7.5 **Counting All Inverse Order Pairs (*Nd*) from Table 7.5**

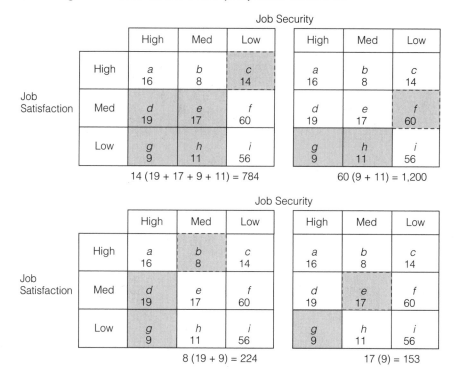

$Nd = 784 + 1{,}200 + 224 + 153 = 2{,}361$

Therefore, we will define these as tied on *Y*. Similarly, if we pair someone from cell *g* with someone from cell *i*, each has a low level of job satisfaction; therefore, we will define these as tied on *Y*.

To count all the pairs that are tied on the dependent variable (*Nty*), multiply the frequency in each cell by the sum of the frequencies in the cells that are both in its row and to its right, and then sum the products. These computations are illustrated in Figure 7.6. Let's sum the total number of pairs tied on the dependent variable (*Nty*):

$Nty = 352 + 112 + 1{,}463 + 1{,}020 + 603 + 616 = 4{,}166$

Before we discuss gamma, let's look again at the meaning of same order and inverse order pairs. (Pairs that are tied on the dependent variable *Nty* are not included in the calculation of gamma; they will be discussed later when we introduce Somers' *d*.) *Same order* pairs are all the pairs of observations formed from Table 7.5 that show a positive relationship between the variables: the higher your job security the higher your job satisfaction, or job satisfaction increases with job security. *Inverse order* pairs are all the pairs of observations that show a negative relationship between the variables: the higher your job security the lower your job satisfaction, or job sat-

Figure 7.6 **Counting All Pairs Tied on the Dependent Variable (Nty) from Table 7.5**

Job Security

		High	Med	Low		High	Med	Low
	High	*a* 16	*b* 8	*c* 14		*a* 16	*b* 8	*c* 14
Job Satisfaction	Med	*d* 19	*e* 17	*f* 60		*d* 19	*e* 17	*f* 60
	Low	*g* 9	*h* 11	*i* 56		*g* 9	*h* 11	*i* 56

16 (8 + 14) = 352   8 (14) = 112

Job Security

		High	Med	Low		High	Med	Low
	High	*a* 16	*b* 8	*c* 14		*a* 16	*b* 8	*c* 14
Job Satisfaction	Med	*d* 19	*e* 17	*f* 60		*d* 19	*e* 17	*f* 60
	Low	*g* 9	*h* 11	*i* 56		*g* 9	*h* 11	*i* 56

19 (17 + 60) = 1,463   17 (60) = 1,020

Job Security

		High	Med	Low		High	Med	Low
	High	*a* 16	*b* 8	*c* 14		*a* 16	*b* 8	*c* 14
Job Satisfaction	Med	*d* 19	*e* 17	*f* 60		*d* 19	*e* 17	*f* 60
	Low	*g* 9	*h* 11	*i* 56		*g* 9	*h* 11	*i* 56

9 (11 + 56) = 603   11 (56) = 616

Nty = 352 + 112 + 1,463 + 1,020 + 603 + 616 = 4,166

isfaction decreases with job security. The purpose of our analysis is to determine whether there is an association between job security and job satisfaction. If there is such an association, we want to be able to determine which of the statements best describes the association—that is, whether the association is *positive* or *negative*. Gamma can help us answer these questions.

■ ■ ■ ■ **Calculating Gamma**

Gamma is a symmetrical measure of association suitable for use with ordinal variables or with dichotomous nominal variables.[6] It can vary from 0.0 to ±1.0 and provides us with an indication of the strength and direction of the association between the variables. When there are more same order pairs than inverse order pairs (Ns is larger than Nd), gamma will be positive. Gamma is calculated using the following formula:

$$\text{gamma} = \frac{Ns - Nd}{Ns + Nd} \qquad (7.2)$$

Using this formula and the pairs calculations we made earlier, let's now find the association between job security and job satisfaction:

$$\text{gamma} = \frac{5,457 - 2,361}{5,457 + 2,361} = .396$$

A gamma of .396 indicates that there is a moderate positive association between job security and job satisfaction. We can conclude that using information on respondents' job security helps us improve the prediction of their job satisfaction by almost 40 percent.

Gamma is a **symmetrical measure of association**. This means that the value of gamma will be the same regardless of which variable is the independent variable or the dependent variable. Thus, if we had wanted to predict job security from job satisfaction rather than the opposite, we would have obtained the same gamma.

Gamma A symmetrical measure of association suitable for use with ordinal variables or with dichotomous nominal variables. It can vary from 0.0 to ±1.0 and provides us with an indication of the strength and direction of the association between the variables. When there are more Ns pairs, gamma will be positive; when there are more Nd pairs, gamma will be negative.

Symmetrical measure of association A measure whose value will be the same when either variable is considered the independent variable or the dependent variable.

---

[6]For 2×2 tables, a measure identical to gamma—Yule's Q—was first introduced by the statistician Udny Yule. However, whereas gamma is suitable for any size of table, Yule's Q is appropriate only for 2×2 tables. When gamma is calculated for 2×2 tables, it is sometimes referred to as Yule's Q.

Positive and Negative Gamma

Gamma can vary from 0.0 to ±1.0. It reflects the proportional reduction in prediction error when incorporating information on the independent variable to predict the dependent variable. Note from Formula 7.2 that the size and the direction of gamma (whether positive or negative) are functions of the relative number of same order (Ns) versus inverse order (Nd) pairs. More Ns pairs makes gamma positive; more Nd pairs makes gamma negative. The larger the difference between Ns and Nd, the larger the size of the coefficient (irrespective of sign). For example, when all the pairs are Ns (Nd = 0), gamma equals 1.00:

$$\text{Gamma} = \frac{Ns}{Ns} = 1.0$$

A gamma of 1.0 indicates that the relationship between the variables is positive, and the dependent variable can be predicted without any error based on the independent variable. When Ns is zero, gamma will be –1.0, indicating a perfect and a negative association between the variables:

$$\text{Gamma} = \frac{0 - Nd}{0 + Nd} = -1.0$$

When $Ns = Nd$, gamma will equal zero:

$$\text{Gamma} = \frac{Ns - Nd}{Ns + Nd} = \frac{0}{Ns + Nd} = 0.0$$

A gamma of zero reflects no association between the two variables; hence, there is nothing to be gained by using order on the independent variable to predict order on the dependent variable.

Gamma as a PRE Measure

Like all PRE measures, gamma is based on two methods of prediction. The first method ignores the relative order of pairs on the independent variable (only untied pairs are included in the computation of gamma), whereas the second method considers it. Suppose that we had tried to predict the rank order of each of the 7,818 pairs (the sum of same and inverse order pairs: $Ns + Nd$ = 5,457 + 2,361 = 7,818) while ignoring information on their relative rank order on job security. If we had used a random method to make these predictions for each of the pairs, chances are that only about 50 percent of our guesses would have been right. Therefore, we would have made errors about half the time, or ($Ns + Nd$) ÷ 2 = (5,457 + 2,361) ÷ 2 = 3,909 errors.

Now let's see if we can improve this prediction by taking the rank order on job security into consideration when predicting the rank order on job satisfaction. The likelihood of improving the prediction depends on the number of Ns versus Nd pairs. When Ns is greater than Nd, as it is in our example, it makes sense to predict that if respondents are higher on job security, they will also be higher on job satisfaction (this is the order displayed by all the *Ns* pairs)—that is, job satisfaction increases with job security. With this prediction we will be making 2,361 errors—the number of *Nd* pairs for which this prediction is not correct—or 1,548 fewer errors than before (3,909 − 2,361 = 1,548). When we divide this number by 3,909 (the original number of errors),

$$\frac{1,548}{3,909} = .396$$

we obtain a proportional reduction of error of .396, which is equal to the gamma coefficient we obtained with Formula 7.2.

Statistics in Practice: Trauma by Social Class

When the number of *Nd* pairs is larger than the number of *Ns* pairs, gamma is negative and the prediction is that if a person has a higher rank on the independent variable, she or he will have a lower rank on the dependent variable. This order is illustrated in Table 7.6, which displays the cross-tabulation of social class and the frequency of traumas of 448 GSS respondents, with social class treated as the independent variable. This table, which was examined earlier in Chapter 6, shows a negative association between the frequency of trauma and social class. Remember that with a negative relationship, higher values of one variable tend to go together with lower values of the other, and vice versa. In this table, for instance, most upper-class respondents (48%) have experienced 0 traumas, whereas most lower-class respondents (47%) have experienced 2+ traumas.

Table 7.6 **Frequency of Trauma by Social Class**

	SOCIAL CLASS			
TRAUMA	Upper	Middle	Lower	Total
2+	32% (58)	17% (38)	47% (22)	43% (118)
1	20% (36)	42% (92)	22% (11)	31% (139)
0	48% (86)	41% (90)	31% (15)	26% (191)
Total (*N*)	100% (180)	100% (220)	100% (48)	100% (448)

Source: General Social Survey, 1987 to 1992.

To calculate gamma we must first count the number of Ns and Nd pairs. The number of same order (Ns) pairs that can be formed from Table 7.6 is

$$Ns = 58(92 + 11 + 90 + 15) + 38(11 + 15) + 36(90 + 15) + 92(15) = 18,212$$

The number of inverse order (Nd) pairs that can be formed from Table 7.6 is

$$Nd = 22(36 + 92 + 86 + 90) + 38(36 + 86) + 11(86 + 90) + 92(86) = 21,172$$

Gamma is

$$\frac{18,212 - 21,172}{18,212 + 21,172} = -.075$$

In this example, the number of inverse order pairs ($Nd = 21,172$) is larger than the number of same order pairs ($Ns = 18,212$). We can predict that if your social class rank is higher than that of the other member of your pair, then your frequency of trauma will be lower—or frequency of trauma decreases with social class. This is the order displayed by all the inverse order pairs. With this prediction we would make 18,212 errors, the number of same order (Ns) pairs for which this prediction is not correct.

A gamma of $-.075$ means that, as expected, the relationship between social class and frequency of trauma is negative; that is, as social class increases, frequency of trauma decreases. However, the relationship between these variables is rather weak: Using social class to predict frequency of trauma results in a proportional reduction of error of only 7.5 percent (.075 × 100 = 7.5%).

■ ■ ■ ■ **Calculating Somers' d**

Somers' d is an asymmetrical measure of association suitable for ordinal variables or for dichotomous nominal variables. This means that its value may be different, depending on which variable is the dependent variable. Because Somers' d is asymmetric—it predicts order of pairs on the dependent variable from their order on the independent variable—it counts as predictive errors all pairs that are tied on the dependent variable (Nty). Like gamma, Somers' d varies from 0.0 to ±1.0 and reflects both the strength and the direction of the association between the variables. The formula for Somers' d is

$$\text{Somers' } d = \frac{Ns - Nd}{Ns + Nd + Nty} \qquad \text{(7.3)}$$

Somers' d An asymmetrical measure of association suitable for ordinal variables or for dichotomous nominal variables.
It varies from 0.0 to ±1.0 and reflects both the strength and the direction of the association between the variables. The greater the number of tied pairs, the smaller the value of Somers' d compared with the value of gamma.

Tied Pairs and Somers' *d*

To illustrate how tied pairs may affect the strength of the association as re-flected in the value of Somers' *d*, let's go back to Figure 7.3, in which we designated the relative rankings of pairs of observations on the two vari-ables of job satisfaction and job security. Notice that four pairs were tied: John and Ruth and Arturo and May had the same rank on job satisfaction, the dependent variable (*Nty*), whereas John and May and Ruth and Arturo had the same rank on job security, the independent variable (*Ntx*). Because Somers' *d* includes only pairs that are tied on the dependent variable (*Nty*), we will consider only the first two pairs.

Comparing the ranking of these two pairs on job satisfaction and job se-curity would lead us to conclude, contradictorily, that higher (or lower) rank on job security is associated with the same level of job satisfaction. This conclusion indicates that there is no relationship between job security and job satisfaction for these pairs. Including such tied pairs in the calcula-tion of Somers' *d* will decrease the magnitude of the coefficient. The larger the number of ties on the dependent variable (*Nty*), the smaller Somers' *d* will be relative to gamma. Based on Table 7.5, we took the following counts of the different types of pairs: $Ns = 5{,}457$, $Nd = 2{,}361$, and $Nty = 4{,}166$. Us-ing this information, we can now calculate Somers' *d* based on Formula 7.3:

$$\text{Somers' } d = \frac{5{,}457 - 2{,}361}{5{,}457 + 2{,}361 + 4{,}166} = .258$$

Somers' *d* Compared with Gamma

Note that for the same data, Somers' *d* is smaller than gamma (gamma was .396). The inclusion of pairs that are tied on *Y* in the denominator of the for-mula for Somers' *d* means that the value of Somers' *d* will never be more than the value of gamma. The larger the number of pairs tied on the depen-dent variable (*Nty*), the smaller will be the size of Somers' *d* relative to gamma.

Like gamma, Somers' *d* is a PRE measure of association. Therefore, the more it departs from 0.0 in either direction, the greater the proportional re-duction of error when information about the independent variable is used to predict values of the dependent variable. A Somers' *d* of .258 indicates that using relative rankings on job security to predict relative ranking on job satisfaction (and considering tied pairs) results in a proportional reduc-tion of error of .258, or in 25.8% ($.258 \times 100 = 25.8\%$) fewer errors.

Because Somers' *d* considers pairs that are tied on the dependent vari-able, a change in the definition of the dependent variable will change the computation of ties. For example, a Somers' *d* in which job security is the dependent variable would include ties on job security rather than on job satisfaction, as calculated in our example.

Since both gamma and Somers' *d* are appropriate to use with ordinal variables or with dichotomous nominal variables, how should we choose between them? Gamma is most often used by social scientists, but Somers' *d* is preferable to gamma in situations when we can clearly distinguish between the independent and dependent variables. For example, let's say that we wanted to look at the relationship between people's family income while growing up and their political views as adults. Clearly, only family income while growing up can influence political views, and *not* vice versa. In this situation, Somers' *d* would be preferable to gamma as a measure of association. In contrast, take the variables *attitude toward gun control* and *attitude toward abortion*. Both could be considered as either the independent or the dependent variable, and therefore gamma should be chosen as a measure of association.

Learning Check.   *Note that we have illustrated how to count tied pairs within rows. If the dependent variable is arranged in the columns, you should calculate ties within columns. Or, if your dependent variable is the column variable, simply switch the table and make it the row variable. You can then follow the procedure suggested here.*

∎ ∎ ∎ ∎ ∎   **Using Ordinal Measures with Dichotomous Variables**

Measures of association for ordinal data are not influenced by the modal category, as is lambda. Consequently, an ordinal measure of association might be preferable for tables when an association cannot be detected by lambda. We can use an ordinal measure for some tables where one or both variables would appear to be measured on a nominal scale. Dichotomous variables (those with only two categories) can be treated as ordinal variables for most purposes. In this chapter we calculated lambda to examine the association between financial satisfaction and home ownership (Tables 7.2, 7.3, and 7.4). Although home ownership might be considered a nominal variable—because it is dichotomized (own/rent)—it might also be treated as an ordinal variable. Thus, the association between home ownership and financial satisfaction (an ordinal variable) might also be examined using an ordinal measure of association.

Let's calculate gamma for Table 7.3. The number of same order pairs that can be formed from Table 7.3 is

$$Ns = 5(11+8) + 5(8) = 135$$

The number of inverse order pairs that can be formed from Table 7.3 is

$$Nd = 2(5+9) + 11(9) = 127$$

Gamma is

$$\frac{135 - 127}{135 + 127} = .03$$

A gamma of .03 confirms our earlier conclusion that for black women, home ownership is a very poor predictor of financial satisfaction. To interpret this negative coefficient, think of ownership as the ordering principle, with "own" as higher and "rent" as lower. Thus, a positive association between home ownership and financial satisfaction means that those who own their homes would express greater financial satisfaction.

■ ■ ■ ■ **Reading the Research Literature:**
Worldview and Abortion Beliefs

Let's conclude this chapter with a typical example of how gamma is presented and interpreted in the social science research literature. The following example is drawn from a study that examines the idea that beliefs about abortion are influenced by a coherent view of the world. Kristin Luker argues that "each side of the abortion debate has an internally coherent and mutually shared view of the world that is tacit, never fully articulated, and most important, completely at odds with the world-view held by their opponents."[7] In general, if a person is religious, views the primary role of women as one of taking care of the home and raising the children, thinks that sex should be practiced for procreation only, and has a conservative political viewpoint, then this person also tends to disapprove of abortion under almost any circumstances. Conversely, a person who is not religious, has an egalitarian view of gender roles, has liberal attitudes toward sexuality, and identifies himself or herself as left of center on the political spectrum probably believes in a woman's right to choose whether or not to have an abortion.[8] The major hypothesis of this study is "the more conservative a person's world view, the greater his or her disapproval of abortion."[9]

To test this hypothesis, Daniel Spicer used data from the 1990 General Social Survey, which is based on a representative sample of all the non-institutionalized adult residents of the continental United States. To measure worldview, Spicer selected six variables: attitudes toward premarital sex, conception of sex roles, religious intensity, political views, fundamentalism, and biblical interpretation. For the purpose of this discussion, we have considered only the first three variables.

Measurement of these variables was based on responses to the following questions:

- *Conception of sex roles* "Do you agree or disagree with this statement? Women should take care of the home and leave running the country up to men."

---

[7]Kristin Luker, *Abortion and the Politics of Motherhood* (Berkeley: University of California Press, 1984), p. 159.

[8]Daniel N. Spicer, "World View and Abortion Beliefs: A Replication of Luker's Implicit Hypothesis," *Sociological Inquiry* 64, no. 1(February 1994): 115. Copyright © 1994 by the University of Texas Press. All rights reserved.

[9]Ibid., pp. 115–116.

- *Premarital sex views* "There's been a lot of discussion about the way morals and attitudes about sex are changing in this country. If a man and a woman have sex relations before marriage, do you think it is always wrong, usually wrong, somewhat wrong, or acceptable?"
- *Religious intensity* "Would you call yourself a strong, moderate, or weak (stated religion)?"

The dependent variable *abortion belief* was constructed from responses to four questions dealing with the following circumstances under which the person believed it acceptable or unacceptable for a woman to have an abortion. A person is said to approve of abortion if he or she said yes to all four of the following items:

If she is married and does not want any more children

If the family has a very low income and cannot afford any more children

If she is not married and does not want to marry the man

The woman wants it for any reason

Based on their response to all four questions, respondents were classified into one of two categories: "approve" or "disapprove" of abortion. The bivariate percentage distributions and the gamma for each of the independent variables and abortion belief are presented in Table 7.7.

Examining the Data

Begin by examining the structure of the table. Note that it is divided into three parts, one for each independent variable. Each part can be read as a

Table 7.7   **Percentage Approval of Abortion by Selected Independent Variables (GSS, 1990)**

Variable	Category	N	Abortion Approval	Gamma
Premarital sex views	Always wrong	92	15%	−.644
	Usually wrong	42	36%	
	Somewhat wrong	75	51%	
	Acceptable	141	71%	
Sex role views	Traditional	58	21%	−.625
	Liberal	292	53%	
Religious intensity	Strong	261	28%	−.433
	Moderate	97	55%	
	Weak	312	59%	

Source: Adapted from Daniel N. Spicer, "World View and Abortion Beliefs: A Replication of Luker's Implicit Hypothesis," *Sociological Inquiry* 64, no. 1 (February 1994): 120. Copyright © 1994 by the University of Texas Press. All rights reserved.

separate table displaying the bivariate percentage and the gamma for each of the independent variables and abortion beliefs. The independent variables and their categories are arrayed in rows of the table; the dependent variable, *abortion approval,* is arrayed in the columns. For each category of the independent variables, the number of people who responded (N) and the percentage who approved of abortion are listed. For example, the variable *sex role views* has two categories, "traditional" and "liberal." Of the 58 traditionals, 21 percent approved of abortion, compared with 53 percent of the 292 liberals.

Learning Check. *The percentages that disapproved of abortion are not listed in the table. However, it is very easy to obtain the numbers: simply subtract the percentage that approved from 100 percent for each category. For example, of the 58 traditionals, 79 percent (100% − 21%) disapproved of abortion. Try to complete the table by calculating the percentages that disapproved of abortion for all the variables.*

Interpreting the Data

Next, interpret the data presented in Table 7.7. In reading the table and interpreting the relationship between each of the independent variables and beliefs about abortion, look at both the percentage differences and the value of gamma. Let's begin by comparing the percentages. Following the rules we learned in Chapter 6, when percentages are calculated within rows (as they are in this table), comparisons are made down the column. For instance, to interpret the relationship between religious intensity and abortion beliefs, compare 28 percent with 55 percent and 59 percent. Similarly, to examine the relationship between sex role views and abortion belief, compare 21 percent with 53 percent. Based on the percentage comparisons for each independent variable, the researcher offers the following interpretation of these findings:

> As shown in [Table 7.7], . . . as the idea of premarital sex becomes more acceptable, the approval of the practice of abortion increases . . . people with traditional sex role conceptions tended to disapprove of abortion and those with a modern, liberal conception of the sexes were in favor of abortion rights. Respondents with a high religious intensity disapproved of abortion, while those with weak religious ties were much more liberal on the question of abortion.[10]

The detailed summary of the relationships between the variables is confirmed by the values of gamma displayed in the table. The gamma values range from −.433 for religious intensity to −.644 for attitudes toward premarital sex. These values indicate a moderate to strong relationship between various aspects of one's worldview and abortion beliefs.

[10]Ibid., pp. 120–121.

Notice that all the gamma values are negative. A negative gamma indicates that low values of one variable "go together" with high values of the other variable, and vice versa.

Often it is tricky to interpret the direction of a relationship between ordinal variables because what is considered "low" or "high" is often a function of arbitrary coding by the researcher. However, a researcher will often specify his or her coding in the text. Spicer makes this statement about the direction of these relationships:

> As high scores on . . . the [independent] variables (*higher* values indicating *"conservative"* social views) were associated with *low* scores (*disapproval*) on the abortion view variable . . . the Gamma values are negative.[11]

In other words, the negative gamma between premarital sex views, sex role views, and religious intensity means that respondents who have more conservative social views tend to disapprove of abortion. Conversely, a large percentage of those who have more liberal views approve of abortion.

MAIN POINTS

- Measures of association are single summarizing numbers that reflect the strength of the relationship between variables, indicate the usefulness of predicting the dependent variable from the independent variable, and often show the direction of the relationship.

- Proportional reduction of error (PRE) underlies the definition and interpretation of several measures of association. PRE measures are derived by comparing the errors made in predicting the dependent variable while ignoring the independent variable with errors made when making predictions that use information about the independent variable.

- PRE measures may range from 0.0 to ±1.0. A PRE of 0.0 indicates that the two variables are not associated and that information about the independent variable will not improve predictions about the dependent variable. A PRE of ±1.0 means that there is a perfect (positive or negative) association between the variables and that information about the independent variable results in a perfect (without any error) prediction of the dependent variable.

- Measures of association may be symmetrical or asymmetrical. When the measure is symmetrical, its value will be the same regardless of which of the two variables is considered the independent or dependent variable. In contrast, the value of asymmetrical measures of association may vary depending on which variable is considered the independent variable and which the dependent variable.

- Lambda is an asymmetrical measure of association suitable for use with nominal variables. It can range from 0.0 to 1.0 and gives an indication of

[11]Ibid., p. 121.

the strength of an association between the independent and the dependent variables.

■ Gamma is a symmetrical measure of association suitable for ordinal variables or for dichotomous nominal variables. It can vary from 0.0 to ±1.0 and reflects both the strength and direction of the association between the variables.

■ Somers' *d* is an asymmetrical measure of association suitable for ordinal variables or for dichotomous nominal variables. Somers' *d* varies from 0.0 to ±1.0 and reflects both the strength and direction of the association between the variables. The greater the number of tied pairs, the smaller the value of Somers' *d* compared with the value of gamma.

KEY TERMS

asymmetrical measure of association

gamma

inverse order pair (Nd)

lambda

measure of association

paired observations

proportional reduction of error (PRE)

same order pair (Ns)

Somers' d

symmetrical measure of association

tied pairs (Ntx, Nty, Ntxy)

SPSS DEMONSTRATION

Demonstration: Producing Nominal Measures of Association for Bivariate Tables

In Chapter 6 we used the Crosstabs procedure in SPSS to create bivariate tables. That same procedure is used to request measures of association. We'll begin by investigating the same relationship we did in Chapter 6: support for legal abortions for married women among various categories of religious affiliation.

Click on *Analyze, Descriptive Statistics*, then *Crosstabs* to get to the Crosstabs dialog box. Put ABNOMORE in the Row(s) box and RELIG in the Column(s) box. Then click on the *Statistics* button. The Statistics dialog box (Figure 7.7) has about a dozen statistics from which to choose. Notice that four statistics are listed in separate categ⌐ ⌐s for "Nominal" and "Ordinal" data. Lambda is listed in the former, and gamma and Somers' *d* in the latter. The other measures of association, such as phi and Cramer's *V*, or Kendall's tau-*b*, will not be discussed in this textbook. The chi-square statistic will be discussed in depth in Chapter 14.

Since religious affiliation is a nominal variable, click on the checkbox for lambda. It is critical that we, as users of statistical programs, understand which statistics to select in any procedure. SPSS, like most programs, can't help us select the appropriate statistic for an analysis. Now click on *Continue*, then *OK* to create the table (Figure 7.8).

Figure 7.7

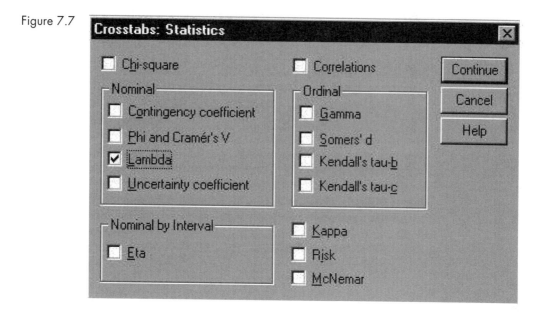

Figure 7.8

ABNOMORE MARRIED--WANTS NO MORE CHILDREN * RELIG RS RELIGIOUS PREFERENCE Crosstabulation

Count

		RELIG RS RELIGIOUS PREFERENCE					
		1 PROTESTANT	2 CATHOLIC	3 JEWISH	4 NONE	5 OTHER	Total
ABNOMORE MARRIED--WANTS NO MORE CHILDREN	1 YES	226	80	18	63	20	407
	2 NO	302	124		45	13	484
Total		528	204	18	108	33	891

Directional Measures

			Value	Asymp. Std. Error[a]	Approx. T[b]	Approx. Sig.
Nominal by Nominal	Lambda	Symmetric	.056	.016	3.433	.001
		ABNOMORE MARRIED--WANTS NO MORE CHILDREN Dependent	.106	.029	3.433	.001
		RELIG RS RELIGIOUS PREFERENCE Dependent	.000	.000	c	c
	Goodman and Kruskal tau	ABNOMORE MARRIED--WANTS NO MORE CHILDREN Dependent	.041	.009		.000[d]
		RELIG RS RELIGIOUS PREFERENCE Dependent	.006	.003		.000[d]

a. Not assuming the null hypothesis.

b. Using the asymptotic standard error assuming the null hypothesis.

c. Cannot be computed because the asymptotic standard error equals zero.

d. Based on chi-square approximation

The cross-tabulation is identical to the one in Chapter 6. Below this table is another table labeled "Directional Measures." For now we will only concern ourselves with the first two columns. Lambda is listed with three values. We've learned that the value of lambda depends on which variable is considered the dependent variable. In our example, *attitude toward abortion for married women* is dependent, so lambda is .106. This indicates a weak relationship between the two variables. We can conclude that knowing the religious affiliation of the respondent increases the ability to predict his or her abortion attitude by just 2.8 percent.

SPSS also calculates a symmetrical lambda for those tables where there is no independent or dependent variable. This calculation goes beyond the scope of this book. (It is not simply an average of the two other values of lambda.) In addition, as a kind of bonus, SPSS provides the Goodman and Kruskal tau statistic, another nominal measure of association, even though it was not requested. This measure will always be produced when lambda is requested.

SPSS PROBLEMS [MODULE A]

1. In Chapter 6, SPSS Problem 3, you examined the relationship between attitudes about women working (FEWORK) and support for women having an abortion for any reason (ABANY).
   a. Study the same relationship, but this time request the appropriate measure of association to describe the relationship. Because these are dichotomous variables, you can use ordinal measures of association.
   b. Add SEX as a control variable and calculate the association measures for each subtable. Is the relationship stronger for women or men? Can you think of reasons why this might be so?

2. Investigate the relationship between the abortion items and various demographic variables (you might begin with gender, age, or race). Examine the relationship of these variables based on the appropriate measures of association. For example, you might examine whether attitude toward each of the abortion items has a similar relationship to gender. That is, if females are supportive of abortion for rape victims, are they also supportive of abortion in other circumstances? Try exploring these relationships further by adding control variables. You might create tables of abortion attitude by race by gender. When you have finished the analysis, write a short report summarizing the findings. Suggest possible causes for the relationships you found.

3. Investigate the relationship between attitudes about affirmative action for blacks (AFFRMACT) and women (FEJOBAFF) with various demographic variables. As you did in Exercise 2, examine the relationship using an appropriate measure of association. Prepare a short report summarizing your findings. What might explain the relationships you've discovered?

CHAPTER EXERCISES

1. In Chapter 6, Exercise 3, we investigated the relationship between whether someone has a close friend or family member who is lesbian or gay and whether she or he believes that homosexuality is a choice. The data from that exercise are displayed again here for convenience.

	Has gay or lesbian close friend or family member?		
Homosexuality is	Yes	No	Total
Choice	81	427	508
Can't change	144	352	496
Don't know	29	121	150
Total	254	900	1,154

a. As before, we will treat whether or not someone has a friend or family member who is lesbian or gay as the independent variable. If we first ignore that variable and try to predict attitude toward homosexuality, how many errors will we make?

b. If we now take into account the independent variable, how many errors of prediction will we make for those who have a gay or a lesbian friend or family member? for those who don't?

c. Combine the answers in (a) and (b) to calculate the proportional reduction in error for this table based on the independent variable. How does this statistic improve our understanding of the relationship between the two variables?

2. In Chapter 6, Exercise 4, we continued the investigation of attitudes toward homosexuality, studying how belief about whether homosexuality is a choice influenced support for gays or lesbians being allowed in the military. That table (minus the "Don't know" responses) is reproduced here.

	Is homosexuality a choice or an orientation that cannot change?		
Should gays and lesbians be allowed in the military?	Choice	Can't Change	Total
Yes	162	268	430
No	276	160	436
Total	438	428	866

a. Treating belief about whether homosexuality is a choice as the independent variable, calculate lambda for the table. How many errors of prediction will be made if the independent variable is ignored? How many fewer errors will be made if the independent variable is taken into account? Use lambda to discuss the relationship between these two beliefs. Why is lambda an appropriate measure of association?

b. Both variables in this table are attitudes, so we could consider belief about whether gays and lesbians should be allowed in the military as the predictor, or independent, variable. If we do, then lambda must be recalculated because it is not a symmetrical measure of association. What is the value for lambda when belief about allowing gays and lesbians in the military is the independent variable? How does it compare with the lambda calculated in (a)?

3. The following table presents data from the 1996 GSS on attitudes toward premarital sex by age of respondent (grouped into intervals of twenty years).

PREMARSX SEX BEFORE MARRIAGE * CAGE categorical age Crosstabulation

Count

		CAGE categorical age			Total
		1.00 20-39	2.00 40-59	3.00 60+	
PREMARSX SEX BEFORE MARRIAGE	1 ALWAYS WRONG	74	77	74	225
	2 ALMST ALWAYS WRG	30	30	32	92
	3 SOMETIMES WRONG	88	77	54	219
	4 NOT WRONG AT ALL	193	170	44	407
Total		385	354	204	943

a. To calculate a measure of association for ordinal data we need to calculate the number of N_s and N_d pairs. Calculate these quantities for this table. (*Hint:* Consider the highest category of attitude toward premarital sex to be "always wrong," and reconstruct the table accordingly.)

b. Using N_s and N_d, calculate gamma for this table. Is gamma positive or negative? Using the value of gamma, interpret the relationship between age and attitude toward premarital sex.

4. Use the table in Exercise 3.

a. Calculate the number of cases tied on the dependent variable (N_{ty}). What is the value of N_{ty}?

b. Use this number to calculate Somers' d for the table. How does it compare with gamma?

c. Would your interpretation of the strength of the relationship between age and attitude toward premarital sex be different if you used Somers' d instead of gamma?

5. Women have increasingly been elected to higher political offices in recent years. Given this fact, is it true that the increasing number of women legislators has, in part, led to a higher percentage of bills being passed on women's or family issues? The following table displays data about the number of bills on women's and family issues that were introduced and then passed by the U.S. Congress during two 2-year periods in the early 1990s.

| | **Bills Introduced on Women's and Family Issues** | |
	1990/91	1992/93
Didn't Pass	231	437
Did Pass	19	64

Source: Data from the Congressional Caucus for Women's Issues.

a. Calculate lambda to assess the strength of the relationship between time period and bill passage. Be sure to use the appropriate independent variable when calculating lambda.

b. Does the calculated value of lambda seem surprising, given the relationship observed in the table? Can you explain why lambda has the value that it does?

6. Are some Americans more integrated into the country than others? What perceptions do blacks and whites have about their "closeness" with the United States? Would there be a difference among social classes in their level of "closeness"? Using SPSS and the 1996 GSS, you decide to investigate level of "closeness" to the United States by race and subjective social class. You obtain the results shown on pages 290–291, with separate tables for whites and blacks. (For ease of illustration and because of low numbers, respondents of lower and upper self-rated social classes are not shown.)

a. Use an appropriate PRE measure, plus percentage differences, to summarize the relationship between level of closeness to the United States and social class, for whites and blacks separately. Can you suggest a reason for any differences you find?

b. Collapse the two tables to create one table that allows you to investigate the relationship between closeness and social class. Calculate an appropriate PRE measure and interpret the relationship.

c. Collapse the two tables to create one table so that you can investigate the relationship between race and level of closeness. Compute the appropriate PRE measure to aid your analysis.

d. Are the results of these three analyses consistent or contradictory? Explain why.

CLSEUSA How close do you feel to America * CLASS SUBJECTIVE CLASS IDENTIFICATION * RACE RACE OF RESPONDENT Crosstabulation

RACE RACE OF RESPONDENT				CLASS SUBJECTIVE CLASS IDENTIFICATION		Total
				2 WORKING CLASS	3 MIDDLE CLASS	
1 WHITE	CLSEUSA How close do you feel to America	1 Very close	Count	89	99	188
			% within CLASS SUBJECTIVE CLASS IDENTIFICATION	39.4%	37.8%	38.5%
		2 Close	Count	101	125	226
			% within CLASS SUBJECTIVE CLASS IDENTIFICATION	44.7%	47.7%	46.3%
		3 Not very close	Count	31	33	64
			% within CLASS SUBJECTIVE CLASS IDENTIFICATION	13.7%	12.6%	13.1%
		4 Not close at all	Count	5	5	10
			% within CLASS SUBJECTIVE CLASS IDENTIFICATION	2.2%	1.9%	2.0%
	Total		Count	226	262	488
			% within CLASS SUBJECTIVE CLASS IDENTIFICATION	100.0%	100.0%	100.0%

2 BLACK

CLSEUSA How close do you feel to America				
1 Very close	Count	16	5	21
	% within CLASS SUBJECTIVE CLASS IDENTIFICATION	33.3%	19.2%	28.4%
2 Close	Count	20	16	36
	% within CLASS SUBJECTIVE CLASS IDENTIFICATION	41.7%	61.5%	48.6%
3 Not very close	Count	8	4	12
	% within CLASS SUBJECTIVE CLASS IDENTIFICATION	16.7%	15.4%	16.2%
4 Not close at all	Count	4	1	5
	% within CLASS SUBJECTIVE CLASS IDENTIFICATION	8.3%	3.8%	6.8%
Total	Count	48	26	74
	% within CLASS SUBJECTIVE CLASS IDENTIFICATION	100.0%	100.0%	100.0%

7. In Exercise 5 we learned about what might seem an oddity in the calculation of lambda that caused it to exactly equal zero. Measures of association for ordinal data are not influenced by the modal category, as is lambda. Consequently, an ordinal measure of association might be preferable for tables like the one in Exercise 5.

a. Calculate an ordinal measure of association for the table in Exercise 5.

b. Use this statistic to discuss the strength and direction of relationship between time period and the passage of bills on women's and family issues.

8. Tolerance of premarital sexual activity is associated with several demographic variables. In this exercise we will explore how well education predicts this attitude. The following table uses data from the 1996 GSS, with education recoded into the five categories displayed in the columns.

PREMARSX SEX BEFORE MARRIAGE * DEGREE RS HIGHEST DEGREE Crosstabulation

			DEGREE RS HIGHEST DEGREE					
			0 LT HIGH SCHOOL	1 HIGH SCHOOL	2 JUNIOR COLLEGE	3 BACHELOR	4 GRADUATE	Total
PREMARSX SEX BEFORE MARRIAGE	1 ALWAYS WRONG	Count	49	129	12	28	8	226
		% within DEGREE RS HIGHEST DEGREE	37.1%	23.9%	21.1%	19.4%	9.9%	23.7%
	2 ALMST ALWAYS WRG	Count	9	61	8	11	5	94
		% within DEGREE RS HIGHEST DEGREE	6.8%	11.3%	14.0%	7.6%	6.2%	9.9%
	3 SOMETIMES WRONG	Count	28	122	15	32	24	221
		% within DEGREE RS HIGHEST DEGREE	21.2%	22.6%	26.3%	22.2%	29.6%	23.2%
	4 NOT WRONG AT ALL	Count	46	228	22	73	44	413
		% within DEGREE RS HIGHEST DEGREE	34.8%	42.2%	38.6%	50.7%	54.3%	43.3%
Total		Count	132	540	57	144	81	954
		% within DEGREE RS HIGHEST DEGREE	100.0%	100.0%	100.0%	100.0%	100.0%	100.0%

Directional Measures

			Value	Asymp. Std. Error[a]	Approx. T[b]	Approx. Sig.
Ordinal by Ordinal	Somers' d	Symmetric	.119	.027	4.318	.000
		PREMARSX SEX BEFORE MARRIAGE Dependent	.125	.029	4.318	.000
		DEGREE RS HIGHEST DEGREE Dependent	.113	.026	4.318	.000

a. Not assuming the null hypothesis.
b. Using the asymptotic standard error assuming the null hypothesis.

Symmetric Measures

		Value	Asymp. Std. Error[a]	Approx. T[b]	Approx. Sig.
Ordinal by Ordinal	Gamma	.181	.042	4.318	.000
N of Valid Cases		954			

a. Not assuming the null hypothesis.
b. Using the asymptotic standard error assuming the null hypothesis.

a. Notice that gamma and Somers' *d* have been calculated. Use these two statistics to discuss the relationship between education and attitude toward premarital sex.

b. Why is Somers' *d* smaller than gamma?

9. Continue your exploration of how education relates to various attitudes by investigating how it influences support for the death penalty for murderers. The following data were also taken from the 1996 GSS file.

CAPPUN FAVOR OR OPPOSE DEATH PENALTY FOR MURDER * DEGREE RS HIGHEST DEGREE Crosstabulation

			DEGREE RS HIGHEST DEGREE					
			0 LT HIGH SCHOOL	1 HIGH SCHOOL	2 JUNIOR COLLEGE	3 BACHELOR	4 GRADUATE	Total
CAPPUN FAVOR OR OPPOSE DEATH PENALTY FOR MURDER	1 FAVOR	Count	144	585	67	153	69	1018
		% within DEGREE RS HIGHEST DEGREE	78.3%	79.3%	77.0%	72.9%	62.2%	76.5%
	2 OPPOSE	Count	40	153	20	57	42	312
		% within DEGREE RS HIGHEST DEGREE	21.7%	20.7%	23.0%	27.1%	37.8%	23.5%
Total		Count	184	738	87	210	111	1330
		% within DEGREE RS HIGHEST DEGREE	100.0%	100.0%	100.0%	100.0%	100.0%	100.0%

a. Calculate Somers' *d* for this table.

b. Use it to describe the relationship between education and support for the death penalty for murder. Is this a strong or weak relationship? What is its direction?

10. Gun ownership is quite common in the United States, but those who own a gun are not necessarily a cross-section of Americans. One possibility is that there might be a difference in gun ownership by marital status (perhaps married individuals are more likely to own a gun to protect their families). Using the GSS1996 data, you construct the following table.

OWNGUN HAVE GUN IN HOME * MARITAL MARITAL STATUS Crosstabulation

			MARITAL MARITAL STATUS					
			1 MARRIED	2 WIDOWED	3 DIVORCED	4 SEPARATED	5 NEVER MARRIED	Total
OWNGUN HAVE GUN IN HOME	1 YES	Count	237	31	51	10	53	382
		% within MARITAL MARITAL STATUS	52.2%	29.5%	34.7%	28.6%	26.1%	40.5%
	2 NO	Count	217	74	96	25	150	562
		% within MARITAL MARITAL STATUS	47.8%	70.5%	65.3%	71.4%	73.9%	59.5%
Total		Count	454	105	147	35	203	944
		% within MARITAL MARITAL STATUS	100.0%	100.0%	100.0%	100.0%	100.0%	100.0%

a. What measure of association is appropriate for this table?

b. Without doing any calculations, you should be able to study this table and provide one possible value for the proper measure of association. What is that value, and why?

11. Exercise 10 in Chapter 6 explored the relationship between attitude toward the effect of pornography on morals and support for women working outside the home. Use the data in the first table (without gender) for this exercise.
    a. Calculate gamma for the table.
    b. Calculate Somers' d for the table.
    c. Use the two measures of association to further characterize the relationship between the two attitudes. Do your conclusions agree with what you said for Exercise 10c in Chapter 6?

12. In a study of adolescent sexual behavior in Ontario, Canada, sexually active teens were asked to report their use of condoms and birth control. The following cross-tabulation of data is presented separately for boys and girls. Results are presented for three age groups.

Males ($N = 2918$)

Frequency of use of protection	Age of teen			
	12–13	14–15	16–17	Total
Always	145	401	532	1078
Intermittent	81	301	213	595
Never	355	557	333	1245
Total	581	1259	1078	2918

Females ($N = 2824$)

Frequency of use of protection	Age of teen			
	12–13	14–15	16–17	Total
Always	75	586	726	1387
Intermittent	48	147	284	479
Never	264	446	248	958
Total	387	1179	1258	2824

Source: Adapted from B. Helen Thomas, Alba DiCenso, and Lauren Griffith, "Adolescent Sexual Behaviour: Results from an Ontario Sample: Part II. Adolescent Use of Protection," *Canadian Journal of Public Health* 89, no. 2 (1998): 94–97, Table 1, p. 95. Used by permission.

Calculate the appropriate measure of association for these tables. Interpret the relationship between age and frequency of use of protection for the sample of sexually active Ontario teens. Is there a difference in the level of association for boys and girls?

13. Is there an association between smoking and school performance among teenagers? The following table reports results from the 1990 California Tobacco Survey. School performance is measured on a 4-point scale; smoking status is designated as nonsmoker, former smoker, or current smoker.

School Performance	Nonsmokers	Former Smokers	Current Smokers	Total
Much better than average	753	130	51	934
Better than average	1439	310	140	1889
Average	1365	387	246	1998
Below average	88	40	58	186
Total	3645	867	495	5007

Source: Adapted from Teh-wei Hu, Zihua Lin, and Theodore E. Keeler, "Teenage Smoking: Attempts to Quit and School Performance," *American Journal of Public Health* 88, no. 6 (1998), Table 1, p. 941. Used by permission of The American Public Health Association.

Calculate the appropriate measure of association for this table. Does a relationship exist between smoking status and school performance for California teenagers? Explain.

8 Bivariate Regression and Correlation

MAIN POINTS
KEY TERMS
SPSS DEMONSTRATIONS
SPSS PROBLEMS
CHAPTER EXERCISES

■ ■ ■ ■ **Introduction**

Many research questions require the analysis of relationships between interval-ratio level variables. Environmental studies, for instance, frequently measure opinions and behavior in terms of quantity, percentages, units of production, and dollar amounts. Let's say that we're interested in the relationship between populations' levels of environmental concern and their wealth. *Bivariate regression analysis* provides us with the tools to express a relationship between two interval-ratio variables in a concise way.[1]

Since 1900 world population has more than tripled and the global economy has expanded twenty times. This growth has resulted in a tremendous increase both in the consumption of oil and natural gas and in the level of environmental pollution worldwide. The decline in environmental resources combined with ecological threat to human security has led to a global environmental movement and considerable support for environmental protection.

In 1992 the Gallup Institute conducted an international survey of environmental concern in twenty-two countries. We have selected eleven of the twenty-two countries included in the original survey for further examination. The survey included a number of questions designed to measure the degree of environmental concern among the public in each country. One of the questions asked respondents whether they would be "willing to pay higher prices to protect the environment." The percentage of respondents who indicated that they would be willing to pay higher prices is presented in Table 8.1. Also presented are the mean, variance, and range for these data.

In examining Table 8.1 and the descriptive statistics, notice the variability in the percentage of citizens who are willing to pay higher prices to protect the environment. The percentage of respondents willing to pay higher prices ranges from a low of 30 percent in the Philippines to a high of 78 percent in Denmark.

One possible explanation for the differences is the economic conditions in these countries. Scholars of the environmental movement have argued that because of limited economic resources, citizens of developing and poorer countries cannot afford to pay for environmental protection. Therefore, there is less support for environmental protection in these countries.

---

[1]Refer to Paul Allison's *Multiple Regression: A Primer* (Thousand Oaks, CA: Pine Forge Press, 1999) for a complete discussion on multiple regression—statistical methods and techniques that consider the relationship between one dependent variable and one or more independent variables.

Table 8.1 **Percentage of Respondents Willing to Pay Higher Prices to Protect the Environment**

Country	Percentages (%)
Denmark	78
Norway	73
Korea	71
Switzerland	70
Chile	64
Canada	61
Ireland	60
Turkey	44
Russia	39
Japan	31
Philippines	30

$$\text{Mean} = \bar{Y} = \frac{\sum Y}{N} = \frac{621}{11} = 56.45$$

$$\text{Variance } Y = S_Y^2 = \frac{\sum (Y - \bar{Y})^2}{N - 1} = \frac{3,032.7}{10} = 303.27$$

Range $Y = 78\% - 30\% = 48\%$

Source: Adapted from Steven R. Brechin and Willett Kempton, "Global Environmentalism: A Challenge to the Postmaterialism Thesis?" *Social Science Quarterly 75*, no. 2 (June 1994): 245–266. Copyright © 1994 by the University of Texas Press. All rights reserved.

One important indicator of economic conditions is the GNP per capita in each country. Table 8.2 displays the GNP per capita, recorded in thousands of dollars, for each of the eleven countries surveyed. Note that GNP per capita ranges widely, from $700 (0.7 × 1,000) in the Philippines to $30,300 (30.3 × 1,000) in Switzerland.

■ ■ ■ ■ **The Scatter Diagram**

Let's examine the possible relationship between the interval-ratio variables *GNP per capita* and the *percentage willing to pay higher prices to protect the environment*. One quick visual method used to display such a relationship between two interval-ratio variables[2] is the **scatter diagram** (or **scatterplot**).

[2]We are interested in examining the *aggregate* relationship between GNP per capita and attitudes toward the environment.

Table 8.2 **GNP per Capita Recorded for Eleven Countries in 1992 (in $1,000)**

Country	GNP per Capita
Denmark	20.0
Norway	22.0
Korea	4.4
Switzerland	30.3
Chile	2.0
Canada	19.0
Ireland	8.0
Turkey	1.4
Russia	3.6
Japan	24.0
Philippines	0.7

$$\overline{X} = \frac{\sum X}{N} = \frac{135.4}{11} = 12.31$$

$$\text{Variance } X = S_X^2 = \frac{\sum (X - \overline{X})^2}{N - 1} = \frac{1,175.3}{10} = 117.52$$

Range $X = \$30.3 - \$0.7 = \$29.6$

Source: Adapted from Steven R. Brechin and Willett Kempton, "Global Environmentalism: A Challenge to the Postmaterialism Thesis?" *Social Science Quarterly* 75, no. 2 (June 1994): 245–266. Copyright © 1994 by the University of Texas Press. All rights reserved.

Often used as a first exploratory step in regression analysis, a scatter diagram can suggest whether two variables are associated.

The scatter diagram showing the relationship between willingness to pay higher prices for environmental protection and per-capita income for the eleven countries is shown in Figure 8.1. In a scatter diagram the scales for the two variables form the vertical and horizontal axes of a graph. Usually the independent variable, X, is arrayed along the horizontal axis and the dependent variable, Y, along the vertical axis. Because differences in GNP per capita are hypothesized to account for differences in the percentage of those willing to pay higher prices, GNP is assumed as the independent variable and is arrayed along the horizontal axis. Willingness to pay higher prices, the dependent variable, is arrayed along the vertical axis. In Figure 8.1, each dot represents a country; its location lies at the exact intersection of that country's GNP per capita and the percentage willing to pay higher prices.

Figure 8.1 **Scatter Diagram of GNP per Capita (in $1,000) and Percentage Willing to Pay More to Protect the Environment**

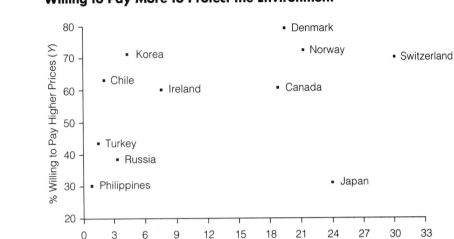

Source: Adapted from Steven R. Brechin and Willett Kempton, "Global Environmentalism: A Challenge to the Postmaterialism Thesis?" *Social Science Quarterly 75*, no. 2 (June 1994): 245–266. Copyright © 1994 by the University of Texas Press. All rights reserved.

> *Scatter diagram (scatterplot)* A visual method used to display a relationship between two interval-ratio variables.

Notice that there is an apparent tendency for countries with lower GNP per capita (for example, the Philippines and Russia) to also have a lower percentage of people willing to pay higher prices for environmental protection, whereas in countries with a higher GNP per capita (for example, Norway and Denmark) a higher percentage of people are willing to pay higher prices. In other words, we can say that GNP per capita and willingness to pay higher prices are *positively associated*. However, there are clearly exceptions to this pattern. For example, with one of the highest GNPs per capita ($24,000), Japan has one of the lowest percentages (31%) of citizens willing to pay higher prices for environmental protection. On the other hand, despite having one of the lowest GNPs per capita ($2,000), a relatively high percentage of Chilean citizens (60%) are willing to pay higher prices for environmental protection.

Scatter diagrams can also illustrate a negative association between two variables. For example, Figure 8.2 displays the association between GNP per capita and the percentage of respondents in fourteen countries who indicated a willingness to volunteer time to help protect the environment. Figure 8.2 suggests that low GNP per capita is associated with a higher

Figure 8.2 **GNP per Capita (in $1,000) and Percentage Willing to Volunteer Time for Environmental Protection**

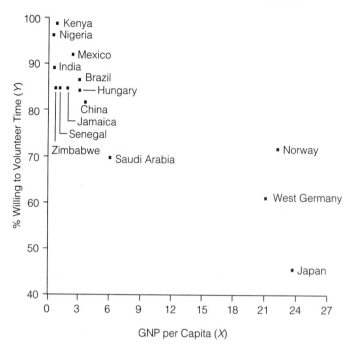

percentage of citizens willing to volunteer time for the environment. Conversely, high GNP per capita seems to be associated with a lower percentage of citizens willing to volunteer time for the environment. Figure 8.2 illustrates a *negative association* between GNP per capita and willingness to volunteer time to protect the environment.

■ ■ ■ ■ **Linear Relations and Prediction Rules**

Scatter diagrams provide a useful but only a preliminary step in exploring a relationship between two interval-ratio variables. We need a more systematic way to express this relationship. Let's examine Figures 8.1 and 8.2 again. Both allow us to see how two sets of measures of environmental concern are related to GNP per capita. The relationships displayed are by no means perfect, but the trends are apparent. In the first case (Figure 8.1), as GNP increases so does the percentage of respondents in most countries who are willing to pay higher prices to protect the environment. In the second case (Figure 8.2), as GNP increases the percentage of respondents who are willing to volunteer time decreases.

One way to evaluate these relationships is by expressing them as *linear relationships*. A **linear relationship** allows us to approximate the observations displayed in a scatter diagram with a straight line. In a perfectly linear relationship all the observations (the dots) fall along a straight line (a perfect relationship is sometimes called a **deterministic relationship**), and the line itself provides a predicted value of Y (the vertical axis) for any value of X (the horizontal axis). For example, in Figure 8.3 we have superimposed a straight line on the scatterplot originally displayed in Figure 8.1. Using this line, we can obtain a predicted value of the percentage willing to pay higher prices for any value of GNP, by reading up to the line from the GNP axis and then over to the percentage axis (indicated by the dotted lines). For example, the predicted value of the percentage willing to pay higher prices for a GNP of $8,000 is 54. Similarly, for a GNP of $24,000 we would get a predicted value of 63 percent willing to pay higher prices.

Linear relationship A relationship between two interval-ratio variables in which the observations displayed in a scatter diagram can be approximated with a straight line.

Deterministic (perfect) linear relationship A relationship between two interval-ratio variables in which all the observations (the dots) fall along a straight line. The line provides a predicted value of Y (the vertical axis) for any value of X (the horizontal axis).

Figure 8.3 **A Straight Line Graph for GNP per Capita (in $1,000) and Percentage Willing to Pay More to Protect the Environment**

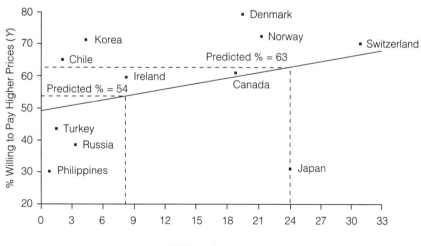

GNP per Capita (X)

As indicated in Figure 8.3, for the eleven countries surveyed, the actual relationship between GNP per capita and the percentage willing to pay higher prices is not perfectly linear. Although some of the countries lie very close to the line, none falls exactly on the line and some deviate from it considerably. Are there other lines that provide a better description of the relationship between GNP per capita and the percentage willing to pay higher prices?

In Figure 8.4 we have drawn two additional lines that approximate the pattern of relationship shown by the scatter diagram. In each case, notice that even though some of the countries lie close to the line, all fall considerably short of perfect linearity. Is there one line that provides the best linear description of the relationship between GNP per capita and the percentage willing to pay higher prices? How do we choose such a line? What are its characteristics? Before we describe a technique for finding the straight line that most accurately describes the relationship between two variables, we first need to review some basic concepts about how straight line graphs are constructed.

Learning Check.   *Use Figure 8.3 to predict the percentage willing to pay higher prices in a country with a GNP of $12,000 and one with a GNP of $27,000.*

Figure 8.4   **Alternative Straight Line Graphs for GNP per Capita (in $1,000) and Percentage Willing to Pay More to Protect the Environment**

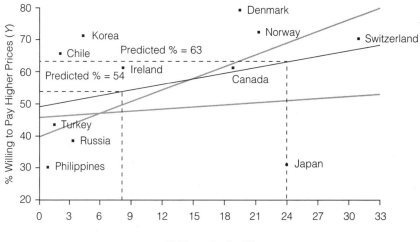

Table 8.3 **Seniority and Salary of Six Teachers (hypothetical data)**

Seniority (in years) X	Salary (in dollars) Y
0	12,000
1	14,000
2	16,000
3	18,000
4	20,000
5	22,000

Constructing Straight Line Graphs

To illustrate the fundamentals of straight line graphs, let's take a simple example. Suppose that in a local school system teachers' salaries are completely determined by seniority. New teachers begin with an annual salary of $12,000, and for each year of seniority their salary increases by $2,000. The seniority and annual salary of six hypothetical teachers are presented in Table 8.3.

Now let's plot the values of these two variables on a graph (Figure 8.5). Because seniority is assumed to determine salary, let it be our independent

Figure 8.5 **A Perfect Linear Relationship Between Seniority (in years) and Annual Salary (in $1,000) of Six Teachers (hypothetical)**

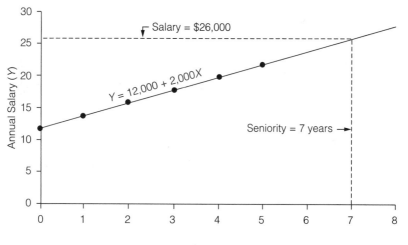

variable (X), and let's array it along the horizontal axis. Salary, the dependent variable (Y), is arrayed along the vertical axis. Connecting the six observations in Figure 8.5 gives us a straight line graph. This graph allows us to obtain a predicted salary value for any value of seniority level simply by reading from the specific seniority level up to the line and then over to the salary axis. For instance, we have marked the lines going up from a seniority of seven years and then over to the salary axis. We can see that a teacher with seven years of seniority makes $26,000.

The relationship between salary and seniority, as depicted in Table 8.3 and Figure 8.5, can also be described with the following algebraic equation:

$$Y = 12,000 + 2,000X$$

where

X = seniority (in years)

Y = salary (in dollars)

This equation allows us to correctly predict salary (Y) for any value for seniority (X) that we plug into the equation. For example, the salary of a teacher with five years of seniority is

$$Y = 12,000 + 2,000(5) = 12,000 + 10,000 = 22,000$$

Note that we can also plug in values of X that are not shown in Table 8.3. For example, the salary of a teacher with ten years of seniority is

$$Y = 12,000 + 2,000(10) = 12,000 + 20,000 = 32,000$$

The equation describing the relation between seniority and salary is an equation for a straight line. The equations for all straight line graphs have the same general form:

$$Y = a + bX \tag{8.1}$$

where

Y = the predicted score on the dependent variable

X = the score on the independent variable

a = the **Y-intercept,** or the point where the line crosses the Y-axis; therefore a is the value of Y when X is 0.

b = the **slope** of the regression line, or the change in Y with a unit change in X. For our example, a = 12,000 and b = 2,000. That is, a teacher will make $12,000 with 0 years of seniority, but then her or his salary will go up by $2,000 with each year of seniority.

Learning Check. *For each of these four lines, as X goes up by one unit, what does Y do? Be sure you can answer this question using both the equation and the line.*

Four Lines: Illustrating the Slope and the Y-Intercept

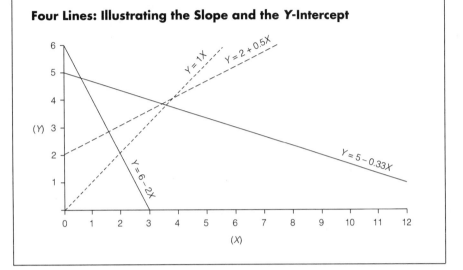

Slope (b) The change in variable Y (the dependent variable) with a unit change in variable X (the independent variable).

Y-intercept (a) The point where the line crosses the Y-axis, or the value of Y when X is 0.

Learning Check. *Use the linear equation describing the relationship between seniority and salary of teachers to obtain the predicted salary of a teacher with twelve (12) years of seniority.*

Finding the Best-Fitting Line

The straight line displayed in Figure 8.5 and the linear equation representing it ($Y = 12,000 + 2,000X$) provide a very simple depiction of the relationship between seniority and salary because salary (the Y variable) is completely determined by seniority (the X variable). When each value of Y is completely determined by X, all of the points (observations) lie on the line, and the relationship between the two variables is called a *deterministic relationship,* or a perfectly linear relationship.

However, most relationships we study in the social sciences are not deterministic, and we are not able to come up with a linear equation that allows us to predict Y from X with perfect accuracy. We are much more likely to find relationships approximating linearity, but in which numerous cases don't follow this trend perfectly. For instance, in reality teachers' salaries are not completely determined by seniority, and therefore knowing years of seniority will not provide us with a perfect prediction of their salary level.

When the dependent variable (Y) is not completely determined by the independent variable (X), not all (sometimes none) of the observations will lie exactly on the line. Look back at Figure 8.4, our example of the percentage willing to pay higher prices to protect the environment in relation to GNP per capital. Though each line represents a linear equation showing us how the percentage of citizens willing to pay higher prices rises with a country's GNP per capita, we do not have a perfect prediction in any of the lines. Although all three lines approximate the linear trend suggested by the scatter diagram, very few of the observations lie exactly on any of the lines and some deviate from them considerably.

Given that none of the lines is perfect, our task is to choose one line—the *best-fitting line*. But which is the best-fitting line?

Defining Error The best-fitting line is the one that generates the least amount of error. Let's think about how this error is defined. Look again at Figure 8.3. For each GNP level, the line (or the equation that this line represents) predicts a value of Y. Ireland, for example, with a GNP of $8,000 gives us a predicted value for Y of 54 percent. But the actual value for Ireland is 60 percent (see also Table 8.1). Thus, we have two values for Y: (1) a predicted Y, which we symbolize as $\hat{Y}$ and which is generated by the prediction equation, also called the *linear regression equation*

$$\hat{Y} = a + bX$$

and (2) the observed Y, symbolized simply as Y. Thus, for Ireland, $\hat{Y} = 54$ percent, whereas $Y = 60$ percent.

We can think of the error as the difference between the observed Y (Y) and the predicted Y ($\hat{Y}$). If we symbolize error as e, then

$$e = Y - \hat{Y}$$

The error for Ireland is $60 - 54$ percent = 6 percentage points.

The Sum of Squared Errors (Σe^2) We want a line or a prediction equation that minimizes e for each individual observation. However, any line we choose will minimize the error for some observations but may maximize it for others. We want to find a prediction equation that minimizes errors over all observations.

There are many mathematical ways of defining errors. For example, we may take the algebraic sum of errors $\Sigma(Y - \hat{Y})$, the sum of the absolute errors $\Sigma(|Y - \hat{Y}|)$, or the sum of the squared errors $\Sigma(Y - \hat{Y})^2$. For mathematical reasons, statisticians prefer to work with the third method—squaring and summing the errors over all observations. The result is the *sum of the squared errors*, or Σe^2. Symbolically, Σe^2 is expressed as

$$\Sigma e^2 = \Sigma \left(Y - \hat{Y} \right)^2$$

The Least-Squares Line   The best-fitting regression line is that line where the sum of the squared errors, or Σe^2, is at a minimum. Such a line is called the **least-squares line**, and the technique that produces this line is called the **least-squares method**. The technique involves choosing a and b for the equation $\hat{Y} = a + bX$ such that Σe^2 will have the smallest possible value. In the next section we use the data from the eleven countries to find the least-squares equation. But before we continue, let's review where we are so far.

Least-squares line (also called the *best-fitting line*)   A line where the sum of the squared errors, or Σe^2, is at a minimum.

Least-squares method   The technique that produces the least-squares line.

Review

1. We examined the relationship between GNP per capita and the percentage willing to pay higher prices to protect the environment, using data collected in eleven countries. We used the *scatter diagram (scatterplot)* to display the relationship between these variables.

2. The scatter diagram indicated that the relationship between these variables might be *linear*; as GNP per capita increases, so does the percentage of citizens willing to pay higher prices to protect the environment.

3. A more systematic way to analyze the relationship is to develop a *straight line equation* to predict the percentage willing to pay higher prices based on GNP. We saw that there are a number of straight lines that can approximate the data.

4. The *best-fitting line* is one that minimizes Σe^2. Such a line is called the *least-squares line*, and the technique that produces this line involves choosing the a and b for the equation $\hat{Y} = a + bX$ that minimize Σe^2.

Computing a and b for the Prediction Equation

Through the use of calculus it can be shown that to figure out the values of a and b in a way that minimizes Σe^2, we need to apply the following formulas:

$$b = \frac{S_{YX}}{S_X} \tag{8.2}$$

$$a = \overline{Y} - b(\overline{X}) \tag{8.3}$$

where

S_{YX} = the covariance of X and Y

S_X = the variance of X

$\overline{Y}$ = the mean of Y

$\overline{X}$ = the mean of X

a = the Y-intercept

b = the slope of the line

These formulas assume that X is the independent variable and Y is the dependent variable.

Before we compute a and b, let's examine these formulas. The denominator for b is the variance of the variable X. It is defined as follows:

$$\text{Variance } (X) = S_X^2 = \frac{\sum(X - \overline{X})^2}{N-1}$$

This formula should be familiar to you from Chapter 5. The numerator (S_{YX}), however, is a new term. It is the covariance of X and Y and is defined as

$$\text{Covariance } (X,Y) = S_{YX} = \frac{\sum(X - \overline{X})(Y - \overline{Y})}{N-1} \tag{8.4}$$

The covariance is a measure of how X and Y vary together. Basically, the covariance tells us to what extent higher values of one variable "go together" with higher values on the second variable (in which case we have a positive covariation) or with lower values on the second variable (which is a negative covariation). Take a look at this formula. It tells us to subtract the mean of X from each X score and the mean of Y from each Y score, and then take the product of the two deviations. The results are then summed for all the cases and divided by $N - 1$.

In Table 8.4 we show the computations necessary to calculate the values of a and b for our eleven countries. The means for GNP per capita and percentage willing to pay higher prices are obtained by summing column 1 and column 2, respectively, and dividing each sum by N. To calculate the covariance we first subtract $\overline{X}$ from each X score (column 3) and $\overline{Y}$ from each Y score (column 5) to obtain the mean deviations. We then multiply these deviations for every observation. The products of the mean deviations are shown in column 7. For example, for the first observation, Denmark, the mean deviation for GNP is 7.69 (20 – 12.31 = 7.69); for the percentage willing to pay higher prices it is 21.55 (78 – 56.45 = 21.55). The product of these deviations, 165.7 (7.69 × 21.55 = 165.7), is shown in column 7. The sum of

Table 8.4 **Worksheet for Calculating a and b for the Regression Equation**

Country	(1) GNP per Capita (X)	(2) % Willing to Pay (Y)	(3) (X − X̄)	(4) (X − X̄)²	(5) (Y − Ȳ)	(6) (Y − Ȳ)²	(7) (X − X̄)(Y − Ȳ)
Denmark	20.0	78	7.69	59.1	21.55	464.4	165.7
Norway	22.0	73	9.69	93.9	16.55	273.9	160.4
Korea	4.4	71	−7.91	62.6	14.55	211.7	−115.1
Switzerland	30.3	70	17.99	323.6	13.55	183.6	243.8
Chile	2.0	64	−10.31	106.3	7.55	57.0	−77.8
Canada	19.0	61	6.69	44.8	4.55	20.7	30.4
Ireland	8.0	60	−4.31	18.6	3.55	12.6	−15.3
Turkey	1.4	44	−10.90	119.0	−12.46	155.2	135.8
Russia	3.6	39	−8.71	75.9	−17.46	304.8	152.1
Japan	24.0	31	11.69	136.7	−25.47	648.7	−297.7
Philippines	.7	30	−11.61	134.8	−26.46	700.1	307.2
	ΣX=135.4	ΣY=621	0.00	1,175.3	0.00	3,032.7	689.5

$$\text{Mean } X = \bar{X} = \frac{\Sigma X}{N} = \frac{135.4}{11} = 12.31$$

$$\text{Mean } Y = \bar{Y} = \frac{\Sigma Y}{N} = \frac{621}{11} = 56.45$$

$$\text{Variance } (Y) = S_Y^2 = \frac{\Sigma (Y - \bar{Y})^2}{N - 1} = \frac{3,032.7}{10} = 303.27$$

$$\text{Standard deviation } (Y) = S_Y = \sqrt{303.27} = 17.41$$

$$\text{Variance } (X) = S_X^2 = \frac{\Sigma (X - \bar{X})^2}{N - 1} = \frac{1,175.3}{10} = 117.53$$

$$\text{Standard deviation } (X) = S_X = \sqrt{117.53} = 10.84$$

$$\text{Covariance } (X, Y) = S_{YX} = \frac{\Sigma (X - \bar{X})(Y - \bar{Y})}{N - 1} = \frac{689.5}{10} = 68.95$$

these products, shown at the bottom of column 7, is 689.5. Dividing it by 10 (N − 1), we get the covariance of 68.95.

The covariance is a measure of the linear relationship between two variables, and its value reflects both the strength and the direction of the relationship. The covariance will be close to zero when X and Y are unrelated;

it will be larger than zero when the relationship is positive, and smaller than zero when the relationship is negative.

Now let's substitute the values for the covariance and the variance from Table 8.4 to calculate b:

$$b = \frac{S_{YX}}{S_X^2} = \frac{68.95}{117.53} = 0.59$$

Once b has been calculated, finding a, the intercept, is simple:

$$a = \overline{Y} - b(\overline{X}) = 56.45 - 0.59(12.31) = 49.19$$

The prediction equation is therefore

$$\hat{Y} = 49.19 + 0.59(X)$$

This equation can be used to obtain a predicted value for the percentage of citizens willing to pay higher prices for environmental protection given a country's GNP per capita. For example, for a country with a GNP per capita of 2 (in $1,000), the predicted percentage is

$$\hat{Y} = 49.19 + 0.59(2) = 50.37\%$$

Similarly, for a country with a GNP per capita of 12 (in $1,000), the predicted value is

$$\hat{Y} = 49.19 + 0.59(12) = 56.27\%$$

Now we can plot the straight line graph corresponding to the regression equation. To plot a straight line we need only two points, where each point corresponds to an X,Y value predicted by the equation. We can use the two points we just obtained: (1) $X = 2$, $Y = 50.37$ and (2) $X = 12$, $Y = 56.27$. In Figure 8.6, the regression line is plotted over the scatter diagram we first displayed in Figure 8.1.

> **Learning Check.** *Use the prediction equation to calculate the predicted values of Y for Chile, Canada, and Japan. Verify that the regression line in Figure 8.6 passes through these points.*

Interpreting a and b

Now let's interpret the coefficients a and b in our equation. The b coefficient is equal to 0.59 percent. This tells us that the percentage of citizens willing to pay higher prices for environmental protection will increase by 0.59 percent for every increment of $1,000 in their country's GNP per capita. Similarly, an increase of $10,000 in a country's GNP corresponds to a 5.9 (0.59 × 10) in-

Figure 8.6 **The Best-Fitting Line for GNP per Capita and Percentage Willing to Pay More to Protect the Environment**

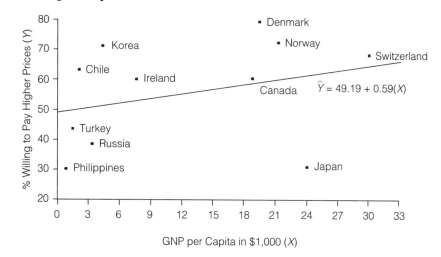

crease in the percentage of citizens willing to pay higher prices for environmental protection.

Note that because the relationships between variables in the social sciences are inexact, we don't expect our regression equation to make perfect predictions for every individual case. However, even though the pattern suggested by the regression equation may not hold for every individual country, it gives us a tool by which to make the best possible guess about how a country's GNP per capita is associated, *on average*, with the willingness of its citizens to pay higher prices for environmental protection. We can say that the slope of 0.59 percent is the estimate of this underlying relationship.

The intercept a is the predicted value of Y when $X = 0$. Thus, it is the point at which the regression line and the Y-axis intersect. With $a = 49.19$, a country with a GNP level equal to zero is predicted to have 49.19 percent of its citizens supporting higher prices for environmental protection. Note, however, that no country has a GNP as low as zero, although the Philippines, with a GNP of $700, comes pretty close to it. As a general rule, be cautious when making predictions for Y based on values of X that are outside the range of the data. Thus, when the lowest value for X is far above zero, the intercept may not have a clear substantive interpretation.

Calculating b Using a Computational Formula

The formula used to calculate b—the slope of the regression equation—can become cumbersome to use as N becomes larger. In most cases you will use a computer program to find a and b, but in situations when you have to

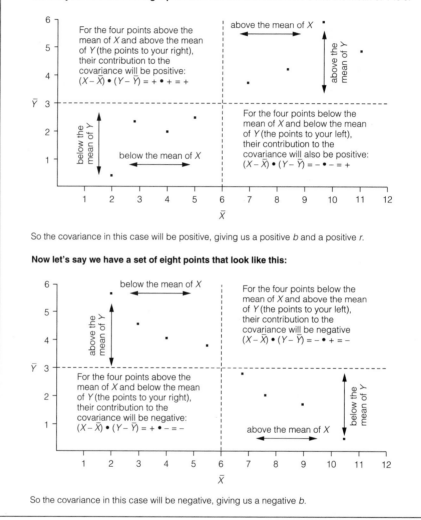

Box 8.1 Understanding the Covariance

Let's say we have a set of eight points for which the mean of *X* is 6 and the mean of *Y* is 3.

For the four points above the mean of *X* and above the mean of *Y* (the points to your right), their contribution to the covariance will be positive:
$(X - \bar{X}) \bullet (Y - \bar{Y}) = + \bullet + = +$

above the mean of *X*

above the mean of *Y*

$\bar{Y}$

below the mean of *Y*

below the mean of *X*

For the four points below the mean of *X* and below the mean of *Y* (the points to your left), their contribution to the covariance will also be positive:
$(X - \bar{X}) \bullet (Y - \bar{Y}) = - \bullet - = +$

$\bar{X}$

So the covariance in this case will be positive, giving us a positive *b* and a positive *r*.

Now let's say we have a set of eight points that look like this:

below the mean of *X*

above the mean of *Y*

For the four points below the mean of *X* and above the mean of *Y* (the points to your left), their contribution to the covariance will be negative
$(X - \bar{X}) \bullet (Y - \bar{Y}) = - \bullet + = -$

$\bar{Y}$

For the four points above the mean of *X* and below the mean of *Y* (the points to your right), their contribution to the covariance will be negative:
$(X - \bar{X}) \bullet (Y - \bar{Y}) = + \bullet - = -$

below the mean of *Y*

above the mean of *X*

$\bar{X}$

So the covariance in this case will be negative, giving us a negative *b*.

use a hand calculator to find *b*, you can use the following computational formula:

$$b = \frac{N\left(\sum XY\right) - \left(\sum X\right)\left(\sum Y\right)}{N\left(\sum X^2\right) - \left(\sum X\right)^2}$$

(8.5)

where

ΣX = the sum of X

ΣY = the sum of Y

ΣXY = the sum of the cross products of the X and Y scores

ΣX^2 = the sum of the squared X scores

Table 8.5 is a worksheet showing the calculations needed to determine b, using our data on GNP per capita and the percentage willing to pay higher prices for environmental protection and the computational formula.

Table 8.5　**Worksheet for Calculating a and b Using a Computational Formula**

Country	(1) GNP (in $1,000) X	(2) X²	(3) Percentage Willing to Pay Y	(4) XY	(5) Y²
Denmark	20.0	400.00	78	1,560.0	6,084
Norway	22.0	484.00	73	1,606.0	5,329
Korea	4.4	19.36	71	312.4	5,041
Switzerland	30.3	918.09	70	2,121.0	4,900
Chile	2.0	4.00	64	128.0	4,096
Canada	19.0	361.00	61	1,159.0	3,721
Ireland	8.0	64.00	60	480.0	3,600
Turkey	1.4	1.96	44	61.6	1,936
Russia	3.6	12.96	39	140.4	1,521
Japan	24.0	576.00	31	744.0	961
Philippines	0.7	0.49	30	21.0	900
	$\Sigma X = 135.4$	$\Sigma X^2 = 2,841.86$	$\Sigma Y = 621$	$\Sigma XY = 8,333.4$	$\Sigma Y^2 = 38,089$

Mean $X = \bar{X} = \dfrac{\Sigma X}{N} = \dfrac{135.4}{11} = 12.31$

Mean $Y = \bar{Y} = \dfrac{\Sigma Y}{N} = \dfrac{621}{11} = 56.45$

$b_{YX} = \dfrac{N\left(\Sigma XY\right) - \left(\Sigma X\right)\left(\Sigma Y\right)}{N\left(\Sigma X^2\right) - \left(\Sigma X\right)^2} = \dfrac{11(8,333.4) - (135.4)(621)}{11(2,841.86) - (135.4)^2}$

$= \dfrac{91,667.4 - 84,083.4}{31,260.46 - 18,333.16} = \dfrac{7,584}{12,927.3} = 0.59$

$a = \bar{Y} - b_{YX}(\bar{X}) = 56.45 - 0.59(12.31) = 56.45 - 7.26 = 49.19$

Box 8.2 A Note on Nonlinear Relationships

In analyzing the relationship between GNP per capita and percentage willing to pay higher prices for environmental protection, we have assumed that the two variables are linearly related. For the most part, social science relationships can be approximated using a linear equation. It is important to note, however, that sometimes a relationship cannot be approximated by a straight line and is better described by some other, nonlinear function. For example, the following scatter diagram shows a nonlinear relationship between age and hours of reading (hypothetical data). Hours of reading increase with age until the twenties, remain stable until the forties, and then tend to decrease with age.

A Nonlinear Relationship Between Age and Hours of Reading per Week

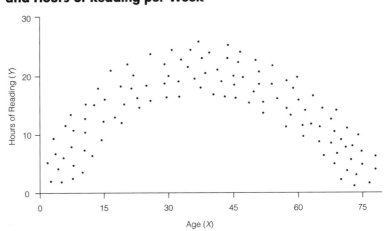

One quick way to find out whether your variables form a linear or a nonlinear pattern of relationship (or whether the variables are related at all!) is to make a scatter diagram of your data. If there is a significant departure from linearity, it would make no sense to fit a straight line to the data. Statistical techniques for analyzing nonlinear relationships between two variables are beyond the scope of this book. Nonetheless, at the very least, you should check for possible departures from linearity when examining your scatter diagram.

■ ■ ■ ■ Statistics in Practice: GNP and Willingness to Volunteer Time for Environmental Protection

In our ongoing example, we have looked at the association between GNP per capita and concern for the environment as measured by the percentage of people willing to pay higher prices to protect the environment. The regression equation we have estimated from data collected by the Gallup organization in eleven countries shows that as a country's GNP per capita rises, more citizens in that country are willing to pay higher prices to protect the environment.

What do these findings suggest? The conventional wisdom has been that citizens of poorer countries do not or cannot care about the environment. Indeed, our findings seem to suggest that people in wealthy countries hold stronger environmental values than those in poorer countries. But, one may ask, is the only reliable measure of concern for the environment the amount of money one is willing to pay to protect it?

In an attempt to challenge the conventional wisdom that people in poor countries lack environmental values, Brechin and Kempton[3] argue that even though few people within the poorest countries would offer monetary payment for anything, even for values they hold highly, they are equally or more likely to agree to commit their labor time. The researchers use the results of a survey collected by the Harris organization to examine this argument.[4] The Harris survey asked respondents if they would be willing to volunteer at least two hours each week for environmental protection. The percentage of citizens in fourteen countries who indicated that they would agree to such weekly labor requirements is presented in Table 8.6, together with the GNP per capita for each country. The scatter diagram for the data was displayed earlier, in Figure 8.2.

Let's examine Figure 8.2 once again. The scatter diagram seems to indicate that the two variables—the percentage agreeing to volunteer time and GNP per capita—are linearly related. It also illustrates that these variables are negatively associated; that is, as GNP per capita rises, the percentage of citizens willing to volunteer time to protect the environment declines.

For a more systematic analysis of the association we need to estimate the least-squares regression equation for these data. Since we want to predict the percentage willing to volunteer time, we treat this variable as our dependent variable (Y). We'll use the shortcut, the computational formula, to calculate the b coefficient.

Table 8.7 shows the calculations necessary to find b for our data on GNP per capita in relation to the percentage willing to volunteer time for environmental protection. Using the computational formula (8.5), let's substitute the values for $\sum XY$, $\sum X$, $\sum Y$, and $\sum X^2$ from Table 8.7 to calculate b:

$$b = \frac{N\left(\sum XY\right) - \left(\sum X\right)\left(\sum Y\right)}{N\left(\sum X^2\right) - \left(\sum X\right)^2} = \frac{14(5,384.07) - (83.55)(1,134.00)}{14(1,529.78) - (83.55)^2}$$

$$= \frac{75,376.98 - 94,745.70}{21,416.92 - 6,980.60} = \frac{-19,368.72}{14,436.32} = -1.34$$

[3]Steven R. Brechin and Willett Kempton, "Global Environmentalism: A Challenge to the Postmaterialism Thesis?" *Social Science Quarterly* 75, no. 2 (June 1994): 245–266.

[4]Both the Harris and the Gallup survey (discussed earlier in this chapter) examine the relationship between GNP and environmental concerns. However, because the two studies used different samples of countries, we need to be cautious about making generalizations based on looking at them jointly.

Table 8.6 **Percentage of Citizens Willing to Volunteer Time and GNP per Capita for Fourteen Countries**

Country	GNP (in $1,000) X	Percentage Willing to Volunteer Time Y
Kenya	.38	98
Nigeria	.25	95
Mexico	1.99	91
India	.35	89
Brazil	2.55	87
Zimbabwe	.64	85
Senegal	.65	85
Jamaica	1.26	85
Hungary	2.56	84
China	.36	83
Norway	21.85	76
Saudi Arabia	6.23	70
West Germany	20.75	62
Japan	23.73	44

The negative slope, −1.34, confirms our earlier impression; the relationship between wealth and willingness to volunteer time is negative. In other words, the higher the GNP per capita, the smaller the percentage of citizens in that country willing to volunteer time to protect its environment. A slope of −1.34 means that an increase of $1,000 in a country's GNP is associated with a decrease of 1.34 percent in the number of people willing to volunteer time for environmental protection.

Now let's find the intercept:

$$a = \bar{Y} - b(\bar{X}) = 81.0 - (-1.34)(5.97) = 81.0 + 8.0 = 89.0$$

The prediction equation is therefore

$$\hat{Y} = 89.0 - 1.34(X)$$

Figure 8.7 shows this regression line plotted over the scatter diagram we first displayed in Figure 8.2.

■ ■ ■ ■ **Methods for Assessing the Accuracy of Predictions**

So far we have developed two regression equations that are helping us to make predictions about people's willingness to contribute money or volunteer their time for environmental protection. But in both cases our predic-

Table 8.7 **GNP per Capita and Percentage Willing to Volunteer Time for Fourteen Countries**

Country	GNP (in $1,000) X	X²	Percentage Willing to Volunteer Time Y	XY	Y²
Kenya	.38	0.14	98	37.24	9,604
Nigeria	.25	0.06	95	23.75	9,025
Mexico	1.99	3.96	91	181.09	8,281
India	.35	0.12	89	31.15	7,921
Brazil	2.55	6.50	87	221.85	7,569
Zimbabwe	.64	0.41	85	54.40	7,225
Senegal	.65	0.42	85	55.25	7,225
Jamaica	1.26	1.59	85	107.10	7,225
Hungary	2.56	6.55	84	215.04	7,056
China	.36	0.13	83	29.88	6,889
Norway	21.85	477.42	76	1,660.60	5,776
Saudi Arabia	6.23	38.81	70	436.10	4,900
West Germany	20.75	430.56	62	1,286.50	3,844
Japan	23.73	563.11	44	1,044.12	1,936
	$\sum X =$ 83.55	$\sum X^2 =$ 1,529.78	$\sum Y =$ 1,134	$\sum XY =$ 5,384.07	$\sum Y^2 =$ 94,476

$$\text{Mean } X = \overline{X} = \frac{\sum X}{N} = \frac{83.55}{14} = 5.97$$

$$\text{Mean } Y = \overline{Y} = \frac{\sum Y}{N} = \frac{1,134}{14} = 81.0$$

tions are far from perfect. If we examine Figures 8.6 and 8.7, we can see that we fail to make accurate predictions in every case. Though some of the countries lie pretty close to the regression line, none lies directly on the line—an indication that some error of prediction was made. You must be wondering by now, "OK, I understand that the model helps us make predictions, but how can I assess the accuracy of these predictions?"

We saw earlier that one way to judge the accuracy of the predictions is to "eyeball" the scatterplot. The closer the observations are to the regression line, the better the "fit" between the predictions and the actual observations. Still we want a more systematic method for making such a judgment. We need a measure that tells us how accurate a prediction the regression model provides. The *coefficient of determination, or r^2,* is such a measure. It tells us how well the bivariate regression model fits the data. Both r^2 and r

Figure 8.7 **Regression Line for GNP per Capita (in $1,000) and Percentage Willing to Volunteer Time for Environmental Protection**

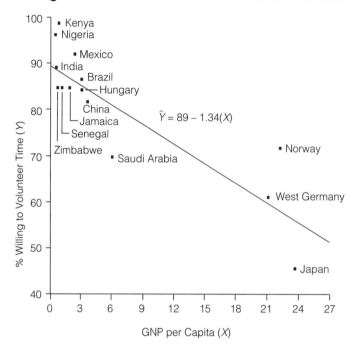

measure the strength of the association between two interval-ratio variables. Before we discuss these measures, let's first examine the notion of prediction errors.

Prediction Errors

Examine Figure 8.8. It displays the regression line for the variables *GNP per capita* (X) and the *percentage willing to pay higher prices to protect the environment* (Y). This regression line and the scatter diagram for the eleven countries surveyed by Gallup were originally presented in Figure 8.6.

In Figure 8.8 we consider the prediction of Y for one country, Norway, out of the eleven countries included in the survey. (The X and Y scores for all eleven countries, including Norway, are presented in Table 8.4.)

Suppose we didn't know the *actual* Y, the percentage of citizens in Norway who agreed to pay higher prices for environmental protection. Suppose further that we did not have knowledge of X, Norway's GNP per capita. Because the mean minimizes the sum of the squared errors for a set of scores, our best guess for Y would be the mean of Y, or = 56.45. The horizontal line in Figure 8.8 represents this mean. Now let's compare the actual Y, 73, with this prediction:

$$Y - \overline{Y} = 73 - 56.45 = 16.55$$

Figure 8.8 **Error Terms for One Observation**

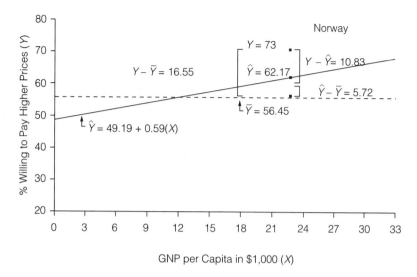

Obviously, with an error of 16.55, our prediction of the average score for Norway is not very accurate (this deviation of $Y - \overline{Y}$ is also illustrated in Figure 8.8).

Now let's see if our predictive power can be improved by using our knowledge of X—the GNP per capita of Norway—and its linear relationship with Y. If we plug Norway's GNP per capita of $22,000 into our prediction equation, as follows,

$$\hat{Y} = 49.19 + 0.59(X)$$
$$\hat{Y} = 49.19 + 0.59(22) = 62.17$$

we obtain a predicted Y of 62.17.

We can now recalculate our new error of prediction by comparing the predicted Y with the actual Y:

$$Y - \hat{Y} = 73 - 62.17 = 10.83$$

Although this prediction is by no means perfect, it is an improvement of 5.72 (16.55 − 10.83 = 5.72) over our earlier prediction.

This improvement is illustrated in Figure 8.8. Note that this improvement of 5.72 is equal to the quantity $Y - \overline{Y}$ (62.17 − 56.45 = 5.72). This quantity represents the improvement in the prediction error resulting from our use of the linear prediction equation.

Let's review what we have done. We have two prediction rules and two measures of error. The first prediction rule is in the absence of information on X, predict $\overline{Y}$. The error of prediction is defined as $Y - \overline{Y}$. The second rule of prediction uses X and the regression equation to predict Y. The error of prediction is defined as $Y - \hat{Y}$.

To calculate these two measures of error for all the cases in our sample, we square the deviations and sum them. Thus, for the deviation from the mean of Y we have

$$\sum\left(Y-\overline{Y}\right)^2$$

and to measure deviation from the regression line, or $\hat{Y}$, we have

$$\sum\left(Y-\hat{Y}\right)^2$$

(We discussed this error term, the sum of squared errors, earlier in the chapter.)

The Coefficient of Determination (r^2) as a PRE Measure   The coefficient of determination, r^2, is a PRE measure of association. We saw in Chapter 7 that all PRE measures adhere to the following formula:

$$\text{PRE} = \frac{E1 - E2}{E1}$$

where

$E1$ = prediction errors made when the independent variable is ignored

$E2$ = prediction errors made when the prediction is based on the independent variable

We have all the elements we need to construct a PRE measure. Because $\sum(Y - \overline{Y})^2$ measures the prediction errors when X is unknown, we can define

$$E1 = \sum(Y - \overline{Y})^2$$

Similarly, because $\sum(Y - \hat{Y})^2$ measures the prediction errors when X is used to predict Y, we can define

$$E2 = \sum(Y - \hat{Y})^2$$

We can now calculate the proportional reduction of error associated with using the linear regression equation as a rule for predicting Y:

$$\text{PRE} = r^2 = \frac{E1 - E2}{E1} = \frac{\sum\left(Y - \overline{Y}\right)^2 - \sum\left(Y - \hat{Y}\right)^2}{\sum\left(Y - \overline{Y}\right)^2} \tag{8.6}$$

The **coefficient of determination**, r^2, measures the proportional reduction of error that results from using the linear regression model. It reflects the proportion of the total variation in the dependent variable, Y, explained by the independent variable, X.

Figure 8.9 **Examples Showing r^2 Near 1.0 and Near 0**

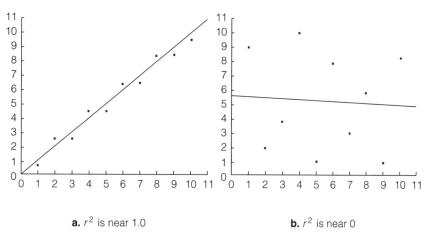

a. r^2 is near 1.0   **b.** r^2 is near 0

The coefficient of determination ranges from 0.0 to 1.0. An r^2 of 1.0 means that by using the linear regression model we have reduced uncertainty by 100 percent. It also means that the independent variable accounts for 100 percent of the variation in the dependent variable. With an r^2 of 1.0, all the observations fall along the regression line, and the prediction error $[\Sigma(Y - \hat{Y})^2]$ is equal to 0.0. An r^2 of 0.0 means that using the regression equation to predict Y does not improve the prediction of Y. Figure 8.9 shows r^2 values near 0.0 and near 1.0. In Figure 8.9a, where r^2 is approximately 1.0, the regression model provides a good fit. In contrast, a very poor fit is evident in Figure 8.9b, where r^2 is near zero. An r^2 near zero indicates either poor fit or a well-fitting line with a b of zero.

> *Coefficient of determination (r^2)*   A PRE measure reflecting the proportional reduction of error that results from using the linear regression model. It reflects the proportion of the total variation in the dependent variable, Y, explained by the independent variable, X.

Calculating r^2 An easier method for calculating r^2 uses the following equation:

$$r^2 = \frac{\left[\text{covariance}(X, Y)\right]^2}{\left[\text{variance}(X)\right]\left[\text{variance}(Y)\right]} = \frac{S_{YX}^2}{S_X^2 S_Y^2} \tag{8.7}$$

This formula tells us to divide the square of the covariance of X and Y by the product of the variance of X and the variance of Y.

To calculate r^2 for our example we can go back to Table 8.4, where the covariance and the variances for the two variables have already been calculated:

Figure 8.10 **A Pie Graph Approach to *r²***

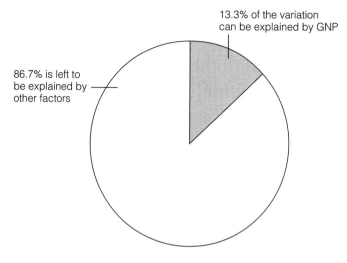

13.3% of the variation can be explained by GNP

86.7% is left to be explained by other factors

Variation in Percentage
Willing to Pay Higher Prices

$$S_{YX} = 68.95$$

$$S_X^2 = 117.53$$

$$S_Y^2 = 303.27$$

Therefore,

$$r^2 = \frac{(68.95)^2}{(117.53)(303.27)} = \frac{4,754.10}{35,643.32} = 0.133$$

An r^2 of 0.133 means that by using GNP per capita and the linear prediction rule to predict Y—the percentage willing to pay higher prices—we have reduced uncertainty of prediction by 13.3 percent (0.133×100). We can also say that the independent variable (GNP per capita) explains 13.3 percent of the variation in the dependent variable (the percentage willing to pay higher prices), as illustrated in Figure 8.10.

Pearson's Correlation Coefficient (*r*)

In the social sciences, it is the square root of r^2, or r—also called **Pearson's *r*** (Pearson's product moment correlation coefficient)—that is most often used as a measure of association between two interval-ratio variables:

$$r = \sqrt{r^2}$$

Pearson's r is usually computed directly[5] by using the following definitional formula:

$$r = \frac{\text{covariance}(X,Y)}{\left[\text{standard deviation}(X)\right]\left[\text{standard deviation}(Y)\right]} = \frac{S_{YX}}{S_X S_Y} \qquad (8.8)$$

Thus, r is defined as the ratio of the covariance of X and Y to the product of the standard deviations of X and Y.

Characteristics of Pearson's r Pearson's r is a measure of relationship or association for interval-ratio variables. It is called the correlation coefficient. Like gamma and Somers' d (introduced in Chapter 7), it ranges from 0.0 to ±1.0, with 0.0 indicating no association between the two variables. An r of +1.0 means that the two variables have a perfect positive association; –1.0 indicates that it is a perfect negative association. The absolute value of r indicates the strength of the linear association between two variables. Thus, a correlation of –0.75 demonstrates a stronger association than a correlation of 0.50. Figure 8.11 illustrates a strong positive relationship, a strong negative relationship, a moderate positive relationship, and a weak negative relationship.

Unlike the b coefficient, r is a symmetrical measure. That is, the correlation between X and Y is identical to the correlation between Y and X. In contrast, b may be different when the variables are switched—for example, when we use Y as the independent variable rather than as the dependent variable.

To calculate r for our example of the relationship between GNP per capita and the percentage of people who support higher prices for environmental protection, let's return to Table 8.4, where the covariance and the standard deviations for X and Y have already been calculated:

$$r = \frac{S_{YX}}{S_X S_Y} = \frac{68.95}{(10.84)(17.41)} = \frac{68.95}{188.72} = 0.365$$

A correlation coefficient of 0.365 indicates that there is a weak-to-moderate positive linear relationship between GNP per capita and the percentage of citizens who support higher prices for environmental protection.

Note that we could have just taken the square root of r^2 to calculate r, because $r = \sqrt{r^2}$, or $\sqrt{0.133} = 0.365$. Similarly, if we first calculate r, we can obtain r^2 simply by squaring r (be careful not to lose the sign of r).

---

[5]If you obtain r simply by taking the square root of r^2, make sure not to lose the sign of r (r^2 is always positive but r can also be negative), which can be ascertained by looking at the sign of S_{YX}.

Figure 8.11 **Scatter Diagrams Illustrating Weak, Moderate, and Strong Relationships as Indicated by the Absolute Value of *r***

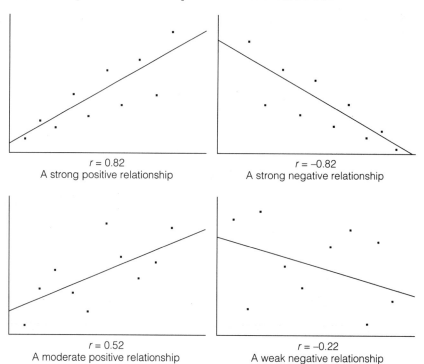

$r = 0.82$
A strong positive relationship

$r = -0.82$
A strong negative relationship

$r = 0.52$
A moderate positive relationship

$r = -0.22$
A weak negative relationship

> ***Pearson's correlation coefficient (r)*** The square root of r^2; it is a measure of association for interval-ratio variables, reflecting the strength of the linear association between two interval-ratio variables. It can be positive or negative in sign.

Calculating r Using a Computational Formula The definitional formula for *r* can become cumbersome to use as *N* becomes larger. In most cases, you will use a computer program to find *r*, but in situations when you have to use a hand calculator, the following computational formula is useful:

$$r = \frac{N\left(\sum XY\right) - \left(\sum X\right)\left(\sum Y\right)}{\sqrt{\left[N\left(\sum X^2\right) - \left(\sum X\right)^2\right]\left[N\left(\sum Y^2\right)\left(\sum Y\right)^2\right]}} \qquad (8.9)$$

where

$\sum X$ = the sum of *X*

ΣY = the sum of Y

ΣXY = the sum of the cross products of the X and Y scores

ΣX^2 = the sum of the squared X scores

ΣY^2 = the sum of the squared Y scores

The data and the calculations needed to calculate r for the variables *GNP per capita* and the *percentage willing to volunteer time for environmental protection* are shown in Table 8.7. The quantities displayed in the table can be substituted directly into Formula 8.9.

$$r = \frac{14(5,384.07) - (83.55)(1,134)}{\sqrt{\left[14(1,529.78) - (83.55)^2\right]\left[14(94,476) - (1,134)^2\right]}}$$

$$= \frac{75,376.98 - 94,745.7}{\sqrt{(21,416.92 - 6,980.60)(1,322,664 - 1,285,956)}}$$

$$= \frac{-19,368.72}{\sqrt{(14,436.32)(36,708)}} = \frac{-19,368.72}{\sqrt{529,928,434.6}} = \frac{-19,368.72}{23,020.17}$$

$$= -0.84$$

An r of -0.84 means that there is a strong negative relationship between GNP per capita and the percentage of citizens who are willing to volunteer time for environmental protection.

■ ■ ■ ■ **Statistics in Practice: Comparable Worth Discrimination**

In Chapter 1 we discussed the dual labor market theory as an explanation for the gender gap in earnings. According to the dual labor market theory, men and women are usually segregated into different types of work, with occupations in which the majority of workers are female usually paying less than occupations in which the majority of workers are male. A related explanation of the gender gap in earnings is the idea of comparable worth discrimination. The concept of comparable worth describes a process in which "employers underpay workers who are doing jobs that are different from predominantly male jobs but are of equal value. . . . As a result of comparable worth discrimination, the more female an occupation, the lower its average pay for both female and male workers, after taking into account such factors as education and experience."[6] The comparable worth hypothesis can be stated more succinctly as follows: In occupations that are of comparable worth, the higher the percentage of female workers, the lower is the average pay for that occupation.

[6]Barbara Reskin and Irene Padavic, *Women and Men at Work* (Thousand Oaks, CA: Pine Forge Press, 1994) p. 119.

Table 8.8   **Sex Composition and Salary of Jobs in a New York County, 1988**

Job Title	Percent Female	Average Salary
Psychologist II (less than full-time)	0	$34,914
Patrol lieutenant	0	32,445
Head groundskeeper II	0	25,140
Respiratory therapy technician	0	21,778
Automotive mechanic	0	21,778
Security aide (part-time)	0	17,205
Public health technician II	14	25,121
Management analyst	33	24,405
Assistant office machine operator	50	14,378
Municipal aide	59	12,716
Mental health worker I	67	14,365
Community service aide	75	14,716
Practical nurse	93	18,082
Community service worker II	100	15,054
Medical technologist	100	24,054
Principal records clerk	100	19,001
Senior accounting clerk/typist	100	15,054
Telephone operator/typist	100	13,100
Accounting clerk/typist	100	13,739

Source: Adapted from Linda J. Ames, "Erase the Bias: A Pay Equity Guide for Eliminating Race and Sex Bias for Wage-Setting Systems" (Washington, DC: National Committee on Pay Equity, 1993).

One way to examine the comparable worth discrimination hypothesis is to compare the pay of predominantly male and female jobs that experts judge to be of comparable worth. Table 8.8 shows the gender composition and average salary of jobs in a New York county in 1988. By eyeballing the data presented in the table, we can see clearly that salaries tend to go down as the percentage of women in the job increases.

Given that both *percent female* and *average salary* are interval-ratio variables, we can use bivariate regression analysis to examine the comparable worth hypothesis. Figure 8.12 shows the scatter diagram for percent female and average salary. Because we are assuming that the percentage of female workers in an occupation can predict the average salary for that occupation, we are going to treat percent female as our independent variable, X. Average salary, then, is our dependent variable, Y. The scatter diagram seems to suggest that the two variables are linearly related. It also illustrates that these variables are negatively associated; that is, as the percentage of women in an occupation rises, the average salary declines.

Figure 8.12   **Scatter Diagram for Percent Female and Average Salary**

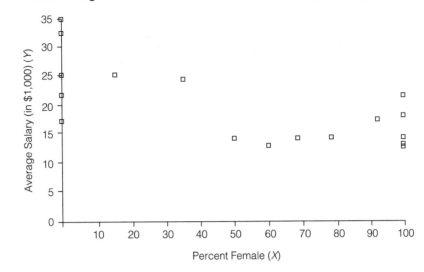

Computing *a* and *b* for the Prediction Equation

For a more systematic analysis of the association, we estimate the following linear regression equation:

$$Y = a + bX$$

where

Y = the average salary for a respective occupation

X = the percentage of women in a respective occupation

Table 8.9 shows the calculations necessary to find b for the New York county data. Using the computational formula for b (formula 8.5), let's substitute the values for ΣXY, ΣX, ΣY, and ΣX^2 from Table 8.9:

$$b = \frac{N\left(\sum XY\right) - \left(\sum X\right)\left(\sum Y\right)}{N\left(\sum X^2\right) - \left(\sum X\right)^2} = \frac{19(16,374.184) - (991)(377.045)}{19(86,029) - (991)^2}$$

$$= \frac{311,109.496 - 373,651.595}{1,634,551 - 982,081} = \frac{-62,542.099}{652,470} = -0.096$$

The negative slope, –0.096, confirms our earlier impression, based on the scatter diagram and Table 8.8, that the relationship between the percentage of women in an occupation and the average salary in that occupation is negative. In other words, the higher the percentage of women in an occupation, the lower the average salary in that occupation. A b equal to –.096 means that every 1 percentage point increase in the representation of

Table 8.9 **Worksheet for Calculating the Regression Equation for Percent Female and Average Salary**

Occupation	Percent Female X	X²	Average Salary (in $1,000) Y	Y²	XY
Psychologist II (less than full-time)	0	0	34.914	1,218.987	0.0
Patrol lieutenant	0	0	32.445	1,052.678	0.0
Head groundskeeper II	0	0	25.140	632.019	0.0
Respiratory therapy technician	0	0	21.778	474.281	0.0
Automotive mechanic	0	0	21.778	474.281	0.0
Security aide (part-time)	0	0	17.205	296.012	0.0
Public health technician II	14	196	25.121	631.064	351.694
Management analyst	33	1,089	24.405	595.604	805.365
Assistant office machine operator	50	2,500	14.378	206.726	718.900
Municipal aide	59	3,481	12.716	161.696	750.244
Mental health worker I	67	4,489	14.365	206.353	962.455
Community service aide	75	5,625	14.716	216.560	1,103.700
Practical nurse	93	8,649	18.082	326.958	1,681.626
Community service worker II	100	10,000	15.054	226.622	1,505.400
Medical technologist	100	10,000	24.054	578.594	2,405.400
Principal records clerk	100	10,000	19.001	361.038	1,900.100
Senior accounting clerk/typist	100	10,000	15.054	226.622	1,505.400
Telephone operator/typist	100	10,000	13.100	171.610	1,310.000
Accounting clerk/typist	100	10,000	13.739	188.760	1,373.900
	$\Sigma X =$ 991	$\Sigma X^2 =$ 86,029	$\Sigma Y =$ 377.045	$\Sigma Y^2 =$ 8,246.465	$\Sigma XY =$ 16,374.184

Mean $X = \overline{X} = \dfrac{\Sigma X}{N} = \dfrac{991}{19} = 52.157$

Mean $Y = \overline{Y} = \dfrac{\Sigma Y}{N} = \dfrac{377.045}{19} = 19.844$

Source: Adapted from Ames, 1993.

Figure 8.13   **Scatter Diagram Showing Regression Line for Percent Female and Average Salary**

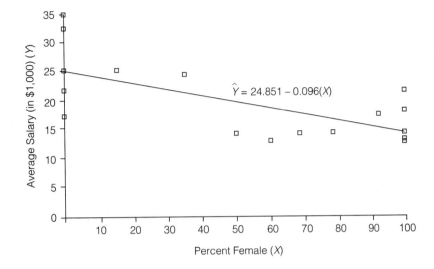

women in an occupation is associated with a decrease of about $96 (−0.096 × $1,000 = −$96) in the average salary for that occupation.

Now let's find the intercept:

$$a = \overline{Y} - b(\overline{X}) = 19.844 - (-.096)(52.157) = 19.844 + 5.007 = 24.851$$

The prediction equation is therefore

$$\hat{Y} = 24.851 - 0.096(X)$$

The regression line corresponding to this linear regression equation is shown in Figure 8.13.

Based on this linear regression equation, we could predict the average salary in an occupation that is, say, 25 percent female, to be

$$\hat{Y} = 24.851 - .096(25) = 24.851 - 2.4 = 22.451 \text{ or } \$22,451$$

In contrast, the average salary for occupations that are 75 percent female would be

$$\hat{Y} = 24.851 - 0.096(75) = 24.851 - 7.2 = 17.651 \text{ or } \$17,651$$

Computing r and r^2

We now find r and r^2 for these data. The quantities needed to calculate r are shown in Table 8.9. These quantities can be substituted directly into the computational formula for r (Formula 8.9):

$$r = \frac{N\left(\sum XY\right) - \left(\sum X\right)\left(\sum Y\right)}{\sqrt{\left[N\left(\sum X^2\right) - \left(\sum X\right)^2\right]\left[N\left(\sum Y^2\right) - \left(\sum Y\right)^2\right]}}$$

$$= \frac{19(16,374.184) - (991)(377.045)}{\sqrt{\left[19(86,029) - (991)^2\right]\left[19(8,246.465) - (377.045)^2\right]}}$$

$$= \frac{311,109.496 - 373,651.595}{\sqrt{(1,634,551 - 982,081)(156,682.835 - 142,162.932)}}$$

$$= \frac{-62,542.099}{\sqrt{(652,470)(14,519.902)}} = \frac{-62,542.009}{97,333.449} = -0.642$$

Next, r^2 can be easily calculated by squaring r:

$$r^2 = (-0.642)^2 = 0.412$$

An r of –0.642 means that the two variables—the percentage of women and average salary—are strongly associated. An r^2 of 0.412 means that by using the percentage of women to predict the average salary of occupations in one New York county, we have reduced the error of prediction by 41.2 percent. We can also say that the percentage of women in an occupation explains 41.2 percent of the variation in the average salary associated with that occupation.

■ ■ ■ ■ **Statistics in Practice: The Marriage Penalty in Earnings**

Among factors commonly associated with earnings are human capital variables (for example, age, education, work experience, and health) and labor market variables (such as the unemployment rate and the structure of occupations). Individual characteristics, such as gender, race, and ethnicity, also explain disparities in earnings. In addition, marital status has been linked to differences in earnings. However, though marriage is associated with higher earnings for men, for women it carries a penalty; married women tend to earn less at every educational level than single women.

The lower earnings of married women have been related to differences in labor force experience. Marriage and the presence of young children tend to limit women's choice of jobs to those that may offer flexible working hours but are generally low paying and offer fewer opportunities for promotion. Moreover, married women tend to be out of the labor market longer and have fewer years on the job than single women. When they re-enter the job market or begin their career after their children are grown, they compete with co-workers with considerably more work experience and on-the-job training. (Women who need to become financially independent after divorce or widowhood may share some of the same liabilities as married women.)

All of this suggests that the returns for formal education will be generally lower for married women. Thus, we would expect single women to earn more for each year of formal education than married women. We explore this issue by analyzing the bivariate relationship between level of education and personal income among single and married females (working full-time) who were included in the 1991 GSS sample. We are assuming that level of education (measured in years) can predict personal income, and therefore we treat education as our independent variable, X. Personal income (measured in dollars), then, is the dependent variable, Y. Since both are interval-ratio variables, we can use bivariate regression analysis to examine the difference in returns for education.

Our bivariate regression equation for single females working full-time is

$$\hat{Y} \text{ (single)} = -\$2,559.10 + \$2,948.47X$$

The regression equation tells us that for every unit increase in education—the unit is one year—we can predict an increase of \$2,948.47 in the annual income of single women in our sample who work full-time.

The bivariate regression equation for married females working full-time is

$$\hat{Y} \text{ (married)} = -\$1,892.02 + \$1,420.54X$$

The regression equation tells us that for every unit increase in education, we can predict an increase of \$1,420.54 in the annual income of married women in our sample who work full-time.

This analysis indicates that, as we suggested, the returns for education are considerably lower for married women. For every year of education, the earnings of single women increase more than twice as much as those of married women!

Let's use these regression equations to predict the difference in annual income between a single woman and a married woman, both with a high school (12 years) education and working full-time:

$$\hat{Y} \text{ (married)} = -\$1,892.02 + \$1,420.54(12) = \$15,154.46$$

$$\hat{Y} \text{ (single)} = -\$2,559.12 + \$2,948.47(12) = \$32,822.52$$

The predicted difference in annual income between a single and a married woman with a high school education, both working full-time, is \$17,668.06 (\$32,822.52 − \$15,154.46).

We also calculated the r and r^2 for these data. For married women $r = 0.336$; for single women, $r = 0.449$. These coefficients indicate that for both groups there is a moderate (the relationship is slightly stronger for single women) positive relationship between education and earnings.

To determine how much of the variation in income can be explained by education we need to calculate r^2. For married women,

$$r^2 \text{ (married)} = (r)^2 = (0.336)^2 = 0.113$$

and for single women,

$$r^2 \text{(single)} = (r)^2 = (0.449)^2 = 0.202$$

Using the regression equation, our prediction of income for married women is improved by 11.3 percent (0.113 × 100) over the prediction we would make using the mean alone. For single women there is a slightly better improvement in prediction, 20.2 percent (0.202 × 100).

Finally, the present analysis deals only with one factor affecting earnings—the level of education. Other important factors associated with earnings (occupation, seniority, race and ethnicity, age, and so on) need to be considered for a complete analysis of the differences in earnings between single and married women.

MAIN POINTS

- A scatter diagram (also called scatterplot) is a quick visual method used to display relationships between two interval-ratio variables. It is used as a first exploratory step in regression analysis and can suggest to us whether two variables are associated.

- Equations for all straight lines have the same general form

  $$Y = a + bX$$

  where

  Y = the predicted score on the dependent variable

  X = the score on the independent variable

  a = the Y-intercept, or the point where the line crosses the Y-axis; therefore, a is the value of Y when X is 0

  b = the slope of the line, or the change in Y with a unit change in X

- The best-fitting regression line is that line where the sum of the squared errors, or Σe^2, is the minimum. Such a line is called the least-squares line, and the technique that produces this line is called the least-squares method.

- The coefficient of determination (r^2) and Pearson's correlation coefficient (r) measure how well the regression model fits the data. Pearson's r also measures the strength of the association between the two variables. The coefficient of determination, r^2, can be interpreted as a PRE measure. It reflects the proportional reduction of error resulting from use of the linear regression model.

KEY TERMS

coefficient of determination (r^2)
deterministic (perfect) linear relationship
least-squares line (best-fitting line)

least-squares method
linear relationship
Pearson's correlation coefficient (r)

scatter diagram (scatterplot)   *Y-intercept (a)*

slope (b)

SPSS DEMONSTRATIONS

Demonstration 1: Producing Scatterplots (Scatter Diagrams)

Do people with more education work longer hours at their jobs? This question can be explored with SPSS using the techniques discussed in this chapter for interval-ratio data because *hours worked* and *education* are both coded at an interval-ratio level in the 1996 GSS file.

We begin by looking at a scatterplot of these two variables. The Scatter procedure can be found under the *Graphs* menu choice. In the opening dialog box, click on *Scatter* (which means we want to produce a standard scatterplot with two variables), then click on *Define*.

The Scatterplot dialog box (Figure 8.14) requires that we specify a variable for both the *X*- and *Y*-axes. We place EDUC (educational attainment) in the *X*-axis because we consider it the independent variable and HRS1 (number of hours worked) in the *Y*-axis because it is the dependent variable. Then click on *OK*.

Figure 8.14

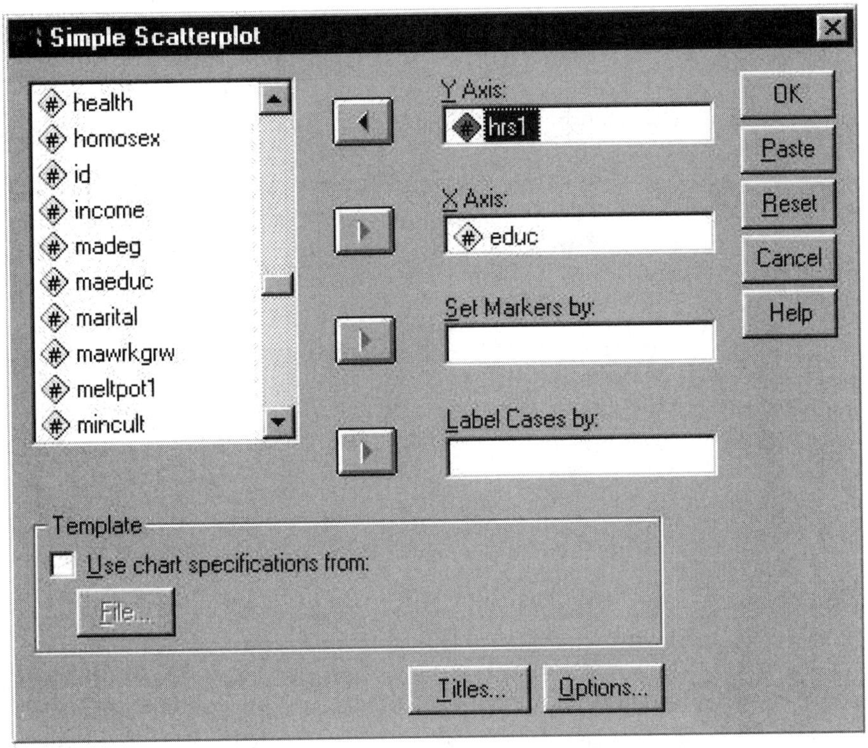

Figure 8.15

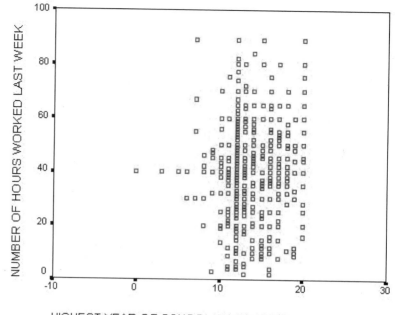

HIGHEST YEAR OF SCHOOL COMPLETED

SPSS creates the requested graph (Figure 8.15) and places it in a window called the Chart Carousel. This is a temporary holding area for graphs. From here charts can be saved, discarded, or edited. It is difficult to tell, by eye, whether or not there is a relationship between the two variables, so we will ask SPSS to place the regression line on the plot.

Clicking on the *Edit* button places a graphic in its own window and changes the menu and toolbar so that choices appropriate for the editing of graphics are available. To add a regression line to the plot, we click on *Chart* from the main menu, then *Options*. After a dialog box opens, we click on *Total* in the Fit Lines section. The result of these actions, plus a minor adjustment of the X-axis and the size of the symbols, is shown in Figure 8.16.

It is now easy to observe that there is a positive relationship between education and number of hours worked each week. However, the relationship doesn't appear to be strong. The predicted value for those with twenty years of education appears to be only a few hours more than for those with eight years of education.

Demonstration 2: Producing Correlation Coefficients

To further quantify the effect of education on hours worked, we request a correlation coefficient. This statistic is available in the Bivariate procedure, which is located by clicking on *Analyze, Correlate,* then *Bivariate* (Figure 8.17). Place the variables you are interested in correlating in the Variable(s) box, then click on *OK*.

Figure 8.16

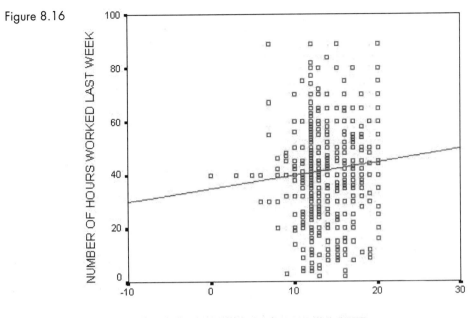

HIGHEST YEAR OF SCHOOL COMPLETED

Figure 8.17

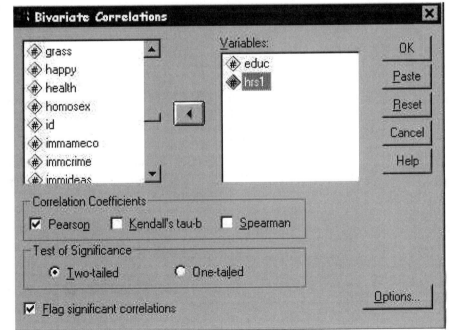

Figure 8.18

Correlations

		EDUC HIGHEST YEAR OF SCHOOL COMPLETED	HRS1 NUMBER OF HOURS WORKED LAST WEEK
EDUC HIGHEST YEAR OF SCHOOL COMPLETED	Pearson Correlation	1.000	.096**
	Sig. (2-tailed)	.	.003
	N	1419	953
HRS1 NUMBER OF HOURS WORKED LAST WEEK	Pearson Correlation	.096**	1.000
	Sig. (2-tailed)	.003	.
	N	953	958

**. Correlation is significant at the 0.01 level (2-tailed).

SPSS produces a matrix of correlations, shown in Figure 8.18. We are interested in the correlation in the bottom left-hand cell, .096. We see that this correlation is closer to 0 than to 1, which tells us that education is not a very good predictor of hours worked, even if it is true that those with more education work longer hours at their job. The number under the correlation coefficient, 953, is the number of valid cases (N)—those respondents who gave a valid response to both questions. The number is reduced because not everyone in the sample is working.

Demonstration 3: Producing a Regression Equation

As a final step, we will use SPSS to calculate the best-fitting regression line and the coefficient of determination. This procedure is located by clicking on *Analyze, Regression,* then *Linear.* The Linear Regression dialog box (Figure 8.19) provides boxes in which to enter the dependent variable and one or more independent variables (regression allows more than one). The Linear Regression dialog box offers many other choices, but the default output from the procedure contains all that we need.

SPSS produces a great deal of output, which is typical for many of the more advanced statistical procedures in the program. We've selected two portions of the output to review here (Figure 8.20). The remaining output from the Regression procedure is beyond the scope of this book. Under the Model Summary, the coefficient of determination is labeled "R Square." Its value is. 009, which is very weak. Educational attainment explains little of the variation in hours worked, only about 1%. This is probably not too sur-

Figure 8.19

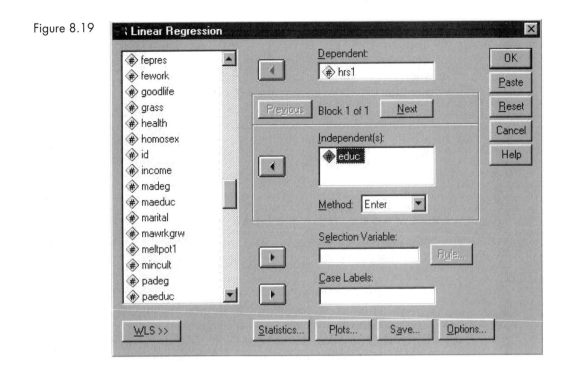

Figure 8.20

Model Summary

Model	R	R Square	Adjusted R Square	Std. Error of the Estimate
1	.096a	.009	.008	13.97

a. Predictors: (Constant), EDUC HIGHEST YEAR OF
   SCHOOL COMPLETED

Coefficientsa

Model		Unstandardized Coefficients		Standardized Coefficients	t	Sig.
		B	Std. Error	Beta		
1	(Constant)	34.905	2.355		14.820	.000
	EDUC HIGHEST YEAR OF SCHOOL COMPLETED	.496	.166	.096	2.982	.003

a. Dependent Variable: HRS1 NUMBER OF HOURS WORKED LAST WEEK

prising; for example, people who own a small business may have no more than a high school degree but work very long hours in their business.

The regression equation results are in the Coefficients section. The regression equation coefficients are listed in the column headed "B." The coefficient for EDUC, or b, is about .496; the intercept term, or a, identified in the "(Constant) " row, is 34.905. Thus, we would predict that every additional year of education increases the number of hours worked each week by a little over a half-hour. Or we could predict that those with a high school education work, on average, 34.905 + 12(.496) hours, or about 40.857 hours.

SPSS PROBLEMS [MODULE A]

1. Use the 1996 GSS data file to study the relationship between years of education (EDUC) and the prestige of the respondent's job (PRESTIGE80).
   a. Construct a scatterplot of these two variables in SPSS and place the best-fit linear regression line on the scatterplot. Describe the relationship between education and prestige.
   b. Have SPSS calculate the regression equation predicting prestige with education. What are the intercept and the slope? What are the coefficient of determination and the correlation coefficient?
   c. What is the predicted job prestige for someone with a college degree (16 years of education)?
   d. What is the predicted job prestige for someone with a graduate degree (20 years of education)?
   e. Can you find a way for SPSS to calculate the error of prediction and predicted value for each respondent and save them as new variables?

2. Use the same variables as in Exercise 1, but do the analysis separately for blacks and whites.
   a. Have SPSS calculate the regression equation for blacks and whites. How similar are they?
   b. What is the predicted job prestige for a black with 14 years of education? For a white respondent with the same amount of education? Which is greater?

3. Use the same variables as in Exercise 1, but do the analysis separately for women and men.
   a. Is there any difference between the regression equations for men and women?
   b. What is the predicted job prestige for women and men with the same amount of education: 12 years? 16 years? 20 years?

4. Use the 1996 GSS file to investigate the relationship between the respondent's education and the education received by his father and mother (PAEDUC and MAEDUC, respectively).

a. Construct scatterplots for these variables, using *mother's education* and *father's education* as predictor variables.
b. Use SPSS to find the correlation coefficient, the coefficient of determination, and the regression equation predicting the respondent's education with father's, then mother's, education. Which variable is the better predictor?
c. Do these same analyses separately for males and females. What differences, if any, do you find? Can male or female education be predicted better by parents' education?

CHAPTER EXERCISES

1. In Chapter 3, Exercise 6, we examined the distribution of pesticides in produce. Now we can use these same data to study the relationship between the percentage of produce with pesticides and the number of pesticides detected in each food type to see whether food types with more pesticides also have more different types of pesticides.
   a. Construct a scatterplot of the two variables, placing *number of types of pesticides detected* on the horizontal or X-axis and *percentage of produce with pesticides* on the vertical or Y-axis.
   b. Does the relationship between the two variables seem linear? Describe the relationship.
   c. Find the value of the Pearson correlation coefficient that measures the association between the two variables, and offer an interpretation.
2. There is often thought to be a relationship between a person's educational attainment and the number of children he or she has. The hypothesis is that as one's educational level increases, he or she has fewer children. Investigate this conjecture with twenty-five cases drawn randomly from the 1994 GSS file. The following table displays educational attainment, in years, and the number of children for each respondent.

EDUC	CHILDS	EDUC	CHILDS
12	0	16	0
12	4	16	2
12	3	18	2
18	2	10	3
12	3	19	1
12	3	17	2
18	0	18	2
12	4	16	3
12	3	12	0
17	0	11	3
16	0	18	2
10	2	12	3
16	0		

    a. Calculate the Pearson correlation coefficient for these two variables. Does its value support the hypothesized relationship?

    b. Calculate the least-squares regression equation using education as a predictor variable. What is the value of the slope, *b*? What is the value of the intercept, *a*?

    c. What is the predicted number of children for a person with a college degree (16 years of education)?

    d. Does any respondent actually have this number of children? If so, what is his or her level of education? If not, is this a problem or an indication that the regression equation you calculated is incorrect? Why or why not?

3. Births out of wedlock have been on the rise in the United States for many years. Discussions of this social phenomenon often focus on the greater number of births to unwed minority mothers. However, births to unwed white mothers have also increased. The following data show the percentage of births to mothers who were not married, separately for whites and nonwhites, over a forty-year period.

Percentage of Unwed Births by Year and Race

Year	Race White	Nonwhite
1950	1.8	18.0
1955	1.9	20.2
1960	2.3	21.6
1965	3.9	26.3
1970	5.7	37.6
1975	7.3	48.8
1980	11.0	55.3
1985	14.5	60.1
1989	19.2	65.7

Source: Data from the National Center for Health Statistics.

    a. Calculate the correlation coefficient between the percentage of unwed births for whites and minorities. What is its value?

    b. Provide an interpretation for the coefficient. Substantively, what does the value of the correlation coefficient imply about the similarity in the rate of increase of births out of wedlock for whites and nonwhites?

4. In Chapter 5, Exercise 9, we studied the variability of the rates of violent crime and incarceration in twenty-one states in the East and the Midwest. We've now been asked to investigate the hypothesis that the number of prisoners is related to the crime rate because states with higher crime rates are likely to have higher rates of incarceration.
   a. Construct a scatter diagram of violent crime rate and incarceration rate, with *crime rate* considered the predictor variable. What can you say about the relationship between these two variables based on the scatterplot?
   b. Find the least-squares regression equation that predicts incarceration rate from the crime rate. What is the slope? What is the intercept?
   c. Calculate the coefficient of determination (r^2), and provide an interpretation.
   d. If the crime rate increased by 100 for a state (that is, 100 more crimes per 100,000 people), by how much would you predict the incarceration rate to increase?
   e. Does it make sense to predict the incarceration rate when the rate of violent crime is equal to zero? Why or why not?

5. Before calculating a correlation coefficient or a regression equation, it is always important to examine a scatter diagram between two variables to see how well a straight line fits the data. If a straight line does not appear to fit, other curves can be used to describe the relationship (this subject is not discussed in our text).

   The following SPSS scatterplot and output display the relationship between education (measured in years) and hours of television viewing (measured in hours) based on GSS1996 data. We can hypothesize that as educational attainment increases, hours of television viewing will decrease, indicating a negative relationship between the two variables.

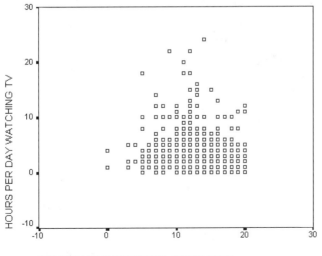

Coefficients[a]

Model		Unstandardized Coefficients		Standardized Coefficients	t	Sig.
		B	Std. Error	Beta		
1	(Constant)	5.647	.248		22.800	.000
	EDUC HIGHEST YEAR OF SCHOOL COMPLETED	-.201	.018	-.244	-11.095	.000

a. Dependent Variable: TVHOURS HOURS PER DAY WATCHING TV

Model Summary[b]

Model	R	R Square	Adjusted R Square	Std. Error of the Estimate
1	.244[a]	.060	.059	2.31

a. Predictors: (Constant), EDUC HIGHEST YEAR OF SCHOOL COMPLETED

b. Dependent Variable: TVHOURS HOURS PER DAY WATCHING TV

   a. Assess the relationship between the two variables based on the scatterplot and output for r^2. Is the relationship between these two variables as hypothesized? Is it a negative or a positive relationship?

   b. Describe the relationship between these two variables, using representative values of years of education and hours of television viewing. For example, if an individual has 16 years of education, what is the predicted hours of television viewing? How can you determine this?

   c. Does a straight line adequately represent the relationship between these two variables? Why or why not?

6. Based on countries in South America (ignoring a few small political entities, such as Surinam in northeast South America), let's analyze the relationship between GNP and infant mortality rate (IMR).

   a. Construct a scatterplot from the following data, predicting IMR from GNP. What is the relationship between GNP and IMR for countries in South America?

   b. Does it appear that a straight line fits these data better than for all countries in the world? Why or why not?

   c. Calculate the correlation coefficient and coefficient of determination. Do these values offer further support for your answer to (b)? How?

   d. If you think a linear relationship is a reasonable description of the scatterplot of GNP and IMR for South American countries, why do you think that is not true for the whole world?

Country	GNP per Capita in 1982 (dollars)	Infant Mortality in 1983 (per 1,000 births)
Argentina	2,520	35.3
Bolivia	570	124.0
Brazil	2,240	71.0
Chile	2,210	23.6
Colombia	1,460	53.0
Ecuador	1,350	70.0
Paraguay	1,610	45.0
Peru	1,310	99.0
Uruguay	2,650	33.2
Venezuela	4,140	39.0

Source: Data from the Population Reference Bureau and the World Bank.

7. Social scientists have long been interested in various civic, labor, and fraternal organizations in the United States and the role they play in the political, civic, and social life of the country. Alexis de Tocqueville, for example, commented in the nineteenth century about the propensity of Americans to join organizations and work together for the common good. The GSS 1994 data set has information on the total number of formal memberships held by each respondent in various organizations. Use this information for the following selected subsample of respondents to see whether those with more education are more likely to have more organizational memberships.

Highest Year of School Completed	Number of Memberships
12	0
16	5
16	1
16	1
11	1
18	0
13	5
10	1
17	1
12	3
14	3
11	2

a. Construct a scatterplot, predicting the number of memberships with education.

b. Calculate the regression equation with *education* as the predictor variable, and draw the regression line on the scatterplot. What is the slope? What is the intercept? Describe how the straight line "fits" the data? Does it fit better for those with more or less education?

c. What is the error of prediction for the second case (the person with 16 years of education and 5 memberships)? What is the error for the person with 10 years of education and 1 membership?

d. What is the predicted number of memberships for someone with 14 years of education? with 4 years of education? Any problems with these predictions?

e. Calculate the mean number of years of education and the mean number of memberships. Plot this point on the scatterplot. Where does it fall? Can you think of a reason why this should be true?

8. In Exercise 6, we investigated the relationship between infant mortality rate and GNP in South America. The birth rates (number of live births per 1,000 inhabitants) in these same countries are shown in the following table:

Country	Birth Rate in 1982
Argentina	24
Bolivia	42
Brazil	31
Chile	24
Colombia	28
Ecuador	41
Paraguay	35
Peru	37
Uruguay	18
Venezuela	33

Source: Data from the Population Reference Bureau.

a. Construct a scatterplot for birth rate and GNP and one for birth rate and infant mortality rate. Do you think each can be characterized by a linear relationship?

b. Calculate the coefficient of determination and correlation coefficient for each relationship.

c. Use this information to describe the relationship between the variables.

9. Minorities are typically arrested at greater rates than whites, given their proportion in the population. Is it possible that the arrest ratio of minori-

ties to whites is related to the percentage of a city's population that is minority? To answer this question, study the following data for eighteen cities with a population over 100,000. Listed in the table is the percentage of the city's population that is black and the arrest ratio for drug crimes (for example, an arrest ratio of 5 means that blacks are five times as likely to be arrested as whites).

a. Construct a scatterplot, predicting the arrest ratio with the percent black population in a city. Does it appear that a straight line relationship will fit the data?

b. Calculate the regression equation with *percent black* as the predictor variable, and draw the regression line on the scatterplot. What is its slope? What is the intercept? Has your opinion changed about whether a straight line seems to fit the data? Are there any cities that fall far from the regression line? Which one(s)?

c. What percentage of the population must be black to obtain a predicted value of 1 for the arrest ratio?

d. Predicting a value that falls beyond the observed range of the two variables in a regression is problematic at best, so your answer in (c) isn't necessarily statistically believable. What is a nonstatistical, or substantive, reason why the prediction for an arrest ratio of 1 may be nonsensical?

	Percent Black Population	Arrest Ratio
Livonia, Michigan	0	43
Warren, Michigan	1	32
Pasadena, Texas	1	27
St. Paul, Minnesota	7	26
Minneapolis, Minnesota	13	22
Madison, Wisconsin	4	21
Alexandria, Virginia	22	18
Columbus, Ohio	23	18
Evansville, Indiana	10	17
Hialeah, Florida	2	17
Sterling Heights, Michigan	0	16
Grand Rapids, Michigan	19	15
Pittsburgh, Pennsylvania	26	14
Little Rock, Arkansas	34	13
Peoria, Illinois	21	13
Seattle, Washington	10	13
Rockford, Illinois	15	12
Lansing, Michigan	19	12

Source: Data from *USA Today,* July 26, 1993; calculated from the FBI's Uniform Crime Report.

10. What can be said about the relationship between family planning utilization and fertility rates? Data from the Population Reference Bureau summarize rates from selected regions in the 1990s.

Region	Total Fertility Rate	Percent Contraceptive Use
Eastern Africa	6.2	17
Western Africa	6.2	14
Northern Asia	4.2	40
Northern Africa	4.0	42
South Central Africa	3.8	41
Southern Africa	3.4	50
Southeast Asia	3.2	51
Latin America and Caribbean	3.0	62
North America	1.9	76
East Asia	1.8	81
Northern Europe	1.7	68
Western Europe	1.5	75
Eastern Europe	1.4	57

Source: Adapted from Population Reference Bureau, "The World's Women: Making Gains but Still Widely Disadvantaged, " Press release for 1998 Women of the World. Used by permission.

a. Construct a scatterplot, predicting the fertility rate based on percentage of contraceptive use in each region. How would you characterize the relationship between the two variables?
b. Calculate the regression equation with *percent contraceptive use* as the predictor variable, and draw the regression line on the scatterplot. What is its slope? What is the intercept?
c. Summarize your findings.

9 Organization of Information and Measurement of Relationships: A Review of Descriptive Data Analysis

■ ■ ■ ■ **Introduction**[1]

In the preceding eight chapters we have introduced you to methods researchers use to organize data and describe relationships between variables. We have presented the chapters sequentially to reflect five cumulative stages of data analysis:

1. Organizing the data using frequency distributions (Chapter 2)
2. Displaying the data using graphic techniques (Chapter 3)
3. Determining what is average or typical about a distribution (Chapter 4)
4. Determining variation within a distribution (Chapter 5)
5. Measuring the association between two variables (Chapters 6, 7, and 8)

Within each chapter, we have described alternative methods used in each stage, depending on the level of data measurement (nominal, ordinal, or interval-ratio) and the purpose of the analysis.

At each of the stages, researchers must decide which technique is appropriate for the data and the research goal. Figure 9.1 presents a flowchart that you can use in this decision-making process. It shows the techniques available at each stage for each of the levels of data—nominal, ordinal, and interval-ratio—and the circumstances that determine the appropriate technique.

The flowchart provides the map for this review chapter. In the sections that follow, we examine real research reports that illustrate the use of one or more of the methods appropriate for the level of data under discussion. The presentation of each application follows the five cumulative steps of descriptive statistical analysis.

■ ■ ■ ■ **Descriptive Data Analysis for Nominal Variables**

We begin our review with the lowest level of measurement—the nominal level. Recall that variables measured at the nominal level are categorized by qualitative differences. *Gender, race, religious preference,* and *political party* are examples of nominal variables. The categories of nominal variables are discrete, and although we can say that each category is different, we cannot measure the difference between the categories quantitatively.

Statistics in Practice: Gender and Local Political Party Activism

The 1992 presidential election campaign brought the role of women in politics to the attention of the public and created renewed interest in the subject among social scientists. In Chapter 1 we suggested that research questions are frequently derived through familiarity with professional literature. In

---

[1]This chapter was co-authored with Pat Pawasarat.

Figure 9.1 **Flowchart of the Systematic Approach to Descriptive Data Analysis**

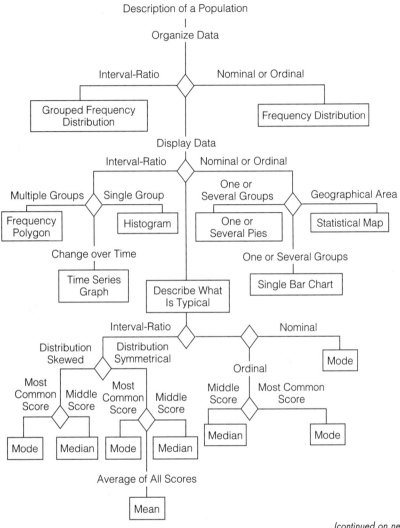

(continued on next page)

1994 Laura van Assendelft and Karen O'Connor[2] examined existing literature on women in politics and discovered the following: Most studies
found that women lack the political ambition to run for public office. The
majority of women who do run for office have been active in local political
party organizations. Male and female party activists differ in many ways,

---

[2]Laura van Assendelft and Karen O'Connor, "Backgrounds, Motivations, and Interests:
A Comparison of Male and Female Local Party Activists," *Women & Politics* 14, no. 3 (1994):
77–91.

Figure 9.1 **Flowchart of the Systematic Approach to Descriptive Data Analysis (continued)**

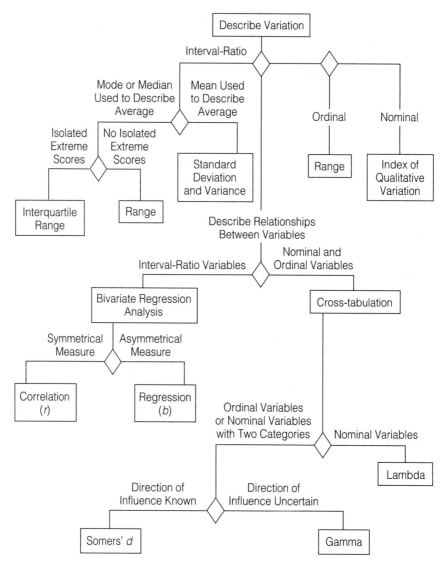

including in political ambition, education, and employment experience. They also found that most of the research in this area was nearly twenty years old and did not take into account the effects of the women's rights movement in recent years.

Van Assendelft and O'Connor reasoned that since local party activity is an apparent stepping-stone to political office for women, it was important to reanalyze women's activity in local party organizations. The purpose of their research was to describe who party activists are, the reasons for their participation, and the similarities and differences between male and female party activists. Based on the existing literature, they hypothesized that female activists differ from male activists in level of education and income, employment experience, and political ambition, as well as in their reasons for participating in political activity. They obtained their data using a mail survey sent to all members of the local Democratic and Republican parties in the metropolitan Atlanta, Georgia, area.

Van Assendelft and O'Connor used eight dependent variables in their study. Two—occupation and marital status—were measured at the nominal level. We will discuss these variables in this section. The other six—level of education, strength of party identification, motivation, years of activism, hours spent on party work, and political ambition—were measured at the ordinal level and will be discussed later. Here we will follow the five steps of descriptive data analysis as we discuss their findings.

Learning Check. *What is the independent variable in van Assendelft and O'Connor's study?*

Organize the Data into a Frequency Distribution Table 9.1 shows the distribution by occupation for men and women. Notice that the table entries are percentages rather than raw frequencies. We use percentages to compare

Table 9.1 **Occupational Differences Among Male and Female Local Party Activists (in percentages)**

Occupation	Males	Females
Small business	21.7	18.6
Large business	31.8	29.4
Education	3.8	13.7
Government	5.7	5.9
Self-employed (business)	13.4	12.7
Self-employed (professional)	21.7	4.9
Homemaker	0.0	12.7
Student	1.9	2.0
Total	100.0	99.9
(N)	(157)	(102)

Source: Laura van Assendelft and Karen O'Connor, "Backgrounds, Motivations, and Interests: A Comparison of Male and Female Local Party Activists," *Women & Politics* 14, no. 3 (1994): 77–91. Used by permission of Haworth Press, Inc., Binghamton, NY.

groups with unequal Ns. Also notice that the total percentage for females adds up to only 99.9 percent. The "missing" 0.1 percent is due to rounding. Small differences such as this frequently occur in real research applications.

An examination of Table 9.1 shows us that, in general, the occupational differences between men and women activists are not great. Van Assendelft and O'Connor note two exceptions: A much higher percentage of women are employed in the educational sector, and women are underrepresented in the self-employed professional sector. The authors point out that the self-employed professional sector includes lawyers—a profession from which many political candidates are drawn—and imply that more women will emerge as candidates as their representation in this sector increases.[3] A third category that shows differences, homemaker, is not discussed by the researchers in this study. Notice that 12.7 percent of the women and 0.0 percent of the men fall into this category.

Display the Data in a Graph Figure 9.2 graphically represents the data shown in Table 9.1. We have chosen to represent occupational differences in a bar graph rather than a pie chart because the number of categories would make a pie chart somewhat difficult to read. The bar graph allows direct comparison of the percentages of males and females in each occupational category, making it easy to identify the categories where the percentage difference is relatively large—education and homemaker (both higher percentages of women) and self-employed professional (a higher percentage of men).

Describe What Is Average or Typical of a Distribution Because these are nominal variables, we can only use the mode to describe the average. Looking at Table 9.1, we can see that the modal occupational sector for both men and women is "large business."

Describe Variability Within a Distribution For nominal variables, we can use the index of qualitative variation (IQV) to describe variation within a distribution and to compare distributions. Table 9.2 shows the distribution by marital status of male and female party activists. Let's calculate the IQVs for this distribution:

$$IQV = \frac{\sum f_i f_j}{\frac{K(K-1)}{2}\left(\frac{N}{K}\right)^2}$$

$$IQV \text{ (men)} = \frac{84.5 \times 15.5}{\frac{2(2-1)}{2}\left(\frac{100}{2}\right)^2} = \frac{1,309.75}{2,500} = 0.52$$

[3]Ibid., pp. 80–81.

Figure 9.2 **Occupational Differences Between Male and Female Local Party Activists**

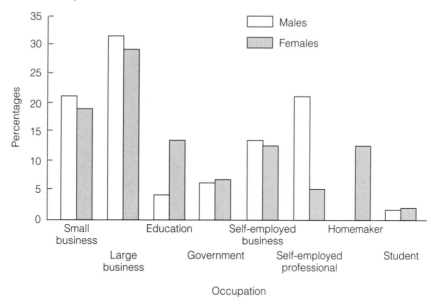

$$\text{IQV (women)} = \frac{68.6 \times 31.4}{\dfrac{2(2-1)}{2}\left(\dfrac{100}{2}\right)^{2}} = \frac{2,154.04}{2,500} = 0.86$$

The comparison of the IQVs for men (0.52) and for women (0.86) shows that there is considerably more variability in marital status among female party activists than among the males.

> **Learning Check.** *Where did the numbers used to calculate the IQVs for marital status come from? The numbers come from the percentage distributions displayed in Table 9.2. Remember that the IQV can be computed using either frequency or percentage distributions. Practice calculating IQVs using Table 9.1, which shows occupational differences between male and female activists. The IQV for occupational differences among males is 0.87; for females it is 0.94. What do these IQVs tell us about variability within and between the distributions?*

Describe the Relationship Between Two Variables Bivariate tables such as Table 9.2 are typically constructed with the independent variables as the column variables and the dependent variables as the row variables. To perform a

Table 9.2 **Marital Status of Male and Female Local Party Activists**

	GENDER		
MARITAL STATUS	Males	Females	Total
Married	84.5% (136)	68.6% (70)	78.3% (206)
Single	15.5% (25)	31.4% (32)	21.7% (57)
Total (N)	100.0% (161)	100.0% (102)	100.0% (263)

Source: Adapted from Laura van Assendelft and Karen O'Connor, "Backgrounds, Motivations, and Interests: A Comparison of Male and Female Local Party Activists," *Women & Politics* 14, no. 3, 1994: 77–91. Used by permission of Haworth Press, Inc., Binghamton, NY.

cross-tabulation, we then compare the rows. Table 9.2 shows us that female party activists are much more likely to be single than male party activists (31.4% of females vs. 15.5% of males).

Notice that the mode in both distributions is "married"; thus, we cannot use lambda to summarize the relationship between the variables. We will rely on cross-tabulation to describe the relationship between gender and marital status.

■ ■ ■ ■ **Descriptive Data Analysis for Ordinal Variables**

The next highest level of measurement is the ordinal level. As with nominal variables, the categories of ordinal variables are discrete; unlike with nominal variables, the categories of ordinal variables can be ranked or ordered from high to low, or vice versa. Though there is a quantitative difference between the categories of an ordinal variable (upper class indicates higher status than lower class), the magnitude of difference is not known.

Gender and Local Political Party Activism: Continuing Our Research Example

Let's return to van Assendelft and O'Connor's study of gender and political party activism and examine some of the variables measured at the ordinal level.

Organize the Data into a Frequency Distribution   Tables 9.3, 9.4, and 9.5 show the frequency distributions for education, strength of party identification, and years active in the political organization by gender. Table 9.3 shows that the percentage of male and female party activists who have completed college is nearly equal (35.8% of males and 35.0% of females). However, a much higher percentage of males than females have completed graduate school (43.2% vs. 29.1%). The cumulative percentages also show that fewer male than female political party activists have a high school education or less (3.7% vs. 12.6%).

Table 9.3   **Educational Differences Between Male and Female Local Party Activists**

Level of Education	Males		Females	
	%	C%	%	C%
Less than high school	1.8	1.8	0.0	0.0
High school	1.9	3.7	12.6	12.6
Some college	17.3	21.0	23.3	35.9
College	35.8	56.8	35.0	70.9
Graduate school	43.2	100.0	29.1	100.0
Total	100.0		100.0	
(N)	(162)		(103)	

Source: Laura van Assendelft and Karen O'Connor, "Backgrounds, Motivations, and Interests: A Comparison of Male and Female Local Party Activists," *Women & Politics* 14, no. 3 (1994): 77–91. Used by permission of Haworth Press, Inc., Binghamton, NY.

Table 9.4   **Strength of Party Identification of Male and Female Local Party Activists (in percentages)**

Strength of Party Identification	Males	Females
Strong	74.4	74.3
Not so strong	15.6	11.9
Weak	8.1	11.9
Total	98.1	98.1
(N)	(160)	(101)

Source: Laura van Assendelft and Karen O'Connor, "Backgrounds, Motivations, and Interests: A Comparison of Male and Female Local Party Activists," *Women & Politics* 14, no. 3 (1994): 77–91. Used by permission of Haworth Press, Inc., Binghamton, NY.

Table 9.5   **Years Active in Local Party Organization (in percentages)**

Years	Males	Females
0–5	31.6	33.3
5–10	20.3	19.6
10–20	18.4	24.5
20+	29.7	22.5
Total	100.0	99.9
(N)	(158)	(102)

Source: Laura van Assendelft and Karen O'Connor, Backgrounds, Motivations, and Interests: A Comparison of Male and Female Local Party Activists," *Women & Politics* 14, no. 3 (1994): 77–91. Used by permission of Haworth Press, Inc., Binghamton, NY.

Table 9.4 shows that men and women are very similar in their identification with their chosen political party; nearly three-quarters of both men and women feel their ties are strong. Notice that the total percentage for both men and women is 98.1 percent rather than 100.0 percent. This difference is too large to be due to rounding. The authors have apparently left out some responses, but the nature of those responses is not apparent from the table or their text.

Look carefully at Table 9.5. Why are we considering the number of years to be an ordinal variable? Measures of time are usually considered interval-ratio level variables. The reason that time is considered as an ordinal variable here is because the intervals of the categories are unequal. The categories "0–5" and "5–10" contain fewer units than the category "10–20," and the category "20+" may contain either fewer or more units than the other categories. Thus, for example, we cannot say how much difference there is between "0–5" and "20+."

Table 9.5 shows that men and women have been involved in party activity for a similar number of years. It is important to note that van Assendelft and O'Connor report that the average age for both male and female activists was 50 years, and a majority of both men and women were older than 50 years.[4] Thus, age differences did not have an effect on the number of years of activity.

Display the Data in a Graph   Recall from Chapter 3 that one of the primary reasons researchers use graphs in presentations is to make it easier for read-

[4]Ibid., p. 80.

Figure 9.3   **Educational Differences Between Male and Female Local Party Activists**

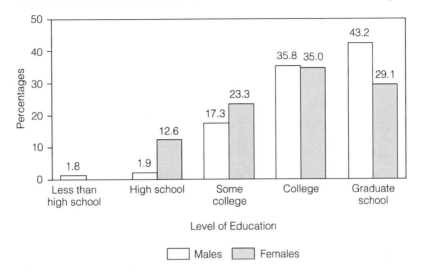

ers to understand data. We can also use graphs when we want to emphasize some aspect of the data. Table 9.3 shows educational differences between male and female party activists. One of van Assendelft and O'Connor's arguments is that fewer women than men run for political office because fewer women are employed in occupations from which many political candidates are drawn. Such occupations—practicing law, for example—generally require a graduate school education. We may want to emphasize the finding that a higher percentage of men than women activists have completed graduate school. We can see from Figure 9.3 that few men or women have less than a high school education. The differences between men and women become apparent at the high school level, with a higher percentage of women than men having only a high school education. At the college levels the percentages become more similar, and the largest percentage difference occurs at the graduate school level.

Table 9.6 shows the hours men and women spent per week on party work during elections; the same information is displayed in Figure 9.4. At the low end of the hours, the percentage of men exceeds the percentage of women; at the high end, the percentage of women exceeds the percentage of men. This distribution has few categories, so we could have chosen to use a pie chart rather than a bar graph to display it.

Learning Check.   *Draw a pie chart (one for men, one for women) and compare it with the bar graph shown in Figure 9.4. Which do you think better emphasizes the data? Remember, you can "explode" slices on a pie chart.*

Table 9.6 **Hours Spent per Week on Party Work During Elections (in percentages)**

Hours	Males	Females
0–5	60.0	42.6
5–10	23.9	23.8
10–20	7.7	15.8
20+	8.4	17.8
Total	100.0	100.0
(N)	(155)	(101)

Source: Laura van Assendelft and Karen O'Connor, Backgrounds, Motivations, and Interests: A Comparison of Male and Female Local Party Activists," *Women & Politics* 14, no. 3 (1994): 77–91. Used by permission of Haworth Press, Inc., Binghamton, NY.

Figure 9.4 **Hours Spent per Week on Party Work During Elections**

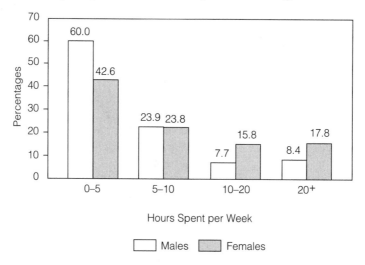

Describe What Is Average or Typical of a Distribution With ordinal data we can choose to report either the median or the mode. The deciding criterion is the purpose of the research. Let's look at the variable *level of education,* shown in Table 9.3, and decide if we will report the median or the mode. The cumulative percentage columns in Table 9.3 show that the 50th percentile for both males and females falls in the "college" category. Thus, the median level of education for both males and females is "college." The modes, however, differ—"graduate school" for men and "college" for women. The purpose of this research is to describe both similarities and differences between male and female party activists, so which do we choose? Our inclination would be to report the mode, for two reasons. First, when we examine the tables for the other variables, we find many areas in which men and women are similar. This makes it all the more important to determine how they differ. Second, reporting the mode supports the argument that women run for political office less frequently than men because—due to lack of educational credentials—they are underrepresented in the professions from which many candidates are chosen. Another option is to report both the mode and the median.

Describe Variability Within a Distribution For most of the variables used in this study, the range of categories for men and women is the same. The similarities in each range may be reported. However, what should be noted is the range in *level of education*, which differs for men and women. As Table 9.3 shows, the range of education for men is less than high school through graduate school; for women it is high school through graduate school. The range shows us that there is more variability among men than among women.

Describe the Relationship Between Two Variables Throughout this discussion we have been performing cross-tabulations of a number of variables. For a more detailed discussion of cross-tabulation as a technique for determining the relationship between variables, let's examine Table 9.7, which shows motivation for party activism among male and female party activists. Before we discuss what this table shows, we need to look at the way the table is constructed.

First, notice that both *gender* and *degree of importance*—the dependent variable—are arrayed in the columns of Table 9.7 instead of in rows as is typical. This suggests that we should compare down the columns. However, if we total the columns, we find that they do not add up to 100 percent. They are not even close. Why? Because in Table 9.7 each row actually represents a separate table, with the independent variable in each case being gender. Table 9.8 shows how a table representing one row (interest in elected office) of Table 9.7 can be constructed. When we compare Table 9.8 with the corresponding row of Table 9.7, we can see that in Table 9.8 we make comparisons across the rows in groups of two.

Table 9.7 shows that generally males and females are similarly motivated to participate in party activity. For both males and females, political philosophy appears to be the strongest motivator (very important for 77.3% of men and 82.1% of women). Supporting a particular candidate (very important for 53.3% of men; 68.8% of women) and specific issues they care about (very important for 35.3% of men, 49.5% of women) are stronger motivators for women than for men.

Table 9.7   **Motivations for Party Activism Among Males and Females (in percentages)**

	DEGREE OF IMPORTANCE					
	Very		Fairly/Not Very		Not at All	
MOTIVATION	(M)	(F)	(M)	(F)	(M)	(F)
Supporting a particular candidate	53.3	68.8	42.1	29.2	4.6	2.1
My political philosophy	77.3	82.1	22.0	17.9	0.6	0.0
Party loyalty	36.4	33.7	57.1	61.0	6.5	5.3
Interest in a government job or political appointment	5.4	8.7	32.6	25.0	61.9	66.3
Interest in elected office	16.4	18.1	43.6	33.0	40.0	48.9
Friendship or opportunities for meeting people	14.0	14.1	72.7	67.4	13.3	18.5
A specific issue that I care about	35.3	49.5	56.8	45.3	7.8	5.3

Source: Laura van Assendelft and Karen O'Connor, Backgrounds, Motivations, and Interests: A Comparison of Male and Female Local Party Activists," *Women & Politics* 14, no. 3 (1994): 77–91. Used by permission of Haworth Press, Inc., Binghamton, NY.

Table 9.8   **Importance of Interest in Elected Office in Motivating Party Activism Among Males and Females (in percentages)**

Degree of Importance	Males	Females
Very	16.4	18.1
Fairly/Not very	43.6	33.0
Not at all	40.0	48.9
Total	100.0	100.0

Source: Laura van Assendelft and Karen O'Connor, Backgrounds, Motivations, and Interests: A Comparison of Male and Female Local Party Activists," *Women & Politics* 14, no. 3 (1994): 77–91. Used by permission of Haworth Press, Inc., Binghamton, NY.

Table 9.9 **Educational Differences Between Male and Female Local Party Activists**

	GENERAL		
	GENDER		
LEVEL OF EDUCATION	Males	Females	Total
Less than high school	3	0	3
High school	3	13	16
Some college	28	24	52
College	58	36	94
Graduate school	70	30	100
Total	162	103	265

Source: Laura van Assendelft and Karen O'Connor, Backgrounds, Motivations, and Interests: A Comparison of Male and Female Local Party Activists," *Women & Politics* 14, no. 3 (1994): 77–91. Used by permission of Haworth Press, Inc., Binghamton, NY.

Gamma and Somers' *d* are appropriate for use with ordinal level variables or with dichotomous nominal variables. However, because the independent variable in this example is always *gender*—a dichotomous nominal level variable—we can use both lambda (in cases where the modes of the distributions differ) and gamma or Somers' *d* to measure the association between gender and any of the dependent variables examined earlier.

For example, let's look at the association between gender and the level of education of local party activists shown in Table 9.3. First, we must convert the percentages into frequencies and construct a bivariate table, as shown in Table 9.9.

Now let's first calculate lambda:

$$\text{lambda} = \frac{E1 - E2}{E1}$$

$$E1 = N - \text{mode} = 265 - 100 = 165$$

$$E2 = (162 - 70) + (103 - 36) = 159$$

$$\text{lambda} = \frac{165 - 159}{165} = .04$$

A lambda of .04 shows us that there is a very weak positive relationship between education and gender. Knowing the gender of an activist will do little to improve our prediction of the level of his or her education.

Next, let's calculate Somers' d. Because we are sure about the direction of influence in this relationship (clearly only gender, and not education, can be the independent variable), we choose Somers' d rather than gamma:

$$Somers'd = \frac{Ns - Nd}{Ns + Nd + Nty}$$

The number of same order (Ns) pairs that can be formed from Table 9.9 is

$$Ns = 3(13 + 24 + 36 + 30) + 3(24 + 36 + 30) + 28(36 + 30) + 58(30) = 4,167$$

The number of inverse order (Nd) pairs that can be formed from Table 9.9 is

$$Nd = 0(3 + 28 + 58 + 70) + 13(28 + 58 + 70) + 24(58 + 70) + 36(70) = 7,620$$

The number of pairs tied on the dependent variable (Nty) that can be formed from Table 9.9 is

$$Nty = 3(0) + 3(13) + 28(24) + 58(36) + 70(30) = 4,899$$

Thus,

$$Somers' \ d = \frac{4,167 - 7,620}{4,167 + 7,620 + 4,899} = -.21$$

A Somers' d of $-.21$ indicates a weak association between gender and education of local party activists. Using gender to predict level of education results in a proportional reduction of error of 21 percent ($.21 \times 100 = 21\%$). If we designate "femaleness" as the ordering principle for gender, then "female" is higher and "male" is lower (using "maleness" as the ordering principle, "female" would be lower and "male" higher). The negative sign of Somers' d can thus be interpreted to mean that male activists tend to have higher levels of education than female activists.

■ ■ ■ ■ **Descriptive Data Analysis for Interval-Ratio Variables**

The highest level of measurement is the interval-ratio level. The categories of interval-ratio variables are continuous and can be ranked from highest to lowest. The measurements for all the cases are expressed in the same units, and the magnitude of difference between categories can be calculated.

Statistics in Practice: Education and Income

The purpose of teaching you how to calculate the statistics presented in this book is to increase your understanding of the statistical procedures and techniques. However, you have also learned that computer software such as SPSS can quickly and accurately provide researchers with frequency distributions, graphs, and statistical output. In this section we use output generated by SPSS to analyze the relationship between education and income for data from the 1996 General Social Survey. This research example will review both SPSS output and the procedures for analyzing interval-ratio data.

Although we will not manually calculate our statistics, we still follow the five basic steps of descriptive data analysis to examine the relationship between education and income for respondents to the General Social Survey.

Organize the Data into a Frequency Distribution Table 9.10 shows the output generated by SPSS when frequency distributions of the variables *income* and *education* are requested. Let's review the parts of this table

Variables are identified by their name and variable label on the top of each table. Next, look at the column headings. The first column indicates values and value labels for the variables. In the case of RINCOME, in the first row the value "1" corresponds to income "LT $1000" (less than $1000). Values of RINCOME do not indicate respondents' actual income; we need to refer to the value label to determine the corresponding income range. For education (EDUC), however, there are no value labels because the values directly represent the number of years of school completed by the respondents (with the exception of 98, which we will discuss in a moment).

The Frequency, Percent, Valid Percent, and Cumulative Percent columns should be familiar. Notice that the categories 0 (NAP), 98 (DK), and 99 (NA) are included in the value labels for RINCOME, as is 98 (DK) for education. Responses coded as NAP (not applicable), DK (don't know) or NA (no answer) are defined as missing cases; these cases are not included in any statistical calculations. The category "refused" for RINCOME remains part of the valid percent calculations. (Generally, SPSS will not omit these cases unless the value is designated as missing.) When a statistical procedure involves more than one variable, cases that are missing data for any of the variables—either refused or NAP—are not included in the calculation.

Display the Data in a Graph To display the data for both education and income we use histograms. SPSS generates histograms as well as a variety of other graphs. However, software graphic packages often allow researchers to produce more professional-looking charts for presentation purposes, and some graphics programs allow the user to import data directly from other sources, such as statistics programs, eliminating the need to enter data manually.

Figure 9.5 shows a histogram for the income of GSS respondents. Notice that we have collapsed the categories into equal-width intervals to make the graph easier to read and to maintain the interval-ratio level of measurement. Note, however, that the income category "$25,000+" is not equal in width to the other categories. The fact that it contains so many cases also indicates that it probably should have been divided into several response categories in the original interview. Figure 9.5 shows that the highest percentage of respondents (44%) earns more than $25,000 per year. However, a substantial percentage (9.5%) earns less than $5,000 per year.

Table 9.10 **Frequency Distributions for Income and Education: GSS 1996**

EDUC HIGHEST YEAR OF SCHOOL COMPLETED

		Frequency	Percent	Valid Percent	Cumulative Percent
Valid	0	1	.1	.1	.1
	3	8	.6	.6	.6
	4	1	.1	.1	.7
	5	7	.5	.5	1.2
	6	8	.6	.6	1.7
	7	7	.5	.5	2.2
	8	45	3.1	3.1	5.3
	9	33	2.3	2.3	7.6
	10	52	3.6	3.6	11.2
	11	72	5.0	5.0	16.2
	12	425	29.3	29.4	45.6
	13	156	10.7	10.8	56.4
	14	171	11.8	11.8	68.2
	15	84	5.8	5.8	74.0
	16	180	12.4	12.4	86.4
	17	58	4.0	4.0	90.5
	18	64	4.4	4.4	94.9
	19	22	1.5	1.5	96.4
	20	52	3.6	3.6	100.0
	Total	1446	99.6	100.0	
Missing	98 DK	6	.4		
Total		1452	100.0		

RINCOME RESPONDENTS INCOME

		Frequency	Percent	Valid Percent	Cumulative Percent
Valid	1 LT $1000	18	1.1	1.6	1.6
	2 $1000 TO 2999	44	3.0	4.3	5.9
	3 $3000 TO 3999	17	1.2	1.7	7.6
	4 $4000 TO 4999	13	.9	1.3	8.9
	5 $5000 TO 5999	21	1.4	2.1	11.0
	6 $6000 TO 6999	25	1.7	2.5	13.4
	7 $7000 TO 7999	17	1.2	1.7	15.1
	8 $8000 TO 9999	37	2.5	3.7	18.8
	9 $10000 - 14999	116	8.0	11.5	30.2
	10 $15000 - 19999	77	5.3	7.6	37.8
	11 $20000 - 24999	121	8.3	12.0	49.8
	12 $25000 OR MORE	445	30.6	44.0	93.8
	13 REFUSED	63	4.3	6.2	100.0
	Total	1012	69.7	100.0	
Missing	0 NAP	320	22.0		
	98 DK	19	1.3		
	99 NA	101	7.0		
	Total	440	30.3		
Total		1452	100.0		

Figure 9.5   **Histogram of Income for GSS Respondents**

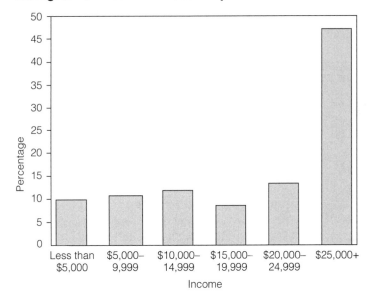

Describe What Is Average or Typical of a Distribution With interval-ratio data we can choose among the mode, the median, and the mean as a measure of central tendency. Whenever possible, the mean is the measure of choice because it allows more advanced statistical analysis than either the mode or the median. Looking at the data for education and income using the frequency distribution for education and the histogram for income, we can see that neither distribution is strongly skewed. For both of these variables, the mean can be used to describe central tendency. However, as we indicated earlier, it is generally a good idea to also compute and interpret the median for interval-ratio data.

Describe Variability Within a Distribution Because we have chosen to use the mean to describe central tendency for each of our two variables, we can use the standard deviation to describe variability. SPSS provides both the mean and the standard deviation for our variables[5] and displays them in the format shown in Table 9.11.

Let's look at this information more closely. The last line [Valid N(listwise)] tells us that in 947 cases the respondents provided usable answers to both survey questions. The column headed "N" indicates the number of usable responses for the individual variables. Verify this by looking at Table 9.10. The valid N should be equal to the total frequency minus REFUSE, NAP, DK, and NA frequencies.

[5]To calculate these statistics, RINCOME was recoded to the midpoint for each category. Recoded values for RRINCOME: 1=500, 2=2000, 3=3500, 4=4500, 5=5500, 6=6500, 7=7500, 8=9000, 9=12500, 10=17500, 11=22500, and 12=27500. The value 13 = Refused was excluded.

Table 9.11   **SPSS Output of Means and Standard Deviations**

Descriptive Statistics

	N	Minimum	Maximum	Mean	Std. Deviation
EDUC HIGHEST YEAR OF SCHOOL COMPLETED	1446	0	20	13.41	2.97
RRINCOME recoded income	949	500.00	27500.00	19715.49	9007.3331
Valid N (listwise)	947				

SPSS has calculated both the mean and the standard deviation for education and income. The mean for education is 13.41 years, with a standard deviation of 2.97 years. The mean indicates that the average respondent has just over a high school education (12 years is equal to completing high school). The standard deviation of 2.97 years, when compared with a mean, shows that a fairly large number of respondents have either less than or more than a high school education. We conclude that this distribution is moderately diverse.

The mean income for this sample is $19,715.49, with a standard deviation of $9,007.33. The standard deviation is very high in comparison with the mean income and indicates that there is a great deal of diversity in the sample distribution.

> **Learning Check.**   *The median is not reported in the SPSS output. However, we can easily determine the median for the variables EDUC and RINCOME from Table 9.10. (Hint: Examine the Cumulative Percent columns.) Compare the median with the mean for each variable. What can you conclude about the shape of the two distributions from this comparison? What can you learn from the medians about the income and education of GSS respondents?*

Describe the Relationship Between Two Variables   We have chosen to explore the relationship between education and income because we know that, in general, better-paying jobs require more education than poorer-paying jobs. Because the relationship between these variables is linear, we can use bivariate regression to make determinations about the relationship between education and income.

SPSS produces the regression calculation shown in Table 9.12. The difference between multiple regression and bivariate regression is that in multiple regression more than one independent variable is used to predict the value of the dependent variable. SPSS uses the multiple regression program to calculate both bivariate and multivariate regression and does not distinguish between the two when it titles the output.

Table 9.12 **SPSS Output for the Bivariate Regression of Recoded Income and Education**

Variables Entered/Removed[b]

Model	Variables Entered	Variables Removed	Method
1	EDUC HIGHEST YEAR OF SCHOOL COMPLETED[a]	.	Enter

a. All requested variables entered.

b. Dependent Variable: RRINCOME recoded income

Model Summary

Model	R	R Square	Adjusted R Square	Std. Error of the Estimate
1	.310[a]	.096	.095	8558.3693

a. Predictors: (Constant), EDUC HIGHEST YEAR OF SCHOOL COMPLETED

ANOVA[b]

Model		Sum of Squares	df	Mean Square	F	Sig.
1	Regression	7.38E+09	1	7.38E+09	100.714	.000[a]
	Residual	6.92E+10	945	73245684		
	Total	7.66E+10	946			

a. Predictors: (Constant), EDUC HIGHEST YEAR OF SCHOOL COMPLETED

b. Dependent Variable: RRINCOME recoded income

Coefficients[a]

Model		Unstandardized Coefficients		Standardized Coefficients	t	Sig.
		B	Std. Error	Beta		
1	(Constant)	5292.607	1465.889		3.611	.000
	EDUC HIGHEST YEAR OF SCHOOL COMPLETED	1038.130	103.445	.310	10.036	.000

a. Dependent Variable: RRINCOME recoded income

Beginning with the first table, SPSS reports what variables were used in the regression analyses. It identifies as a footnote (b) that our dependent variable is the recoded respondent income variable (RRINCOME); our independent variable is EDUC (under variables entered).

The SPSS program for multiple regression produces information on a variety of tests not required for a simple bivariate regression. So where in all those numbers do we find what we need to produce a bivariate regression equation? Look under the section labeled "Coefficients." Producing the regression equation is actually very simple. The value of b is listed in column B and row EDUC. Thus, the value of b is about $1,038.13. The value of a is found in column B and row (Constant). Thus, our bivariate regression equation is

$$\hat{Y} = \$5,292.61 + \$1,038.13(EDUC)$$

The regression equation tells us that for every unit change in education—the unit is one year—we can predict an increase of $1038.13 in annual income for the respondents in our sample. Let's use the regression equation to predict the difference in annual income between respondents with a high school (12 years) and a college (16 years) education:

$$\hat{Y} \text{ (high school)} = \$5,292.61 + \$1,038.13(12) = \$17,750.17$$

$$\hat{Y} \text{ (college)} = \$5,292.61 + \$1,038.13(16) = \$21,902.69$$

Based on the regression equation, we predict that individuals with a college degree make an annual income of $4,152.52 more than those with just a high school education ($21,902.69 − $17,750.17 = $4,152.52). Over the course of a normal lifetime of work, we would expect a college graduate to earn almost a quarter of a million dollars more than a high school graduate.

SPSS can also calculate Pearson's correlation coefficient, r. The output produced by SPSS is shown in Table 9.13. The output is arranged in a matrix that resembles a bivariate table. Each variable is listed in a row and a column. To find the value of r, we need to locate the intersection of the two variables. Since r is a symmetrical measure, it does not matter which variable we consider the independent variable, so we can read either down the columns or across the rows. The correlation coefficient for each pair of variables is the first number in the cell. For EDUC and RRINCOME, r is equal to 0.310. Remember, r can range from 0.0 to ±1.0, with 0.0 indicating no relationship between the variables and ±1.0 indicating a perfect positive or negative relationship between the variables. An r of 0.310 indicates a moderate positive relationship between education and income. To determine how much of the variation in income can be explained by education we need to calculate r^2:

$$r^2 = (r)^2 = (0.310)^2 = 0.096$$

Table 9.13　**SPSS Output of Correlation Coefficient, Recoded Income and Education**

Correlations

		EDUC HIGHEST YEAR OF SCHOOL COMPLETED	RRINCOME Recoded Rincome
EDUC HIGHEST YEAR OF SCHOOL COMPLETED	Pearson Correlation	1.000	.310**
	Sig. (2-tailed)	.	.000
	N	1446	947
RRINCOME Recoded Rincome	Pearson Correlation	.310**	1.000
	Sig. (2-tailed)	.000	.
	N	947	949

**. Correlation is significant at the 0.01 level (2-tailed).

Using the regression equation, our prediction of income is improved by 9.6 percent (0.096×100) over the prediction we would make using the mean of income alone. We can also say that education explains 9.6 percent of the variation in income in our sample.

∎　∎　∎　∎　**A Final Note**

Many people find that the most difficult part of statistics is determining the proper procedures and techniques to use with the data. We've designed this chapter as a reference tool for you to use in making those decisions when performing descriptive data analysis.

In the remaining chapters of this book, we will introduce you to inferential statistics—procedures researchers use to make predictions about a population using data collected from a sample of the population. We close our book with a final review chapter (Chapter 15), which will provide you with a reference guide when performing inferential statistics.

SPSS PROBLEMS [MODULE B]

1. The 1996 GSS file contains two questions asking how satisfied a respondent is with his or her finances and job (called SATFIN and SATJOB, respectively). Each is measured on a scale of 1 to 3, where low scores mean more satisfaction. Use SPSS to answer these questions.
   a. Characterize the distribution of these variables, using measures of central tendency and variability.

b. What percentage of the sample said they were "satisfied" with their financial situation? How about with their job?

c. Create a chart to display SATFIN by SEX. Describe the resulting graph.

2. Continue your exploration of the satisfaction variables by examining their relationship to each other and to other items.

a. Create a cross-tabulation of SATFIN by SEX and SATJOB by SEX. Have SPSS calculate appropriate percentages, statistics, and measures of association. Use these results to describe the relationships between SEX and the two variables.

b. Calculate the correlation coefficient between the satisfaction variables. Is there a positive or a negative relationship between these variables? Can you suggest why? Although these variables are measured on a three-point ordinal scale, it is quite common in the social sciences to use interval-ratio techniques with such data.

3. Is a respondent's age related to satisfaction with his/her health?

a. To investigate this question, first create a scatterplot of the two variables. Do you find this graphic display useful? Why or why not? Have SPSS place a regression line on the plot.

b. Now have SPSS calculate the regression equation to predict health satisfaction with age. What is the regression equation? What is the coefficient of determination? Describe the relationship between the two variables. Is age a strong predictor of self-rated health?

c. What is the predicted health satisfaction rating for a person who is 30 years old? 70 years old?

4. Although the satisfaction variables and the health variable are measured on an ordinal scale, it is much easier to study their relationship to age with interval-ratio techniques. Why is that? What problem would you encounter if you tried to use a cross-tabulation instead?

CHAPTER EXERCISES

1. Indicate whether the following types of data are nominal, ordinal, or interval-ratio.

a. The eye color of people

b. The number of general elections in which a person has voted in his or her lifetime

c. Ranking of school quality as below average, average, or above average

d. Political party membership

e. The time at which people have dinner each night

f. The classification of burns as first, second, or third degree

2. A question from the 1996 GSS asks about whether the income differentials in the United States are too big. Following is a portion of the output from the SPSS Frequencies procedure.

INCGAP INCOME DIFFERENTIALS IN USA TOO BIG

		Frequency	Percent	Valid Percent	Cumulative Percent
Valid	0 NAP	738	50.8	50.8	50.8
	1 STRONGLY AGREE	235	16.2	16.2	67.0
	2 AGREE	211	14.5	14.5	81.5
	3 NEITHER	85	5.9	5.9	87.4
	4 DISAGREE	91	6.3	6.3	93.7
	5 STRONGLY DISAGREE	66	4.5	4.5	98.2
	8 CANT CHOOSE	22	1.5	1.5	99.7
	9 NA	4	.3	.3	100.0
	Total	1452	100.0	100.0	

a. On what scale of measurement is INCGAP measured?

b. The values of 0, 8, and 9 should be recoded as missing because they correspond to "not applicable," "can't choose," and "no answer." Taking this recoding into account, calculate the correct values for the columns labeled Percent, Valid Percent, and Cumulative Percent.

c. Is cumulative percentage proper to calculate for INCGAP? Why or why not?

3. Construct an appropriate graph to display the categories of INCGAP.

4. For the variable INCGAP:

a. Calculate the mode and the median. (*Reminder:* Don't use the missing data.)

b. Calculate the IQV. Calculate the range.

c. Use this information to describe the distribution of responses to this question.

5. The following table shows the relationship between INCGAP and race of the respondent (RACE).

INCGAP INCOME DIFFERENTIALS IN USA TOO BIG * RACE RACE OF RESPONDENT Crosstabulation

				RACE RACE OF RESPONDENT			
				1 WHITE	2 BLACK	3 OTHER	Total
INCGAP INCOME DIFFERENTIALS IN USA TOO BIG	1 STRONGLY AGREE		Count	188	30	17	235
			% within RACE RACE OF RESPONDENT	34.0%	31.6%	42.5%	34.2%
	2 AGREE		Count	171	28	12	211
			% within RACE RACE OF RESPONDENT	30.9%	29.5%	30.0%	30.7%
	3 NEITHER		Count	69	9	7	85
			% within RACE RACE OF RESPONDENT	12.5%	9.5%	17.5%	12.4%
	4 DISAGREE		Count	73	15	3	91
			% within RACE RACE OF RESPONDENT	13.2%	15.8%	7.5%	13.2%
	5 STRONGLY DISAGREE		Count	52	13	1	66
			% within RACE RACE OF RESPONDENT	9.4%	13.7%	2.5%	9.6%
Total			Count	553	95	40	688
			% within RACE RACE OF RESPONDENT	100.0%	100.0%	100.0%	100.0%

a. Describe the relationship you observe in the table, using percentages.
b. What is the strength of the relationship between race and the belief that income differentials are too big? Calculate the appropriate measure of association. Is the value you calculated consistent with your description in (a)?

6. A poll of white and black Americans in 1993 asked, "Is racial discrimination against blacks where you live serious?" Here are the results:

Race	Serious	Not Serious
Whites	33%	67%
Blacks	68%	32%

Source: Data from a USA Today/CNN/Gallup Poll of 840 adults on February 8–9, 1993. Missing data have been removed.

a. Use these data to construct a clustered bar graph of the seriousness of racial discrimination against blacks.
b. Describe the relationship between race and belief about discrimination against blacks. Why might the two variables be related?
c. Calculate an appropriate measure of association for this table to measure the strength of the relationship. If you don't think this can be done, explain why.

7. The following table displays the amount that 284 people in a large city paid in sales tax in 1995.

Amount Paid	Frequency
0 to $99.99	34
$100 to $199.99	52
$200 to $499.99	121
$500 to $1,000	77

a. Calculate the median amount of sales tax paid.
b. Calculate the mean sales tax paid.
c. Calculate the IQR for the sales tax data.

8. You and a friend are discussing measures of central tendency (you are both very diligent statistics students). Your friend says, "No distribution of an interval-ratio variable is ever truly symmetrical. All of them are skewed to a certain extent." You admit this is probably true. Your friend goes on to make this claim: "Therefore, given Figure 9.1 in the textbook, I think we should never use the mean as a measure of central tendency for interval-ratio data. Instead, it's safer to use the median." Do you agree or disagree with your friend? Provide reasons for your answer.

Figure 9.6

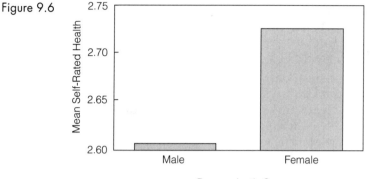

Respondent's Sex

9. The bar graph in Figure 9.6 displays mean self-rated physical health by gender. (Health is measured on a scale of 1 to 7, where lower numbers mean better health.) Explain what error has been made in creating this chart and what its effect might be on an unsuspecting reader.

10. In 1990 the United States had 248,709,873 residents, according to the U.S. Census Bureau. Also in 1990, the U.S. Department of Justice reported that there were 292 federal and state prisoners per 100,000 persons in the United States.
    a. How many total federal and state prisoners were there in 1990?
    b. What was the proportion of the U.S. population in federal and state prison?

11. In the National Election Study of 1994, respondents were asked their feelings toward eventual 1996 presidential contenders Bill Clinton and Bob Dole on a scale of 0 to 100, where a higher score means warmer, more positive feelings. A score of –99 means a non-response. A sub-sample of the responses follows.

Clinton	Dole	Clinton	Dole
30	70	70	70
15	70	100	70
0	85	100	–99
40	50	80	0
30	60	60	60
30	–99	70	70
85	–99	90	0
0	85	70	50
75	20	85	0
70	70	85	15
50	50	85	70
70	50	60	70

a. Calculate the mean feelings toward Clinton and toward Dole.

b. Calculate the variance and standard deviation for the feelings toward Clinton.

c. Calculate the IQV for feelings toward Dole.

12. What is the relationship between feelings toward Bill Clinton and those toward Bob Dole?

a. Investigate this question by calculating the correlation coefficient between feelings toward each politician.

b. Use this statistic to describe how the two variables are related. Does the result make sense?

13. The 1994 National Election Study included a question about approval of President Bill Clinton on a four-point scale. The following contingency table shows the relationship between approval of Clinton and whether a respondent classified himself or herself as middle or working class.

Approval of Clinton	Working Class	Middle Class	Total
Strongly disapprove	175	303	478
Not strongly disapprove	150	166	317
Not strongly approve	280	250	530
Strongly approve	174	136	310

a. Assume that class can be used to predict approval rating. Calculate appropriate percentages, and describe the relationship in the table.

b. Calculate a measure of association to further characterize this relationship.

14. A child psychologist measured the amount of time parents spent talking to their young children each day and then, several years later, gave the students a standardized achievement test (the test is measured on a scale from 0 to 50). She obtained the following results.

Time Spent Talking to Child (in minutes)	Test Score
15	34
45	47
9	27
60	49
22	29
30	39
5	18
20	26
25	33
40	42
30	36

    a. Construct a scatterplot of time spent talking and test score. Use it to describe the relationship between these two variables.

    b. Calculate the coefficient of determination between these two variables.

    c. Calculate the regression equation.

    d. What is the predicted test score for a child whose parents talked with her for 50 minutes per day?

15. Has there been a decline in the poverty rates since the Clinton Administration's welfare reform programs? Data released in January 1999 by the U.S. Census Bureau are presented in the following table, which lists the percentage of people of all ages in poverty by state for 1993 and 1995.

State	1995	1993	State	1995	1993
Alabama	17.6	18.8	Montana	15.8	15.2
Alaska	10.1	11.2	Nebraska	9.8	10.7
Arizona	16.3	18.5	Nevada	10.5	11.5
Arkansas	18.2	18.9	New Hampshire	6.9	8.6
California	16.5	17.4	New Jersey	8.7	10.0
Colorado	10.4	11.7	New Mexico	20.2	21.6
Connecticut	8.3	9.2	New York	15.8	16.3
Delaware	9.7	11.1	North Carolina	13.1	14.1
Florida	15.2	16.0	North Dakota	11.9	12.5
Georgia	15.6	16.3	Ohio	11.3	13.7
Hawaii	9.9	9.6	Oklahoma	18.2	18.3
Idaho	12.6	12.5	Oregon	12.5	13.2
Illinois	11.3	13.4	Pennsylvania	11.4	12.9
Indiana	9.8	11.9	Rhode Island	11.6	12.4
Iowa	9.8	11.1	South Carolina	15.7	16.6
Kansas	11.0	12.2	South Dakota	13.7	14.3
Kentucky	17.9	19.7	Tennessee	14.7	17.8
Louisiana	21.2	23.9	Texas	18.5	19.6
Maine	12.3	13.7	Utah	9.7	11.2
Maryland	9.2	10.2	Vermont	11.2	12.2
Massachusetts	9.7	11.1	Virginia	11.3	12.0
Michigan	12.6	15.0	Washington	10.8	12.0
Minnesota	8.7	10.8	West Virginia	19.9	21.7
Mississippi	21.4	24.6	Wisconsin	8.9	10.9
Missouri	13.4	15.1	Wyoming	11.5	11.9

Source: Data adapted from U.S. Bureau of the Census, *Small Area Income and Poverty Estimates Program*, 1999, Tables A93-00 and A95-00.

    a. What would be the best way to graphically present the data for 1995?

    b. What would be the best way to graphically present the *difference* in rates (comparing each state's 1993 and 1995 figures)?

    c. Calculate the mean, median, and standard deviation for 1993 and 1995. Has there been a change in poverty rates? Justify your answer based on your calculations and your analysis of the data.

16. In Chapter 5, Box 5.2, we reviewed student diversity at the University of California, Berkeley, using IQV to demonstrate the difference in student body enrollment in 1984 versus 1994. In 1996, Californians approved a referendum that banned the use of race and ethnicity in college admissions. What was its impact on entering freshmen in 1998?

    The *New York Times* (April 1, 1998) reported the racial breakdown of freshmen admitted to Berkeley, comparing 1997 and 1998. (Data for those who did not report their ethnicity are not presented here.)

Number of Accepted Freshmen	Black	Asian	American Indian	Hispanic	White/ Other
1997	598	3866	77	1411	3831
1998	255	3861	47	852	3635

    a. Based on the frequencies, describe the differences in freshmen admissions, comparing 1997 to 1998.

    b. Calculate the IQV for both years. Based on the IQV calculations, what can you conclude? Is there a difference in ethnic variability between the freshmen cohorts?

10   The Normal Distribution

■ ■ ■ ■ **Introduction**

In the preceding chapters we have learned some important things about distributions: how to organize them into frequency distributions; how to display them using graphs; and how to describe their central tendencies and variation using measures such as the mean and the standard deviation. We have also learned that distributions can have different shapes. Some distributions are symmetrical; others are negatively or positively skewed. The distributions we have described so far are all *empirical distributions;* that is, they are all based on real data.

The distribution we describe in this chapter—known as the *normal curve* or the **normal distribution**—is a theoretical distribution. A *theoretical distribution* is similar to an empirical distribution in that it can be organized into frequency distributions, displayed using graphs, and described by its central tendency and variation using measures such as the mean and the standard deviation. However, unlike an empirical distribution, a theoretical distribution is based on theory rather than on real data. The value of the theoretical normal distribution lies in the fact that many empirical distributions we study seem to approximate it. We can often learn a lot about the characteristics of these empirical distributions based on our knowledge of the theoretical normal distribution.

■ ■ ■ ■ **Properties of the Normal Distribution**

The normal curve (Figure 10.1) looks like a bell-shaped frequency polygon. Because of this property it is sometimes called the *bell-shaped curve.* One of the most striking characteristics of the normal distribution is its perfect symmetry. Notice that if you fold Figure 10.1 exactly in the middle, you have two equal halves, each the mirror image of the other. This means that precisely half the observations fall on each side of the middle of the distribution. In addition, the midpoint of the normal curve is the point having the maximum frequency. This is also the point at which three measures coincide: the mode (the point of the highest frequency), the median (the point that divides the distribution into two equal halves), and the mean (the average of all the scores). Notice also that most of the observations are clustered around the middle, with the frequencies gradually decreasing at both ends of the distribution.

Figure 10.1 **The Normal Curve**

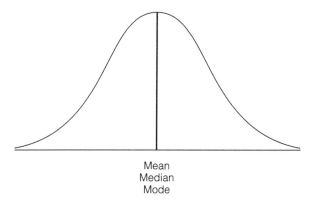

Mean
Median
Mode

> *Normal distribution* A bell-shaped and symmetrical theoretical
> distribution, with the mean, the median, and the mode all coincid-
> ing at its peak and with the frequencies gradually decreasing at both
> ends of the curve.

Empirical Distributions Approximating
the Normal Distribution

The normal curve is a theoretical ideal, and real-life distributions never
match this model perfectly. However, researchers study many variables (for
example, standardized tests such as the SAT, ACT, or GRE; height; athletic
ability; and numerous social and political attitudes) that closely resemble
this theoretical model. When we say that a variable is "normally distributed," we mean that a graphic display will reveal an approximately bell-
shaped and symmetrical distribution closely resembling the idealized
model shown in Figure 10.1. This property makes it possible for us to describe many empirical distributions based on our knowledge of the normal
curve.

An Example: Final Grades in Statistics

It is easier to understand the properties of a normal curve if we think in
terms of a real distribution that is near normal. Let's examine the frequencies and the bar chart presented in Table 10.1. These data are the final scores
of 1,200 students who took Professor Frankfort-Nachmias' social statistics
class at the University of Wisconsin–Milwaukee between 1983 and 1993. To
convince you that the variable *final score in statistics* is normally distributed,

Table 10.1   **Final Grades in Social Statistics of 1,200 Students (1983–1993): A Near Normal Distribution**

Frequency Bar Chart

Midpoint Score		Freq	Cum Freq	%	Cum %
40	★	4	4	.33	.33
50	★★★★★★★	78	82	6.50	6.83
60	★★★★★★★★★★★★★★★★	275	357	22.92	29.75
70	★★★★★★★★★★★★★★★★★★★★★★★★★	483	840	40.25	70.00
80	★★★★★★★★★★★★★★★	274	1,114	22.83	92.83
90	★★★★★★★	81	1,195	6.75	99.58
100	★	5	1,200	.42	100.00

	10	50	100	200	300	400	500

Mean $(\bar{Y})$ = 70.07   Median = 70.00   Mode = 70.00

Standard deviation (S_Y) = 10.27

we overlaid a normal curve on the distribution shown in Table 10.1. Notice how closely our empirical distribution of statistics scores approximates the normal curve!

Notice that 70 is the most frequent score obtained by the students, and therefore it is the mode of the distribution. Because about half the students are either above (49.99%) or below (50.01%) this score (based on raw frequencies), both the mean (70.07) and the median (70) are approximately 70. Also shown in Table 10.1 is the gradual decrease in the number of students who scored either above or below 70. Very few students scored higher than 90 or lower than 50.

When we use the term *normal curve*, we are not referring to identical distributions. The shape of a normal distribution varies, depending on the mean and standard deviation of the particular distribution. For example, in Figure 10.2 we present two normally shaped distributions with identical means (μ_Y = 12) but with different standard deviations (σ_{Y1} = 3; σ_{Y2} = 5). Notice that the distribution with the larger standard deviation appears relatively wider and flatter.

Figure 10.2 **Two Normal Distributions with Equal Means but Different Standard Deviations**

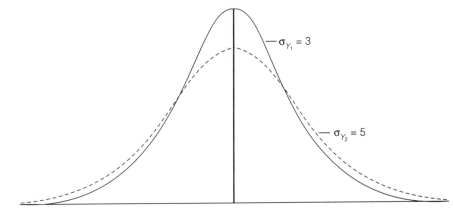

$\sigma_{Y_1} = 3$

$\sigma_{Y_2} = 5$

$\mu_Y = 12$

Areas Under the Normal Curve

Regardless of the precise shape of the distribution, in all normal or nearly normal curves we find a constant proportion of the area under the curve lying between the mean and any given distance from the mean when measured in standard deviation units. The area under the normal curve may be conceptualized as a proportion or percentage of the number of observations in the sample. Thus, the entire area under the curve is equal to 1.00, or 100 percent (1.00 × 100) of the observations. Because the normal curve is perfectly symmetrical, exactly 0.5000 or 50 percent of the observations lie above or to the right of the center, which is the mean of the distribution, and 50 percent lie below or to the left of the mean.

In Figure 10.3, note the percentage of cases that will be included between the mean and 1, 2, and 3 standard deviations above and below the mean. The mean of the distribution divides it exactly in half: 34.13 percent is included between the mean and 1 standard deviation to the right of the mean; the same percentage is included between the mean and 1 standard deviation to the left of the mean. The plus signs indicate standard deviations above the mean; the minus signs denote standard deviations below the mean. Thus, between the mean and ±1 standard deviations, 68.26 percent of all the observations in the distribution occur; between the mean and ±2 standard deviations, 95.46 percent of all observations in the distribution occur; and between the mean and ±3 standard deviations, 99.72 percent of the observations occur.

Figure 10.3   **Percentages Under the Normal Curve**

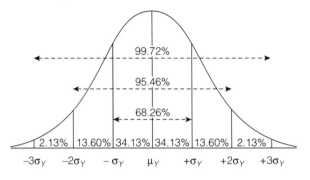

2.13% | 13.60% | 34.13% | 34.13% | 13.60% | 2.13%

$-3\sigma_Y$ $-2\sigma_Y$ $-\sigma_Y$ μ_Y $+\sigma_Y$ $+2\sigma_Y$ $+3\sigma_Y$

> **Learning Check.**   *Review and confirm the properties of the normal curve. What is the area underneath the curve equal to? What percentage of the distribution is within 1 standard deviation? within 2 and 3 standard deviations? Verify the percentage of cases by summing the percentages in Figure 10.3.*

Interpreting the Standard Deviation

The fixed relationship between the distance from the mean and the areas under the curve represents a property of the normal curve that has highly practical applications. As long as a distribution is normal and we know the mean and the standard deviation, we can determine the relative frequency (proportion or percentage) of cases that fall between any score and the mean.

This property provides an important interpretation for the standard deviation of empirical distributions that are approximately normal. For such distributions, when we know the mean and the standard deviation, we can determine the percentage of scores that are within any distance, measured in standard deviation units, from that distribution's mean. For example, we know that college entrance tests such as the SAT and ACT are normally distributed. The SAT, for instance, has a mean of 500 and a standard deviation of 100. This means that approximately 68 percent of the students who take the test obtain a score between 400 (1 standard deviation below the mean) and 600 (1 standard deviation above the mean). We can also anticipate that approximately 95 percent of the students who take the test will score between 300 (2 standard deviations below the mean) and 700 (2 standard deviations above the mean).

Not every empirical distribution is normal. We've learned that the distributions of some common variables, such as income, are skewed and therefore not normal. The fixed relationship between the distance from the mean and the areas under the curve applies *only* to distributions that are normal or approximately normal.

■ ■ ■ ■ Standard (Z) Scores

We can express the difference between any score in a distribution and the mean in terms of *standard scores*, also known as *Z scores*. A **standard (Z) score** is the number of standard deviations that a given raw score (or the observed score) is above or below the mean. A raw score can be transformed into a Z score to find how many standard deviations it is above or below the mean.

> *Standard (Z) score*   The number of standard deviations that a given raw score is above or below the mean.

Transforming a Raw Score into a Z Score

To transform a raw score into a Z score, we divide the difference between the score and the mean by the standard deviation. For instance, to transform a final score in the statistics class into a Z score, we subtract the mean of 70.7 from that score and divide the difference by the standard deviation of 10.27. Thus, the Z score of 80 is

$$\frac{80 - 70.07}{10.27} = 0.97$$

or 0.97 standard deviations above the mean. Similarly, the Z score of 60 is

$$\frac{60 - 70.07}{10.27} = -0.98$$

or 0.98 standard deviations below the mean; the negative sign indicates that this score is below the mean.

This calculation, in which the difference between a raw score and the mean is divided by the standard deviation, gives us a method of standardization known as *transforming a raw score into a Z score* (also known as a standard score). The Z score formula is

$$Z = \frac{Y - \overline{Y}}{s_Y} \tag{10.1}$$

A Z score allows us to represent a raw score in terms of its relationship to the mean and to the standard deviation of the distribution. It represents

Table 10.2   **Final Social Science Statistics Scores Converted to Z Scores**

Final Score	Z Score
40	$Z = \dfrac{40 - 70.07}{10.27} = \dfrac{-30.07}{10.27} = -2.93$
50	$Z = \dfrac{50 - 70.07}{10.27} = \dfrac{-20.07}{10.27} = -1.95$
60	$Z = \dfrac{60 - 70.07}{10.27} = \dfrac{-10.07}{10.27} = -0.98$
70	$Z = \dfrac{70 - 70.07}{10.27} = \dfrac{-.07}{10.27} = -0.01$
80	$Z = \dfrac{80 - 70.07}{10.27} = \dfrac{9.93}{10.27} = 0.97$
90	$Z = \dfrac{90 - 70.07}{10.27} = \dfrac{19.93}{10.27} = 1.94$
100	$Z = \dfrac{100 - 70.07}{10.27} = \dfrac{29.93}{10.27} = 2.91$
$\bar{Y} = 70.07$	$S_Y = 10.27$

how far a given raw score is from the mean in standard deviation units. A positive Z indicates that a score is larger than the mean, and a negative Z indicates that it is smaller than the mean. The larger the Z score, the larger the difference between the score and the mean.

Learning Check.   *To go back to our example of the final scores in statistics, we can convert the students' final scores into Z scores using Formula 10.1, as shown in Table 10.2. How many standard deviations above the mean is a score of 90? Below is a visual interpretation of what this question is asking.*

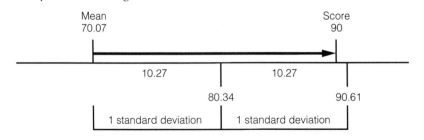

> From the figure, we can see that a score of 90 is almost 2 standard deviations above the mean. When we use the formula, we find that the Z score for a score of 90 is 1.94.

Transforming a Z Score into a Raw Score

For some normal curve applications, we need to reverse the process, transforming a Z score into a raw score instead of transforming a raw score into a Z score. A Z score can be converted to a raw score to find the score associated with a particular distance from the mean when this distance is expressed in standard deviation units. For example, suppose we are interested in finding out the final score in the statistics class that lies 1 standard deviation above the mean. To solve this problem we begin with the Z-score formula:

$$Z = \frac{Y - \overline{Y}}{S_Y}$$

Note that for this problem we have the values for Z ($Z = 1$), the mean ($\overline{Y} = 70.07$), and the standard deviation ($S_Y = 10.27$), but we need to determine the value of Y:

$$1 = \frac{Y - 70.07}{10.27}$$

Through simple algebra we solve for Y:

$$Y = 70.07 + 1(10.27) = 70.07 + 10.27 = 80.34$$

The score of 80.34 lies 1 standard deviation (or 1 Z score) above the mean of 70.07.

The general formula for transforming a Z score into a raw score is

$$Y = \overline{Y} + Z(S_Y) \tag{10.2}$$

Thus, to transform a Z score into a raw score, multiply the Z score by the standard deviation and add the product to the mean.

Now, what statistics score lies 1.5 standard deviations below the mean? Because the score lies below the mean, the Z score is negative. Thus,

$$Y = 70.07 + (-1.5)(10.27) = 70.07 - 15.41 = 54.66$$

The score of 54.66 lies 1.5 standard deviations below the mean of 70.07.

> **Learning Check.** *Transform the Z scores in Table 10.3 back into raw scores. Your answers should agree with the raw scores listed in the table.*

Figure 10.4 **The Standard Normal Distribution**

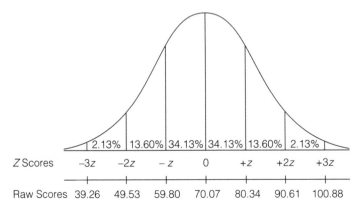

Z Scores   −3z   −2z   − z   0   +z   +2z   +3z

Raw Scores   39.26   49.53   59.80   70.07   80.34   90.61   100.88

■ ■ ■ ■ **The Standard Normal Distribution**

When a normal distribution is represented in standard scores (Z scores), we call it the **standard normal distribution**. Standard scores, or Z scores, are numbers that tell us the distance between an actual score and the mean in terms of standard deviation units. The standard normal distribution has a mean of 0.0 and a standard deviation of 1.0.

> *Standard normal distribution*   A normal distribution represented in standard (Z) scores.

Figure 10.4 shows a standard normal distribution with areas under the curve associated with 1, 2, and 3 standard scores above and below the mean. To help you understand the relationship between raw scores of a distribution and standard Z scores, we also show the raw scores in the statistics class that correspond to these standard scores. For example, notice that the mean for the statistics score distribution is 70.07; the corresponding Z score—the mean of the standard normal distribution—is 0. The score of 80.34 is 1 standard deviation above the mean (70.07 + 10.27 = 80.34), and therefore its corresponding Z score is +1. Similarly, the score of 59.80 is 1 standard deviation below the mean (70.07 − 10.27 = 59.80), and its Z-score equivalent is −1.

> **Learning Check.** *Can you explain why the mean of the standard normal curve is 0 and the standard deviation is equal to 1?*

■ ■ ■ ■ ■ The Standard Normal Table

We can use Z scores to determine the proportion of cases that are included between the mean and any Z score in a normal distribution. The areas or proportions under the standard normal curve, corresponding to any Z score or its fraction, are organized into a special table called the **standard normal table**. The table is presented in Appendix B. In this section we will discuss how to use this table.

The Structure of the Standard Normal Table

Table 10.3 reproduces a small part of the standard normal table. Note that the table consists of three columns.

Column A lists positive Z scores. Because the normal curve is symmetrical, the proportions that correspond to positive Z scores are identical to the proportions corresponding to negative Z scores.

Column B shows the area included between the mean and the Z score listed in column A. Note that when Z is positive the area is located on the right side of the mean (see Figure 10.5a), whereas for a negative Z score the same area is located left of the mean (Figure 10.5b).

Column C shows the proportion of the area that is beyond the Z score listed in column A. Areas corresponding to positive Zs are on the right side of the curve (see Figure 10.5). Areas corresponding to negative Z scores are identical except that they are on the left side of the curve (Figure 10.5b).

> *Standard normal table* A table showing the area (as a proportion, which can be translated into a percentage) under the standard normal curve corresponding to any Z score or its fraction.

Transforming Z Scores into Proportions (or Percentages)

We illustrate how to use Appendix B with some simple examples, using our data on students' final statistics scores (see Table 10.1). The examples in this section are applications that require the transformation of Z scores into proportions (or percentages).

Finding the Area Between the Mean and a Specified Positive Z Score Use the standard normal table to find the area between the mean and a specified positive Z score. To find the percentage of students whose scores range between the mean (70.07) and 85, follow these steps.

1. Convert 85 to a Z score:

$$Z = \frac{85 - 70.07}{10.27} = 1.45$$

Table 10.3 **The Standard Normal Table**

(A) Z	(B) Area Between Mean and Z	(C) Area Beyond Z	(A) Z	(B) Area Between Mean and Z	(C) Area Beyond Z
0.00	0.0000	0.5000	0.21	0.0832	0.4168
0.01	0.0040	0.4960	0.22	0.0871	0.4129
0.02	0.0080	0.4920	0.23	0.0910	0.4090
0.03	0.0120	0.4880	0.24	0.0948	0.4052
0.04	0.0160	0.4840	0.25	0.0987	0.4013
0.05	0.0199	0.4801	0.26	0.1026	0.3974
0.06	0.0239	0.4761	0.27	0.1064	0.3936
0.07	0.0279	0.4721	0.28	0.1103	0.3897
0.08	0.0319	0.4681	0.29	0.1141	0.3859
0.09	0.0359	0.4641	0.30	0.1179	0.3821
0.10	0.0398	0.4602			
0.11	0.0438	0.4562	0.31	0.1217	0.3783
0.12	0.0478	0.4522	0.32	0.1255	0.3745
0.13	0.0517	0.4483	0.33	0.1293	0.3707
0.14	0.0557	0.4443	0.34	0.1331	0.3669
0.15	0.0596	0.4404	0.35	0.1368	0.3632
0.16	0.0636	0.4364	0.36	0.1406	0.3594
0.17	0.0675	0.4325	0.37	0.1443	0.3557
0.18	0.0714	0.4286	0.38	0.1480	0.3520
0.19	0.0753	0.4247	0.39	0.1517	0.3483
0.20	0.0793	0.4207	0.40	0.1554	0.3446

Figure 10.5 **Areas Between Mean and Z (B) and Beyond Z (C)**

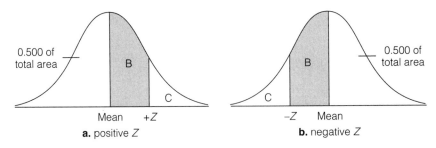

Figure 10.6 **Finding the Area Between the Mean and a Specified Positive Z Score**

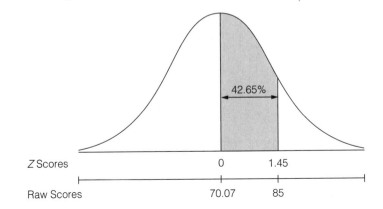

Z Scores   0   1.45

Raw Scores   70.07   85

2. Look up 1.45 in column A (in Appendix B) and find the corresponding area in column B, 0.4265. We can translate this proportion into a percentage (0.4265 × 100 = 42.65%) of the area under the curve included between the mean and a Z of 1.45 (see Figure 10.6).

3. Thus, 42.65 percent of the students scored between 70.07 and 85.

To find the actual number of students who scored between 70.07 and 85, multiply the proportion 0.4265 by the total number of students. Thus, approximately 512 students (0.4265 × 1,200 = 512) obtained a score between 70.07 and 85.

Finding the Area Between the Mean and a Specified Negative Z Score What is the percentage of students whose scores ranged between 65 and 70.07? We can use the standard normal table and the following steps to find out.

1. Convert 65 to a Z score:

$$Z = \frac{65 - 70.07}{10.27} = -0.49$$

2. Because the proportions that correspond to positive Z scores are identical to the proportions corresponding to negative Z scores, we ignore the negative sign of Z and look up 0.49 in column A. The area corresponding to a Z score of 0.49 is 0.1879. This indicates that 0.1879 of the area under the curve is included between the mean and a Z of −0.49 (see Figure 10.7). We convert this proportion to 18.79 percent (0.1879 × 100 = 18.79%).

3. Thus, approximately 225 (0.1879 × 1,200 = 225) students obtained a score between 65 and 70.07.

Figure 10.7 **Finding the Area Between the Mean and a Specified Negative Z Score**

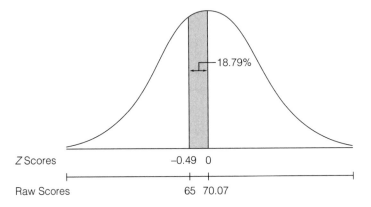

Finding the Area Between Two Z Scores on the Same Side of the Mean Suppose the grade of B was assigned to students who scored between 74 and 84. What is the percentage of students who obtained a B during the ten-year period for which the data were collected?

1. First, find the Z scores corresponding to 74 and 84:

$$Z = \frac{74 - 70.07}{10.27} = 0.38 \qquad Z = \frac{84 - 70.07}{10.27} = 1.36$$

2. Look up the areas corresponding to the Z scores. We find that 0.38 corresponds to an area of 0.1480 (14.80%) and 1.36 corresponds to an area of 0.4131 (41.31%). The area in which we are interested is shown in Figure 10.8.

3. To find the area highlighted in Figure 10.8, subtract the smaller area (the area corresponding to a Z of 0.38, or 14.80%) from the larger area (the

Figure 10.8 **Finding the Area Between Two Z Scores on the Same Side of the Mean**

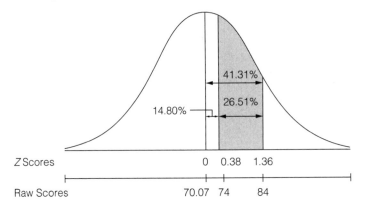

area corresponding to a Z of 1.36, or 41.31%). The area included between the scores of 74 and 84 is 41.31% − 14.80% = 26.51%.

4. Thus, 26.51 percent of all students scored between 74 and 84.

Finding the Area Between Two Z Scores on Opposite Sides of the Mean   When the scores we are interested in lie on opposite sides of the mean, we add the areas together rather than subtract one from the other. For example, suppose we want to find the number of students who scored between 62 and 72.

1. First, find the Z scores corresponding to 62 and 72:

$$Z = \frac{72 - 70.07}{10.27} = 0.19 \qquad Z = \frac{62 - 70.07}{10.27} = -0.79$$

2. Look up the areas corresponding to these Z scores. We find that 0.19 corresponds to an area of 0.0753 (or 7.53%) and −0.79 corresponds to an area of 0.2852 (28.52%) (see Figure 10.9).

3. Because the scores are on opposite sides of the mean, add the two areas obtained in step 2. The total area between these scores is 7.53% + 28.52% = 36.05%.

4. The number of students who scored between 62 and 72 is 433 (1,200 × 0.3605).

Figure 10.9   **Finding the Area Between Two Z Scores on Opposite Sides of the Mean**

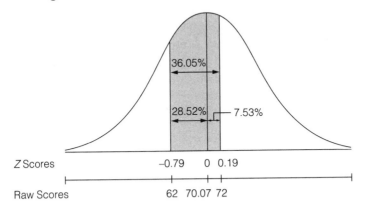

Finding the Area Above a Positive Z Score or Below a Negative Z Score   We can compare students who have done very well or very poorly to get a better idea of how they compare with other students in the class.

To identify students who did very well, we selected all students who scored above 85. To find how many students scored above 85, first convert 85 to a Z score:

Figure 10.10 **Finding the Area Above a Positive Z Score or Below a Negative Z Score**

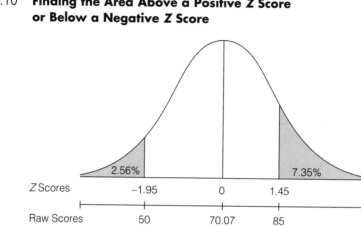

$$Z = \frac{85 - 70.07}{10.27} = 1.45$$

Thus, the Z score corresponding to a final score of 85 in statistics is equal to 1.45.

The area beyond a Z of 1.45 includes all students who scored above 85. This area is shown in Figure 10.10. To find the proportion of students whose scores fall into this area, refer to the entry in column C that corresponds to a Z of 1.45, 0.0735. This means that 7.35 percent (0.0735 × 100 = 7.35%) of the students scored above 85. To find the actual number of students in this group, multiply the proportion 0.0735 by the total number of students. Thus, there were 1,200 × 0.0735, or about 88 students, who scored above 85 over the ten-year period.

A similar procedure can be applied to identify the number of students who did not do well in the class. The cutoff point for poor performance in this class was the score of 50. To determine how many students did poorly, we first converted 50 to a Z score:

$$Z = \frac{50 - 70.07}{10.27} = -1.95$$

The Z score corresponding to a final score of 50 is equal to –1.95. The area beyond a Z of –1.95 includes all students who scored below 50. This area is also shown in Figure 10.10. Locate the proportion of students in this area in column C, in the entry corresponding to a Z of 1.95. (Remember the proportions corresponding to positive or negative Zs are identical.) This proportion is equal to 0.0256. Thus, 2.56 percent (0.0256 × 100 = 2.56%) of the group, or about 31 (0.0256 × 1,200) students, performed poorly in statistics.

Transforming Proportions (or Percentages) into Z Scores

The examples in this section are applications that require transforming proportions (or percentages) into Z scores.

Finding a Z Score Bounding an Area Above It   Assuming an A is assigned to the top 10 percent of the students, what would it take to get an A in the class? To answer this question we need to identify the cutoff point for the top 10 percent of the class. This problem involves two steps:

1. Find the Z score that bounds the top 10 percent, or 0.1000 (0.1000 × 100 = 10%), of all the students who took statistics (see Figure 10.11).

    Refer to the areas under the normal curve, shown in Appendix B. First, look for an entry of 0.1000 (or the value closest to it) in column C. The entry closest to 0.1000 is 0.1003. Then locate the Z in column A that corresponds to this proportion. The Z score associated with the proportion 0.1003 is 1.28.

2. Find the final score associated with a Z of 1.28.

    This step involves transforming the Z score into a raw score. We learned earlier in this chapter (Formula 10.2) that to transform a Z score into a raw score we multiply the Z score by the standard deviation and add that product to the mean. Thus,

$$Y = 70.07 + 1.28(10.27) = 70.07 + 13.15 = 83.22$$

The cutoff point for the top 10 percent of the class is a score of 83.22.

Finding a Z Score Bounding an Area Below It   Now let's assume that an F was assigned to the bottom 5 percent of the class. What would be the cutoff point for a failing score in statistics? Again, this problem involves two steps:

Figure 10.11   **Finding a *Z* Score Bounding an Area Above It**

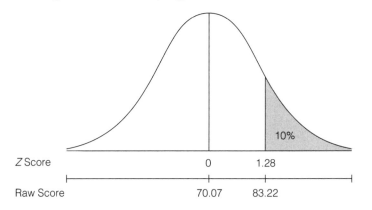

Figure 10.12   **Finding a Z Score Bounding an Area Below It**

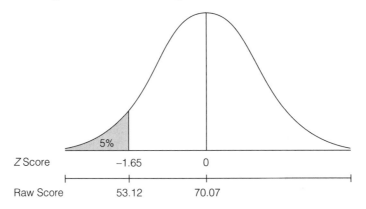

1. Find the Z score that bounds the lowest 5 percent, or 0.0500, of all the students who took the class (see Figure 10.12).

   Refer to the areas under the normal curve, and look for an entry of 0.0500 (or the value closest to it) in column C. The entry closest to 0.0500 is 0.0495. Then locate the Z in column A that corresponds to this proportion, 1.65. Because the area we are looking for is on the left side of the curve—that is, below the mean—the Z score is negative. Thus, the Z associated with the lowest 0.0500 (or 0.0495) is –1.65.

2. To find the final score associated with a Z of –1.65, convert the Z score to a raw score:

   $$Y = 70.07 + (-1.65)(10.27) = 70.07 - 16.95 = 53.12$$

   The cutoff for a failing score in statistics is 53.12.

Learning Check.   *Can you find the number of students who got a score of at least 90 in the statistics course? How many students got a score below 60?*

Working with Percentiles

In Chapter 4 we defined percentiles as scores below which a specific percentage of the distribution falls. For example, the 95th percentile is a score that divides the distribution so that 95 percent of the cases are below it and 5 percent are above it. How are percentile ranks determined? How do you convert a percentile rank to a raw score? To determine the percentile rank of a raw score requires transforming Z scores into proportions or percentages. Converting percentile ranks to raw scores is based on transforming proportions or percentages into Z scores. In the following examples, we illustrate both procedures based on our statistics scores example.

Figure 10.13 **Finding the Percentile Rank of a Score Higher Than the Mean**

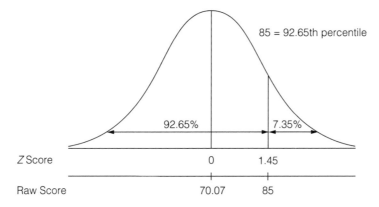

Finding the Percentile Rank of a Score Higher Than the Mean Suppose you are one of the 1,200 students who took the statistics course. Your final score in the course was 85. How well did you do relative to the other students who took the class? To evaluate your performance you must translate your raw score into a percentile rank. Figure 10.13 illustrates this problem. To find the percentile rank of a score higher than the mean, follow these steps.

1. Convert the raw score to a Z score:

$$Z = \frac{85 - 70.07}{10.27} = 1.45$$

The Z score corresponding to a raw score of 85 is 1.45.

2. Find the area beyond Z in Appendix B, column C. The area beyond a Z score of 1.45 is 0.0735.

3. Subtract the area from 1.00 and multiply by 100 to obtain the percentile rank:

percentile rank = $(1.0000 - 0.0735 = 0.9265)100 = 92.65\%$

Being in the 92.65th percentile means that 92.65 percent of all the students enrolled in social statistics scored lower than 85 and 7.35 percent scored higher than 85.

Finding the Percentile Rank of a Score Lower Than the Mean Now let's say that you were unfortunate enough to obtain a score of 65 in the class. What is your percentile rank? Again, to evaluate your performance you must translate your raw score into a percentile rank. Figure 10.14 illustrates this problem. To find the percentile rank of a score lower than the mean, follow these steps.

1. Convert the raw score to a Z score:

Figure 10.14 **Finding the Percentile Rank of a Score Lower Than the Mean**

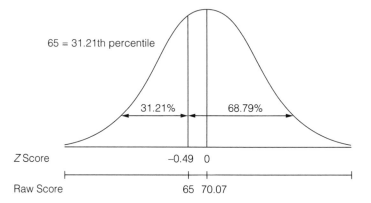

$$Z = \frac{65 - 70.07}{10.27} = -0.49$$

The Z score corresponding to a raw score of 65 is –0.49.

2. Find the area beyond Z in Appendix B, column C. The area beyond a Z score of –0.49 is 0.3121.

3. Multiply the area by 100 to obtain the percentile rank:

percentile rank = 0.3121(100) = 31.21%

The 31.21th percentile rank means that 31.21 percent of all the students enrolled in social statistics did worse than you (that is, 31.21% scored lower than 65 but 68.79% scored higher than 65).

Learning Check. *In Chapter 4 we learned to identify percentiles using cumulative percentages in a distribution. Examine Table 10.1 and find the 92nd percentile. Does your answer differ from the results we obtained earlier (finding the percentile rank of a score higher than the mean)? If it does, explain why.*

Finding the Raw Score Associated with a Percentile Higher Than 50 Now let's assume that our graduate program in sociology will accept only students who scored at the 95th percentile. What is the cutoff point required for admission? Figure 10.15 illustrates this problem. To find the score associated with a percentile higher than 50, follow these steps.

1. Divide the percentile by 100 to find the area below the percentile rank:

$$\frac{95}{100} = 0.9500$$

Figure 10.15 **Finding the Raw Score Associated with a Percentile Higher Than 50**

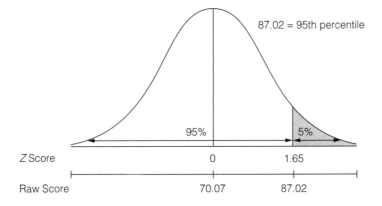

2. Subtract the area below the percentile rank from 1.00 to find the area above the percentile rank:

   $1.0000 - 0.9500 = 0.0500$

3. Find the Z score associated with the area above the percentile rank.
   Refer to the area under the normal curve, shown in Appendix B. First, look for an entry of 0.0500 (or the value closest to it) in column C. The entry closest to 0.0500 is 0.0495. Now locate the Z in column A that corresponds to this proportion, 1.65.

4. Convert the Z score to a raw score:

   $Y = 70.07 + 1.65(10.27) = 70.07 + 16.95 = 87.02$

   The final statistics score associated with the 95th percentile is 87.02. This means that you will need a score of 87.02 or higher to be admitted to the graduate program in sociology.

> **Learning Check.** *In a normal distribution, how many standard deviations from the mean is the 95th percentile? If you can't answer this question, review the material in this section.*

Finding the Raw Score Associated with a Percentile Lower Than 50 Finally, what is the score associated with the 40th percentile? To find the percentile rank of a score lower than 50 follow these steps (see Figure 10.16).

1. Divide the percentile by 100 to find the area below the percentile rank:

   $\dfrac{40}{100} = 0.4000$

2. Find the Z score associated with this area.

Figure 10.16 **Finding the Raw Score Associated with a Percentile Lower Than 50**

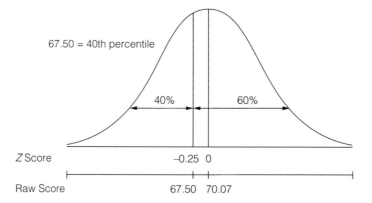

Refer to the area under the normal curve, shown in Appendix B. First, look for an entry of 0.4000 (or the value closest to it) in column C. The entry closest to 0.4000 is 0.4013. Now locate the Z in column A that corresponds to this proportion. The Z score associated with the proportion 0.4013 is –0.25.

3. Convert the Z score to a raw score:

$$Y = 70.07 + (-0.25)(10.27) = 70.07 - 2.568 = 67.50$$

The final statistics score associated with the 40th percentile is 67.50. This means that 40 percent of the students scored below 67.50 and 60 percent scored above it.

Learning Check.   *What is the raw score in statistics associated with the 50th percentile?*

■ ■ ■ ■ **A Final Note**

In this chapter we have learned how the properties of the theoretical normal curve can be applied to describe important characteristics of empirical distributions that are approximately normal. The normal curve has other practical applications as well, however, beyond the description of "real life" distributions. In subsequent chapters we will see that the normal distribution also enables us to describe the characteristics of a theoretical distribution—the sampling distribution—of great significance in inferential statistics. The techniques learned in this chapter—transforming scores and finding areas under the normal curve—will be used in many of the procedures described in subsequent chapters. Make sure you understand these techniques before you proceed to the next chapter.

MAIN POINTS

- The normal distribution is central to the theory of inferential statistics. It also provides a model for many empirical distributions that approximate normality.

- In all normal or nearly normal curves, we find a constant proportion of the area under the curve lying between the mean and any given distance from the mean when measured in standard deviation units.

- The standard normal distribution is a normal distribution represented in standard scores, or Z scores. Z scores express the number of standard deviations that a given score is above or below the mean. The proportions corresponding to any Z score or its fraction are organized into a special table called the standard normal table.

KEY TERMS

normal distribution   *standard normal table*
standard normal distribution   *standard (Z) score*

SPSS DEMONSTRATIONS

Demonstration 1: Producing Z Scores with SPSS
[Module A]

In this chapter we have discussed the theoretical normal curve, Z scores, and the relationship between raw scores and Z scores. The SPSS Descriptives procedure can calculate Z scores for any distribution. We'll use it to study the distribution of occupational prestige in the 1996 GSS file. Locate the Descriptives procedure in the *Analyze* menu, under *Descriptive Statistics*, then *Descriptives*. We can select one or more variables to place in the Variable(s) box; for now, we'll just place PRESTG80 in this box (see Figure 10.17). A checkbox in the bottom left corner tells SPSS to create standardized

Figure 10.17

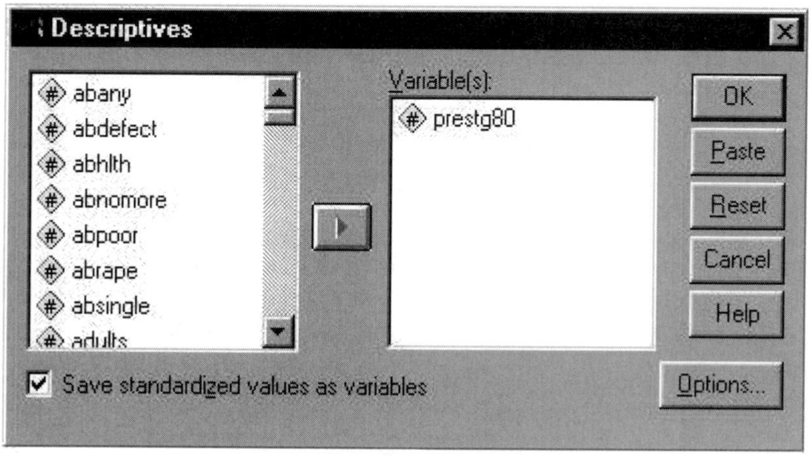

values, or Z scores, as new variables. Any new variable is placed in a new column in the Data Editor Window and will then be available for additional analyses. Click on *OK* to run the procedure.

The output from Descriptives (Figure 10.18) is brief, listing the mean and standard deviation for PRESTG80, plus the minimum and maximum values and the number of valid cases.

Figure 10.18

Descriptive Statistics

	N	Minimum	Maximum	Mean	Std. Deviation
PRESTG80 RS OCCUPATIONAL PRESTIGE SCORE (1980)	1370	17	86	43.34	14.33
Valid N (listwise)	1370				

Though not indicated in the Output1 window, SPSS has created a new Z score variable for PRESTG80. To see this new variable, switch to the Data Editor window by clicking on *Window* from the main menu, then on the file GSS96.SAV. Then go to the last column by pressing the End key (see Figure 10.19). By default, SPSS appends a Z to the variable name, so the new variable is called ZPRESTG8. Notice that the last character of the old name (0) had to be dropped because SPSS has a limit of eight characters for any variable name.

Figure 10.19

racesex	relig	sex	sphmewrk	zprestg8
3.00	3	1	0	-.93124
4.00	1	2	0	-.65201
2.00	1	2	0	.95356
2.00	1	2	0	1.51201
2.00	4	2	0	.18568
1.00	4	1	0	-.79162
1.00	4	1	0	.18568

The first case in the file has a Z score of –.93124, so the prestige score for this person must be below the mean of 43.34. If we locate the respondent's PRESTG80 score, we see that the score for this person was 30 (not pictured), below the mean as we expected. Anywhere SPSS has placed a system missing value in the column for ZPRESTG8, it means that the original prestige score is missing.

If the data file is saved, the new Z score variable will be saved along with the original data and then can be used in analyses. In addition, if we have SPSS calculate the mean and standard deviation of ZPRESTG8, we find that they are equal to 0 and 1.00, respectively.

SPSS PROBLEMS [MODULE A]

1. The majority of variables that social scientists study are not normally distributed. This doesn't typically cause problems in analysis when the goal of a study is to calculate means and standard deviations—as long as sample sizes are greater than about 50. (This will be discussed in later chapters.) However, when characterizing the distribution of scores in *one* sample, or in a complete population (if this information is available), a nonnormal distribution can cause complications. We can illustrate this point by examining the distribution of age in the GSS96 file.
   a. Access this file, then create a histogram for AGE with a superimposed normal curve. [Click on the option *Display normal curve.*] How does the distribution of AGE deviate from the theoretical normal curve?
   b. Calculate the mean and standard deviation for AGE in this sample, using either the Frequencies or Descriptives procedure.
   c. Assuming the distribution of AGE is normal, calculate the number of people who should be 25 years of age or less.
   d. Use Frequencies to get a table of the percentage of cases at each value of AGE. Compare the theoretical calculation in (c) with the actual distribution of age in the sample. What percentage of people in the sample are 25 years old or less? Is this value close to what you calculated? Why might there be a discrepancy?

2. SPSS will calculate standard scores for any distribution. Examine the distribution of HRS1 (number of hours worked last week).
   a. Access the 1996 GSS file. Have SPSS calculate Z scores for HRS1.
   b. What is the equivalent Z score for someone who worked 60 hours last week?
   c. Use the Frequencies procedure to find the percentile rank, in this sample, for a score of 60.
   d. Does the percentile rank you found from Frequencies correspond to the Z score for a value of 60? In other words, is the distribution of hours worked last week normal? If so, then the Z score SPSS calculates should be very close, after transforming it into an appropriate area, to the percentile rank for that same score.
   e. Create histograms for HRS1 and the new variable ZHRS1. Explain why they have the same shape.

3. Repeat the procedure in Exercise 2, this time running separate analyses for men versus women and blacks versus whites based on the variable PRESTG80. Remember, you can run separate analyses using the *Data–*

Split File command. Is there a difference in PRESTG80 among men/women and blacks/whites in the GSS sample? How would you describe the distribution of PRESTG80 for the four groups?

CHAPTER EXERCISES

1. It is increasingly true that government agencies, on all levels, use examinations to screen applicants and remove bias from the hiring process. Consider a police department in a large midwestern city that uses such an examination to hire new officers. The mean score for all applicants this year on the exam is 98, with a standard deviation of 13. The distribution of scores for the applicants is approximately normal.
   a. Assume that only 12 percent of all applicants can be accepted this year. Will an applicant be accepted if his or her score on the exam is 115?
   b. What is the cutoff score of this year's test? In other words, what score is above 88 percent of all scores in the distribution?
   c. What is the Z value for this score?

2. If a particular distribution you are studying is not normal, it may be difficult to determine the area under the curve of the distribution or translate a raw score into a Z value. Is this statement true? Why or why not?

3. In 1990 the average population of all the countries in the world (192 countries) was 27.3 million, with a standard deviation of 105.7.
   a. China's population in 1990 was 1,101 million (that is, 1 billion, 101 million). If the population distribution is approximately normal, convert the value of China's population to a Z score.
   b. For a normal distribution, what percentage of cases should fall less than 1 standard deviation below the mean, or equivalently, below a Z score of –1? How many countries would fall below this value in 1990? (*Hint:* You don't need a listing of each country's population to answer this question.)
   c. What does your answer in (b) imply about the shape of the distribution of population for the 192 countries? When a distribution isn't normal, what statistic is a better measure of central tendency than the mean?

4. A social psychologist has developed a test to measure gregariousness. The test is normed so that it has a mean of 70 and a standard deviation of 20, and the gregariousness scores are normally distributed in the population of college students used to develop the test.
   a. What is the percentile rank of a score of 40?
   b. What percentage of scores falls between 35 and 90?
   c. What is the standard score for a test score of 65?
   d. What proportion of students should score above 115?
   e. What is the cutoff score below which 87 percent of all scores fall?

5. The 1991 General Social Survey provides the following statistics for the average income of males and females, and their associated standard deviations.

	Mean	**Standard Deviation**	**N**
Males	$22,052.51	$17,734.92	434
Females	$14,331.21	$12,165.89	448

a. Assuming that income is normally distributed in the population, what proportion of males have incomes between $30,000 and $40,000? What proportion of females have incomes in the same range?

b. What is the probability that a male, drawn at random from the population, will have an income over $50,000? What is the equivalent probability for a female drawn at random?

c. What is the probability that a male or female will have an income below $15,000?

d. Find the upper and lower income limits, centered around the mean, that will include 50 percent of all females.

e. If income is actually positively skewed in the population, how would that change your other answers?

6. The following table displays information for each U.S. state (and the District of Columbia) on two variables. Use the first variable, concerning the living conditions of children, in this problem. (Including the District of Columbia, there are 51 scores.)

Severely distressed neighborhoods have been defined as having at least four of these five characteristics: a poverty rate above 27.5 percent; at least 39.6 percent of families headed by females; a high school dropout rate above 23.3 percent; more than 17 percent of families on welfare; and more than 46.5 percent of males out of the labor force (not working or seeking work). The percentage of children living in such neighborhoods, by state, is shown in the table.

	% of Children in Distressed Neighborhoods	**% of Eligible Voters Who Voted in the 1992 Election**
Alabama	9.6	54
Alaska	0.5	51
Arizona	5.2	51
Arkansas	7.6	53
California	5.1	45
Colorado	2.5	62
Connecticut	5.7	63
Delaware	1.7	41
D.C.	25.2	47
Florida	4.6	49
Georgia	6.2	46

(continued on next page)

	% of Children in Distressed Neighborhoods	% of Eligible Voters Who Voted in the 1992 Election
Hawaii	0.5	41
Idaho	0.0	53
Illinois	9.5	58
Indiana	3.0	54
Iowa	1.0	64
Kansas	2.0	62
Kentucky	7.1	53
Louisiana	17.2	59
Maryland	6.2	51
Maine	0.6	71
Massachusetts	5.1	60
Michigan	11.5	62
Minnesota	2.0	70
Mississippi	17.4	52
Missouri	5.2	62
Montana	2.3	69
Nebraska	1.3	62
Nevada	3.3	48
New Hampshire	0.1	62
New Jersey	4.9	54
New Mexico	4.5	51
New York	12.8	48
North Carolina	2.6	49
North Dakota	2.1	67
Ohio	8.1	60
Oklahoma	3.1	60
Oregon	1.1	55
Pennsylvania	6.9	54
Rhode Island	5.1	54
South Carolina	4.4	44
South Dakota	3.6	67
Texas	4.8	49
Tennessee	7.3	52
Utah	0.3	62
Vermont	0.6	64
Virginia	2.8	52
Washington	2.4	51
West Virginia	2.9	50
Wisconsin	5.4	68
Wyoming	0.2	62

Source: Data on children in distressed neighborhoods come from the *1994 Kids Count Data Book.*

a. What are the mean and the standard deviation for the percentage of children in distressed neighborhoods for all states?

b. Using the information from (a), how many states fall more than 1 standard deviation above the mean? How does this number compare with the number expected from the theoretical normal curve distribution? Can you suggest anything these states have in common that might cause them to have more children in distressed neighborhoods?

c. How many states fall more than 1 standard deviation below the mean? Is this number greater or lower than the expected value from the theoretical normal curve? Again, can you suggest any characteristics these states have in common that might cause them to have fewer children in distressed neighborhoods?

d. Create a histogram of the percentages of children in distress. Does the distribution appear to be normal? Use this information to further explain why the number of states falling more than 1 standard deviation below the mean differs from the expected value.

7. Refer to the data in Chapter 5, Exercise 6, on the occupational prestige of whites and blacks. Assume that occupational prestige is normally distributed in each population.

a. What percentage of whites should have occupational prestige scores above 60?

b. What percentage of blacks should have occupational prestige scores above 60?

c. What proportion of whites have prestige scores between 30 and 70?

d. Given that the black sample size is 73, how many blacks in the sample have an occupational prestige score between 50 and 60?

8. SAT scores are normed so that, in any year, the mean of the verbal or math test should be 500 and the standard deviation 100. Assuming this is true (it is only approximately true, both because of variation from year to year and because scores have decreased since the SAT tests were first developed), answer the following questions.

a. What percentage of students score above 625 on the math SAT in any given year?

b. What percentage of students score between 400 and 600 on the verbal SAT?

c. A college decides to liberalize its admission policy. As a first step, the admissions committee decides to exclude only those applicants scoring below the 20th percentile on the verbal SAT. Translate this percentile into a Z score. Then calculate the equivalent SAT verbal test score.

9. The Chicago police department was asked by the mayor's office to estimate the cost of crime to citizens of Chicago. The police began their study with the crime of burglary, relying on a random sample of 500 files (there is too much crime to calculate statistics for all the crimes

committed). They found the average dollar loss in a burglary was $678, with a standard deviation of $560, and that the dollar loss was normally distributed.

a. What proportion of burglaries had dollar losses above $1,000?

b. What percentage of burglaries had dollar losses between $200 and $300?

c. What is the probability that any one burglary had a dollar loss above $400?

d. What proportion of burglaries had dollar losses below $500?

10. The number of hours people work each week varies widely for many reasons. Using the 1996 General Social Survey, you find that the mean number of hours worked last week was 42.36 with a standard deviation of 14.27 hours, based on a sample size of 954.

a. Assume that *hours worked* is approximately normally distributed in the sample. What is the probability that someone in the sample will work 60 or more hours in a week? How many people in the sample of 954 should have worked 60 or more hours?

b. What is the probability that someone will work 30 or fewer hours in a week (that is, work part-time)? How many people does this represent in the sample?

c. What number of hours worked per week corresponds to the 60th percentile?

11. We discovered that 405 GSS respondents watched television for an average of 2.91 hours a day, with a standard deviation of 1.62. Answer the following questions assuming the distribution of the number of television hours is normal.

a. What is the Z score for a person who watches more than 8 hours per day?

b. What proportion of people watch television less than 5 hours a day? How many does this correspond to in the sample?

c. What number of television hours per day corresponds to a Z score of −1.3?

e. What is the percentage of people who watch between 1 and 6 hours of television per day?

12. A company tests applicants for a job by giving writing and software proficiency tests. The means and standard deviations for each exam follow, along with the scores for two applicants, Bill and Ted. Assume test scores are normally distributed.

Exam	Mean	Standard Deviation	Bill	Ted
Writing	56.4	9.3	65	67
Software use	68.7	5.6	70	75

a. On which test did Bill do better, relative to the other applicants? Calculate appropriate statistics to answer this question.

b. On which test did Ted do better, relative to the other applicants? Calculate statistics to answer this question.

c. What proportion of applicants scored below Bill's Software Use test score?

d. What is the percentile rank of Ted's Writing score of 67?

13. What is the value of the mean for any standard normal distribution? What is the value of the standard deviation for any standard normal distribution? Explain why this is true for any standard normal distribution.

14. You are asked to do a study of shelters for abused and battered women to determine the necessary capacity in your city to provide housing for most of these women. After recording data for a whole year, you find that the mean number of women in shelters each night is 250, with a standard deviation of 75. Fortunately, the distribution of the number of women in the shelters each night is normal, so you can answer the following questions posed by the city council.

a. If the city's shelters have a capacity of 350, will that be enough places for abused women on 95 percent of all nights? If not, what number of shelter openings will be needed?

b. The current capacity is only 220 openings because some shelters have closed. What is the percentage of nights that the number of abused women seeking shelter will exceed current capacity?

15. Based on the chapter discussion:

a. What are the properties of the normal distribution? Why is it called "normal"?

b. What is the meaning of a positive (+) Z score? What is the meaning of a negative (–) Z score?

11 Sampling and Sampling Distributions

SPSS DEMONSTRATION

SPSS PROBLEM

CHAPTER EXERCISES

■ ■ ■ ■ **Introduction**

Until now we have ignored the question of who or what should be observed when we collect data or whether the conclusions based on our observations can be generalized to a larger group of observations. The truth is that we are rarely able to study or observe everyone or everything we are interested in. Though we have learned about various methods to analyze observations, remember that these observations represent only a tiny fraction of all the possible observations we might have chosen. Consider the following examples.

> *Example 1* The student union on your campus is trying to find out how it can better address the needs of commuter students and has commissioned you to conduct a needs assessment survey. You have been given enough money to survey about 500 students. Given that there are nearly 15,000 commuters on your campus, is this an impossible task?

> *Example 2* Your chancellor has appointed a task force to investigate issues of concern to the lesbian, gay, and bisexual community at the university. The task force has been charged with assessing the campus climate for members of these university communities and studying the coverage of lesbian, gay, and bisexual subjects in the curriculum. There are about 30,000 students, faculty, and staff on your campus and about 2,000 courses offered every year. How should the task force proceed?

What do these problems have in common? In both situations the major problem is that there is too much information and not enough resources to collect and analyze all of it.

■ ■ ■ ■ **Aims of Sampling**[1]

Researchers in the social sciences almost never have enough time or money to collect information about the entire group that interests them. Known as the **population**, this group includes all the cases (individuals, objects, or groups) in which the researcher is interested. For example, in our first illustration the population is all 15,000 commuter students; the population in the second illustration consists of all 30,000 faculty, staff, and students and 2,000 courses.

---

[1]This discussion has benefited from a more extensive presentation on the aims of sampling, in Richard Maisel and Caroline Hodges Persell, *How Sampling Works* (Thousand Oaks, CA: Pine Forge Press, 1996).

Fortunately, we can learn a lot about a population if we carefully select a subset of it. This subset is called a **sample**. Through the process of *sampling*—selecting a subset of observations from the population of interest—we attempt to generalize to the characteristics of the larger group (population) based on what we learn from the smaller group (the sample). This is the basis of *inferential statistics*—making predictions or inferences about a population from observations based on a sample.

The term **parameter**, associated with the population, refers to measures used to describe the distribution of the population we are interested in. For instance, the average commuting time for *all* 15,000 students on your campus is a population parameter because it refers to a population characteristic. In previous chapters we have learned many ways of describing a distribution, such as a proportion, a mean, or a standard deviation. When used to describe the population distribution, these measures are referred to as parameters. Thus, a population mean, a population proportion, and a population standard deviation are all population parameters.

Population A group that includes all the cases (individuals, objects, or groups) in which the researcher is interested.

Sample A relatively small subset selected from a population.

We use the term **sample statistic** when referring to a corresponding characteristic calculated for the sample. For example, the average commuting time for a *sample* of commuter students is a sample statistic. Similarly, a sample mean, a sample proportion, and a sample standard deviation are all sample statistics.

In this and the following chapter we will discuss some of the principles involved in generalizing results from samples to the population. In our discussion we will use different notations when referring to sample statistics and population parameters. Table 11.1 presents the sample notation and the corresponding population notation.

The distinctions between a sample and a population and between a parameter and a statistic are illustrated in Figure 11.1. We've included for illustration the population parameter of .60—the proportion of white

Table 11.1   **Sample and Population Notations**

Measure	Sample Notation	Population Notation
Mean	$\bar{Y}$	μ_Y
Proportion	p	π
Standard deviation	S_Y	σ_Y
Variance	S_Y^2	σ_Y^2

Figure 11.1   **The Proportion of White Respondents in a Population and in a Sample**

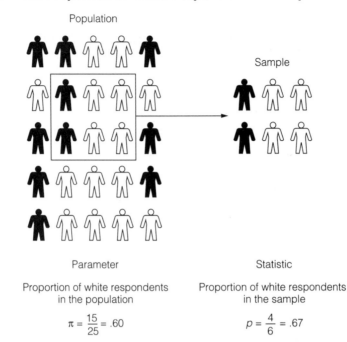

Population

Sample

Parameter

Proportion of white respondents
in the population

$$\pi = \frac{15}{25} = .60$$

Statistic

Proportion of white respondents
in the sample

$$p = \frac{4}{6} = .67$$

respondents in the population. However, since we almost never have enough resources to collect information about the population, it is rare that we know the value of a parameter. The goal of most research is to find the population parameter. Researchers usually select a sample from the population to obtain an estimate of the population parameter. Thus, the major objective of sampling theory and statistical inference is to provide estimates of unknown parameters from sample statistics that can be easily obtained and calculated.

Parameter A measure (for example, mean or standard deviation) used to describe the population distribution.

Statistic A measure (for example, mean or standard deviation) used to describe the sample distribution.

Learning Check.   *It is important that you understand what the terms* population, sample, parameter, *and* statistic *mean. Use your own words so the meaning makes sense to you. If you cannot clearly define these terms, review the preceding material. You will see these sample and population notations over and over again. If you memorize them, you will find it much easier to understand the formulas used in inferential statistics.*

■ ■ ■ ■ **Some Basic Principles of Probability**

In the following sections we will discuss a variety of techniques adopted by social scientists to select samples from populations. The techniques all follow a general approach called *probability sampling*. Before we discuss these techniques, we will review some basic principles of probability.

We all use the concept of probability in everyday conversation. We might ask, "What is the probability that it will rain tomorrow?" or "What is the likelihood that we will do well on a test?" In everyday conversations our answers to these questions are rarely systematic, but in the study of statistics *probability* has a far more precise meaning.

Probability theory applies to situations in which we can specify the possible outcomes resulting from a certain situation or task. For example, the task of rolling a die has six possible outcomes: 1, 2, 3, 4, 5, 6. The task of flipping a coin results in either of two outcomes: heads or tails. For any such situation or task, the probability of an outcome occurring is defined as the ratio (over the long run) of the number of times the desired outcome can occur relative to the total number of times all the outcomes can occur. Probability estimates outcomes as the number of trials becomes infinite. For instance, over the long run, the probability of getting heads when flipping an evenly balanced coin is 1 to 2, or ½, because heads can occur only once within a total of two possible outcomes (heads and tails). Similarly, over the long run, the probability of rolling a 3 on a die is ⅙ because the outcome 3 can occur only once within a total of 6 possible equally likely outcomes.

Probabilities are usually measured in terms of proportions. A ratio can be converted to a proportion by dividing its numerator by the denominator. Thus, the probability of ½ is equivalent to .500 (½ = .500); similarly, .167 corresponds to the ratio ⅙ (⅙ = .167). Probabilities can range from 0 to 1. The closer the probability is to 1, the more likely it is that the event will occur. A probability that is close to 0 means that an event is highly unlikely.

> **Learning Check.** *What is the probability of drawing an ace out of a normal deck of 52 playing cards? It's not ¹/₅₂. There are four aces, so the probability of drawing one of them is ⁴/₅₂ or ¹/₁₃. The proportion is .077. The probability of drawing the ace of spades is ¹/₅₂.*

■ ■ ■ ■ **Probability Sampling**

Social researchers are usually much more systematic in their effort to obtain samples that are representative of the population than we are when we gather information in our everyday life. Such researchers have adopted a number of approaches for selecting samples from populations. Only one general approach, *probability sampling,* allows the researcher to use the principles of statistical inference to generalize from the sample to the population.

Probability sampling is a method that enables the researcher to specify for each case in the population the probability of its inclusion in the sample. The purpose of probability sampling is to select a sample that is as representative as possible of the population. The sample is selected in such a way as to allow use of the principles of probability to evaluate the generalizations made from the sample to the population. A probability sample design enables the researcher to estimate the extent to which the findings based on one sample are likely to differ from what would be found by studying the entire population.

> *Probability sampling*   A method of sampling that enables the researcher to specify for each case in the population the probability of its inclusion in the sample.

Although accurate estimates of sampling error can be made only from probability samples, social scientists often use nonprobability samples because they are more convenient and cheaper to collect. Nonprobability samples are useful under many circumstances for a variety of research purposes. Their main limitation is that they do not allow the use of the method of inferential statistics to generalize from the sample to the population. Because in this and the next chapter we deal only with inferential statistics, we do not discuss nonprobability sampling. In the following sections we will learn about three sampling designs that follow the principles of probability sampling: the simple random sample, the stratified random sample, and the systematic random sample.[2]

The Simple Random Sample

The *simple random sample* is the most basic probability sampling design, and it is incorporated into even more elaborate probability sampling designs. A **simple random sample** is a sample design chosen in such a way as to ensure that (1) every member of the population has an equal chance of being chosen and (2) every combination of N members has an equal chance of being chosen.

Let's take a very simple example to illustrate. Suppose we are conducting a cost-containment study of the ten hospitals in our region, and we want to draw a sample of two hospitals to study intensively. We can put into a hat ten slips of paper, each representing one of the ten hospitals and mix the slips carefully. We select one slip out of the hat and identify the hospital it represents. We then make the second draw and select another slip out of the hat and identify it. The two hospitals we identified on the two draws become the two members of our sample. The sample is a simple ran-

---

[2]The discussion in these sections is based on Chava Frankfort-Nachmias and David Nachmias, *Research Methods in the Social Sciences* (New York: St. Martin's Press, 1996), pp. 183–194.

dom sample because—assuming we made sure the slips were really well mixed—(1) pure chance determined which hospital was selected, (2) every hospital had the same chance of being selected as a member of our sample of two, and (3) every combination of (N = 2) hospitals was equally likely to be chosen.

> **Simple random sample** A sample designed in such a way as to ensure that (1) every member of the population has an equal chance of being chosen and (2) every combination of N members has an equal chance of being chosen.

Researchers usually use computer programs or tables of random numbers in selecting random samples. An abridged table of random numbers is reproduced in Appendix A. To use a random number table, list each member of the population and assign the member a number. Begin anywhere on the table and read each digit that appears in the table in order—up, down, or sideways; the direction does not matter, as long as it follows a consistent path. Whenever we come across a digit in the table of random digits that corresponds to the number of a member in the population of interest, that member is selected for the sample. Continue this process until the desired sample size is reached.

Suppose now that, in your job as a hospital administrator, you are planning to conduct a cost-containment study by examining patients' records. Out of a total of 300 patients' records, you want to draw a simple random sample of 5. You follow these steps:

1. Number the patient accounts, beginning with 001 for the first account and ending with 300, which represents the 300th account.

2. Use some random process to enter Appendix A (you might close your eyes and point a pencil). For our illustration, let's start with the first column of numbers. Notice that each column lists five-digit numbers. Because your population contains only three-digit numbers (001–300), drop the last two digits of each number and read only the first three digits in each group of numbers. (Alternatively, you could choose any other group of three-digit numbers in this block—for example, the last three digits in the block.)

3. Dropping the last two digits of each five-digit block and proceeding down the column, you obtain the following three-digit numbers:

104*	375	963	289*
223*	779	895	635
241*	995	854	094*
421			

Among these numbers, five correspond to numbers within the range of numbers assigned to the patient records. They are starred. The last number listed is 094 from line 13. You do not need to list more numbers because you already have five different numbers that qualify for inclusion in the sample. The starred numbers represent the records you will choose for your sample because these are the only ones that fall between 001 and 300, the range you specified.

4. We now have five records in our simple random sample. Let's list them: 104, 223, 241, 289, and 094.

The Systematic Random Sample

Now let's look at a sampling method that is easier to implement than a simple random sample. The *systematic random sample,* although not a true probability sample, provides results very similar to those obtained with a simple random sample. It uses a ratio, K, obtained by dividing the population size by the desired sample size:

$$K = \frac{\text{population size}}{\text{sample size}}$$

Systematic random sampling is a method of sampling in which every Kth member in the total population is chosen for inclusion in the sample after the first member of the sample is selected at random from among the first K members in the population.

Recall our example in which we had a population of 15,000 commuting students and our sample was limited to 500. In this example,

$$K = \frac{15,000}{500} = 30$$

Using a systematic random sampling method, we first choose any one student at random from among the first 30 students on the list of commuting students. Then we select every 30th student after that until we reach 500, our desired sample size. Suppose that our first student selected at random happens to be the 8th student on the list. The second student in our sample is then 38th on the list ($8 + 30 = 38$). The third would be $38 + 30 = 68$, the fourth, $68 + 30 = 98$, and so on. The systematic random sample is illustrated in Figure 11.2.

> *Systematic random sampling* A method of sampling in which every Kth member (K is a ratio obtained by dividing the population size by the desired sample size) in the total population is chosen for inclusion in the sample after the first member of the sample is selected at random from among the first K members in the population.

> **Learning Check.** *How does a systematic random sample differ from a simple random sample?*

Figure 11.2 **Systematic Random Sampling**

From a population of 40 students, let's select a systematic random sample of 8 students. Our skip interval will be 5 (40 ÷ 8 = 5). Using a random number table, we choose a number between 1 and 5. Let's say we choose 4. We then start with student 4 and pick every 5th student:

Our trip to the random number table could have just as easily given us a 1 or a 5, so all the students do have a chance to end up in our sample.

The Stratified Random Sample

A third type of probability sampling is the *stratified random sample*. We obtain a **stratified random sample** by (1) dividing the population into subgroups based on one or more variables central to our analysis and (2) then drawing a simple random sample from each of the subgroups. We could stratify by race/ethnicity, for example, by dividing the population into different racial/ethnic groups and then drawing a simple random sample from each group. For instance, suppose we want to compare the attitudes of Hispanics toward abortion with the attitudes of white and black respondents. Our population of interest consists of 1,000 individuals, with 700 (or 70%) whites, 200 (20%) blacks, and 100 (10%) Hispanics. Because we know the proportion of each subgroup in the population, we may want to draw a stratified sample that would reflect these exact proportions. For instance, we can draw a stratified sample size of $N = 180$ that includes 126 (70%) whites, 36 (20%) blacks, and 18 (10%) Hispanics. In such a **proportionate stratified sample**, the size of the sample selected from each subgroup is proportional to the size of that subgroup in the entire population.

In a **disproportionate stratified sample**, the size of the sample selected from each subgroup is deliberately made disproportional to the size of that subgroup in the population. For instance, for our example we could select a

---

Box 11.1 Disproportionate Stratified Samples and Diversity

Disproportionate stratified sampling is especially useful given the increasing diversity of American society. In a diverse society factors such as race, ethnicity, class and gender, as well as other categories of experience such as age, religion, and sexual orientation, become central in shaping our experiences and defining the differences among us. These factors are an important dimension of the social structure, and they not only operate independently but also are experienced simultaneously by all of us.* For example, if you are a white woman, you may share some common experiences with a woman of color based on your gender, but your racial experiences are going to be different. Moreover, your experiences within the race/gender system are further conditioned by your social class. Similarly, if your are a man, your experiences are shaped as much by your class, race, and sexual orientation as they are by your gender. If you are a black gay man, for instance, you might not benefit equally from patriarchy compared with a classmate who is a white heterosexual male.

What are the research implications of an inclusive approach that emphasizes social differences? Such an approach will include women and men in a study of race, Hispanics and people of color when considering class, and women and men of color when studying gender. Such an approach makes the experience of previously excluded groups more visible and central because it puts those who have been excluded at the center of the analysis so that we can better understand the experience of all groups, including those with privilege and power.

What are the sampling implications of such an approach? Let's think of an example. Suppose you are looking at the labor force experiences of black and Hispanic women who are over 50 years of age, and you want to compare these experiences with those of white women in the same age group. Both Hispanic and black women compose a small proportion of the population. A proportional sample probably would not include enough Hispanic or black women to provide an adequate basis for comparison with white women. To make such comparisons, it would be desirable to draw a disproportionate stratified sample that deliberately overrepresents both Hispanic and black women so that these subsamples will be of sufficient size (see Figure 11.3).

*Margaret L. Andersen and Patricia Hill Collins, *Race, Class, and Gender* (Belmont, CA: Wadsworth, 1992), pp. 1–6.

---

sample ($N = 180$) consisting of 90 whites (50%), 45 blacks (25%), and 45 Hispanics (25%). In such a sampling design, although the sampling probabilities for each population member are not equal (they vary between groups), they are *known,* and therefore we can make accurate estimates of error in the inference process.[3] Disproportionate stratified sampling is especially

[3]We discuss more on sampling error in the next section.

Figure 11.3 **A Random Sample Stratified by Race/Ethnicity**

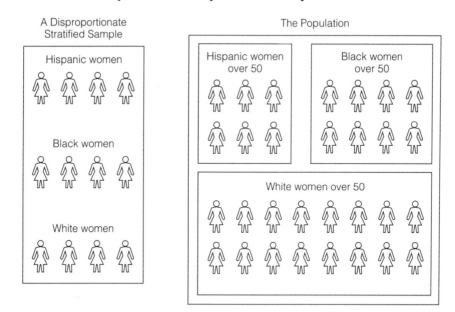

useful when we want to compare subgroups with each other, and when the size of some of the subgroups in the population is relatively small. Proportionate sampling can result in the sample having too few members from a small subgroup to yield reliable information about them.

Stratified random sample A method of sampling obtained by (1) dividing the population into subgroups based on one or more variables central to our analysis and (2) then drawing a simple random sample from each of the subgroups.

Proportionate stratified sample The size of the sample selected from each subgroup is proportional to the size of that subgroup in the entire population.

Disproportionate stratified sample The size of the sample selected from each subgroup is disproportional to the size of that subgroup in the population.

Learning Check. *Can you think of some research questions that could best be studied using a disproportionate stratified random sample? When might it be important to use a proportionate stratified random sample?*

■ ■ ■ ■ **The Concept of the Sampling Distribution**

We began this chapter with a few examples illustrating why researchers in the social sciences almost never collect information on the entire population that interests them. Instead, they usually select a sample from that population and use the principles of statistical inference to estimate the characteristics, or parameters, of that population based on the characteristics, or statistics, of the sample. In this section we describe one of the most important concepts in statistical inference—*sampling distribution.* The sampling distribution helps estimate the likelihood of our sample statistics and, therefore, enables us to generalize from the sample to the population.

The Population

To illustrate the concept of the sampling distribution, let's consider as our population the twenty individuals listed in Table 11.2.[4] Our variable, Y, is the income (in dollars) of these twenty individuals, and the parameter we are trying to estimate is the mean income.

We use the symbol μ_Y to represent the population mean; the Greek letter mu (μ) stands for the mean, and the subscript Y identifies the specific variable, income. Using Formula 4.1, we can calculate the population mean:

$$\mu_Y = \frac{\sum Y}{N} = \frac{Y_1 + Y_2 + Y_3 + Y_4 + Y_5 + \ldots + Y_{20}}{20}$$

$$= \frac{11,350 + 7,859 + 41,654 + 13,445 + 17,458 + \ldots + 25,671}{20}$$

$$= 22,766$$

Using Formula 5.3, we can also calculate the standard deviation for this population distribution. We use the Greek symbol sigma (σ) to represent the population's standard deviation and the subscript Y to stand for our variable, income:

$$\sigma_Y = 14,687$$

Of course, most of the time we do not have access to the population. So instead we draw one sample, compute the mean—the statistic—for that sample, and use it to estimate the population mean—the parameter.

[4]The population of the twenty individuals presented in Table 11.2 is considered a finite population. A finite population consists of a finite (countable) number of elements (observations). Other examples of finite populations include all women in the labor force in 1996 and all public hospitals in New York City. A population is considered infinite when there is no limit to the number of elements it can include. Examples of infinite populations include all women in the labor force, in the past or future. Most samples studied by social scientists come from finite populations. However, it is also possible to sample from an infinite population.

Table 11.2 **The Population: Personal Income for Twenty Individuals (hypothetical data)**

Individual	Income (Y)
Case 1	11,350 (Y_1)
Case 2	7,859 (Y_2)
Case 3	41,654 (Y_3)
Case 4	13,445 (Y_4)
Case 5	17,458 (Y_5)
Case 6	8,451 (Y_6)
Case 7	15,436 (Y_7)
Case 8	18,342 (Y_8)
Case 9	19,354 (Y_9)
Case 10	22,545 (Y_{10})
Case 11	25,345 (Y_{11})
Case 12	68,100 (Y_{12})
Case 13	9,368 (Y_{13})
Case 14	47,567 (Y_{14})
Case 15	18,923 (Y_{15})
Case 16	16,456 (Y_{16})
Case 17	27,654 (Y_{17})
Case 18	16,452 (Y_{18})
Case 19	23,890 (Y_{19})
Case 20	25,671 (Y_{20})
Mean (μ_Y) = 22,766	Standard deviation (σ_Y) = 14,687

The Sample

Let's pretend that μ_Y is unknown and that we estimate its value by drawing a random sample of three individuals ($N = 3$) from the population of twenty individuals and calculate the mean income for that sample. The incomes included in that sample are as follows:

Case 8	18,342
Case 16	16,456
Case 17	27,654

Now let's calculate the mean for that sample:

$$\overline{Y} = \frac{18,342 + 16,456 + 27,654}{3} = 20,817$$

Notice that our sample mean ($\overline{Y}$), \$20,817, differs from the actual population parameter, \$22,766. This discrepancy is due to sampling error. **Sampling error** is the discrepancy between a sample estimate of a population parameter and the real population parameter. By comparing the sample statistic with the population parameter, we can determine the sampling error. The sampling error for our example is 1,949 (22,766 – 20,817 = 1,949).

Now let's select another random sample of three individuals. This time the incomes included are:

Case 15	18,923
Case 5	17,458
Case 17	27,654

The mean for this sample is

$$\overline{Y} = \frac{18,923 + 17,458 + 27,654}{3} = 21,345$$

The sampling error for this sample is 1,421 (22,766 – 21,345 = 1,421), somewhat less than the error for the first sample we selected.

The Dilemma

Although comparing the sample estimates of the average income with the actual population average is a perfect way to evaluate the accuracy of our estimate, in practice we rarely have information about the actual population parameter. If we did, we would not need to conduct a study! Moreover, few if any sample estimates correspond exactly to the actual population parameter. This, then, is our dilemma: If sample estimates vary and if most estimates result in some sort of sampling error, how much confidence can we place in the estimate? On what basis can we infer from the sample to the population?

Sampling error The discrepancy between a sample estimate of a population parameter and the real population parameter.

The Sampling Distribution

The answer to this dilemma is to use a device known as the *sampling distribution*. The **sampling distribution** is a theoretical probability distribution of all possible sample values for the statistic in which we are interested. If we were to draw all possible random samples of the same size from our population of interest, compute the statistic for each sample, and plot the frequency distribution for that statistic, we would obtain an approximation of the sampling distribution. Every statistic—for example, a proportion, a mean, or a variance—has a sampling distribution. Because it includes all possible sample values, the sampling distribution enables us to compare our sample result with other sample values and determine the likelihood associated with that result.[5]

> *Sampling distribution* A theoretical probability distribution of all possible sample values for the statistic in which we are interested.

■ ■ ■ ■ **The Sampling Distribution of the Mean**

Sampling distributions are theoretical distributions, which means that they are never really observed. Constructing an actual sampling distribution would involve taking all possible random samples of a fixed size from the population. This process would be very tedious because it would involve a very large number of samples. However, to help grasp the concept of the sampling distribution, let's illustrate how one could be generated from a limited number of samples.

An Illustration

For our illustration, we use one of the most common sampling distributions—the sampling distribution of the mean. The **sampling distribution of the mean** is a theoretical distribution of sample means that would be obtained by drawing from the population all possible samples of the same size.

Let's go back to our example in which our population is made up of twenty individuals and their incomes. From that population (Table 11.2) we now randomly draw fifty possible samples of size 3, computing the mean income for each sample and replacing it before drawing another.

In our first sample of size 3 we draw three incomes: $8,451, $41,654, and $18,923. The mean income for this sample is

$$\overline{Y} = \frac{8,451 + 41,654 + 18,923}{3} = 23,009$$

Now we restore these individuals to the original list and select a second sample of three other individuals. The mean income for this sample is

$$\overline{Y} = \frac{15,436 + 25,345 + 16,456}{3} = 19,079$$

We repeat this process forty-eight more times, each time computing the sample mean and restoring the sample to the original list. Table 11.3 lists the

[5]Here we are using an idealized example in which the sampling distribution is actually computed. However, please bear in mind that in practice one never computes a sampling distribution because it is also infinite.

Table 11.3   **Mean Income of Fifty Samples of Size 3**

Sample	Mean ($\overline{Y}$)
First	23,009
Second	19,079
Third	18,873
Fourth	26,885
Fifth	21,847
.	.
.	.
.	.
Fiftieth	26,645
Total (M) = 50	$\sum \overline{Y} = 1,237,482$

means of the first five and the fiftieth samples of $N = 3$ that were drawn from the population of twenty individuals. (Please note that $\sum \overline{Y}$ refers to the sum of all the means computed for each of the samples and M refers to the total number of samples that were drawn.)

The grouped frequency distribution for all fifty sample means ($M = 50$) is displayed in Table 11.4; Figure 11.4 is a histogram of this distribution. This distribution is an example of a sampling distribution of the mean. Notice that in its structure the sampling distribution resembles a frequency distribution of raw scores, except that here each score is a sample mean and the corresponding frequencies are the number of samples with that particular mean value. For example, the third interval in Table 11.4

Table 11.4   **Sampling Distribution of Sample Means for Sample Size N = 3 Drawn for the Population of Twenty Individuals' Incomes**

Sample Mean Intervals	Frequency	Percentage (%)
11,500–15,500	6	12
15,500–19,500	7	14
19,500–23,500	14	28
23,500–27,500	4	8
27,500–31,500	9	18
31,500–35,500	7	14
35,500–39,500	1	2
39,500–43,500	2	4
Total (M)	50	100

Figure 11.4 **Sampling Distribution of Sample Means for Sample Size $N = 3$ Drawn from the Population of Twenty Individuals' Incomes**

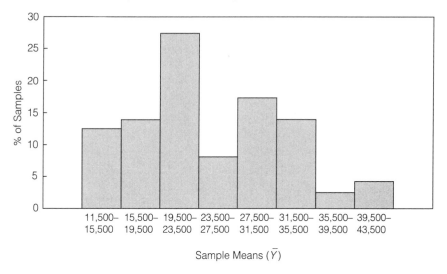

ranges from $19,500 to $23,500, with a corresponding frequency of 14, or 28 percent. This means that we drew 14 samples (28%) with means ranging between $19,500 and $23,500.

Remember the distribution depicted in Table 11.4 and Figure 11.4 is an empirical distribution, whereas the sampling distribution is a theoretical distribution. In reality, we never really construct the sampling distribution. However, even this simple empirical example serves to illustrate some of the most important characteristics of the sampling distribution.

> *Sampling distribution of the mean* A theoretical probability distribution of sample means that would be obtained by drawing from the population all possible samples of the same size.

Review

Before we continue, let's take a moment to review the three distinct types of distribution.

The Population We began with the *population distribution* of twenty individuals. This distribution actually exists. It is an empirical distribution that is usually unknown to us. We are interested in estimating the mean income for this population.

The Sample We drew a sample from that population. The *sample distribution* is an empirical distribution that is known to us and is used to help us estimate the mean of the population. We selected fifty samples of $N = 3$ and calculated the mean income. We usually use the sample mean ($\overline{Y}$) as an estimate of the population mean (μ_y).

The Sampling Distribution of the Mean For illustration, we generated an approximation of the sampling distribution of the mean, consisting of fifty samples of $N = 3$. The *sampling distribution of the mean* does not really exist. It is a theoretical distribution.

To help you understand the relationship among the population, the sample, and the sampling distributions, we have illustrated in Figure 11.5 the process of generating an empirical sampling distribution of the mean. From a population of raw scores (*Y*s), we draw *M* samples of size *N* and calculate the mean of each sample. The resulting sampling distribution of the mean, based on *M* samples of size *N*, shows the values that the mean could take and the frequency (number of samples) associated with each value. Make sure you understand these relationships. The concept of the sampling distribution is crucial to understanding statistical inference. In this and the next chapter, we learn how to employ the sampling distribution to draw inferences from the sample to the population.

The Mean of the Sampling Distribution

Like the sample and population distributions, the sampling distribution can be described in terms of its mean and standard deviation. We use the symbol $\mu_{\overline{Y}}$ to represent the mean of the sampling distribution. The subscript $\overline{Y}$ indicates that the variable of this distribution is the mean. To obtain the mean of the sampling distribution, add all the individual sample means ($\sum \overline{Y} = 1,237,482$) and divide by the number of samples ($M = 50$). Thus, the mean of the sampling distribution of the mean is actually the mean of means:

$$\mu_{\overline{Y}} = \frac{\sum \overline{Y}}{M} = \frac{1,237,482}{50} = 24,750$$

The Standard Error of the Mean

The standard deviation of the sampling distribution is also called the **standard error of the mean**. The standard error of the mean, $\sigma_{\overline{Y}}$, describes how much dispersion there is in the sampling distribution, or how much variability there is in the value of $\overline{Y}$ from sample to sample:

$$\sigma_{\overline{Y}} = \frac{\sigma_Y}{\sqrt{N}}$$

This formula tells us that the standard error of the mean is equal to the standard deviation of the population (σ_Y) divided by the square root of the

Figure 11.5 **Generating the Sampling Distribution of the Mean**

From a population (with a population mean of μ_Y) we start drawing samples and calculating the means for those samples:

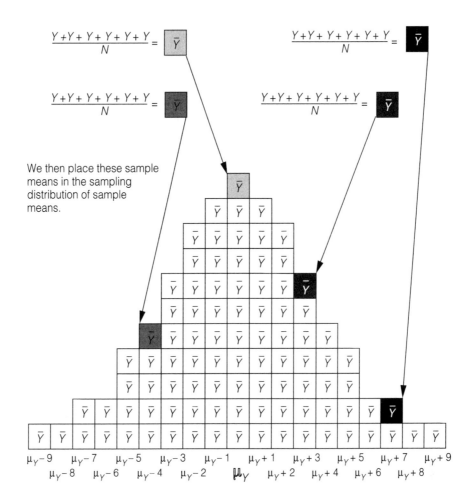

Once we lay all of the sample means out onto the sampling distribution, we notice a few things:

1. The mean of this distribution is μ_Y, the mean for the population of scores.
2. Most of the sample means fall fairly near μ_Y: the probability of pulling such sample means is high.
3. As we move farther and farther away from μ_Y on either side, we have fewer and fewer sample means: the probability of pulling such sample means is low.

sample size (N). For our example, because the population standard deviation is 14,687 and our sample size is 3, the standard error of the mean is

$$\sigma_{\bar{Y}} = \frac{14,687}{\sqrt{3}} = 8,480$$

Standard error of the mean The standard deviation of the sampling distribution of the mean. It describes how much dispersion there is in the sampling distribution of the mean.

Box 11.2 Population, Sample, and Sampling Distribution Symbols

In the discussions that follow we make frequent references to the mean and standard deviation of the three distributions. To distinguish among the different distributions, the symbols that refer to the means and standard deviations for the sample, population, and sampling distributions follow. Notice that we use Greek letters to refer to both the sampling and the population distributions.

	Mean	Standard Deviation
Sample distribution	$\bar{Y}$	S_Y
Population distribution	μ_Y	σ_Y
Sampling distribution of $\bar{Y}$	$\mu_{\bar{Y}}$	$\sigma_{\bar{Y}}$

■ ■ ■ ■ **The Central Limit Theorem**

In Figures 11.6a and 11.6b, we compare the histograms for the population and sampling distributions of Tables 11.2 and 11.4. Figure 11.6a shows the population distribution of twenty incomes, with a mean $\mu_Y = 22,766$ and a standard deviation $\sigma_Y = 14,687$. Figure 11.6b shows the sampling distribution of the means from fifty samples of $N = 3$ with a mean $\mu_{\bar{Y}} = 24,749$ and a standard deviation (the standard error of the mean) $\sigma_{\bar{Y}} = 8,479$. These two figures illustrate some of the basic properties of sampling distributions in general and the sampling distribution of the mean in particular.

First, as can be seen from Figures 11.6a and 11.6b, the shapes of the two distributions differ considerably. Whereas the population distribution is skewed to the right, the sampling distribution of the mean is less skewed—that is, closer to symmetry and a normal distribution.

Figure 11.6 **Three Income Distributions**

a. Population distribution of personal income for twenty individuals (hypothetical data)

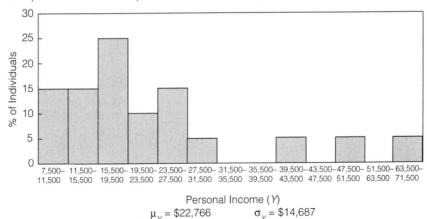

Personal Income (Y)

$\mu_Y = \$22,766$ $\sigma_Y = \$14,687$

b. Sampling distribution of sample means for sample size $N = 3$ drawn from the population of twenty individuals' incomes

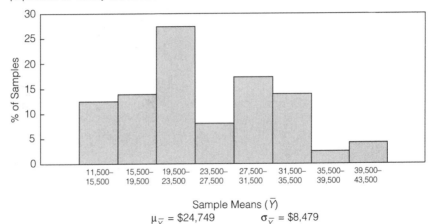

Sample Means ($\bar{Y}$)

$\mu_{\bar{Y}} = \$24,749$ $\sigma_{\bar{Y}} = \$8,479$

c. Sampling distribution of sample means for sample size $N = 6$ drawn from the population of twenty individuals' incomes

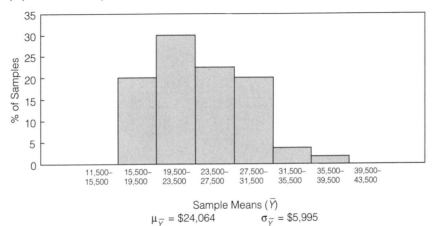

Sample Means ($\bar{Y}$)

$\mu_{\bar{Y}} = \$24,064$ $\sigma_{\bar{Y}} = \$5,995$

Second, whereas only a few of the sample means coincide exactly with the population mean, $22,766, the sampling distribution centers around this value. The mean of the sampling distribution is a pretty good approximation of the population mean.

Third, the variability of the sampling distribution is considerably smaller than the variability of the population distribution. Notice that the standard deviation for the sampling distribution ($\sigma_{\bar{Y}} = 8,479$) is almost half that for the population ($\sigma_Y = 14,687$).

These properties of the sampling distribution are even more striking as the sample size increases. To illustrate the effect of a larger sample on the shape and properties of the sampling distribution, we went back to our population of twenty individual incomes and drew fifty additional samples of $N = 6$. We calculated the mean for each sample and constructed another sampling distribution. This sampling distribution is shown in Figure 11.6c. It has a mean $\mu_{\bar{Y}} = 24,064$ and a standard deviation $\sigma_{\bar{Y}} = 5,995$. Notice that as the sample size increased, the sampling distribution became more compact. This decrease in the variability of the sampling distribution is reflected in a smaller standard deviation: with an increase in sample size from $N = 3$ to $N = 6$ the standard deviation of the sampling distribution decreased from 8,479 to 5,995. Furthermore, with a larger sample size the sampling distribution of the mean is an even better approximation of the normal curve.

These properties of the sampling distribution of the mean are summarized more systematically in one of the most important statistical principles underlying statistical inference. It is called the **Central Limit Theorem**, and it states: If all possible random samples of size N are drawn from a population with a mean μ_Y and a standard deviation σ_Y, then as N becomes larger, the sampling distribution of sample means becomes approximately normal, with mean $\mu_{\bar{Y}}$ and standard deviation $\sigma_Y/\sqrt{N}$.

The significance of the Central Limit Theorem is that it tells us that with *sufficient sample size* the sampling distribution of the mean will be normal regardless of the shape of the population distribution. Therefore, even when the population distribution is skewed, we can still assume that the sampling distribution of the mean is normal, given random samples of large enough size. Furthermore, the Central Limit Theorem also assures us that (1) as the sample size gets larger, the mean of the sampling distribution becomes equal to the population mean; and (2) as the sample size gets larger, the standard error of the mean (the standard deviation of the sampling distribution of the mean) decreases in size. The standard error of the mean tells how much variability in the sample estimates there is from sample to sample. The smaller the standard error of the mean, the closer (on average) the sample means will be to the population mean. Thus, the larger the sample, the more closely the sample statistic clusters around the population parameter.

Central Limit Theorem An important principle in statistical inference that relates the normal distribution and the sampling distribution of the means with a sufficient sample size.

> **Learning Check.**   *Make sure you understand the difference between the number of samples that can be drawn from a population and the sample size. Whereas the number of samples is infinite in theory, the sample size is under the control of the investigator.*

The Size of the Sample

Although there is no hard and fast rule, a general rule of thumb is that when N is 50 or more, the sampling distribution of the mean will be approximately normal regardless of the shape of the distribution. However, we can assume that the sampling distribution will be normal even with samples as small as 30 if we know that the population distribution approximates normality.

> **Learning Check.**   *What is a normal population distribution? If you can't answer this question, go back to Chapter 10. You must understand the concept of a normal distribution before you can understand the techniques involved in inferential statistics.*

The Significance of the Sampling Distribution and the Central Limit Theorem

In the preceding sections we have covered a lot of abstract material. You may have a number of questions at this time. Why is the concept of the sampling distribution so important? What is the significance of the Central Limit Theorem? To answer these questions, let's go back and review our twenty incomes example.

In order to estimate the mean income of a population of twenty individuals, we drew a sample of 3 cases and calculated the mean income for that sample. Our sample mean, $\overline{Y} = 20{,}817$, differs from the actual population parameter, $\mu_Y = 22{,}766$. When we selected different samples, we found that each time the sample mean differed from the population mean. These discrepancies are due to sampling errors. Had we taken a number of additional samples, we probably would have found that the mean was different each time because every sample differs slightly. Few, if any, sample means would correspond exactly to the actual population mean. Usually we have only one sample statistic as our best estimate of the population parameter.

So now let's restate our dilemma: If sample estimates vary and if most result in some sort of sampling error, how much confidence can we place in the estimate? On what basis can we infer from the sample to the population?

The solution lies in the sampling distribution and its properties. Because the sampling distribution is a theoretical distribution that includes all possible sample outcomes, we can compare our sample outcome with it and estimate the likelihood of its occurrence.

Since the sampling distribution is theoretical, how can we know its shape and properties so that we can make these comparisons? Our knowledge is based on what the Central Limit Theorem tells us about the properties of the sampling distribution of the mean. We know that if our sample size is large enough (at least 50 cases), most sample means will be quite close to the true population mean. It is highly unlikely that our sample mean would deviate much from the actual population mean.

In Chapter 10 we saw that in all normal curves, a constant proportion of the area under the curve lies between the mean and any given distance from the mean when measured in standard deviation units, or Z scores. We can find this proportion in the standard normal table (Appendix B).

Knowing that the sampling distribution of the means is approximately normal, with a mean $\mu_{\bar{Y}}$ and a standard deviation $\sigma_Y / \sqrt{N}$ (the standard error of the mean), we can use Appendix B to determine the probability that a sample mean will fall within a certain distance—measured in standard deviation units, or Z scores—of $\mu_{\bar{Y}}$ or μ_Y. For example, we can expect approximately 68 percent (or we can say the probability is approximately .68) of all sample means to fall within ±1 standard error ($\sigma_Y / \sqrt{N}$, or the standard deviation of the sampling distribution of the mean) of $\mu_{\bar{Y}}$ or μ_Y. Similarly, the probability is about .95 that the sample mean will fall within ±2 standard errors of $\mu_{\bar{Y}}$ or μ_Y. In the next chapter we will see how this information helps us evaluate the accuracy of our sample estimates.

Learning Check. *Suppose a population distribution has a mean $\mu_Y =$ 150 and a standard deviation $\sigma_Y = 30$ and you draw a simple random sample of N = 100 cases. What is the probability that the mean is between 147 and 153? What is the probability that the sample mean exceeds 153? Would you be surprised to find a mean score of 159? Why? (Hint: To answer these questions you need to apply what you learned in Chapter 10 about Z scores and areas under the normal curve [Appendix B].) Remember, to translate a raw score into a Z score we used this formula:*

$$Z = \frac{Y - \bar{Y}}{S_Y}$$

However, because here we are dealing with a sampling distribution, replace Y with the sample mean $\bar{Y}$, $\bar{Y}$ with the sampling mean $\mu_{\bar{Y}}$, and S_Y with the standard error of the mean.

$$Z = \frac{\bar{Y} - \mu_{\bar{Y}}}{\sigma_Y / \sqrt{N}}$$

MAIN POINTS

■ Through the process of sampling, researchers attempt to generalize to the characteristics of a large group (the population) from a subset (sample) selected from that group. The term *parameter,* associated with the population, refers to the information we are interested in finding out. *Statistic* refers to a corresponding calculated sample statistic.

■ A probability sample design allows us to estimate the extent to which the findings based on one sample are likely to differ from what we would find by studying the entire population.

■ A simple random sample is chosen in such a way as to ensure that every member of the population and every combination of N members have an equal chance of being chosen.

■ In systematic sampling, every Kth member in the total population is chosen for inclusion in the sample after the first member of the sample is selected at random from the first K members in the population.

■ A stratified random sample is obtained by (1) dividing the population into subgroups based on one or more variables central to our analysis and (2) then drawing a simple random sample from each of the subgroups.

■ The sampling distribution is a theoretical probability distribution of all possible sample values for the statistic in which we are interested. The sampling distribution of the mean is a frequency distribution of all possible sample means of the same size that can be drawn from the population of interest.

■ According to the Central Limit Theorem, if all possible random samples of size N are drawn from a population with a mean μ_Y and a standard deviation σ_Y, then as N becomes larger, the sampling distribution of sample means becomes approximately normal, with mean $\mu_{\overline{Y}}$ and standard deviation $\sigma_Y / \sqrt{N}$.

■ The Central Limit Theorem tells us that with sufficient sample size, the sampling distribution of the mean will be normal regardless of the shape of the population distribution. Therefore, even when the population distribution is skewed, we can still assume that the sampling distribution of the mean is normal, given a large enough randomly selected sample size.

KEY TERMS

Central Limit Theorem

disproportionate stratified sample

parameter

population

probability sampling

proportionate stratified sample

sample

sampling distribution

sampling distribution of the mean

sampling error

simple random sample

standard error of the mean

statistic

stratified random sample

systematic random sampling

SPSS DEMONSTRATION

Demonstration 1: Selecting a Random Sample
[Module A]

In this chapter we've discussed various types of samples and the definition of the standard error of the mean. Usually, data entered into SPSS have already been sampled from some larger population. However, SPSS does have a sampling procedure that can take random samples of data. Systematic samples and stratified samples can also be drawn with SPSS, but they require the use of the SPSS command language.

When might it be worthwhile to use the SPSS Sample procedure? One instance is when doing preliminary analysis of a very large data set. For example, if you worked for your local hospital and had complete data records for all patients (tens of thousands), there would be no need to use *all* the data during initial analysis. You could select a random sample of individuals and use the subset of data for preliminary analysis. Later, the complete patient data set could be used for completing your final analyses.

To use the Sample procedure, click on *Data* from the main menu, then on *Select Cases.* The opening dialog box (Figure 11.7) has four choices that will select a subset of cases via various methods. By default, the *All cases* button is checked. We click on the *Random sample of cases* button, then on the *Sample* button to give SPSS our specification.

Figure 11.7

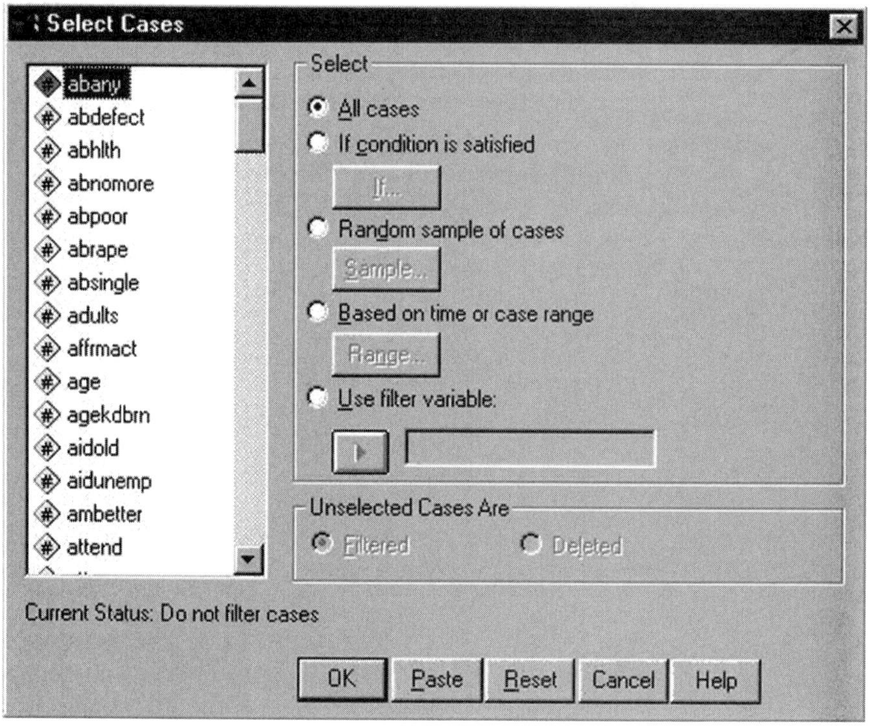

⑤ The next dialog box (Figure 11.8) provides two options to create a random sample. The most convenient is normally the first, where we tell SPSS what percentage of cases to select from the larger file. Alternatively, we can tell SPSS to take an exact number of cases. The second option is available because SPSS will only take approximately the percentage specified in the first option.

Figure 11.8

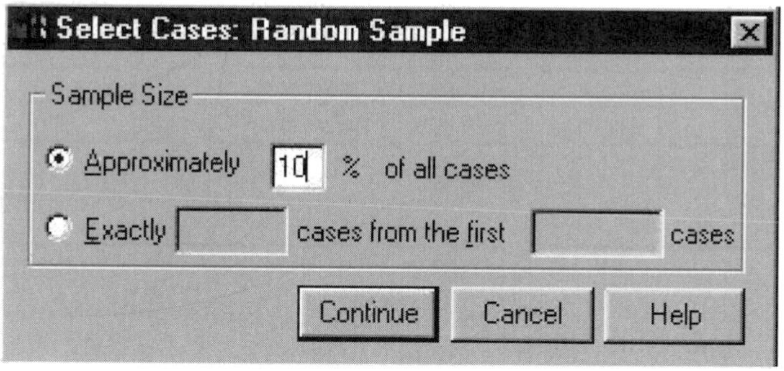

We type "10" in the box to ask for 10 percent of the original sample of 1,427 respondents from the GSS. Then click on *Continue* and *OK*, as usual, to process the request.

SPSS does not delete the cases from the active data file that aren't selected for the sample. Instead, they are filtered out (you can identify them in the Data Editor window by the slash across their row number). This means that we can always return to the full data file by going back to the Select Cases dialog box and selecting the *All cases* button.

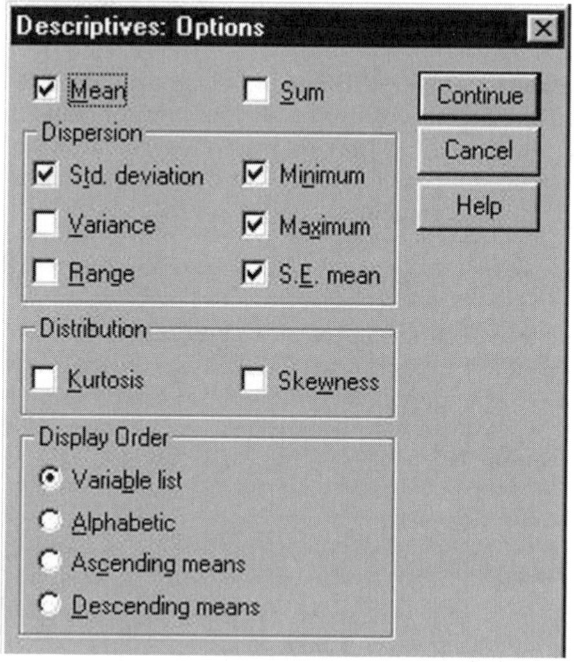

When SPSS processes our request, it tells us that the data have been filtered by putting the words "Filter On" in the status area at the bottom of the SPSS window (the status area has many helpful messages from SPSS).

To demonstrate the effect of sampling, we ask for univariate statistics for the variable CHILDS, measuring the number of children a respondent has. Click on *Statistics, Summarize,* and then *Descriptives* to open this dialog box and place CHILDS in the variable list. Click on the *Options* button to obtain the box shown in Figure 11.9. Select the mean, standard deviation, minimum, and maximum values. In

Figure 11.9

addition, we'll add the standard error of the mean by clicking the *S.E. mean* box. Then click *Continue* and *OK* to put SPSS to work.

The results (Figure 11.10) show that the number of valid cases is exactly 146, or 10 percent of the whole file. The mean of CHILDS is 1.95, and the standard error of the mean is .13.

Figure 11.10

Descriptive Statistics

	N	Minimum	Maximum	Mean		Std.
	Statistic	Statistic	Statistic	Statistic	Std. Error	Statistic
CHILDS NUMBER OF CHILDREN	146	0	8	1.95	.13	1.53
Valid N (listwise)	146					

If we repeat the process again, this time asking for a 25 percent sample, we obtain the results shown in Figure 11.11.

Figure 11.11

Descriptive Statistics

	N	Minimum	Maximum	Mean		Std.
	Statistic	Statistic	Statistic	Statistic	Std. Error	Statistic
CHILDS NUMBER OF CHILDREN	328	0	8	1.77	.0885	1.60
Valid N (listwise)	328					

How closely does the mean for CHILDS from these two random samples match that of the full file? The mean for all 1,427 respondents is 1.82, so it certainly appears that SPSS did take a random sample of this larger file. Both samples produced means and standard deviations that are within the range of the population parameters.

SPSS PROBLEM

Repeat the SPSS demonstration, selecting 25 percent, 50 percent, and 75 percent samples and requesting descriptives for *two* interval variables of your choice. The variable CHILDS in the previous demonstration is an example of an interval-ratio variable. Compare your descriptive statistics with descriptives for the entire sample. What can you say about the accuracy of your random samples?

CHAPTER EXERCISES

1. Explain which of the following is a statistic and which is a parameter.
   a. The mean age of Americans from the 1990 census

b. The unemployment rate for the population of U.S. adults, estimated by the government from a large sample

c. The percentage of Texans opposed to abortion from a poll of 1,000 residents

d. The mean salaries of employees at your school (for example, administrators, faculty, maintenance, etc.)

2. The mayor of your city has been talking about the need for a tax hike. The city's newspaper uses letters sent to the editor to judge public opinion about this possible hike, reporting on their results in an article.

a. Do you think that these letters represent a random sample? Why or why not?

b. What alternative sampling method would you recommend to the mayor?

3. The following four common situation scenarios involve selecting a sample and understanding how a sample relates to a population.

a. A friend interviews every tenth shopper that passes by her as she stands outside one entrance of a major department store in a shopping mall. What type of sample is she selecting? How might you define the population from which she is selecting the sample?

b. A political polling firm samples fifty potential voters from a list of registered voters in each county in a state to interview for an upcoming election. What type of sample is this? Do you have enough information to tell?

c. Another political polling firm in the same state selects potential voters from the same list of registered voters with a very different method. First, they alphabetize the list of last names, then pick the first twenty names that begin with an A, the first twenty that begin with a B, and so on until Z (the sample size is thus 20×26, or 520). Is this a probability sample?

d. A social scientist gathers a carefully chosen group of twenty people whom she has selected to represent a broad cross section of the population in New York City. She interviews them in depth for a study she is doing on race relations in the city. Is this a probability sample? What type of sample has she chosen?

4. An upper-level sociology class at a large urban university has 120 students, including 34 seniors, 57 juniors, 22 sophomores, and 7 freshmen.

a. Imagine that you choose one student at random from the classroom (perhaps by using a random number table). What is the probability that the student will be a junior?

b. What is the probability that the student will be a freshman?

c. If you are asked to select a proportionate stratified sample of size 30 from the classroom, stratified by class level (senior, junior, and so on), how many students from each group will be in the sample?

d. If instead you are to select a disproportionate sample of size 20 from the classroom, with equal numbers of students from each class level in the sample, how many freshmen will be in the sample?

5. Can the standard error of a variable ever be larger than, or even equal in size to, the standard deviation for the same variable? Justify your answer by means of both a formula and a discussion of the relationship between these two concepts.

6. When taking a random sample from a very large population, how does the standard error of the mean change when
   a. the sample size is increased from 100 to 1,600?
   b. the sample size is decreased from 300 to 150?
   c. the sample size is multiplied by 4?

7. Many television news shows conduct "instant" polls by providing an 800 number and asking an interesting question of the day for viewers to call and answer.
   a. Is this poll a probability sample? Why or why not?
   b. Specify the population from which the sample of calls is drawn.

8. Use the data from Chapter 10, Exercise 6 (repeated below), concerning the percentage of eligible adults who voted in the 1992 election, by state.

	% of Children in Distressed Neighborhoods	% of Eligible Voters Who Voted in the 1992 Election
Alabama	9.6	54
Alaska	0.5	51
Arizona	5.2	51
Arkansas	7.6	53
California	5.1	45
Colorado	2.5	62
Connecticut	5.7	63
Delaware	1.7	41
D.C.	25.2	47
Florida	4.6	49
Georgia	6.2	46
Hawaii	0.5	41
Idaho	0.0	53
Illinois	9.5	58
Indiana	3.0	54
Iowa	1.0	64
Kansas	2.0	62
Kentucky	7.1	53
Louisiana	17.2	59
Maryland	6.2	51
Maine	0.6	71
Massachusetts	5.1	60
Michigan	11.5	62
Minnesota	2.0	70

	% of Children in Distressed Neighborhoods	% of Eligible Voters Who Voted in the 1992 Election
Mississippi	17.4	52
Missouri	5.2	62
Montana	2.3	69
Nebraska	1.3	62
Nevada	3.3	48
New Hampshire	0.1	62
New Jersey	4.9	54
New Mexico	4.5	51
New York	12.8	48
North Carolina	2.6	49
North Dakota	2.1	67
Ohio	8.1	60
Oklahoma	3.1	60
Oregon	1.1	55
Pennsylvania	6.9	54
Rhode Island	5.1	54
South Carolina	4.4	44
South Dakota	3.6	67
Texas	4.8	49
Tennessee	7.3	52
Utah	0.3	62
Vermont	0.6	64
Virginia	2.8	52
Washington	2.4	51
West Virginia	2.9	50
Wisconsin	5.4	68
Wyoming	0.2	62

Source: Data on children in distressed neighborhoods come from the *1994 Kids Count Data Book.*

a. Calculate the mean and standard deviation for the population.
b. Now take ten samples of size 5 from the population. Use either simple random sampling or systematic sampling with the help of the table of random numbers in Appendix A. Calculate the mean for each sample.
c. After selecting the samples, calculate the mean and standard deviation for the ten sample means. How does the standard deviation of the sample means compare with that for the population of fifty states?
d. Now select ten samples of size 10 from the population of states, calculating the mean for each. Then, as before, calculate the mean and

standard deviation for the ten sample means. How does the standard deviation here compare with that in (c)? Why?

e. Now construct a histogram of the distribution of values in the population, and of the sampling distribution of means for the two sample sizes. Describe and explain any differences between the three distributions you observe.

9. You've been asked to determine the percentage of students who would support increased ethnic diversity on your campus. You want to take a random sample of fellow students to make the estimate. Explain whether each of the following scenarios describes a random sample.

   a. You ask all students eating lunch in the cafeteria on a Tuesday at 12:30 P.M.

   b. You ask every tenth student from the list of enrolled students.

   c. You ask every tenth student passing by the student union.

   d. What sampling procedure would you recommend to complete your study?

10. For the total population of a large southern city, mean family income is $34,000, with a standard deviation (for the population) of $5,000.

    a. Imagine that you take a sample of 200 city residents. What is the probability that your sample mean is between $33,000 and $34,000?

    b. For this same sample size, what is the probability that the sample mean exceeds $37,000?

11. A small population of $N = 10$ has values of 4, 7, 2, 11, 5, 3, 4, 6, 10, and 1.

    a. Calculate the mean and standard deviation for the population.

    b. Take ten simple random samples of size 3, and calculate the mean for each.

    c. Calculate the mean and standard deviation of all these sample means. How closely does the mean of all sample means match the population mean? How is the standard deviation of the means related to the standard deviation for the population?

12   Estimation

Introduction
Estimation Defined
Reasons for Estimation
Point and Interval Estimation
Box 12.1 Estimation as a Type of Inference
Procedures for Estimating Confidence Intervals for Means
Calculating the Standard Error of the Mean
Deciding on the Level of Confidence and Finding the Corresponding Z Value
Calculating the Confidence Interval
Interpreting the Results
Reducing Risk
Estimating Sigma
Calculating the Standard Error of the Mean
Deciding on the Level of Confidence and Finding the Corresponding Z Value
Calculating the Confidence Interval
Interpreting the Results
Sample Size and Confidence Intervals
Box 12.2 What Affects Confidence Interval Width? A Summary
Statistics in Practice: Hispanic Migration and Earnings
Confidence Intervals for Proportions
Procedures for Estimating Proportions
Calculating the Standard Error of the Proportion
Deciding on the Level of Confidence and Finding the Corresponding Z Value
Calculating the Confidence Interval
Interpreting the Results
Increasing the Sample Size
Statistics in Practice: Should Affirmative Action
Programs Be Increased?
Calculating the Standard Error of the Proportion
Deciding on the Level of Confidence and Finding the Corresponding Z Value
Calculating the Confidence Interval
Interpreting the Results

■　■　■　■　**Introduction**

In this chapter we discuss the procedures involved in estimating population means and proportions. These procedures are based on the principles of sampling and statistical inference discussed in Chapter 11. Knowledge about the sampling distribution allows us to estimate population means and proportions from sample outcomes and to assess the accuracy of these estimates.

> *Example 1* An article published in *Time* magazine (May 20, 1996) reports the results of a survey conducted by a national polling organization for *Time*/CNN. Based on a sample of 826 registered voters, this poll estimated that 50 percent of all registered voters are Clinton supporters.

> *Example 2* Each month the Bureau of Labor Statistics interviews a sample of about 50,000 adult Americans to determine job-related activities. Based on these interviews, monthly estimates are made of vital statistics, such as the unemployment rate (the proportion who are unemployed), average earnings, the percentage of the workforce working part-time, and the percentage collecting unemployment benefits. These estimates are considered so vital that they cause fluctuations in the stock market and influence economic policies of the federal government.

> *Example 3* The Gallup Organization conducted a telephone poll of 1,269 blacks and 1,680 whites in 1997. Individuals were asked of their perceptions on current black/white relations in the United States. When asked whether affirmative action programs should be increased, 53% of blacks and 22% of white respondents agreed.

Each year the National Opinion Research Center (NORC) conducts the General Social Survey (GSS) on a representative sample of about 1,500 respondents. The GSS, from which many of the examples in this book are selected, is designed to provide social science researchers with a readily accessible database of socially relevant attitudes, behaviors, and attributes of a cross section of the U.S. adult population. For example, in analyzing the responses to the 1991 GSS, researchers found the average income was $18,130. This average probably differs from the average of the population from which the GSS sample was drawn. However, we can establish that in most cases the sample mean (in this case $18,130) is fairly close to the actual true average in the population.

As you read these examples, you may have questioned the reliability of some of the numbers. Is it possible to establish the voting preferences of millions of Americans, to determine their opinion regarding affirmative action, or to find their average income based on a sample of about 1,000 respondents? If elections were held the day the poll was conducted by the national polling organization, would 50 percent of all registered voters in the United States really vote for Bill Clinton? What is the actual percentage of black and white Americans in the United States who favor increasing affirmative action programs?

■ ■ ■ ■ **Estimation Defined**

The average income of all adult Americans, the percentage of all registered voters who would vote for Bill Clinton, and the percentage of all Americans who favor increasing affirmative action programs are *population parameters.* The average income calculated from the GSS, the percentage who stated that they would vote for Bill Clinton in the *Time*/CNN poll, and the percentage who favored increasing affirmative action programs in the Gallup survey are all *sample estimates of population parameters.* Thus, the responses to the *Time*/CNN political poll were used to estimate the percentage of registered voters in favor of Bill Clinton; the mean income of $18,130 calculated from the GSS sample can be used to estimate the mean income of all adults in the United States. Similarly, based on a national sample of adult Americans, the Gallup polling organization estimated the percentage of black and white adults in the United States who favor increasing affirmative action programs.

These are all illustrations of *estimation.* **Estimation** is a process whereby we select a random sample from a population and use a sample statistic to estimate a population parameter. We can use sample proportions as estimates of population proportions, sample means as estimates of population means, or sample variances as estimates of population variances.

> *Estimation* A process whereby we select a random sample from a population and use a sample statistic to estimate a population parameter.

Reasons for Estimation

Why estimate? The goal of most research is to find the population parameter. Yet we hardly ever have enough resources to collect information about the entire population. We rarely know the value of the population parameter. On the other hand, we can learn a lot about a population by randomly selecting a sample from that population and obtaining an estimate of the population parameter. The major objective of sampling theory and statistical inference is to provide estimates of unknown population parameters from sample statistics. *Sample theory*

Point and Interval Estimation *(ie confidence inter points interval estimates)*

Estimates of population characteristics can be divided into two types: point estimates and interval estimates. **Point estimates** are sample statistics used to estimate the exact value of a population parameter. When *Time* magazine projected that 50 percent of all registered voters are Clinton supporters, it was using a point estimate. Similarly, if we reported the average income of the population of adult Americans to be exactly $18,130, we would be using a point estimate.

The problem with point estimates is that sample estimates usually vary, and most result in some sort of sampling error. As a result, when we use a sample statistic to estimate the exact value of a population parameter, we never really know how accurate it is.

One method of improving accuracy is to use an *interval estimate* rather than a point estimate. In interval estimation, we identify a range of values within which the population parameter may fall. This range of values is called a **confidence interval**. Instead of using a single value, $18,130, as an estimate of the mean earnings of adult Americans, we could say that the population mean is somewhere between $17,500 and $19,100.

When we use confidence intervals to estimate population parameters, such as the mean earnings, we can also evaluate the accuracy of this estimate by assessing the likelihood that any given interval will contain the mean. This likelihood, expressed as a percentage or a probability, is called a **confidence level**. Confidence intervals are defined in terms of confidence levels. Thus, by selecting a 95 percent confidence level, we are saying that there is a .95 probability—or 95 chances out of 100—that a specified interval will contain the population mean. Confidence intervals can be constructed for any level of confidence, but the most common ones are the 90 percent, 95 percent, and 99 percent levels.

Confidence intervals can be constructed for many different parameters based on their corresponding sample statistics. In this chapter we describe the rationale and the procedure for the construction of confidence intervals for means and proportions.

Point estimate   A sample statistic used to estimate the exact value of a population parameter.

Confidence interval (interval estimate)   A range of values defined by the confidence level within which the population parameter is estimated to fall.

Confidence level   The likelihood, expressed as a percentage or a probability, that a specified interval will contain the population parameter.

Box 12.1 *Estimation as a Type of Inference*

The goal of inferential statistics is to say something meaningful about the population, based entirely on information from a sample of that population. A confidence interval attempts to do just that: By knowing a sample mean, sample size,
and sample standard deviation, we are able to say something about the population from which that sample was drawn.

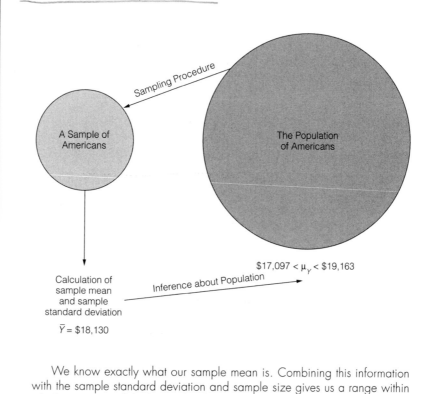

We know exactly what our sample mean is. Combining this information
with the sample standard deviation and sample size gives us a range within
which we can confidently say that the population mean falls.

Learning Check.   *What is the difference between a point estimate and
a confidence interval?*

■ ■ ■ ■ Procedures for Estimating Confidence Intervals for Means

To illustrate the procedure of establishing confidence intervals for means,
we'll reintroduce one of the research examples discussed in Chapter 11—
assessing the needs of commuting students on our campus.

Recall that we have been given enough money to survey a random sample of 500 students. One of our tasks is to estimate the average commuting time of all 15,000 commuters on our campus—the population parameter. To obtain this estimate we calculate the average commuting time for the sample. Suppose the sample average is $\overline{Y}$ = 7.5 hours per week, and we want to use it as an estimate of the true average commuting time for the entire population of commuting students.

Because it is based on a sample, this estimate is subject to sampling error. We do not know how close it is to the true population mean. However, based on what the Central Limit Theorem tells us about the properties of the sampling distribution of the mean, we know that with a large enough sample size, most sample means will tend to be close to the true population mean. Therefore, it is unlikely that our sample mean, $\overline{Y}$ = 7.5, deviates much from the true population mean.

We know that the sampling distribution of the means is approximately normal with a mean $\mu_{\overline{Y}}$, equal to the population mean μ_Y, and a standard error $\sigma_{\overline{Y}}$ (standard deviation of the sampling distribution) as follows:

$$\sigma_{\overline{Y}} = \frac{\sigma_Y}{\sqrt{N}} \qquad (12.1)$$

Standard error of the mean = Standard deviation of the sampling distribution

This information allows us to use the normal distribution to determine the probability that a sample mean will fall within a certain distance—measured in standard deviation (standard error) units, or Z scores—of $\mu_{\overline{Y}}$ or μ_Y. We can make the following assumptions:

- 68 percent of all random sample means will fall within ±1 standard error of the true population mean

- 95 percent of all random sample means will fall within ±1.96 standard errors of the true population mean

- 99 percent of all random sample means will fall within ±2.58 standard errors of the true population mean

Based on these assumptions and the value of the standard error, we can establish a range of values—a confidence interval—that is likely to contain the actual population mean. We can also evaluate the accuracy of this estimate by assessing the likelihood that this range of values will actually contain the population mean.

The general formula for constructing a confidence interval (CI) for any level is

$$CI = \overline{Y} \pm Z(\sigma_{\overline{Y}}) \qquad (12.2)$$

Notice that to calculate a confidence interval, we take the sample mean and add to or subtract from it the product of a Z value and the standard error.

The Z score we choose depends on the desired confidence level. For example, to obtain a 95 percent confidence interval we would choose a Z of 1.96 because we know (from Appendix B) that 95 percent of the area under

Figure 12.1 **Relationship Between Confidence Level and Z
for 95 and 99 Percent Confidence Intervals**

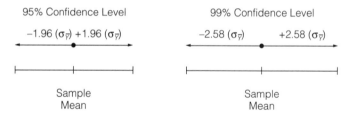

Source: Adapted from David Freedman, Robert Pisani, Roger Purves, and
Ani Adhikari, *Statistics*, 2nd ed., p. 348. Copyright © 1991 by W. W.
Norton & Co., Inc. Reprinted by permission of W. W. Norton & Co., Inc.

the curve is included between ±1.96. Similarly, for a 99 percent confidence
level we would choose a Z of 2.58. The relationship between the confidence
level and Z is graphically illustrated in Figure 12.1 for the 95 percent and 99
percent confidence levels.

(98 ÷ 2)
look up .4900
on C column
+ standard
Normal Table
= 2.33

> **Learning Check.** *If you don't understand the relationship between
> the confidence level and Z, review the material in Chapter 10. What would
> be the appropriate Z value for a 98 percent confidence interval?* *2.33*

To determine the confidence interval for means, follow these steps:

1. Calculate the standard error of the mean.
2. Decide on the level of confidence, and find the corresponding Z value.
3. Calculate the confidence interval.
4. Interpret the results.

Let's return to the problem of estimating the mean commuting time of the
population of students on our campus. How would you find the 95 percent
confidence interval?

Calculating the Standard Error of the Mean Let's suppose that the standard
deviation for our population of commuters is $\sigma_Y = 1.5$. We calculate the
standard error for the sampling distribution of the mean:

$$\sigma_{\bar{Y}} = \frac{\sigma_Y}{\sqrt{N}} = \frac{1.5}{\sqrt{500}} = 0.07$$

Deciding on the Level of Confidence and Finding the Corresponding Z Value We
decide on a 95 percent confidence level. The Z value corresponding to a 95
percent confidence level is 1.96.

Figure 12.2   **95 Percent Confidence Interval for the Mean Commuting Time (N = 500)**

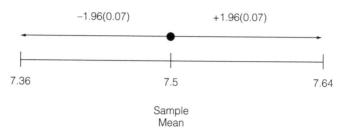

7.36   7.5   7.64

Sample
Mean

Calculating the Confidence Interval   The confidence interval is calculated by adding and subtracting from the observed sample mean the product of the standard error and Z:

$$95\% \ CI = 7.5 \pm 1.96(0.07)$$
$$= 7.5 \pm 0.14$$
$$= 7.36 \ to \ 7.64$$

The 95 percent CI for the mean commuting time is illustrated in Figure 12.2.

Interpreting the Results   We can be 95 percent confident that the actual mean commuting time—the true population mean—is not less than 7.36 hours and not greater than 7.64 hours. In other words, if we collected a large number of samples (N = 500) from the population of commuting students, 95 times out of 100 the true population mean would be included within our computed interval. With a 95 percent confidence level, there is a 5 percent risk that we are wrong. Five times out of 100, the true population mean will not be included in the specified interval.

Remember that we can never be sure whether the population mean is actually contained within the confidence interval. Once the sample is selected and the confidence interval defined, the population mean either does or does not contain the population mean—but we will never be sure.

> **Learning Check.**   *What is the 90 percent confidence interval for the mean commuting time? (Hint: First, find the Z value associated with a 90 percent confidence level.)* $Z = 1.645$

[handwritten notes in margin: $90 \div 2 = 45$ / look up .4500 / on Standard Z / table C column / Z = 1.645 / $CI = 7.5 \pm 1.645(0.07)$]

To further illustrate the concept of confidence intervals, let's suppose that we draw ten different samples (N = 500) from the population of commuting students. For each sample mean, we construct a 95 percent confidence interval. Figure 12.3 displays these confidence intervals. Each horizontal line represents a 95 percent confidence interval constructed around a sample mean (marked with a circle).

Figure 12.3 **95 Percent Confidence Intervals for Ten Samples**

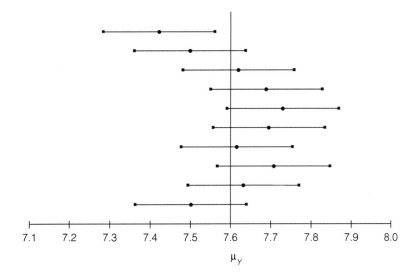

The vertical line represents the population mean. Note that the horizontal lines that intersect the vertical line are the intervals that contain the true population mean. Only 1 out of the 10 confidence intervals does not intersect the vertical line, meaning it does not contain the population mean. What would happen if we continued to draw samples of the same size from this population and constructed a 95 percent confidence interval for each sample? For about 95 percent of all samples the specified interval would contain the true population mean, but for 5 percent of all samples it would not.

Reducing Risk *but less precision*

One way to reduce the risk of being incorrect is by increasing the level of confidence. For instance, we can increase our confidence level from 95 to 99 percent. The 99 percent confidence interval for our commuting example is

$$99\% \ CI = 7.5 \pm 2.58(0.07)$$
$$= 7.5 \pm 0.18$$
$$= 7.32 \ to \ 7.68$$

When using the 99 percent confidence interval, there is only a 1 percent risk that we are wrong and the specified interval does not contain the true population mean. We can be almost certain that the true population mean is included in the interval ranging from 7.32 to 7.68 hours per week. Notice that by increasing the confidence level we have also increased the width of the confidence interval from 0.28 (7.36 – 7.64) to 0.36 hours (7.32 – 7.68), thereby making our estimate less precise.

Figure 12.4   **95 Percent Versus 99 Percent Confidence Intervals**

Among the population of black Americans, what is the mean number of children in a family? Using the 1996 GSS sample, we can construct the following confidence intervals:

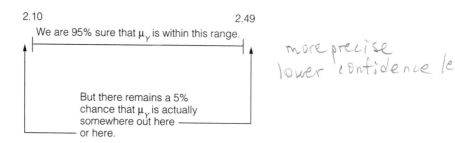

A 95% confidence interval

2.10   2.49

We are 95% sure that μ_Y is within this range.

But there remains a 5% chance that μ_Y is actually somewhere out here —— or here.

more precise
lower confidence le

A 99% confidence interval

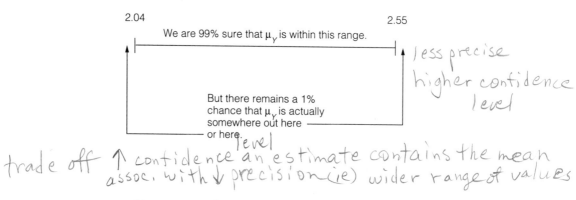

2.04   2.55

We are 99% sure that μ_Y is within this range.

But there remains a 1% chance that μ_Y is actually somewhere out here —— or here.

less precise
higher confidence level

trade off ↑ confidence level an estimate contains the mean
assoc. with ↓ precision (ie) wider range of values

You can see that there is a trade-off between achieving greater confidence in an estimate and the precision of that estimate. Although using a higher level of confidence (such as 99 percent) increases our confidence that the true population mean is included in our confidence interval, the estimate becomes less precise as the width of the interval increases. Although we are only 95 percent confident that the interval ranging between 7.36 and 7.64 hours includes the true population mean, it is a more precise estimate than the 99 percent interval ranging from 7.32 to 7.68 hours. The relationship between the confidence level and the precision of the confidence interval is illustrated in Figure 12.4.

Estimating Sigma

To calculate confidence intervals, we need to know the standard error of the sampling distribution, $\sigma_{\bar{Y}}$. The standard error is a function of the population standard deviation and the sample size:

$$\sigma_{\bar{Y}} = \frac{\sigma_Y}{\sqrt{N}}$$

μ_Y = population mean
σ_Y = population standard deviation

In our commuting example we have been using a hypothetical value, σ_Y = 1.5, for the population standard deviation. Typically, both the mean (μ_Y) and the standard deviation (σ_Y) of the population are unknown to us. When $N \geq 50$, however, the sample standard deviation S_Y is a good estimate of σ_Y. The standard error is then calculated as follows:

$$S_{\bar{Y}} = \frac{S_Y}{\sqrt{N}}$$

S_Y sample standard deviation
S_Y is good estimate of μ_Y when $N > 50$ (12.3)

As an example, we'll estimate the mean income for all adult Americans based on the 1991 GSS survey. The mean income for a sample of $N = 880$ is $\bar{Y} = \$18,130$, and the standard deviation is $S_Y = \$15,639$. Let's determine the 95 percent confidence interval for these data.

Calculating the Standard Error of the Mean   The standard error for the sampling distribution of the mean is

$\wedge$ sample means

$$S_{\bar{Y}} = \frac{S_Y}{\sqrt{N}} = \frac{15,639}{\sqrt{880}} = 527.19$$

Deciding on the Level of Confidence and Finding the Corresponding Z Value   We decide on a 95 percent confidence level, estimating a 5% risk of error. The Z value corresponding to a 95 percent confidence level is 1.96.

Calculating the Confidence Interval   The confidence interval is calculated by adding to and subtracting from the observed sample mean the product of the standard error and Z:

95% CI = 18,130 ± 1.96(527.19)
    = 18,130 ± 1,033
    = 17,097 to 19,163

Interpreting the Results   We can be 95 percent confident that the actual mean income of the population of adult Americans from which the GSS sample was taken is not less than $17,097 and not greater than $19,163. In other words, if we drew a large number of samples ($N = 880$) from this population, then 95 times out of 100 the true population mean would be included within our computed interval.

Sample Size and Confidence Intervals

Researchers can increase the precision of their estimate by increasing the sample size. In Chapter 11 we learned that larger samples result in smaller standard errors and, therefore, in sampling distributions that are more clustered around the population mean (Figure 11.6). A more tightly clustered sampling distribution means that our confidence intervals will be narrower and more precise. To illustrate the relationship between sample size and the standard error, and thus the confidence interval, let's calculate the 95 percent confidence interval for our GSS data with (1) a sample of $N = 440$ and (2) a sample of $N = 1,760$.

With a sample size $N = 440$, the standard error for the sampling distribution is

$$S_{\overline{Y}} = \frac{15,639}{\sqrt{440}} = 745.56$$

The 95 percent confidence interval is

$$95\% \text{ CI} = 18,130 \pm 1.96(745.56)$$
$$= 18,130 \pm 1,461$$
$$= 16,669 \text{ to } 19,591$$

With a sample size $N = 1,760$, the standard error for the sampling distribution is

$$S_{\overline{Y}} = \frac{15,639}{\sqrt{1760}} = 372.78$$

and the 95 percent confidence interval is

$$95\% \text{ CI} = 18,130 \pm 1.96(372.78)$$
$$= 18,130 \pm 731$$
$$= 17,399 \text{ to } 18,861$$

↑ sample size →
more narrow C.I,
ie ↑ precision

In Table 12.1 we summarize the 95 percent confidence intervals for the mean income for three sample sizes: $N = 440$, $N = 880$, and $N = 1,760$.

Notice that there is an inverse relationship between sample size and the width of the confidence interval. The increase in sample size is linked with increased precision of the confidence interval. The 95 percent confidence interval for the GSS sample of 440 cases is $2,922. But the interval widths de-

Table 12.1 **95 Percent Confidence Interval and Width for Mean Income for Three Different Sample Sizes**

Sample Size	Confidence Interval	Interval Width	S_Y	$S_{\overline{Y}}$
$N = 440$	$16,669–$19,591	$2,922	$15,639	745.56
$N = 880$	$17,097–$19,163	$2,066	$15,639	527.19
$N = 1,760$	$17,399–$18,861	$1,462	$15,639	372.78

crease to $2,066 and $1,462, respectively, as the sample sizes increase to $N = 880$ and then to $N = 1,760$. We had to quadruple the size of the sample (from 440 to 1,760) to reduce the confidence interval by half[1] (from $2,922 to $1,462). In general, although the precision of estimates increases steadily with sample size, the gains are rather modest after N reaches about 400. An important factor to keep in mind is the increased cost associated with a larger sample. Researchers have to consider at what point the increase in precision is too small to justify the additional cost associated with a larger sample.

Learning Check.   *Why do smaller sample sizes produce wider confidence intervals? (See Figure 12.5.)* (Hint: *Compare the standard errors of the mean for the three sample sizes.)*

Box 12.2   What Affects Confidence Interval Width?
A Summary

"Holding other factors constant . . . "

If the sample size goes up	↑	the confidence interval becomes more precise. → ←
If the sample size goes down	↓	the confidence interval becomes less precise. ← →
If the value of the sample standard deviation goes up	↑	the confidence interval becomes less precise. ← →
If the value of the sample standard deviation goes down	↓	the confidence interval becomes more precise. → ←
If the level of confidence goes up (from 95% to 99%)	↑	the confidence interval becomes less precise. ← →
If the level of confidence goes down (from 99% to 95%)	↓	the confidence interval becomes more precise. → ←

---

[1]The slight variation is due to rounding.

Figure 12.5 **The Relationship Between Sample Size and Confidence Interval Width**

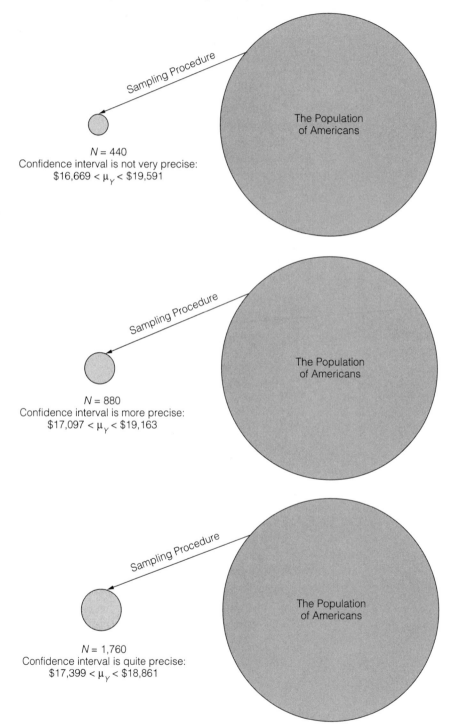

$N = 440$
Confidence interval is not very precise:
$\$16,669 < \mu_Y < \$19,591$

$N = 880$
Confidence interval is more precise:
$\$17,097 < \mu_Y < \$19,163$

$N = 1,760$
Confidence interval is quite precise:
$\$17,399 < \mu_Y < \$18,861$

■ ■ ■ ■ **Statistics in Practice: Hispanic Migration and Earnings**

Tienda and Wilson have investigated the relationship between migration and the earnings of Hispanic men.[2] Based on a sample of the 1980 census that included 5,726 Mexicans, 5,908 Puerto Ricans, and 3,895 Cubans, Tienda and Wilson argue that these three Hispanic groups vary markedly in socioeconomic characteristics because of differences in the timing and circumstances of their immigration to the United States. The authors claim that the period of entry and the circumstances prompting migration have affected the geographical distribution and the employment opportunities of each group. For example, Puerto Ricans are disproportionately located in the Northeast, where the labor market is characterized by the highest unemployment rates, whereas the majority of Cuban immigrants reside in the Southeast, where the unemployment rate is the lowest in the United States.

The contemporary profiles of Hispanic men also reveal persistent differences in educational levels among Mexicans and Puerto Ricans compared with Cubans. About 60 percent of Mexicans and Puerto Ricans had not completed high school, compared with 42 percent of Cuban men. At the other extreme, 17 percent of Cuban men were college graduates, compared with about 4 percent of Mexican men and Puerto Rican men.

These differences in migrant status and socioeconomic characteristics are likely to be evident in disparities in earnings among the three groups. Tienda and Wilson anticipated that the earnings of Cubans would be higher than the earnings of Mexicans and Puerto Ricans. As hypothesized, with average earnings of $16,368 ($S_Y$ = $3,069) Cubans are at the top of the income hierarchy. Puerto Ricans are at the bottom of the income hierarchy, with earnings averaging $12,587 ($S_Y$ = $8,647). Mexican men are intermediate among the groups, with average annual earnings of $13,342 ($S_Y$ = $9,414).

Although Tienda and Wilson did not calculate confidence intervals for their estimates, we will use the data they present to calculate a 95 percent confidence interval for the mean income for the three groups of Hispanic men.

To find the 95 percent confidence interval for Cuban income, we first calculate the standard error:

$$S_{\bar{Y}} = \frac{3,069}{\sqrt{3,895}} = 49.17$$

Then we calculate the confidence interval:

95% CI = 16,368 ± 1.96(49.17)
      = 16,368 ± 96
      = 16,272 to 16,464

[2]Adapted from Marta Tienda and Franklin D. Wilson, "Migration and the Earnings of Hispanic Men," *American Sociological Review* 57 (1992): 661–678.

For Puerto Rican income, the standard error is

$$S_{\bar{Y}} = \frac{8,647}{\sqrt{5,908}} = 112.50$$

and the confidence interval is

$$95\% \text{ CI} = 12,587 \pm 1.96(112.50)$$
$$= 12,587 \pm 220$$
$$= 12,367 \text{ to } 12,807$$

Finally, for Mexican income the standard error is

$$S_{\bar{Y}} = \frac{9,414}{\sqrt{5,726}} = 124.41$$

and the confidence interval is

$$95\% \text{ CI} = 13,342 \pm 1.96(124.41)$$
$$= 13,342 \pm 244$$
$$= 13,098 \text{ to } 13,586$$

The confidence intervals for mean annual income of Cuban, Puerto Rican, and Mexican immigrants are illustrated in Figure 12.6. We can say with 95 percent confidence that the true income mean for each Hispanic

Figure 12.6 **The 95 Percent Confidence Intervals for the Mean Income of Puerto Ricans, Mexicans, and Cubans**

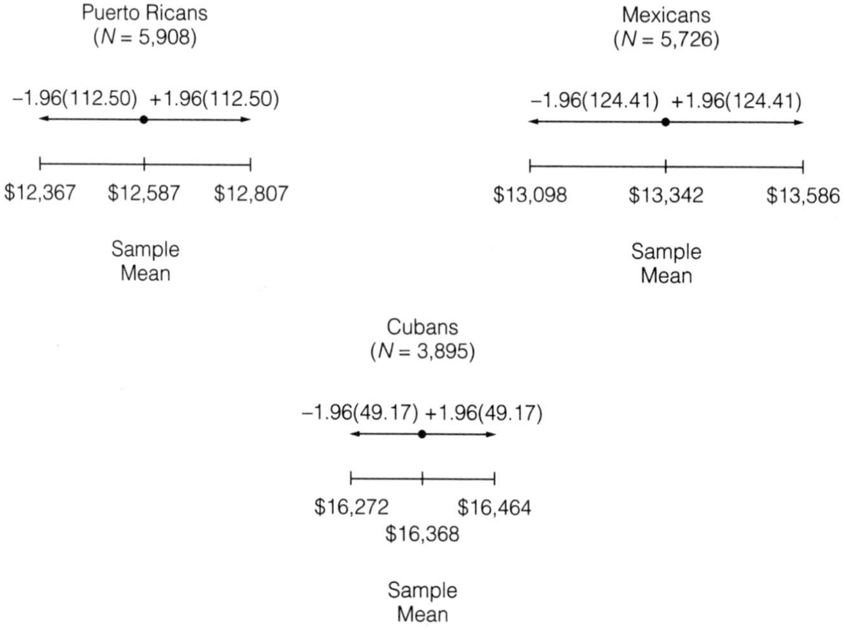

group lies somewhere within the corresponding confidence interval. Notice that the confidence intervals do not overlap, revealing great disparities in earnings among the three groups. As noted earlier, highest interval estimates are for Cubans, followed by Mexicans and then Puerto Ricans.

Confidence Intervals for Proportions

Confidence intervals can also be computed for sample proportions or percentages in order to estimate population percentages. The procedures for estimating proportions and percentages are identical. Any of the formulas presented for proportions can be applied to percentages, and vice versa. We can obtain a confidence interval for a percentage by calculating the confidence interval for a proportion and then multiplying the result by 100.

The same conceptual foundations of sampling and statistical inference that are central to the estimation of population means—the selection of random samples and the special properties of the sampling distribution—are also central to the estimation of population proportions.

Earlier we saw that the sampling distribution of the means underlies the process of estimating population means from sample means. Similarly, the *sampling distribution of proportions* underlies the estimation of population proportions from sample proportions. Based on the Central Limit Theorem, we know that with sufficient sample size the sampling distribution of proportions is approximately normal, with mean μ_p equal to the population proportion π and with a standard error of proportions (the standard deviation of the sampling distribution of proportions) equal to

$$\sigma_p = \sqrt{\frac{(\pi)(1-\pi)}{N}}$$ (12.4)

[handwritten annotation: μ_p = mean of population proportion; π = population proportion; N = pop size]

where

  σ_p = the standard error of proportions
  π = the population proportion
  N = the population size

However, since the population proportion, π, is unknown to us (that is what we are trying to estimate), we can use the sample proportion, p, as an estimate of π. The estimated standard error then becomes

$$S_p = \sqrt{\frac{(p)(1-p)}{N}}$$ (12.5)

where

  S_p = the estimated standard error of proportions
  p = the sample proportion
  N = the sample size

assumptions of normality for population proportions
① *when p + 1-p are about .5 N of 50 OK*
when p >.5 or <.5 we need N of 100 or more

As an example, let's calculate the estimated standard error for the 1996 *Time*/CNN poll. Based on a random sample of 826 registered voters, the percentage favoring Bill Clinton was estimated to be 50 percent. Based on Formula 12.5, with $p = .50$, $1 - p = (1 - .50) =. 50$, and $N = 826$, the standard error is

$$S_p = \sqrt{\frac{(.50)(1-.50)}{826}} = .017$$

We will have to consider two factors to meet the assumption of normality with the sampling distribution of proportions: (1) the sample size N and (2) the sample proportions p and $1 - p$. When p and $1 - p$ are about .50, a sample size of at least 50 is sufficient. But when $p > .50$ (or $1 - p < .50$), a larger sample is required to meet the assumption of normality. Usually, a sample of 100 or more is adequate for any single estimate of a population proportion.

Procedures for Estimating Proportions

Because the sampling distribution of proportions is approximately normal, we can use the normal distribution to establish confidence intervals for proportions in the same manner that we used the normal distribution to establish confidence intervals for means.

The general formula for constructing confidence intervals for proportions for any level of confidence is

$$CI = p \pm Z(S_p)$$

where

CI = the confidence interval
p = the observed sample proportion
Z = the Z corresponding to the confidence level
S_p = the standard error of proportions

Let's examine this formula in more detail. Notice that to obtain a confidence interval at a certain level we take the sample proportion and add to or subtract from it the product of a Z value and the standard error. The Z value we choose depends on the desired confidence level. We want the area between the mean and the selected $\pm Z$ to be equal to the confidence level.

For example, to obtain a 95 percent confidence interval we would choose a Z of 1.96 because we know (from Appendix B) that 95 percent of the area under the curve is included between ± 1.96. Similarly, for a 99 percent confidence level we would choose a Z of 2.58. (The relationship between confidence level and Z values was graphically illustrated earlier, in Figure 12.1.)

To determine the confidence interval for a proportion, we follow the same steps used to find confidence intervals for means:

1. Calculate the standard error of the proportion.
2. Decide on the desired level of confidence, and find the corresponding Z value.
3. Calculate the confidence interval.
4. Interpret the results.

To illustrate these steps we use the results of the *Time*/CNN presidential preference poll.

Calculating the Standard Error of the Proportion The standard error of the proportion .50 (50%) with a sample $N = 826$ is .017.

Deciding on the Desired Level of Confidence and Finding the Corresponding Z Value We choose the 95 percent confidence level. The Z corresponding to a 95 percent confidence level is 1.96.

Calculating the Confidence Interval We calculate the confidence interval by adding to and subtracting from the observed sample proportion the product of the standard error and Z:

$$95\% \text{ CI} = .50 \pm 1.96(.017)$$
$$= .50 \pm .033$$
$$= .467 \text{ to } .533$$

Interpreting the Results We are 95 percent confident that the true population proportion is somewhere between .467 and .533. In other words, if we drew a large number of samples from the population of registered voters, then 95 times out of 100 the confidence interval we obtained would contain the true population proportion. We can also express this result in percentages and say that we are 95 percent confident that the true population percentage of support for President Clinton is included somewhere within our computed interval of 46.7 percent to 53.3 percent.

> **Learning Check.** *Calculate the confidence interval for the presidential preference poll using percentages rather than proportions. Your results should be identical with ours except that they are expressed in percentages.*

Note that with a 95 percent confidence level there is a 5 percent risk that we are wrong. If we continued to draw large samples from this population, in 5 out of 100 samples the true population proportion would not be included in the specified interval.

We can decrease our risk by increasing the confidence level from 95 to 99 percent.

$$99\% \text{ CI} = .50 + 2.58(.017)$$
$$= .50 \pm .044$$
$$= .456 \text{ to } .544$$

When using the 99 percent confidence interval we can be almost certain (99 times out of 100) that the true population proportion is included in the interval ranging from .456 (45.6%) to .544 (54.4%). However, as we saw earlier, there is a trade-off between achieving greater confidence in making an estimate and the precision of that estimate. Although using a 99 percent level increased our confidence level from 95 percent to 99 percent (thereby reducing our risk of being wrong from 5% to 1%), the estimate became less precise as the width of the interval increased.

Increasing the Sample Size

The relationship between sample size and interval width when estimating means also holds true for sample proportions. When the sample size increases, the standard error of the proportion decreases, and therefore the width of the confidence interval decreases as well. For instance, when we increase the sample of registered voters from $N = 826$ to $N = 1,500$, our standard error becomes

$$S_p = \sqrt{\frac{(.50)(1 - .50)}{1,500}} = .013$$

and the 95 percent confidence interval is

$$95\% \text{ CI} = .50 \pm 1.96(.013)$$
$$= .50 \pm .025$$
$$= .475 \text{ to } .525$$

Notice that the width of the confidence interval decreased from .066 (.467 to .533) with $N = 826$ to .05 (.475 to .525) with $N = 1,500$. Thus, to increase the precision of our estimate by .016, we had to almost double our sample size! In a research project, an increase in sample size is often associated with increased cost, so we need to consider whether the increased accuracy of the estimate justifies the increased cost of the project.

■ ■ ■ ■ **Statistics in Practice: Should Affirmative Action Programs Be Increased?**

Poll or survey results may be limited to a single estimate of a parameter. For instance, the *Time*/CNN poll only reported the estimated percentage of Clinton supporters among registered voters in the United States. Most survey studies, however, are not limited to single estimates for the overall population. Often, separate estimates are reported for subgroups within the overall population of interest. The 1997 Gallup survey, for example, compared support for affirmative action programs among black and white respondents.

When estimates are reported for subgroups, the confidence intervals are likely to vary from subgroup to subgroup. Each confidence interval is based on the confidence level, the standard error of the proportion (which can be estimated from p), and the sample size. Even when a confidence interval is reported only for the overall sample, we can easily compute separate confidence intervals for each of the subgroups if the confidence level and the size of each of the subgroups are included.

To illustrate this, let's calculate the 95 percent confidence intervals for the proportions of whites and blacks that support increasing affirmative action programs. Out of 1,680 whites included in the overall sample, .22 (or 22%) favor increasing affirmative action programs. In contrast, .53 (or 53%) of the 1,269 black respondents agree.

Calculating the Standard Error of the Proportion   The standard error for the proportion of whites is

$$S_p = \sqrt{\frac{.22(1 - .22)}{1,680}} = .010$$

The standard error for the proportion of blacks that support affirmative action programs is

$$S_p = \sqrt{\frac{.53(1 - .53)}{1,269}} = .014$$

Deciding on the Desired Level of Confidence and Finding the Corresponding Z Value   We choose the 95 percent confidence level, with a corresponding Z value of 1.96.

Calculating the Confidence Interval   For whites,

$$95\% \text{ CI} = .22 \pm 1.96(.010)$$
$$= .22 \pm .020$$
$$= .200 \text{ to } .240$$

and for blacks,

$$95\% \text{ CI} = .53 \pm 1.96(.014)$$
$$= .53 \pm .027$$
$$= .503 \text{ to } .557$$

The 95 percent confidence interval for the proportion of white and black respondents who support affirmative action is illustrated in Figure 12.7.

Interpreting the Results   We are 95 percent confident that the true population proportion supporting increases in affirmative action programs is somewhere between .200 and .240 (or between 20.0% and 24.0%) for whites, and

Figure 12.7 **The 95 Percent Confidence Interval for the Proportion of Whites and Blacks Supporting Affirmative Action**

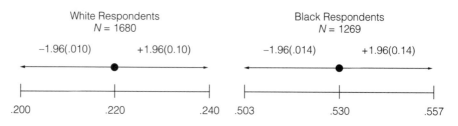

somewhere between .503 and .557 (or between 50.3% and 55.7%) for blacks. Based on the Gallup sample, a clear majority of black Americans support increases in affirmative action programs, whereas less than a quarter of white Americans agree.

MAIN POINTS

- The goal of most research is to find population parameters. The major objective of sampling theory and statistical inference is to provide estimates of unknown parameters from sample statistics.

- Researchers make point estimates and interval estimates. Point estimates are sample statistics used to estimate the exact value of a population parameter. Interval estimates are ranges of values within which the population parameter may fall.

- Confidence intervals can be used to estimate population parameters such as means or proportions. Their accuracy is defined with the confidence level. The most common confidence levels are 90, 95, and 99 percent.

- To establish a confidence interval for a mean or a proportion, add or subtract from the mean or the proportion the product of the standard error and the Z value corresponding to the confidence level.

KEY TERMS

confidence interval (interval estimate)   estimation

confidence level   point estimate

SPSS DEMONSTRATION

Producing Confidence Intervals Around a Mean [Module A]

SPSS calculates confidence intervals around a sample mean or proportion with the Explore procedure. Let's investigate the average number of ideal children for men and women.

Activate the Explore procedure by selecting the *Analyze* menu, *Descriptive Statistics,* then *Explore.* The opening dialog box has spaces for both dependent and independent variables. Place CHLDIDEL in the Dependent List box, and SEX in the Factor List box, as shown in Figure 12.8.

Figure 12.8

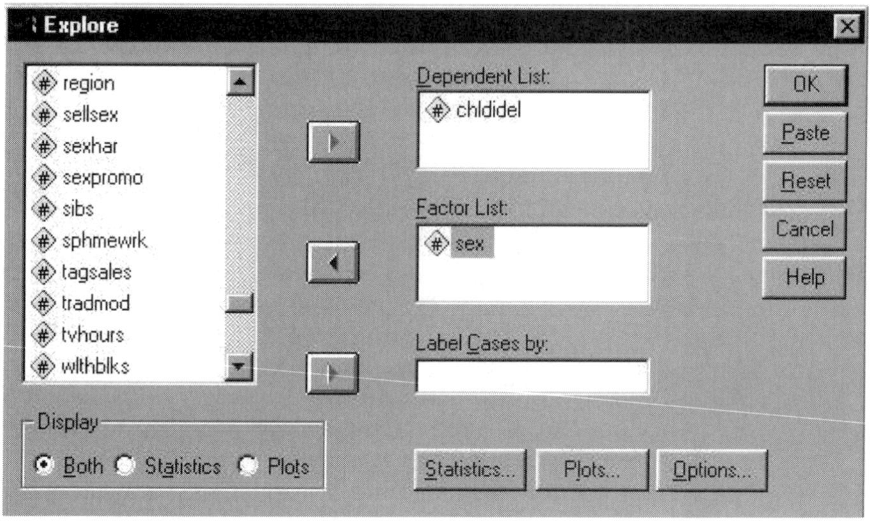

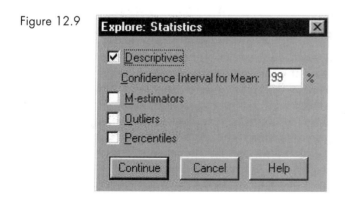

Click on the *Statistics* button. Notice that the Descriptives choice also includes the confidence interval for the mean, which by default is calculated at the 95 percent confidence level. Let's change that to the 99 percent level by erasing the "95" and substituting "99" (see Figure 12.9).

Figure 12.9

Click on *Continue* to return to the main dialog box. Recall that Explore produces several statistics and plots by default. For this example, we don't need to view the graphics, so click on the *Statistics* button in the *Display* section. Your screen should now look like Figure 12.10.

Figure 12.10

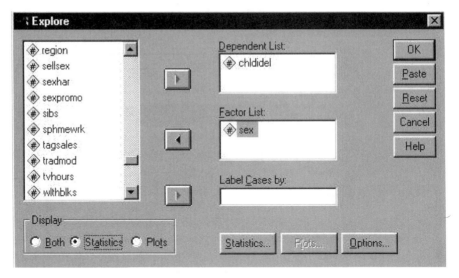

Click on *OK* to run the procedure.

The output from the Explore procedure (Figure 12.11) is broken into two parts, one for males and one for females. The mean number of ideal children for males is 2.74; for females, it's 2.78. Our data indicate that, on average, women and men are nearly identical in their ideal number of children.

The 99 percent confidence interval for males runs from about 2.54 to 2.95 children. One way to interpret this result is to state that, in 100 samples of size 396 of males (the number of males in this sample) from the U.S. adult population, we would expect the confidence interval to include the true population value for the mean ideal number of children 99 times out of those 100. We can never be sure that in *this* particular sample the confidence interval includes the population mean. As explained in this chapter, any one sample's confidence interval either does or does not contain the (unknown) population mean, so no probability value can be associated with a particular confidence interval. Still, our best estimate for the mean ideal number of children falls within a narrow range of only about .41.

For females, the 99 percent confidence interval is even narrower, varying from about 2.60 to 2.97 children, or only .37. This is because we have a larger sample of females, 533. Notice that the standard deviation of CHLDIDEL is very similar for males and females.

SPSS PROBLEMS

1. Recall that the GSS sample includes men and women from 18 to 89 years of age. Does it matter that we may have responses from men and women of diverse ages? Would our results change if we selected a younger sample of men and women? [Module A]

Figure 12.11

Case Processing Summary

	SEX RESPONDENTS SEX	Cases					
		Valid		Missing		Total	
		N	Percent	N	Percent	N	Percent
CHLDIDEL IDEAL NUMBER OF CHILDREN	1 MALE	396	63.6%	227	36.4%	623	100.0%
	2 FEMALE	533	66.3%	271	33.7%	804	100.0%

Descriptives

SEX RESPONDENTS				Statistic	Std. Error
CHLDIDEL IDEAL NUMBER OF CHILDREN	1 MALE	Mean		2.74	7.94E-02
		99% Confidence Interval for Mean	Lower Bound	2.54	
			Upper Bound	2.95	
		5% Trimmed Mean		2.57	
		Median		2.00	
		Variance		2.499	
		Std. Deviation		1.58	
		Minimum		0	
		Maximum		8	
		Range		8	
		Interquartile Range		1.00	
		Skewness		2.234	.123
		Kurtosis		5.100	.245
	2 FEMALE	Mean		2.78	7.06E-02
		99% Confidence Interval for Mean	Lower Bound	2.60	
			Upper Bound	2.97	
		5% Trimmed Mean		2.60	
		Median		2.00	
		Variance		2.654	
		Std. Deviation		1.63	
		Minimum		0	
		Maximum		8	
		Range		8	
		Interquartile Range		1.00	
		Skewness		2.162	.106
		Kurtosis		4.375	.211

a. To take the SPSS demonstration one step further, use the *Select Cases* procedure to do that analysis. Select respondents based on the variable AGE, less than or equal to 35 years.

b. Repeat the Explore procedure we just completed in the demonstration based on CHLDIDEL. What differences exist between men and women in the under 35 sample? How does this compare with our results based on the entire sample? As we discussed in Chapter 11, what impact may it have on our sample findings?

2. Calculate the 95 percent confidence interval for the following variables, comparing lower, working, middle, and upper classes (CLASS) in the GSS sample. Make a summary statement of your findings. [Module A]
   a. CHILDS (Number of children in the household)
   b. EDUC (Respondent's highest year of school completed)
   c. PAEDUC (Father's highest year of school completed)
   d. HRS1 (Hours at work per week)

CHAPTER EXERCISES

1. In a study of crime, the FBI found that 13.2 percent of all Americans had been victims of crime during a one-year period. This result was based on a sample of 1,105 adults.
   a. Estimate the percentage of U.S. adults who were victims at the 90 percent confidence level. State in words the meaning of the result.
   b. Estimate the percentage of victims at the 99 percent confidence level.
   c. Imagine that the FBI doubles the sample size in a new sample but finds the same value of 13.2 percent for the percentage of victims in the second sample. By how much would the 90 percent confidence interval shrink? By how much would the 99 percent confidence interval shrink?
   d. Considering your answers to (a), (b), and (c), can you suggest why national surveys, such as those by Gallup, Roper, or the *New York Times,* typically take samples of size 1,000 to 1,500?

2. Use the data in Chapter 10, Exercise 5 about income.
   a. Construct the 95 percent confidence interval for the mean income of males.
   b. Construct the 99 percent confidence interval for the mean income of males.
   c. As our confidence in the result increases, how does the size of the confidence interval change? Explain why this is true.

3. The United States has often adopted an isolationist foreign policy designed to stay out of foreign entanglements (as George Washington advised his fellow citizens two hundred years ago). In 1995 the Times Mirror organization polled a random sample of Americans, asking, "Please tell me whether you agree or disagree that the United States should mind its own business internationally and let other countries get along as best they can on their own." The poll found that .41 of the 1,007 respondents agreed with this statement.
   a. Estimate the proportion of all adult Americans who agree with the statement at the 95 percent confidence level.
   b. Estimate the proportion of all adult Americans who agree with the statement at the 99 percent confidence level.
   c. If you were going to write a report on this poll result, would you prefer to use the 99 percent or 95 percent confidence interval? Explain why.

4. Use the data in Chapter 5, Exercise 6, about occupational prestige for blacks and whites.
   a. Construct the 95 percent confidence interval for occupational prestige for blacks ($N = 73$).
   b. Construct the 95 percent confidence interval for occupational prestige for whites ($N = 455$). State in words the meaning of the result.
   c. Use these statistics to discuss differences in occupational prestige for blacks and whites. Does it appear that whites do have greater job prestige than blacks? Why?

5. A newspaper does a poll to determine the likely vote for the incumbent mayor, Ann Johnson, in the upcoming election. They find that 52 percent of the voters favor Johnson in a sample of 500 registered voters. The newspaper asks you, their statistical consultant, to tell them whether they should declare Johnson the likely winner of the election. What is your advice? Why?

6. The police department in your city was asked by the mayor's office to estimate the cost of crime. The police began their study with burglary records, taking a random sample of 500 files since there were too many crime records to calculate statistics for all the crimes committed.
   a. If the average dollar loss in a burglary, for this sample of size 500, is $678, with a standard deviation of $560, construct the 95 percent confidence interval for the true average dollar loss in burglaries.
   b. An assistant to the mayor, who claims to understand statistics, complains about your confidence interval calculation. She asserts that the dollar losses from burglaries are not normally distributed, which in turn makes the confidence interval calculation meaningless. Assume that she is correct about the distribution of money loss. Does that imply that the calculation of a confidence interval is not appropriate? Why or why not?

7. From the 1996 GSS we find that the mean number of hours worked the previous week in the U.S. population was 42.36, with a standard deviation of 14.27. A total of 954 adults answered this question.
   a. What is the 95 percent confidence interval for the mean number of hours worked the previous week in the U.S. population?
   b. Does the size of this confidence interval seem compatible with the fact that a sizable proportion of adults work only part-time? Why or why not?

8. A social service agency plans to conduct a survey to determine the mean income of its clients. The director of the agency prefers that you measure the mean income very accurately, to within ±$500. From a sample taken two years ago, you estimate that the standard deviation of income for this population is about $5,000. Your job is to figure out the necessary sample size to reduce sampling error to ±$500.
   a. Do you need to have an estimate of the current mean income to answer this question? Why or why not?

b. What sample size should be drawn to meet the director's requirement at the 95 percent level of confidence? (*Hint:* Use the formula for a confidence interval and solve for *N*, the sample size.)

c. What sample size should be drawn to meet the director's requirement at the 99 percent level of confidence?

9. Data from the 1996 General Social Survey show that the mean number of children per respondent was 1.86, with a standard deviation of 1.68. A total of 1,445 people answered this question. Estimate the population mean number of children per adult using a 90 percent confidence interval.

10. A finance company took a random sample of its records to determine the proportion of auto loans they financed that were not repaid. In their sample of 500 they found that .12 of the loans defaulted (and the car had to be repossessed). Estimate at the 99 percent confidence level the proportion of all auto loans financed by the company that eventually default.

11. The cost of health care continues to be a serious concern of Americans, especially those who are without health insurance for any reason. To gauge the magnitude of this problem, the Gallup Organization conducted a poll in May 1993 that asked whether a person had been without insurance at some point in the past. Out of the sample of 1,011, 41 percent answered yes to this question. What is the 95 percent confidence interval for the percentage of adult Americans who have been without insurance?

12. The Social Security system in the United States may encounter serious financial difficulties as baby boomers begin to retire in the future. Several polls have asked Americans their opinion about the financial condition of Social Security. In one poll, taken in 1992 by CBS News and the *New York Times*, 53 percent of a sample of 1,281 adults said that they did not think "the Social Security system will have the money available to provide the benefits you expect for your retirement."

a. Calculate the 95 percent confidence interval to estimate the percentage of Americans who don't think Social Security will be able to provide for them.

b. Calculate the 99 percent confidence interval.

c. Are both these results compatible with the view that *less* than 50 percent of Americans believe that the Social Security system will not be able to pay their benefits after retirement?

13. Is homosexuality a biological trait or based on one's environment and socialization? In 1998 the nature versus nurture debate was reopened as several religious groups proclaimed their ability to "cure" gays of their sexual orientation. What do Americans attribute homosexuality to? In July 1998 the Gallup Organization surveyed 1,016 adults. (Data from Frank Newport, July 25, 1998. Used by permission of the Gallup Poll.) They found that 31 percent believed it was something homosexuals are

born with, while 47 percent attributed homosexuality to upbringing/environment.

a. For each reported percentage, calculate the 95 percent confidence interval.

b. To the question, "Do you personally believe homosexual behavior is morally wrong or is not morally wrong?" 59 percent responded "Yes, morally wrong," while 35 percent answered "No, not morally wrong." For each percentage, calculate the 95 and 99 percent confidence intervals.

c. What conclusions can you draw about the public's opinions of homosexual behavior based on your calculations?

14. Based on telephone interviews with 1,009 adults, the Gallup Organization reported on the current status of religious faith in the United States. (Data from Frank Newport and Lydia Saad, March 29, 1997. Used by permission of the Gallup Poll.) Among the respondents, 67 percent reported being a member of a church or synagogue. While 61 percent felt that religion was very important in their lives, 57 percent thought that religion was losing its influence on American life.

a. Calculate the 95 percent confidence interval for each proportion reported.

b. What can you conclude about the current status of religion based on these Gallup Poll results?

15. Poll results released by the Gallup Organization in 1997 revealed differences in the perceptions of blacks and whites on the status of race relations in the United States ("Black/White Relations in the U.S.," *Gallup Poll Archives*, June 10, 1997). Among a series of questions posed to black and white respondents, Gallup asked individuals to assess their level of satisfaction on a number of aspects of their personal life. The percentage satisfied in selected life categories are presented in the following table.

	Blacks	Whites
Way things are going in U.S.	44%	51%
Way things are going in personal life	74%	87%
Income	53%	72%
Education, Preparing for Job	68%	75%
Total (*N*)	1269	1680

Source: Data from *Special Report*, June 10, 1997. Used by permission of the Gallup Poll.

a. Estimate the 95 percent confidence interval for each percentage.

b. Prepare a paragraph summarizing the results of your calculations.

13 Testing Hypotheses About Two Samples

Reading the Research Literature: Reporting the Results of Statistical Hypothesis Testing

MAIN POINTS

KEY TERMS

SPSS DEMONSTRATION

SPSS PROBLEMS

CHAPTER EXERCISES

■ ■ ■ ■ **Introduction**

When researchers make comparisons between groups, they are usually interested in identifying a relationship between variables. For instance, in Chapter 3 we compared elderly men and women on their housing arrangements. As reported in the 1993 census, 15.5 percent of elderly men and 40.8 percent of elderly women live alone. Can we conclude that there is an association between gender and elderly living arrangements?

In Chapter 4 we compared the average number of traumatic events experienced by working-class black males and black females. We learned that the average is higher for women (1.76) than for men (1.04). Does the difference in the average number of traumatic events experienced by working-class black males and black females imply that gender is associated with the frequency of trauma among black Americans?

Because most comparisons are based on data from a sample and not from the entire population, we need to consider that the differences we observe—between elderly men and women or between black males and females—may not be "real" differences. It may be that if we took more samples, we would get different results.

For example, based on the 1996 GSS, men reported an average of 13.59 years of education, while women had an average of 13.26 years. Data for years of education are reported in Table 13.1. These sample averages could mean either: (1) the average number of years of education for men is higher than the average for women or (2) the average for women is actually about the same as for men; but our sample just happens to indicate a higher average for men.

Table 13.1 **Years of Education for Men and Women, GSS1996**

	Men Sample 1	Women Sample 2
Mean ($\bar{Y}$)	13.59	13.26
Standard Deviation (S)	3.03	2.91
Variance (S^2)	9.18	8.47
N	635	811

How can we decide which of these explanations makes more sense? Because most estimates are based on single samples and different samples may result in different estimates, sampling results cannot be used directly to make statements about a population. We need a procedure that allows us to evaluate hypotheses about population parameters based on sample statistics. In Chapter 12 we saw that population parameters can be estimated from sample statistics. In this chapter we will learn how to use sample statistics to make decisions about population parameters. This procedure is called **statistical hypothesis testing**.

■ ■ ■ ■ Assumptions of Statistical Hypothesis Testing

Statistical hypothesis testing requires several assumptions. These assumptions include considerations of the level of measurement of the variable, the method of sampling, the shape of the population distribution, and the sample size. The specific assumptions may vary, depending on the test or the conditions of testing. However, without exception, *all* statistical tests assume random sampling. Tests of hypotheses about means also assume interval-ratio level of measurement and require that the population under consideration be normally distributed or that the sample size be larger than 50.

Based on our data, we test the hypothesis that average years of completed education are higher for men than for women. The test we are considering meets these conditions:

1. The sample is a subgroup in the GSS sample, which is a national probability sample, randomly selected.
2. The variable *years of education* is an interval-ratio level of measurement.
3. We cannot assume that the population is normally distributed. However, because our sample sizes are sufficiently large (*each N* > 50), we know, based on the Central Limit Theorem, that the sampling distribution of the mean will be approximately normal.
4. The samples are independent of each other. The choice of sample members from one population has no effect on the choice of sample members from the second population. In our comparison of men and women, we are assuming that the selection of men is independent of the selection of women. (The requirement of independence is also satisfied by selecting one sample randomly, then dividing the sample into appropriate subgroups. For example, we could randomly select a sample and then divide it into groups based on gender, religion, income, or any other attribute we are interested in.)

■ ■ ■ ■ Stating the Research and Null Hypotheses

Hypotheses are usually defined in terms of interrelations between variables and are often based on a substantive theory. Earlier, we defined *hypotheses* as tentative answers to research questions. They are tentative because they can be verified only after being empirically tested. The testing of hypotheses is an important step in this verification process.

The Research Hypothesis (H_1)

Our first step is to formally express the hypothesis in a way that makes it amenable to a statistical test. The substantive hypothesis is called the **research hypothesis** and is symbolized by H_1. Research hypotheses are always expressed in terms of population parameters because we are interested in making statements about population parameters based on our sample statistics.

In our research hypothesis (H_1), we believe that the average years of education is higher for men than women. We are stating a hypothesis about the relationship between gender and education in the general population by comparing the mean educational attainment of men with the mean educational attainment of women. Symbolically, we use μ to represent the population mean; the subscript 1 refers to our first sample (men) and 2 to our second sample (women). Our hypothesis can be expressed as

$H_1: \mu_1 > \mu_2$

In general, the research hypothesis (H_1) specifies one of the following:

1. The population parameters are not equal to each other: $\mu_1 \neq \mu_2$.
2. One parameter is greater than the other: $\mu_1 > \mu_2$.
3. One parameter is less than the other: $\mu_1 < \mu_2$.

> *Research hypothesis (H_1)* A statement reflecting the substantive hypothesis. It is always expressed in terms of population parameters, but its specific form varies from test to test.

The Null Hypothesis (H_0)

Is it possible that there is no real difference between the mean education of men and women in the population and that the observed difference of .33 years (13.59 – 13.26) is actually due to the fact that this particular sample happened to contain men with more years of education? Since statistical inference is based on probability theory, it is not possible to prove or disprove the research hypothesis directly. We can, at best, estimate the likelihood that it is true or false.

As a solution, statisticians set up a hypothesis that is counter to the research hypothesis. The **null hypothesis**, symbolized as H_0, contradicts the research hypothesis and usually states that there is no difference between the population means. It is also referred to as the hypothesis of "no difference." Our null hypothesis can be stated symbolically as

$H_0: \mu_1 = \mu_2$

Rather than directly testing the substantive hypothesis (H_1) that men have higher years of education on average than women, we test the null hypothesis (H_0) that there are no differences between the years of education for both groups. In hypothesis testing, we hope to reject the null hypothesis in order to provide support for the research hypothesis. The rejection of the null hypothesis will strengthen our belief in the research hypothesis and increase our confidence in the importance and utility of the broader theory from which the research hypothesis was derived.

> *Null hypothesis (H_0)* A statement of "no difference," which contradicts the research hypothesis and is always expressed in terms of population parameters.

More About Research Hypotheses: One- and Two-Tailed Tests

In a **one-tailed test,** the research hypothesis is directional; that is, it specifies that a population mean is either less (<) or greater (>) than another population mean. We can express our research hypothesis as either:

$$H_1: \mu_1 < \mu_2 \text{ or } \mu_1 > \mu_2$$

The research hypothesis we've stated for the average years of education for men and women is a one-tailed test.

> *One-tailed test* A type of hypothesis test that involves a directional hypothesis. It specifies that the values of one group are either larger or smaller than those of another.

When a one-tailed test specifies that one population value is greater than another, we call it a **right-tailed test** because we will evaluate the difference between the values at the right tail of the sampling distribution. Our example is a right-tailed test because the research hypothesis states that men's average years of education is greater than women's.

In contrast, the research hypothesis can also specify that the value of one population is less than another. This is called a **left-tailed test** because the difference between values will be evaluated at the left tail of the sampling distribution.

> *Right-tailed test* A one-tailed test in which the difference between values is hypothesized to be at the right tail of the sampling distribution.
>
> *Left-tailed test* A one-tailed test in which the critical region of rejection is hypothesized to be at the left tail of the sampling distribution.

Sometimes, however, although we have some theoretical basis to assume differences between groups, we cannot anticipate the direction of those differences. For example, we may have reason to believe that the average education of men is *different* from that of women, but we may not have enough research or support to predict whether it is *higher* or *lower*.

When we have no theoretical reason for specifying a direction to the research hypothesis, we conduct a **two-tailed test**. The research hypothesis specifies that the population means are not equal. For example, we can express the research hypothesis about the mean education as

$H_1: \mu_1 \neq \mu_2$

Two-tailed test A type of hypothesis test that involves a nondirectional research hypothesis. We are equally interested in whether the values are less than or greater than one another.

With both one-tailed and two-tailed tests, our null hypothesis of no difference remains the same. It can be expressed as

$H_0: \mu_1 = \mu_2$

Learning Check. *For the following research situations, state your research and null hypotheses:*

■ *There is a difference between the mean statistics grades of social science majors and the mean statistics grades of non–social science majors.*

■ *The average number of children in two-parent black families is lower than the average number of children in two-parent nonblack families.*

■ *There is a difference in years of education between women whose mothers have college degrees and those whose mothers do not.*

■ *Grade point averages are higher among girls who participate in organized sports than girls who do not.*

■ ■ ■ ■ **The Sampling Distribution of the Differences Between Means**

The sampling distribution allows us to compare our sample results with all possible sample outcomes and estimate the likelihood of their occurrence. Tests about differences between two sample means are based on the **sampling distribution of the difference between means.** The sampling distribution of the difference between two sample means is a theoretical probability distribution that would be obtained by calculating all the possible mean differences $(\overline{Y}_1 - \overline{Y}_2)$ by drawing all possible independent random samples of size N_1 and N_2 from two populations.

The properties of the sampling distribution of the differences between two sample means are determined by a corollary to the Central Limit Theorem. This theorem assumes that our samples are independently drawn from normal populations, but that with sufficient sample size ($N_1 > 50$, $N_2 > 50$) the sampling distribution of the difference between means will be approximately normal, even if the original populations are not normal. This sampling distribution has a mean $\mu_{Y_1} - \mu_{Y_2}$ and a standard deviation (standard error)

$$\sigma_{\overline{Y}_1 - \overline{Y}_2} = \sqrt{\frac{\sigma_{Y_1}^2}{N_1} + \frac{\sigma_{Y_2}^2}{N_2}} \tag{13.1}$$

which is based on the variances in each of the two populations ($\sigma_{Y_1}^2$ and $\sigma_{Y_2}^2$).

We can use the normal distribution as our model of the sampling distribution of the difference between two sample means. We are able to use the probabilities associated with the normal distribution and the Z statistic to test the null hypothesis that $\mu_1 = \mu_2$, as long as the populations are normally distributed or the sample sizes are sufficiently large.

> *Sampling distribution of the difference between means* A theoretical probability distribution that would be obtained by calculating all the possible mean differences ($\overline{Y}_1 - \overline{Y}_2$) that would be obtained by drawing all the possible independent random samples of size N_1 and N_2 from two populations where N_1 and N_2 are both greater than or equal to 50.

To test the null hypothesis about differences between means, the sample means need to be transformed into a Z statistic. The obtained Z statistic is calculated using the following formula:

$$Z = \frac{(\overline{Y}_1 - \overline{Y}_2) - (\mu_1 - \mu_2)}{\sigma_{\overline{Y}_1 - \overline{Y}_2}} \tag{13.2}$$

where $\sigma_{\overline{Y}_1 - \overline{Y}_2}$ is the standard error of the sampling distribution of the difference between means. Notice that the second term in the numerator contains the term $\mu_1 = \mu_2$, which is assumed to be zero under the null hypothesis. Therefore, the actual formula we use to calculate Z is

$$Z = \frac{(\overline{Y}_1 - \overline{Y}_2)}{\sigma_{\overline{Y}_1 - \overline{Y}_2}} \tag{13.3}$$

Estimating the Standard Error

The Z ratio in Formula 13.3 assumes that the population variances are known and we can calculate the standard error $\sigma_{\overline{Y}_1 - \overline{Y}_2}$ (the standard deviation of the sampling distribution) by using Formula 13.1. However, in most situations the only data we have are based on sample data, and we do not know the true value of the population variances, $\sigma_{Y_1}^2$ and $\sigma_{Y_2}^2$. Thus, we need to estimate the standard error from the sample variances, $S_{Y_1}^2$ and $S_{Y_2}^2$. The estimated standard error of the difference between means is symbolized by $S_{\overline{Y}_1 - \overline{Y}_2}$ (instead of $\sigma_{\overline{Y}_1 - \overline{Y}_2}$).

The *t* Statistic

Since we are estimating the standard error, instead of the Z statistic we use the **t statistic**. The *t* value we calculate is called the **obtained *t***. The obtained *t* represents the number of standard deviation units (or standard error units) that our mean difference $(\overline{Y}_1 - \overline{Y}_2)$ is from the hypothesized value of $\mu_1 - \mu_2$, assuming the null hypothesis is true.

The formula for computing the *t* statistic is

$$t = \frac{\overline{Y}_1 - \overline{Y}_2}{S_{\overline{Y}_1 - \overline{Y}_2}}$$

(13.4)

where $S_{\overline{Y}_1 - \overline{Y}_2}$ is the estimated standard error.

---

t *statistic (obtained)* The test statistic computed for determining the difference between two means when the population variance is unknown and the standard error must be estimated.

■ ■ ■ ■ The *t* Distribution and the Degrees of Freedom

In order to understand the *t* statistic, we should first be familiar with its distribution. The **t distribution** is actually a family of curves, each determined by its *degrees of freedom.* The concept of degrees of freedom is used in calculating several statistics, including the *t* statistic. The **degrees of freedom (df)** define the number of scores that are free to vary in calculating each statistic.

To calculate the degrees of freedom, we must know the sample size and whether there are any restrictions from calculating that statistic. The number of restrictions is then subtracted from the sample size to determine the degrees of freedom. When calculating the *t* statistic for the two-sample test, we start with N_1 and N_2 and lose 2 degrees of freedom, one for every population variance we estimate.[1] Notice that degrees of freedom will increase as

---

[1]The simplest way to think about it is in terms of subtracting one df for each prior estimate used to get the variance. In other words, to compute $S_{\overline{Y}_1}$ we first had to compute $\overline{Y}_1$. Since the sum of the deviations about the mean must equal zero, only $N - 1$ of the deviation scores are free to vary with each variance estimate.

the sample sizes increase. For a difference between means test, when population variances are assumed equal or the size of both samples is greater than 50, the df is calculated as follows:

$$df = (N_1 + N_2) - 2 \qquad\qquad (13.5)$$

When the population variances are unequal and the size of one or both samples is equal to or less than 50, we use another formula to calculate the degrees of freedom associated with the t statistic:[2]

$$df = \frac{(S_{\bar{Y}_1}^2 - S_{\bar{Y}_2}^2)^2}{(S_{\bar{Y}_1}^2)^2(N_1 - 1) + (S_{\bar{Y}_2}^2)^2(N_2 - 1)} \qquad\qquad (13.6)$$

Comparing the t and Z Statistics

Notice the similarities between the formulas for the t and Z statistics. The only apparent difference is in the denominator. The denominator of Z is $\sigma_{\bar{Y}_1 - \bar{Y}_2}$—the standard error based on the population variances, $\sigma_{\bar{Y}_1}^2$ and $\sigma_{\bar{Y}_2}^2$. For the denominator of t we replace $\sigma_{\bar{Y}_1 - \bar{Y}_2}$ with $S_{\bar{Y}_1 - \bar{Y}_2}$, the estimated standard error based on the sample variances.

However, there is another important difference between the Z and t statistics: Because it is estimated from sample data, the denominator of the t statistic is subject to sampling error. The sampling distribution of the test statistic is not normal, and the standard normal distribution cannot be used to determine probabilities associated with it.

In Figure 13.1 we present the t distribution for several df's. Like the standard normal distribution, the t distribution is bell shaped. The t statistic, similar to the Z statistic, can have positive and negative values. A positive t statistic corresponds to the right tail of the distribution; a negative value corresponds to the left tail. Notice that when the df is small, the t distribution is much flatter than the normal curve. But as the degrees of freedom increase, the shape of the t distribution gets closer to the normal distribution, until the two are almost identical when df is greater than 120.

---

t distribution A family of curves, each determined by its degrees of freedom (df). It is used when the population variance is unknown and the standard error is estimated from the sample variances.

Degrees of freedom (df) The number of scores that are free to vary in calculating a statistic.

---

[2]Degrees of freedom formula based on Dennis Hinkle, Willian Wiersma, and Stephen Jurs, *Applied Statistics for the Behavioral Sciences* (Boston: Houghton Mifflin, 1998), p. 268.

Figure 13.1 **The Normal Distribution and *t* Distributions for 1, 5, 20, and ∞ Degrees of Freedom**

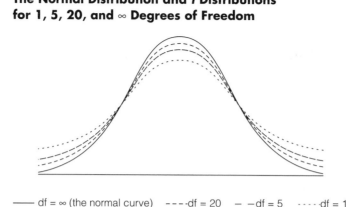

——— df = ∞ (the normal curve)   - - - -df = 20   — —df = 5   · · · ·df = 1

Calculating the Estimated Standard Error

To calculate the *t* statistic, we use one of two formulas for the calculation of the estimated standard error, depending on whether or not we can assume that the two population variances are equal. As a rule of thumb, when either sample variance is more than *twice* as large as the other, we can no longer assume that the two population variances are equal.

The Population Variances Are Assumed Equal   When we can assume that the two population variances are equal, we combine information from the two sample variances to estimate the standard deviation of the sampling distribution:

$$S_{\bar{Y}_1-\bar{Y}_2} = \sqrt{\frac{(N_1-1)S^2_{Y_1}+(N_2-1)S^2_{Y_2}}{(N_1+N_2)-2}}\sqrt{\frac{N_1+N_2}{N_1N_2}} \tag{13.7}$$

where $S_{\bar{Y}_1-\bar{Y}_2}$ is the estimated standard error of the difference between means, and $S^2_{Y_1}$ and $S^2_{Y_2}$ are the variances of the two samples. To calculate the *t* statistic, substitute this computed value of $S_{\bar{Y}_1-\bar{Y}_2}$ into the denominator of Formula 13.4.

The Population Variances Are Assumed Unequal   If the variances of the two samples $\left(S^2_{Y_1} \text{ and } S^2_{Y_2}\right)$ are very different, the formula for the estimated standard error becomes

$$S_{\bar{Y}_1-\bar{Y}_2} = \sqrt{\frac{S^2_{Y_1}}{N_1}+\frac{S^2_{Y_2}}{N_2}} \tag{13.8}$$

The computed value of $S_{\bar{Y}_1-\bar{Y}_2}$ can then be substituted into the denominator of Formula 13.4 to calculate t.

Calculating the Degrees of Freedom, Standard Error, and t Statistic

Based on our example in Table 13.1, variances for both samples are approximately equal—that is, one variance is not twice as large as the other. We thus calculate degrees of freedom (df) based on Formula 13.5. With $N_1 = 635$ and $N_2 = 811$,

$$df = (635 + 811) - 2 = 1,444$$

Next we estimate our standard error using Formula 13.7:

$$S_{\bar{Y}_1-\bar{Y}_2} = \sqrt{\frac{(635-1)3.03^2 + (811-1)2.91^2}{(635+811)-2}} \sqrt{\frac{635+811}{635(811)}} = 0.16$$

Finally, our obtained t statistic, based on Formula 13.4, is

$$t = \frac{13.59 - 13.26}{.16} = 2.06$$

■ ■ ■ ■ **Determining What Is Sufficiently Improbable: Probability Values and Alpha**

Now let's put all our information together. We're assuming that our null hypothesis ($\mu_1 = \mu_2$) is true, but want to determine whether our sample evidence casts doubt on that assumption (suggesting that actually our research hypothesis, $\mu_1 > \mu_2$, is correct). What are the chances that we would have randomly collected two samples of men and women such that the average education for men is this much higher than for women? We can determine the chances or probability based on the t statistic that we've just calculated.

Appendix C lists the corresponding values of the t distribution for various degrees of freedom and levels of significance; we have reproduced a small part of this appendix in Table 13.2. The t statistics listed in this table, also referred to as the values of t, are a function of (1) the degrees of freedom, (2) the level of significance, and (3) whether the test is one- or two-tailed. The significance levels are arrayed across the top of the table in two rows, the first for one-tailed and the second for two-tailed tests. The first column on the left side of the table shows the degrees of freedom.

The levels of significance in Table 13.2 can also be interpreted as probability or *P* **values**. A *P* value can be defined as the actual probability associated with the obtained value of t. It is a measure of how unusual or rare our obtained test statistic is compared to what is predicted in our null hypothesis. The smaller the *P* value, the more evidence we have that the null hypothesis is not true.

Table 13.2 **Values of the *t* Distribution**

	Level of Significance for One-Tailed Test					
	.10	.05	.025	.01	.005	.0005
	Level of Significance for Two-Tailed Test					
df	.20	.10	.05	.02	.01	.001
1	3.078	6.314	12.706	31.821	63.657	636.619
2	1.886	2.920	4.303	6.965	9.925	31.598
3	1.638	2.353	3.182	4.541	5.841	12.941
4	1.533	2.132	2.776	3.747	4.604	8.610
5	1.476	2.015	2.571	3.365	4.032	6.859
10	1.372	1.812	2.228	2.764	3.169	4.587
15	1.341	1.753	2.131	2.602	2.947	4.073
20	1.325	1.725	2.086	2.528	2.845	3.850
25	1.316	1.708	2.060	2.485	2.787	3.725
30	1.310	1.697	2.042	2.457	2.750	3.646
40	1.303	1.684	2.021	2.423	2.704	3.551
60	1.296	1.671	2.000	2.390	2.660	3.460
120	1.289	1.658	1.980	2.358	2.617	3.373
∞	1.282	1.645	1.960	2.326	2.576	3.291

Source: Abridged from R. A. Fisher and F. Yates, *Statistical Tables for Biological, Agricultural and Medical Research* (London: Longman, 1974), Table 111. Used by permission of Addison Wesley Longman Ltd.

Before we determine the probability of our obtained *t* statistic, let's determine whether it is consistent with our research hypothesis. Recall that we defined a right-tailed test ($\mu_1 > \mu_2$) as our research hypothesis, predicting that the difference would assessed on the right tail of the *t* distribution. The positive value of our obtained *t* statistic confirms that we will be evaluating the difference on the right tail. (If we had a negative obtained *t* statistic, it would mean the difference between the two means would have to be evaluated at the left tail of the distribution, contrary to our research hypothesis.)

To determine the probability of observing a *t* value of 2.06 with 1,444 degrees of freedom, let's refer to Table 13.2. From the first column, we can see that 1,444 degrees of freedom is not listed, so we'll have to use the last row, df = ∞, to locate our obtained *t* statistic.

Though our obtained *t* statistic of 2.06 is not listed in the last row of *t* statistics, it lies somewhere between 1.960 and 2.326. The *t* statistic of 1.960 corresponds to the .025 level of significance and 2.326 to the .01 level of significance for one-tailed tests. Restated, we can say that our obtained t statistic of 2.06 falls between 1.960 and 2.326 (1.960 < 2.06 < 2.326), or the probability of obtaining a t statistic of 2.06 is between .025 and .01 (.025 > $P(2.06)$ > .01).

Researchers usually define in advance what is a sufficiently improbable t value by specifying a cutoff point below which P must fall to reject the null hypothesis. This cutoff point, called **alpha** and denoted by the Greek letter α, is customarily set at .05, .01 or .001. Let's say that we decide to reject the null hypothesis if $P \le .05$. The value of .05 is referred to as alpha (α), and it defines for us what result is sufficiently improbably to allow us to take the risk and reject the null hypothesis. An alpha of .05 means that even if the obtained t statistic is due to sampling error, so the null hypothesis is true, we would have a 5 percent (or less) risk of rejecting it. Alpha values of .01 or .001 are more cautionary levels of risk. The difference between P and alpha is that P is the *actual probability* associated with the obtained value of t, whereas alpha is the level of probability *determined in advance* at which the null hypothesis is rejected. The null hypothesis is rejected when $P \le \alpha$.

We've already determined that our obtained t of 2.06 has a probability value between .025 and .01. The value of .025 means that fewer than 2.5 out of 100 samples drawn from this population are likely to have a mean difference of .33 (13.59 – 13.26). Another way we can say it is: There are only 2.5 chances out of 100 (or 2.5 percent) that we would have drawn a random sample with $t = 2.06$ if the mean education of men was in fact equal to mean education for women. The same can be said for the P value of .01: There is less than 1 chance out of 100 that we would have drawn a random sample with a t statistic of 2.06 if the null hypothesis were correct. Since both P values are less than the alpha level of .05, we can reject the null hypothesis of no difference.

Based on the P value(s), we can also make a statement regarding the "significance" of the difference between the two samples. If the P value is equal to or less than our alpha level, our obtained t statistic is considered *statistically significant*—that is to say, it is very unlikely to have occurred by random chance or sampling error. We can state that the difference between men and women's educational attainment is significant at the .05 level, or we can specify the level of significance by saying that the "level of significance falls between .025 and .01."

---

t *statistic*   The t statistic that refers to a particular level of significance and degrees of freedom.

P *value*   The probability associated with the obtained value of t.

***Alpha* (α)**   The level of probability at which the null hypothesis is rejected. It is customary to set alpha at the .05, .01, or .001 level.

■ ■ ■ ■ **The Five Steps in Hypothesis Testing: A Summary**

Regardless of the particular application or problem, statistical hypothesis testing can be organized into five basic steps. Let's summarize these steps:

1. Making assumptions
2. Stating the research and the null hypotheses and selecting alpha
3. Selecting the sampling distribution and specifying the test statistic
4. Computing the test statistic
5. Making a decision and interpreting the results

Making Assumptions Statistical hypothesis testing involves making several assumptions regarding the level of measurement of the variable, the method of sampling, the shape of the population distribution, and the sample size. In our example we made the following assumptions:

1. Independent random samples are used.
2. The variable *years of education* is measured on an interval-ratio level of measurement.
3. Because $N_1 > 50$ and $N_2 > 50$, the assumption of normal population is not required.
4. The population variances are assumed equal.

Stating the Research and the Null Hypotheses and Selecting Alpha The substantive hypothesis is called the *research hypothesis* and is symbolized by H_1. Research hypotheses are always expressed in terms of population parameters because we are interested in making statements about population parameters based on sample statistics. Our research hypothesis was

H_1: $\mu_1 > \mu_2$

The *null hypothesis,* symbolized as H_0, contradicts the research hypothesis in a statement of no difference between the population means. For our example, the null hypothesis was stated symbolically as

H_0: $\mu_1 = \mu_2$

We'll set alpha at .05, meaning that we will reject the null hypothesis if the probability of our obtained t is less than or equal to .05.

Selecting the Sampling Distribution and Specifying the Test Statistic The t distribution and the t statistic are used to test the significance of the difference between the two sample means.

Computing the Test Statistic We first calculate the df associated with our test:

$$df = (N_1 + N_2) - 2 = (635 + 811) - 2 = 1{,}444$$

We then estimate our standard error based on Formula 13.7:

$$S_{\bar{Y}_1 - \bar{Y}_2} = \sqrt{\frac{(635-1)3.03^2 + (811-1)2.91^2}{(635+811)-2}} \sqrt{\frac{635+811}{635(811)}} = 0.16$$

Based on Formula 13.4, our t statistic is:

$$t = \frac{\bar{Y}_1 - \bar{Y}_2}{S_{\bar{Y}_1 - \bar{Y}_2}}$$
$$= \frac{13.59 - 13.26}{.16} = 2.06$$

Making a Decision and Interpreting the Results We confirm that our obtained t is on the right tail of the distribution, consistent with our research hypothesis. Based on our obtained t statistic of 2.06, we determine that its P value is between .01 and .025, less than our .05 alpha level. We can reject the null hypothesis of no difference between men and women's mean education. We conclude that men, on average, have significantly higher years of education than women do.

■ ■ ■ ■ **Errors in Hypothesis Testing**

We should emphasize that because our conclusion is based on sample data, we will never really know if the null hypothesis is true or false. In fact, as we have seen, there is a 1–2.5 percent chance that the null hypothesis is true and that we are making an error by rejecting it.

The null hypothesis can be either true or false, and in either case it can be rejected or not rejected. If the null hypothesis is true and we reject it nonetheless, we are making an incorrect decision. This type of error is called a **Type I error**. Conversely, if the null hypothesis is false but we fail to reject it, this incorrect decision is a **Type II error**.

In Table 13.3 we show the relationship between the two types of errors and the decisions we make regarding the null hypothesis. The probability of a Type I error—rejecting a true hypothesis—is equal to the chosen alpha level. For example, when we set alpha at the .05 level, we know that if the null hypothesis is in fact true, the probability of rejecting it is .05 (or 5%).

We can control the risk of rejecting a true hypothesis by manipulating alpha. For example, by setting alpha at .01, we are reducing the risk of making a Type I error to 1 percent. Unfortunately, however, Type I errors and Type II errors are inversely related; thus, by reducing alpha and lowering the risk of making a Type I error, we are increasing the risk of making a Type II error.

Table 13.3 **Type I and Type II Errors**

	Null Hypothesis Is True	Null Hypothesis Is False
Reject null hypothesis	Type I error (α)	Correct decision
Do not reject null hypothesis	Correct decision	Type II error

As long as we base our decisions on sample statistics and not population parameters, we have to accept a degree of uncertainty as part of the process of statistical inference.

> *Type I error* The probability associated with rejecting a null hypothesis when it is true.
>
> *Type II error* The probability associated with failing to reject a null hypothesis when it is false.

> **Learning Check.** *The implications of research findings are not created equal. For example, researchers might hypothesize that eating spinach increases the strength of weight lifters. Little harm will be done if the null hypothesis that eating spinach has no effect on the strength of weight lifters is rejected in error. The researchers would most likely be willing to risk a high probability of a Type I error, and all weight lifters would eat spinach. However, when the implications of research have important consequences, the balancing act between Type I and Type II errors becomes more important. Can you think of some examples where researchers would want to minimize Type I errors? When might they want to minimize Type II errors?*

■ ■ ■ ■ Hypotheses About Differences Between Means: Illustrations

In this section we present two research examples illustrating the process of testing hypotheses about differences between means with the *t* statistic. These examples illustrate how to test for differences between sample means based on two types of population variances—when they are assumed equal and unequal.

The Population Variances Are Assumed Equal:
The Earnings of Asian American Men

Because of their socioeconomic achievement in American society, Asian Americans have been cited as a "model minority."[3] According to the 1980 census, the level of education and the average family income of Asian Americans are the highest among minority groups in the United States.[4] Among Asian Americans, the achievements of Chinese and Japanese have been particularly impressive. The 1980 census reports that the median years of schooling were 13.7 for Chinese Americans and 12.7 for Japanese Americans, as compared with 12.5 years of schooling for non-Hispanic whites. Similar trends have been observed in the relative earnings of these groups. For example, in 1990 the median income of Chinese Americans was 99 percent that of white workers; the median income of Japanese Americans was 112 percent that of white workers.[5]

The success of Chinese and Japanese Americans and their image as a "model minority" challenge the predominant view that being nonwhite is an inherent liability to achievement in American society. It seems to reinforce the notion advanced by human capital theorists that the economic success of immigrant group members is determined solely by individual human capital (credentials and skills) and not by race or national origin. Thus, according to the human capital perspective, we would expect to find earning parity between Chinese and Japanese Americans and non-Hispanic whites who have similar credentials.

Other researchers argue that the "model minority" image diverts attention from problems such as employment discrimination and economic marginality confronting Asian Americans. They suggest that the earnings parity between Asian Americans and white workers may be due to overachievement in educational attainment, longer working hours, and regional concentration of Asians in states such as California, where earnings are generally higher than in other states.[6] Thus, when factors such as education, work experience, and job training are controlled for, the earnings of Chinese and Japanese are lower than the earnings of white Americans. It is suggested that racial discrimination is the most likely explanation for these earning differentials.

To examine these competing explanations for the socioeconomic status of Chinese and Japanese Americans, Zhou and Kamo analyzed the earning patterns of Chinese and Japanese in the United States, relative to the earnings of

[3]Roger Daniels, *Asian Americans: Chinese and Japanese in the United States Since 1850* (Seattle: University of Washington Press, 1988).

[4]Zhou Min and Yoshinori Kamo, "An Analysis of Earnings Patterns for Chinese, Japanese, and Non-Hispanic White Males in the United States," *The Sociological Quarterly* 35, no. 4 (1994): 581–602.

[5]Zhou and Kamo; William P. O'Hare and Judy C. Felt, *Asian Americans: America's Fastest Growing Minority Group* (Washington, DC: Population Reference Bureau, Inc., 1991).

[6]Zhou and Kamo.

Table 13.4   **Means and Standard Deviations for Earnings of College-Educated Chinese, Japanese, and Non-Hispanic White Americans (in California)**

	Chinese	Japanese	Non-Hispanic White
Mean	$21,439	$22,907	$24,891
Standard deviation	10,289	11,120	14,225
N	471	758	2,123

Source: Adapted from M. Zhou and Y. Kamo, *The Sociological Quarterly* 35, no. 4 (November 1994), Table 2, p. 591. © 1994 by The Midwest Sociological Quarterly. Used by permission.

non-Hispanic whites. Zhou and Kamo argue that, as a group, Chinese and Japanese Americans have not achieved earnings parity with whites with identical credentials. They based their analysis on random samples from the 1980 census.[7] Their sample is limited to male workers between the ages of 25 and 64 who were in the labor force.[8]

Table 13.4 shows the mean earnings and standard deviations for U.S.-born Chinese, Japanese, and non-Hispanic whites who reside in California and who have a college degree.[9] The results in Table 13.4 show that Chinese and Japanese Americans have not achieved earnings parity with whites, despite similar credentials (college degree) and employment in the same labor market (California). The mean earnings for whites ($\overline{Y}$ = $24,891) are higher than the earnings of either Japanese ($\overline{Y}$ = $22,907) or Chinese American ($\overline{Y}$ = $21,439) workers.

Does the gap of $3,452 ($24,891 – $21,439) or of $1,984 ($24,891 – $22,907) between the earnings of Chinese and Japanese Americans, respectively, and the earnings of whites provide support for Zhou and Kamo's argument? We can use the data shown in Table 13.4 and the procedure of hypothesis testing about differences between means to answer this question. We limit our discussion to a test of the differences in mean earnings between Chinese Americans and non-Hispanic whites. We calculate the *t* statistic to test whether the observed differences in earnings between these two groups are large enough for us to conclude that the populations from which these samples are drawn are different as well.

[7]U.S. Bureau of the Census, *Public-Use Microdata Samples*, 1983.

[8]To justify the exclusion of females from the analysis, Zhou and Kamo argue that the nature of female employment differs from that of male employment and that patterns of female labor participation vary among racial and ethnic groups.

[9]Zhou and Kamo analyzed the earnings of additional subsamples of Chinese, Japanese, and non-Hispanic whites. We are focusing on U.S.-born persons who reside in California and who have a college degree.

Making Assumptions Our assumptions are as follows:

1. Independent random samples are selected.
2. Because $N_1 > 50$ and $N_2 > 50$, the assumption of normal population is not required.
3. The level of measurement of the variable *income* is interval-ratio.
4. The population variances are assumed equal.

Stating the Research and the Null Hypotheses and Selecting Alpha If Chinese Americans as a group have not achieved earnings parity with whites with identical credentials, as suggested by Zhou and Kamo, we would expect the earnings of non-Hispanic white men to be higher than the earnings of Chinese American men. The research hypothesis we will test is that the mean earnings for the population of non-Hispanic white men are greater than the mean earnings of the population of Chinese American men.

Symbolically, the hypothesis is expressed as

$$H_1: \mu_1 > \mu_2$$

with μ_1 representing the mean earnings of white men and μ_2 the mean earnings of Chinese American men. Note that because H_1 specifies that the mean earnings for white men are greater than the mean earnings for Chinese men, it is a directional hypothesis. Thus, our test is a right-tailed test.

The null hypothesis states that there are no differences between the two population means, or

$$H_0: \mu_1 = \mu_2$$

We are interested in rejecting the null hypothesis of no difference so that we have sufficient support for our research hypothesis that white men's earnings are higher than Chinese men's. We will reject the null hypothesis if the probability of t (obtained) is less than or equal to .05.

Selecting the Sampling Distribution and Specifying the Test Statistic We use the t distribution and the t statistic to test the significance of the difference between the two sample means.

Computing the Test Statistic To test the null hypothesis about the differences between the mean earnings of Chinese Americans and non-Hispanic whites, we need to translate the ratio of the observed differences to its standard error into a t statistic. The obtained t statistic is calculated following Formula 13.4:

$$t = \frac{\overline{Y}_1 - \overline{Y}_2}{S_{\overline{Y}_1 - \overline{Y}_2}}$$

where $S_{\overline{Y}_1 - \overline{Y}_2}$ is the estimated standard error of the sampling distribution. Because the population variances are assumed equal, df is $N_1 + N_2 - 2 = 2{,}123 + 471 - 2 = 2{,}592$.

Because we assume that the population variances are equal, we can combine information from the two sample variances to estimate the standard error (Formula 13.7):

$$S_{\bar{Y}_1 - \bar{Y}_2} = \sqrt{\frac{(N_1 - 1)S_{Y_1}^2 + (N_2 - 1)S_{Y_2}^2}{(N_1 + N_2) - 2}} \sqrt{\frac{N_1 + N_2}{N_1 N_2}}$$

For our example, the estimate for $S_{\bar{Y}_1 - \bar{Y}_2}$ is

$$S_{\bar{Y}_1 - \bar{Y}_2} = \sqrt{\frac{2,122(14,225)^2 + 470(10,289)^2}{(2,123 + 471) - 2}} \sqrt{\frac{2,123 + 471}{(2,123)(417)}} = 692.48$$

We substitute this value into the denominator for the t statistic (Formula 13.4):

$$t = \frac{24,891 - 21,439}{692.48} = 4.98$$

Making a Decision and Interpreting the Results Consistent with our research hypothesis, the obtained t is positive, on the right tail of the distribution. Since the obtained t of 4.98 is greater than $t = 3.291$ (df = ∞, one-tailed test; see Appendix C), we can state that its probability is less than .0005. This P value is below our .05 alpha level. The probability of obtaining this difference between non-Hispanic white and Chinese American men, if the null hypothesis were true, is extremely low. We have sufficient evidence to reject the null hypothesis and conclude that the average earnings of non-Hispanic white males are significantly higher than the average earnings of Chinese American men. The difference of $3,462 is significant at the .0005 level.

> **Learning Check.** *Would you change your decision in the previous example if we selected an alpha of .01?*

> **Learning Check.** *Using the data presented in Table 13.4, test the null hypothesis that the mean earnings of Japanese Americans are equal to the mean earnings of non-Hispanic whites. Set alpha at .05. What is your conclusion?*

The Population Variances Are Assumed Unequal: Earnings of Immigrant and Native-Born Men

Alejandro Portes and Min Zhou (1996) examined the economic benefits of self-employment among immigrant and native-born minorities.[10] According

[10]Alejandro Portes and Min Zhou, "Self-Employment and the Earnings of Immigrants," *American Sociological Review* 61 (1996): 219–230.

to Portes and Zhou, there is a lot of sociological and economic debate on whether there are any benefits from immigrant entrepreneurial activity. The authors analyzed differences in annual income among self-employed and salaried workers for samples of native-born black, Japanese, and non-Hispanic white American men, and immigrant Cuban, Chinese, and Korean men. They based their analysis on random samples from the 1980 Public Use Microdata Sample from the U.S. Census. Their samples are limited to male workers aged 25 to 64 who worked at least 160 hours in the preceding year. Table 13.5 shows annual income for 1979 for the four groups.

For illustration, we'll use data for black, non-Hispanic white, Cuban, and Korean men. The data in Table 13.5 show that among all four groups, self-employed men reported higher average earnings than salaried workers. However, notice that the variability (based on standard deviation) in earnings is larger among those who were self-employed. Korean immigrant males reported the highest average self-employment earnings, followed closely by native non-Hispanic white males. Both salaried and self-employed black males reported the lowest average earnings of the four groups.

We'll focus on the samples of Cuban men, self-employed and salaried. In comparing the variances for the two groups, notice that the variance for self-employed Cuban men is almost 3.5 times the size of the variance among salaried Cuban men ($351,450,009 \div 101,646,724 = 3.46$). Thus, we will assume unequal variances.

Making Assumptions   Our assumptions are as follows:

1. Independent random samples are used.
2. Because $N_1 > 50$ and $N_2 > 50$, the assumption of normal population is not required.
3. The level of measurement of the variable *income* is interval-ratio.
4. The population variances are assumed unequal.

Stating the Research and the Null Hypotheses and Selecting Alpha   The research hypothesis is that the mean annual income of salaried Cuban immigrant men is different from that of self-employed Cuban immigrant men. This hypothesis can be stated as a two-tailed test. A two-tailed test indicates that we will be assessing the difference between the two means at either the left (negative) or right (positive) tail of the distribution.

$H_1: \mu_1 \neq \mu_2$

The null hypothesis states that there are no differences between the two population means:

$H_0: \mu_1 = \mu_2$

We will set our alpha for this test at .01.

Table 13.5 **Annual Earnings for Black, Non-Hispanic White, Cuban, and Korean Males, 1980 PUMS Sample**

Variable	Black		Non-Hispanic White		Cuban		Korean	
	Salaried	Self-Employed	Salaried	Self-Employed	Salaried	Self-Employed	Salaried	Self-Employed
Mean Annual Earnings	$13,145	$14,808	$19,002	$22,773	$15,008	$20,990	$16,593	$23,387
S	7,995	15,211	10,848	19,436	10,082	18,747	12,346	19,803
S^2	63,920,025	231,374,521	117,679,104	377,758,096	101,646,724	351,450,009	152,423,716	392,158,809
N	9,096	420	8,214	1,334	6,877	1,338	1,984	697

Source: Adapted from A. Portes and M. Zhou, "Self-Employment and the Earnings of Immigrants," *American Sociological Review* 61 (April 1996), Table 1, p. 222. Used by permission.

Selecting the Sampling Distribution and Specifying the Test Statistic The *t* distribution and the *t* statistic are used to test the significance of the difference between the two sample means.

Computing the Test Statistic Although the population variances are unequal, since both samples are greater than 50 we use the standard equation, Formula 13.5, for calculating the degrees of freedom:

$$df = (6{,}877 + 1{,}338) - 2 = 8{,}213$$

We have assumed that the population variances are unequal, so we'll use Formula 13.8 to estimate the standard error:

$$S_{\overline{Y}_1 - \overline{Y}_2} = \sqrt{\frac{101{,}646{,}724}{6{,}877} + \frac{351{,}450{,}009}{1{,}338}} = 526.73$$

We substitute this value into the denominator for the obtained *t* statistic:

$$t = \frac{15{,}008 - 20{,}990}{526.73} = -11.36$$

Making a Decision and Interpreting the Results Notice that we have obtained a negative *t* statistic because the mean annual earnings of salaried Cuban males are lower than the earnings of self-employed Cuban males. We will assess the difference between these two means at the left tail of the distribution. However, when we determine the probability of our obtained *t*, we will ignore its negative sign, as Appendix C lists only positive values of *t*.

Based on Appendix C, given the calculated degrees of freedom (8,213) and our two-tailed research hypothesis, we can determine that the probability of obtaining a *t* statistic of 11.36 is less than .001. Our obtained *t* statistic far exceeds the highest *t* statistic value of 3.291 (df = ∞, two-tailed test). Its probability of occurrence is less than our stated alpha of .01. We have evidence to reject the null hypothesis of no difference between the two groups of Cuban immigrants. We can conclude that self-employed Cuban immigrants make, on average, higher annual earnings than salaried Cuban immigrant males. The corollary statement can also be made: Salaried Cuban immigrant males make less than those who are self-employed. The annual earnings difference of $5,982 is significant at the .001 level.

> **Learning Check.** *Calculate the* t *statistics for blacks, non-Hispanic whites, and Korean men. Is there an economic benefit in being self-employed? Based on your obtained* t *statistics and their probability levels, does one group demonstrate greater earning differences than the others?*

■ ■ ■ ■ **Testing the Significance of the Difference Between Two Sample Proportions**

In the preceding sections, we have learned how to test for the significance of the difference between two population means when the variable is measured on an interval-ratio level of measurement. Yet numerous variables in the social sciences are measured on a nominal or an ordinal level. These variables are often described in terms of proportions. For example, we might be interested in comparing the proportion of blacks and whites who support President Bill Clinton or the proportion of men and women who support federal funding for abortion. In this section we present statistical inference techniques to test for significant differences between two sample proportions.

Hypothesis testing with two sample proportions follows the same structure as the statistical tests presented earlier. The assumptions of the test are stated; the research and the null hypotheses are formulated; the sampling distribution and the test statistic are specified; and a decision is made whether or not to reject the null hypothesis following the calculation of the test statistic.

Illustration: Public Opinion About the President

To illustrate hypothesis testing with proportions, let's begin with a recent example. In the face of President Bill Clinton's sex scandal, do Americans approve of the way he is handling his job as president? Do black Americans and white Americans feel the same way about President Clinton? Ninety-two blacks (N_1) and 788 whites (N_2) responded to a *New York Times/CBS News* poll conducted in August 1998. The proportion of black respondents who reported that they "approve of the way Bill Clinton is handling his job as President" was 0.94 (p_1); the proportion of whites indicating approval was 0.60 (p_2).

Making Assumptions Our assumptions are as follows:

1. Independent random samples of $N_1 + N_2 > 100$ are used.
2. The level of measurement of the variable is nominal.

Stating the Research and the Null Hypotheses and Selecting Alpha We assume a two-tailed test, that the population proportions for blacks and whites are not equal.

$$H_1: \pi_1 \neq \pi_2$$
$$H_0: \pi_1 = \pi_2$$

In this case, we set alpha at .05.

Selecting the Sampling Distribution and Specifying the Test Statistic The population distributions of dichotomies are not normal. However, based on the

Central Limit Theorem, we know that the sampling distribution of the difference between sample proportions is normally distributed when the sample size is large (when $N_1 + N_2 > 100$), with mean $\mu_{p_1-p_2}$ and the estimated standard error $S_{p_1-p_2}$. Therefore, we can follow the same procedure applied when testing the differences between means, using the normal distribution as the sampling distribution and calculating Z as the test statistic.[11]

To calculate the test statistic Z, we follow Formula 13.3 for testing the difference between means. Substituting p_1 and p_2 for $\overline{Y}_1$ and $\overline{Y}_2$ in the numerator and $S_{p_1-p_2}$ for $\sigma_{\overline{Y}_1-\overline{Y}_2}$ in the denominator, we get

$$Z = \frac{p_1 - p_2}{S_{p_1-p_2}} \tag{13.9}$$

where p_1 and p_2 are the sample proportions for blacks and whites, and $S_{p_1-p_2}$ is the estimated standard error of the sampling distribution of the differences between sample proportions.

The estimated standard error is calculated using the following formula:

$$S_{p_1-p_2} = \sqrt{\frac{p_1(1-p_1)}{N_1} + \frac{p_2(1-p_2)}{N_2}} \tag{13.10}$$

Calculating the Test Statistic We calculate the standard error using Formula 13.10:

$$S_{p_1-p_2} = \sqrt{\frac{.94(1-.94)}{92} + \frac{.60(1-.60)}{788}} = .03$$

Substituting this value into the denominator for Z, we get

$$Z = \frac{.94 - .60}{.03} = 11.33$$

Making a Decision and Interpreting the Results Our obtained Z of 11.33 indicates that the difference between the two proportions will be evaluated at the right tail of the Z distribution. To determine the probability of observing a Z value of 11.33 if the null hypothesis is true, look up the value in Appendix B to find the area to the right of (below) the obtained Z. Recall from Chapter 10 when we calculated Z scores and their probability, that the Z values are located in column A. The P value is the probability to the right of the obtained Z, or the "area beyond Z" in column C.

Notice that our obtained Z statistic is not listed in the table. It exceeds the highest listed Z value of 4.00. We can state that the P value of 11.33 is

---

[11]The sample proportions are unbiased estimates of the corresponding population proportions. Therefore, we can use the Z statistic although our standard error is estimated from the sample proportions.

less than .0001 for a one-tailed test. However, for a two-tailed test we'll have to multiply P by 2 (.0001 × 2 = .0002). The probability of 11.33 for a two-tailed test is less than .0002, indicating an extremely rare occurrence if the null hypothesis were true. In addition, the probability of 11.33 is less than our stated alpha (.0002 < .05).

Thus, we reject the null hypothesis of no difference and conclude that there is a significant difference in the approval of President Clinton between black and white Americans. Black Americans are more likely to approve of President Clinton's job performance than white Americans.

Illustration: Gender and Abortion Attitudes

Do men and women hold different attitudes about abortion? Some people suggest that because women are more directly affected by abortion, they will be more supportive of legal abortion. However, other studies[12] show that there is no significant difference between men's and women's views on abortion. We have randomly selected a subsample from the 1996 GSS survey and compared the proportion of men and women who reported that they "approve of abortion for any reason":

Men	Women
$p_1 = 0.46$	$p_2 = 0.44$
$N_1 = 399$	$N_2 = 505$

Making Assumptions   Our assumptions are as follows:

1. Independent random samples of $N_1 + N_2 > 100$ are used.
2. The level of measurement of the variable is nominal.

Stating the Research and the Null Hypotheses and Selecting Alpha   We assume that the population proportions are not equal.

$$H_1: \pi_1 \neq \pi_2$$
$$H_0: \pi_1 = \pi_2$$

We will set alpha at .05.

Selecting the Sampling Distribution and Specifying the Test Statistic   The sampling distribution is the normal distribution, and the test statistic is Z.

Calculating the Test Statistic   We calculate the standard error using Formula 13.10:

$$S_{p_1-p_2} = \sqrt{\frac{.46(1-.46)}{399} + \frac{.44(1-.44)}{505}} = .03$$

---

[12]See, for example, Elizabeth Addel Cook, Ted G. Jelen, and Clyde Wilcox, *Between Two Absolutes: Public Opinion and the Politics of Abortion* (Boulder, CO: Westview Press, 1992).

Substituting this value into the denominator for Z, we get

$$Z = \frac{.46 - .44}{.03} = .67$$

Making a Decision and Interpreting the Results The positive Z obtained of .67 indicates that the difference between the proportions will be evaluated at the right tail of the distribution. The two-tailed probability of .67 is .2514 × 2 = .5028. This is a very large P value—much larger than our alpha of .05. We do not reject the null hypothesis of no difference. We conclude that the observed differences in the opinions of men and women regarding abortion probably *do not* reflect a difference that would have been seen had the entire population been measured.

■ ■ ■ ■ ■ **Reading the Research Literature: Reporting the Results of Statistical Hypothesis Testing**

Let's conclude with an example of how the results of statistical hypothesis testing are presented in the social science research literature. Keep in mind that the research literature does not follow the same format or the degree of detail that we've presented in this chapter. For example, most research articles do not include a formal discussion of the null hypothesis or the sampling distribution. The presentation of statistical analyses and detail will vary based on the journal's editorial policy or the standard format for the discipline.

It is not uncommon for a single research article to include the results of ten to twenty statistical tests. Results have to be presented succinctly and in summary form. An author's findings are usually presented in a summary table that may include the sample statistics (for instance, the sample means), the obtained test statistics (t or Z), the alpha level, and an indication of whether or not the results are statistically significant.[13]

Kenneth Ferraro (1996) examined the link between fear and the perceived risk of violent crime among men and women.[14] According to Ferraro, gender is the most important predictor of the fear of crime. Women are more likely to report fear of crime than men. His sample is based on the Fear of Crime in America Survey 1990, with 1,101 respondents. Given a list of ten types of crime, respondents were asked to rate their fear of being a victim on a 10-point scale; 1 indicates they were "not afraid at all" and 10 means that they were "very afraid." They were also asked to rate the chance that each of the ten crimes would happen to them in the coming year on a 10-point scale; 1 means that they consider "it's not at all likely" and 10

---

[13]Similar discussion is presented in Joseph F. Healey, *Statistics: A Tool for Social Research*, 3d ed. (Belmont, CA: Wadsworth, 1999), pp. 216–217.

[14]Kenneth F. Ferraro, "Women's Fear of Victimization: Shadow of Sexual Assault?" *Social Forces* 75, no. 2 (1996): 667–690.

Table 13.6 **Means and Standard Deviations for Victimization Fear and Perceived Risk by Sex**

	Men	Women	Mean Difference
Type of Fear			
Sexual assault	2.21+2.47	6.09+3.36	3.88**
Murder	3.48+3.05	5.30+3.67	1.82**
Robbery	3.66+2.54	5.05+3.16	1.39**
Assault	4.31+2.92	5.69+3.40	1.38**
Burglary/home	3.85+2.97	5.90+3.42	2.05**
Car theft	4.25+2.79	4.76+2.90	.51**
Burglary/away	5.18+2.82	6.18+2.98	1.00**
Cheat/con	3.40+2.56	3.89+2.90	.49**
Vandalism	4.31+2.58	4.89+2.95	.58**
Panhandler	2.36+1.86	3.36+2.57	1.00**
Fear (total)	37.02+18.41	51.22+23.25	14.20**
Type of Risk			
Sexual assault	1.38+1.03	2.98+2.33	1.60**
Murder	1.80+1.64	2.27+2.19	.47**
Robbery	2.66+1.98	3.21+2.51	.55**
Assault	2.38+1.85	2.76+2.34	.38**
Burglary/home	2.07+1.77	2.79+2.21	.72**
Car theft	3.60+2.43	3.79+2.66	.19
Burglary/away	3.55+2.29	4.20+2.57	.65**
Cheat/con	4.08+2.90	3.60+2.78	−.48**
Vandalism	3.37+2.34	3.72+2.75	.35*
Panhandler	4.54+3.48	3.83+3.31	−.71**
Risk (total)	29.34+13.84	33.26+17.98	3.92**

*$p < .05$ **$p < .01$ *Note:* Differences assessed by *t*-test of means.
Source: Adapted from K. Ferraro, "Women's Fear of Victimization: Shadow of Sexual Assault?" *Social Forces* 72, no. 2 (1996), Table 2, p. 676. Copyright © The University of North Carolina Press. Used by permission.

indicates that "it's very likely" to occur. In addition, Ferraro calculated the total scores for all fear and risk items. Ferraro's results are summarized in Table 13.6.

Let's examine this table carefully. Each row represents the mean responses and standard deviation for men and women for a particular type of fear or risk. In the last column of the table, Ferraro presents the mean difference between the scores for men and women. Obtained *t*-test statistics are not presented. Yet the footnote informs us the differences were assessed by *t*-test of means. The asterisks in the last column indicate which differences are statistically significant at the .05 (*) and .01 (**) levels. Note that with the exception of one risk comparison (car theft), all mean comparisons are significant.

Ferraro concludes:

> For each battery of victimizations, the personal offenses are listed first. Consistent with most previous research, women in the Fear of Crime in America survey are more fearful of all the ten offenses considered. The differences are greater for the personal crimes including rape, burglary, and robbery, but the difference in fear of sexual assault is particularly dramatic. Women in this sample were more afraid of rape than murder. . . . While the difference in total victimization fear is over 14, the difference in the total across types of perceived risk is less than 4. Men are more likely than women to perceive high risk of being approached by a panhandler and cheated or conned out of money. There is no significant gender difference in perceived risk of car theft.[15]

MAIN POINTS

- Statistical hypothesis testing is a decision-making process that enables us to determine whether a particular sample result falls within a range that can occur by an acceptable level of chance. The process of statistical hypothesis testing consists of five steps: making assumptions; stating the research and the null hypotheses and selecting alpha; selecting a sampling distribution and a test statistic; computing the test statistic; and making a decision and interpreting the results.

- The most common type of statistical hypothesis testing involves a comparison between two sample means. If we knew the population variances when testing for differences between sample means, we would use the Z statistic and the normal distribution. However, in practice, we are unlikely to have this information.

- When testing for differences between sample means when the population variances are unknown, we use the *t* statistic and the *t* distribution.

- Tests involving differences between proportions follow the same procedure as tests for differences between means when population variances are known. The test statistic is Z, and the sampling distribution is approximated by the normal distribution.

[15]Ibid, p. 675.

KEY TERMS

alpha (α)

degrees of freedom (df)

left-tailed test

null hypothesis (H$_0$)

one-tailed test

P *value*

research hypothesis (H$_1$)

right-tailed test

sampling distribution of the difference between means

t *distribution*

t *statistic*

t *statistic (obtained)*

two-tailed test

Type I error

Type II error

SPSS DEMONSTRATION

Producing a Test of Mean Differences

In this chapter we have discussed methods of testing differences in means or proportions between two samples (or groups). The two-sample T Test procedure can be found under the *Analyze* menu choice, then under *Compare Means*, where it is labeled *Independent-Samples T Test*.

The opening dialog box requires that you specify various test variables (the dependent variable) and one independent or grouping variable (see Figure 13.2).

Figure 13.2

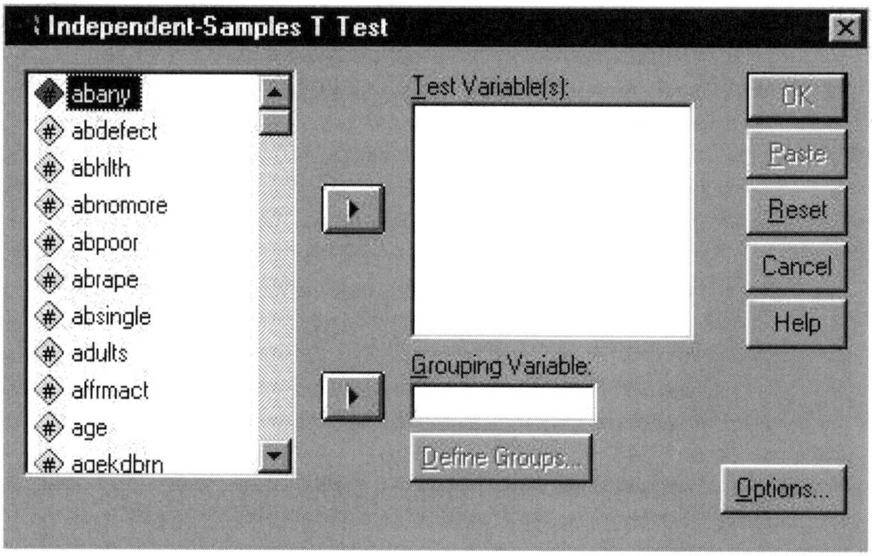

We'll test the null hypothesis that females and males work the same number of hours each week by using the variable HRS1. Place that variable in the Test Variable(s) box and SEX in the Grouping Variable box. When you do so, question marks appear next to SEX, indicating that you must supply two values to define the two groups (independent samples). Click on *Define Groups*. Then put "1" in the first box and "2" in the second box (recall that 1 = males and 2 = females), as shown in Figure 13.3. Then click on *Continue* and *OK* to run the procedure.

Figure 13.3

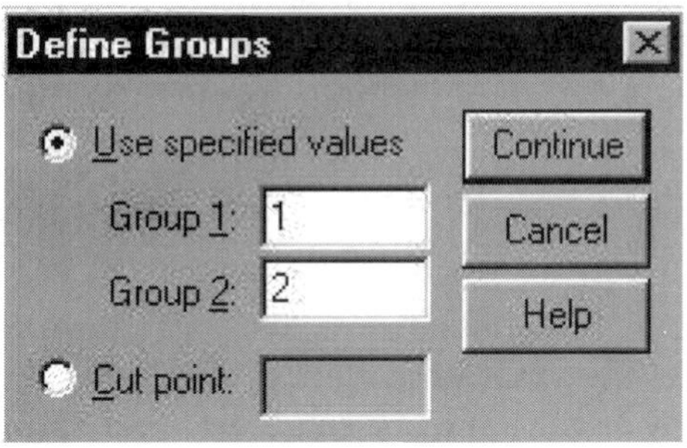

The output from T Test (Figure 13.4) is detailed and contains more information than we have reviewed in this chapter. The first part of the output displays the mean number of hours worked for males and females, the standard deviation (SD), the number of respondents in each group, and the standard error of the mean. We see that males worked about 5.95 hours more per week than females (44.79 versus 38.84 hours).

Recall from the chapter that an important decision of the *t* statistic calculation is whether the variances of the two groups are equal. If the variances are assumed equal, you can use a simple formula to calculate the number of

Figure 13.4

Group Statistics

	SEX RESPONDENTS SEX	N	Mean	Std. Deviation	Std. Error Mean
HRS1 NUMBER OF HOURS WORKED LAST WEEK	1 MALE	476	44.79	14.16	.65
	2 FEMALE	482	38.84	13.18	.60

Independent Samples Test

		Levene's Test for Equality of Variances		t-test for Equality of Means				
		F	Sig.	t	df	Sig. (2-tailed)	Mean Difference	Std. Error Difference
HRS1 NUMBER OF HOURS WORKED LAST WEEK	Equal variances assumed	6.898	.009	6.722	956	.000	5.94	.88
	Equal variances not assumed			6.719	949.279	.000	5.94	.88

degrees of freedom. However, SPSS can take account of those times when the variances are unequal and still calculate an appropriate t statistic and degrees of freedom. To do so, SPSS does a direct test of whether or not the variances for hours worked are identical for men and women (we did not discuss this test in the chapter). This is the Levene Test in the middle of the output screen. The test has a null hypothesis that the variances are equal. SPSS reports a probability of .009 for this test, so, for example, at the .01 level, we would conclude that the variances are not equal. That is, we reject the null hypothesis and conclude that our variances are unequal.

Based on our decision, we know which of the t tests to use in the bottom section of the output. Because the Levene Test suggests that the variances are *not equal*, we use the second line (this allows us to pool the variances). The actual t value is 6.719, with 949.279 degrees of freedom. The two-tailed exact significance (2-Tail Sig) is listed as .000, which means it is less than .0005. SPSS calculates the exact significance, so there is no need to look in Appendix C. This probability is very small and well below the .05 or .01 level, so we conclude that there is a mean difference in hours worked between males and females. We reject the null hypothesis.

What if we wanted to do a one-tailed test instead? SPSS does not directly list the probability for a one-tailed test, but it is easy to calculate. If we had specified a directional research hypothesis—such as that men work longer hours than women—we would simply take the probability reported by SPSS and divide it in half for a one-tailed test. Because the probability is so small in this case, our conclusion will be the same whether we do a one- or two-tailed test.

The last bit of output on each line is the 95 percent confidence interval for the mean difference in hours worked between the two groups. (We do not show this output in Figure 13.4.) You should be able to understand this based on the discussion of confidence intervals in Chapter 12. It is helpful information when testing mean differences because the actual mean difference measured (here 5.95 hours) is a sample mean, which will vary from sample to sample. The 95 percent confidence interval gives us a range over which the sample mean differences are likely to vary.

SPSS PROBLEMS [MODULE B]

1. Use the 1996 GSS file to investigate differences in educational attainment by class and race. The SPSS T Test procedure is the appropriate tool to test for any differences because SPSS doesn't directly calculate a Z statistic for the two-sample case. Race and class have more than two categories each, but they can be used in the T Test procedure by comparing two categories at a time (for example, blacks with whites, or the middle class with the lower class). Do all your tests at the .05 significance level as two-tailed tests. Write a report based on your findings. If you find any differences in educational level, do they seem to be large enough to make a critical difference in the real world?

2. Do males and females report the same average hours of television watching per day (TVHOURS)? Use the 1996 GSS to investigate the question, doing a two-tailed test at the .05 alpha level. Write a brief statement describing your findings and suggesting several causes for the result.

3. Extend your analysis in Exercise 2, this time by comparing men and women with the same educational degree. For example, is there a difference in television viewing between high school males and females? Using the variable DEGREE, use the *Select Cases* option to select a particular degree group.

4. Is there a difference in television watching between married men and married women? Between single men and single women? Select MARITAL=1 (married) as your *Select Cases* option and repeat your analysis in Exercise 2. Examine how the *t*-test statistics and significance levels have changed from the results in Exercise 2. Repeat the analysis with MARITAL=5 (never married).

CHAPTER EXERCISES

1. Consider the problem facing security personnel at a military facility in the Southwest. Their job is to detect infiltrators (spies trying to break in). The facility has an alarm system to assist the security officers. However, sometimes the alarm doesn't work properly, and sometimes the officers don't notice a real alarm. In general, the security personnel must decide between these two alternatives at any given time:

H_0: Everything is fine; no one is attempting an illegal entry.

H_1: There are problems; someone is trying to break into the facility.

Based on this information, fill in the blanks in these statements:

A "missed alarm" is a Type _____ error, and its probability of occurrence is denoted by _____.

A "false alarm" is a Type _____ error.

2. For each of the following situations determine whether a one- or a two-tailed test is appropriate. Also, state the research and the null hypotheses.
   a. A sociologist believes that the average income of elderly women is lower than the average income of elderly men.
   b. Is there a difference in the amount of study time on-campus and off-campus students devote to their schoolwork during an average week? You prepare a survey to determine the average number of study hours for each group of students.
   c. Reading scores for a group of third-graders enrolled in an accelerated reading program is predicted to be higher than the scores for nonenrolled third-graders.

    d. Stress (measured on an ordinal scale) is predicted to be lower for adults who own dogs (or other pets) than for non–pet owners.

3. For each situation in Exercise 2:
   a. Describe the Type I and Type II errors that could occur.
   b. What are the general implications of making a Type I error? of making a Type II error?
   c. When would you want to minimize Type I error? Type II error?

4. Fill in the blanks in these statements.
   a. The process of using sample statistics to make decisions about population parameters is called _____ .
   b. The _____ is a statement reflecting the substantive hypothesis. It is always expressed in terms of population parameters.
   c. Tests of hypotheses about means require that the population under consideration be normally distributed or _____ .
   d. A _____ is a type of hypothesis test in which the region of rejection is located on one side of the sampling distribution.

5. A social psychologist has been studying the stress levels of people who live in cities compared with those who live in the suburbs. She had the people in her survey record the number of times during one week that they experienced a stressful situation, such as a driver cutting them off, a rude sales clerk, and so forth. Her research hypothesis is that suburban residents will have fewer stressful experiences or interactions than city dwellers. The data show that those who live in cities had an average of 14.2 stressful experiences; suburban residents had 12.5 stressful interactions, on the average. From many past studies, she estimates that the standard deviation of the variable of interest is 2.3 interactions for both rural and city residents. She included 80 city dwellers and 80 suburbanites in the study.
   a. State the appropriate research and null hypotheses.
   b. Based on an alpha of .01, what can you conclude?

6. Do whites have jobs of higher prestige than nonwhites? Use data from the 1996 GSS to investigate this question. The variable PRESTG80 records job prestige for each respondent. The SPSS Means procedure was used to display means for job prestige, broken down into the racial classifications of white, black, and other. The standard deviation and number of respondents in each group are also listed. Assume that the standard deviations are unequal across the races.

Report

PRESTG80 RS OCCUPATIONAL PRESTIGE SCORE (1980)

RACE RACE OF RESPONDENT	Mean	N	Std. Deviation
1 WHITE	44.25	455	13.60
2 BLACK	38.08	73	12.51
3 OTHER	40.19	32	15.32
Total	43.21	560	13.72

    a. Which two-sample test is appropriate to determine whether whites have higher job prestige than the other racial groups? Why?

    b. Use this test to conduct a one-tailed test of the null hypothesis, alpha = .05, comparing whites with blacks and whites with others. What did you discover?

    c. Would your conclusions have been different if you had used a two-tailed test?

7. A social worker has been investigating the effect of residence on the lives of her clients. She randomly selected 35 people who live in public housing and 35 people who live in standard housing (a house, apartment, etc.), then studied several aspects of their lives. In particular, she was interested in whether people stay longer in a job if they are out on their own rather than living in public housing (for this study, she selected from only those people who had a job). She found that those who live in public housing have been working in their current job an average of 1.3 years with a standard deviation of 4.3 years. Those living in standard housing have been working in their current job an average of 2.4 years with a standard deviation of 3.6 years. Assume that the standard deviations are equal.

    a. State the research and the null hypotheses.

    b. Test the hypothesis at the .05 level. What conclusion can you draw?

8. You have previously examined the relationship between education and support for the busing of students to reduce segregation. Data from the 1996 GSS show that 37.5 percent (21 out of 56) of those with less than a high school education favor busing for this reason, whereas 43.2 percent (19 out of 44) of those with a graduate degree favor busing. You wonder whether there is any difference in support, in the population, between people with these two levels of education. Use a test of the difference between proportions when answering these questions. (Set alpha at .05.)

    a. What is the research hypothesis? Should you conduct a one- or a two-tailed test? Why?

    b. What do you conclude?

9. A friend suggests that people who are happier in life have more children than those who are unhappy. You think that she might be correct but would prefer to investigate the question rather than speculate. You and she collect data from a random sample of people in the town in which you live, asking about happiness (happy or not happy) and the number of children. You find that those who are happy have an average of 1.81 children, with a variance of 2.21 (33 people in this group); those who are unhappy have an average of 2.12 children, with a variance of 2.99 (31 people in this group). Assume the variances are unequal.

    a. What is the appropriate test statistic? Why?

    b. Test the null hypothesis with a one-tailed test; alpha = .01. What do you conclude about your friend's conjecture?

10. A study is done to see whether employees of a large manufacturing company support a new plan to allow flexible working hours. The company is so large that a random sample is taken to answer this question. The results of the study show that 45 percent of the employees who belong to a union at the company support the new plan (276 union employees were included in the sample). You've also been told by the researcher who did the study that 28 percent of the nonunion employees, chiefly management, support the plan (this figure also comes from a random sample).
   a. Do you have enough information to test the null hypothesis that there is no difference in support between union and nonunion employees? If not, why not?
   b. What if you learned that there are 19,546 employees at the company? Do you now have enough information? Why or why not?
   c. What if you learned that the sample size for nonunion employees is 202? Can you now do the test? If so, test the hypothesis at the .05 alpha level.

11. The gender gap—differences between men and women in their political attitudes and behavior—was a central issue in the 1996 presidential election. The gender gap is evident in the tendency of women to hold liberal views and to vote Democratic more often than men. A pre-election poll found that among 371 women, 53 percent supported Bill Clinton for president, whereas only 42 percent of 402 men did so. Do these differences reflect a real gender gap in the population of voters?
   a. What are the research and the null hypotheses to test for a gender gap? Would you conduct a one- or a two-tailed test? Test the null hypothesis at the .01 alpha level.
   b. What do you conclude?

12. Social class is an important sociological variable in many circumstances. Social class membership is related to many attitudes and behaviors, including number of children. The 1996 GSS contains the variables CHILDS and CLASS, which allow a test of this hypothesis. The Means procedure in SPSS was used to produce the following output:

Report

CHILDS NUMBER OF CHILDREN

CLASS SUBJECTIVE CLASS IDENTIFICATION	Mean	N	Std. Deviation
1 LOWER CLASS	2.11	35	1.51
2 WORKING CLASS	1.88	241	1.69
3 MIDDLE CLASS	1.83	276	1.65
4 UPPER CLASS	1.87	23	1.36
Total	1.87	575	1.65

   a. Use this information to see whether persons of self-perceived lower-class status have more children than those in the middle class. Assume equal variances. Test at the .01 alpha level.

b. Test whether the upper class has significantly fewer children than the lower class. Assume unequal variances. Test at the .05 alpha level.

c. Imagine that there were 150 respondents from the upper class in the sample rather than the 23 actually in the study. Redo your calculations in (b). Does your conclusion change? What does this tell you about the effect of sample size on statistical tests?

13. In Chapter 12 we discussed a study by Tienda and Wilson on the relationship between migration and earnings of Hispanic men. Using a sample of the 1980 census, Tienda and Wilson found that the earnings of Cuban men are higher than the earnings of either Puerto Rican or Mexican men. In Chapter 12 we calculated the 95 percent confidence interval for the mean earnings of the three groups and found that the interval estimate for Cubans was considerably higher than the interval estimates for Puerto Ricans and Mexicans. A more direct way to investigate the question of whether the earnings of Cuban men differ from the earnings of Puerto Rican and Mexican men is by testing directly for the differences between means. Use the data presented in the following table to conduct these tests.

Hispanic Group	N	$\bar{Y}$	S_Y
Mexicans	5,726	$13,342	$9,414
Puerto Ricans	5,980	$12,587	$8,647
Cubans	3,895	$16,368	$3,069

a. Which two-sample test is appropriate to determine whether Cuban men earn more than either Puerto Rican or Mexican men? Why?

b. Use this test to conduct one-tailed tests of the null hypothesis that the earnings of Cuban men are equal to those of the other two groups. Use the alpha level of .05, and interpret the results.

c. Compare your results with the results we obtained in Chapter 12 using confidence intervals. Do your results confirm the conclusions we drew earlier?

14. Do men and women have different beliefs on the ideal number of children in a family? Based on the following GSS data and obtained t statistic, what would you conclude? (Assume a two-tailed test; alpha = .01.)

	Men	Women
Mean ideal number of children	2.64	2.94
Standard deviation	1.45	1.67
N	409	543
Obtained t statistic	**−2.851**	

15. Do women consider themselves more successful in their family life than men? Using an ordinal scale measurement, the GSS1996 asked 311 men

and 419 women to rate their level of success in their family life. On a scale of 1 to 5, 1 means "not at all successful," 5 means "completely successful." Women reported an average score of 3.57 ($S = .85$) and men, 3.50 ($S = .86$). We obtained a t statistic of 1.035. Based on an alpha of .05, what can you conclude?

16. We recalculated our comparison of ideal number of children, this time only for men and women over age 35. Our results are presented below. What conclusions can you draw, based on the same alpha of .01? How would you compare these results to those in Exercise 14?

	Men	Women
Mean ideal number of children	2.65	3.04
Standard deviation	1.46	1.78
N	294	372
Obtained t statistic	−3.068	

14 The Chi-Square Test

■ ■ ■ ■ **Introduction**

Figures collected by the U.S. Department of Justice suggest that violent crime is not an equal opportunity offender. Your chances of being a victim of a violent crime are strongly influenced by your age, race, gender, and neighborhood. For example, you are far more likely to be a victim of crime if you live in a city rather than in a suburb or in the country; if you are a young black male rather than a middle-aged white male; or if you are a black woman between the ages of 16 and 24 rather than a white woman of the same age.

As we learned in the previous chapter, the fear of being a crime victim—regardless of actual victimization—is greater for women at every age and of every race than for men.[1] We now extend Ferraro's analysis with an analysis of two GSS variables: *fear of walking alone at night* and *gender*. Based on a random sample taken from the 1996 General Social Survey data set, these data confirm the observation that fear of crime differs according to gender: 55.7 percent of the women surveyed compared with only 24.5 percent of the men are afraid to walk alone in their neighborhoods at night.

The percentage differences in the perceptions of safety between males and females, shown in Table 14.1, suggest that there is a relationship between gender and fear in our sample. In inferential statistics we base our statements about the larger population on what we observe in our sample. How do we know that the gender differences in Table 14.1 reflect a real difference in the perception of safety among the larger population? How can we be sure that these differences are not just a quirk of sampling? If we took another sample, would these differences be wiped out or even reversed?

Let's assume that men and women are equally likely to be afraid to walk alone at night, that in the population from which this sample was drawn there are no real differences between them. What would be the expected percentages of men and women who would be afraid to walk alone at night?

If gender and fear were not associated, we would expect the same percentage of men and women to be fearful. Similarly, we would expect to see the same percentage of men and women who are not fearful. These percentages should be equal to the percentage of "fearful" and "not fearful" respondents in the sample as a whole. The last column of Table 14.1—the row marginals—displays these percentages: 42.3 percent of all respondents were afraid to walk alone at night, whereas 57.7 percent were not afraid. Therefore, if there were no association between gender and fear, we would expect to see 42.3 percent of the men and 42.3 percent of the women in the sample afraid to walk alone at night. Similarly, 57.7 percent of the men and 57.7 percent of the women would not be afraid to do so.

Table 14.2 shows these hypothetical expected percentages. Because the percentage distributions of the variable *fear* are identical for men and

---

[1]Kenneth F. Ferraro, "Women's Fear of Victimization: Shadow of Sexual Assault?" *Social Forces* 75, no. 2 (1996): 667–690.

Table 14.1 **Percentage of Men and Women Afraid to Walk Alone in Their Neighborhood at Night: GSS1996**

Afraid	Men	Women	Total
No	75.5%	44.3%	57.7%
	(120)	(93)	(213)
Yes	24.5%	55.7%	42.3%
	(39)	(117)	(156)
Total (N)	100.00%	100.00%	100.00%
	(159)	(210)	(369)

women, we can say that Table 14.2 demonstrates a perfect model of "no association" between the variable *fear* and the variable *gender.*

If there is an association between gender and fear, then at least some of the observed percentages in Table 14.1 should differ from the hypothetical expected percentages shown in Table 14.2. On the other hand, if gender and fear are not associated, the observed percentages should approximate the expected percentages shown in Table 14.2. In a cell-by-cell comparison of Tables 14.1 and 14.2, you can see that there is quite a disparity between the observed percentages and the hypothetical percentages. For example, in Table 14.1, 75.5 percent of the men reported that they were not afraid, whereas the corresponding cell for Table 14.2 shows that only 57.7 percent of the men report no fear. The remaining three cells reveal similar discrepancies.

Are the disparities between the observed and expected percentages large enough to convince us that there is a genuine pattern in the population? The *chi-square* statistic helps answer this question. It is obtained by

Table 14.2 **Percentage of Men and Women Afraid to Walk Alone in Their Neighborhood at Night: Hypothetical Data Showing No Association**

Afraid	Men	Women	Total
No	57.7%	57.7%	57.7%
			(213)
Yes	42.3%	42.3%	42.3%
			(156)
Total (N)	100.00%	100.00%	100.00%
	(159)	(210)	(369)

comparing the actual observed frequencies in a bivariate table with the frequencies that are generated under an assumption that the two variables in the cross-tabulation are not associated with each other. If the observed and expected values are very close, the chi-square statistic will be small. If the disparities between the observed and expected values are large, the chi-square statistic will be large. In the following sections, we will learn how to compute the chi-square statistic in order to determine whether the differences between men's and women's fear of walking alone in their neighborhood at night could have occurred simply by chance.

The Concept of Chi-Square as a Statistical Test

The **chi-square test** (pronounced kai-square and written as χ^2) is an inferential statistics technique designed to test for significant relationships between two variables organized in a bivariate table. The test has a variety of research applications and is one of the most widely used tests in the social sciences. Chi-square requires no assumptions about the shape of the population distribution from which a sample is drawn. It can be applied to nominally or ordinally measured variables.

> *Chi-square test* An inferential statistics technique designed to test for significant relationships between two variables organized in a bivariate table.

The Concept of Statistical Independence

When two variables are not associated (as in Table 14.2), one can say that they are **statistically independent**. That is, an individual's score on one variable is independent of his/her score on the second variable. We identify statistical independence in a bivariate table by comparing the distribution of the dependent variable in each category of the independent variable. When two variables are statistically independent, the percentage distributions of the dependent variable within each category of the independent variable are identical. The hypothetical data presented in Table 14.2 illustrate the notion of statistical independence. The distributions of the variable *fear* were identical within each category of the independent variable *gender*: 42.3 percent of the men and women were afraid to walk alone at night, and 57.7 percent were not afraid. Based on Table 14.2, we can say that level of fear is independent of one's gender.[2]

---

[2]Because statistical independence is a symmetrical property, the distribution of the independent variable within each category of the dependent variable will also be identical. That is, if gender and fear were statistically independent, we would also expect to see the distribution of gender identical in each category of the variable fear.

Learning Check. *The data we will use to practice calculating chi-square are taken from the General Social Survey 1996. They examine the relationship between* preferred family size *(the independent variable) and* support for abortion *(the dependent variable), as shown in the following bivariate table:*

Preferred Family Size and Support for Abortion

| | **PREFERRED FAMILY SIZE** | | |
SUPPORT	Large	Small	
Yes	25%	50%	41%
	(38)	(127)	(165)
No	75%	50%	59%
	(112)	(126)	(238)
Total	100%	100%	100%
(N)	(150)	(253)	(403)

Construct a bivariate table (in percentages) showing no association between preferred family size *and* support for abortion.

Independence (statistical) The absence of association between two cross-tabulated variables. The percentage distributions of the dependent variable within each category of the independent variable are identical.

■ ■ ■ ■ **The Structure of Hypothesis Testing with Chi-Square**

The chi-square test follows the same five basic steps as the statistical tests presented in Chapter 13: (1) the assumptions of the test are stated; (2) the research and the null hypotheses are formulated and an alpha level is selected; (3) the sampling distribution and the test statistic are specified; (4) the test statistic is calculated; and (5) a decision is made whether or not to reject the null hypothesis. Before we apply the five-step model to a specific example, let's discuss some of the elements that are specific to the chi-square test.

The Assumptions

The chi-square test requires no assumptions about the shape of the population distribution from which the sample was drawn. However, like all inferential techniques it assumes random sampling. It can be applied to variables measured on a nominal and/or an ordinal level of measurement.

Stating the Research and the Null Hypotheses

The research hypothesis (H_1) proposes that the two variables are related in the population.

H_1: The two variables are related in the population. (Gender and fear of walking alone at night are statistically dependent.)

Like all other tests of statistical significance, the chi-square is a test of the null hypothesis. The null hypothesis (H_0) states that no association exists between two cross-tabulated variables in the population, and therefore the variables are statistically independent.

H_0: There is no association between the two variables. (Gender and fear of walking alone at night are statistically independent.)

Learning Check. *Refer to the data in the previous Learning Check. Are the variables* preferred family size *and* support for abortion *statistically independent? Write out the research and the null hypotheses for your practice data.*

The Concept of Expected Frequencies

Assuming that the null hypothesis is true, we compute the cell frequencies we would expect to find if the variables are statistically independent. These frequencies are called **expected frequencies** (and are symbolized as f_e). The chi-square test is based on cell-by-cell comparisons between the expected frequencies (f_e) and the frequencies actually observed (**observed frequencies** are symbolized as f_o).

Expected frequencies (f_e) The cell frequencies that would be expected in a bivariate table if the two variables in a bivariate table were statistically independent.

Observed frequencies (f_o) The cell frequencies actually observed in a bivariate table.

Calculating the Expected Frequencies

The difference between f_o and f_e will determine the likelihood that the null hypothesis is true and that the variables are, in fact, statistically independent. When there is a large difference between f_o and f_e, it is unlikely that the two variables are independent, and we will probably reject the null hypothesis. On the other hand, if there is little difference between f_o and f_e, the variables are probably independent of each other, as stated by the null hypothesis (and therefore we will not reject the null hypothesis).

The most important element in using chi-square to test for the statistical significance of cross-tabulated data is the determination of the expected frequencies. Because chi-square is computed on actual frequencies instead of on percentages, we need to calculate the expected frequencies based on the null hypothesis.

In practice, expected frequencies are more easily computed directly from the row and column frequencies than from percentages. We can calculate the expected frequencies by following this formula:

$$f_e = \frac{(\text{column marginal})(\text{row marginal})}{N} \tag{14.1}$$

To obtain the expected frequencies for any cell in any cross-tabulation in which the two variables are assumed independent, multiply the row and column totals for that cell and divide the product by the total number of cases in the table.

Let's use this formula to recalculate the expected frequencies for our data on gender and fear as displayed in Table 14.1. Consider the men who were not afraid to walk alone at night (the upper left cell). The expected frequency for this cell is the product of the row total (213) and the column total (159) divided by all the cases in the table (369):

$$f_e = \frac{159 \times 213}{369} = 91.78$$

For men who are afraid to walk alone at night (the lower left cell) the expected frequency is

$$f_e = \frac{159 \times 156}{369} = 67.22$$

Next, let's compute the expected frequencies for women who are not afraid to walk alone at night (the upper right cell):

$$f_e = \frac{210 \times 213}{369} = 121.22$$

Finally, the expected frequency for women who are afraid to walk at night (the lower right cell) is

$$f_e = \frac{210 \times 156}{369} = 88.78$$

These expected frequencies are displayed in Table 14.3.

Note that the table of expected frequencies contains the identical row and column marginals as the original table (Table 14.1). Although the expected frequencies usually differ from the observed frequencies (depending on the degree of relationship between the variables), the row and column marginals must always be identical with the marginals observed in the original table.

Table 14.3 **Expected Frequencies of Men and Women Afraid to Walk Alone in Their Neighborhood at Night**

Afraid	Men	Women	Total
No	91.78	121.22	213
Yes	67.22	88.78	156
Total (N)	159	210	369

Learning Check. *Calculate the expected frequencies for* preferred family size *and* support for abortion *and construct a bivariate table. Are your column and row marginals the same as in the original table?*

Calculating the Obtained Chi-Square

The next step in calculating chi-square is to compare the differences between the expected and observed frequencies across all cells in the table. In Table 14.4, the expected frequencies are shown in the shaded area in each cell below the corresponding observed frequencies. Note that the difference between the observed and expected frequencies in each cell is quite large. Is it large enough to be significant? The way we decide is by calculating the **obtained chi-square** statistic:

$$\chi^2 = \sum \frac{(f_o - f_e)^2}{f_e}$$ (14.2)

where

f_o = observed frequencies
f_e = expected frequencies

According to this formula, for each cell subtract the expected frequency from the observed frequency, square the difference, and divide by the expected frequency. After performing this operation for every cell, sum the results to obtain the chi-square statistic.

Let's follow these procedures using the observed and expected frequencies from Table 14.4. Our calculations are displayed in Table 14.5. The obtained χ^2 statistic, 36.07, summarizes the differences between the observed frequencies and the frequencies we would expect to see if the null hypothesis were true and the variables—gender and fear of walking alone at night—were not associated. Next, we need to interpret our obtained chi-square statistic and decide whether it is large enough to allow us to reject the null hypothesis.

Table 14.4 **Observed and Expected Frequencies of Men and Women Afraid to Walk Alone in Their Neighborhood at Night**

	Men (f_o)	Women (f_o)	Total
Afraid	(f_e)	(f_e)	
No	120	93	213
	91.78	121.22	
Yes	39	117	156
	67.22	88.78	
Total (N)	159	210	369

Table 14.5 **Calculating Chi-Square for *Fear***

Fear	f_o	f_e	$f_o - f_e$	$(f_o - f_e)^2$	$\chi^2 = \sum \dfrac{(f_o - f_e)^2}{f_e}$
Men not afraid	120	91.78	28.22	796.37	8.68
Men afraid	39	67.22	−28.22	796.37	11.85
Women not afraid	93	121.22	−28.22	796.37	6.57
Women afraid	117	88.78	28.22	796.37	8.97

$$\chi^2 = \sum \frac{(f_o - f_e)^2}{f_e} = 36.07$$

> **Learning Check.** *Using the format of Table 14.5, construct a table to calculate chi-square for* preferred family size *and* support for abortion.

> *Chi-square (obtained)* The test statistic that summarizes the differences between the observed (f_o) and the expected (f_e) frequencies in a bivariate table.

The Sampling Distribution of Chi-Square

In Chapters 11 through 13, we've learned that test statistics such as Z and t have characteristic sampling distributions that tell us the probability of obtaining a statistic, assuming the null hypothesis is true. In the same way, the sampling distribution of chi-square tells the probability of getting values of chi-square, assuming no relationship exists in the population.

Figure 14.1   **Chi-Square Distributions for 1, 5, and 9 Degrees of Freedom**

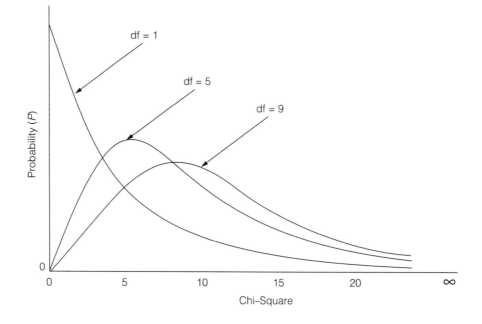

Like other sampling distributions, the chi-square sampling distributions depend on the degrees of freedom. In fact, the χ^2 sampling distribution is not one distribution, but—like the t distribution—is a family of distributions. The shape of a particular chi-square distribution depends on the number of degrees of freedom. This is illustrated in Figure 14.1, which shows chi-square distributions for 1, 5, and 9 degrees of freedom. Here are some of the main properties of the chi-square distributions that can be observed in this figure:

■ The distributions are positively skewed. The research hypothesis for the chi-square is always a one-tailed test.

■ Chi-square values are always positive. The minimum possible value is zero, with no upper limit to its maximum value. A χ^2 of zero means that the variables are completely independent; the observed frequencies in every cell are equal to the corresponding expected frequencies.

■ As the number of degrees of freedom increases, the χ^2 distribution becomes more symmetrical and, with df greater than 30, begins to resemble the normal curve.

Determining the Degrees of Freedom

In Chapter 13 we defined degrees of freedom (df) as the number of values that are free to vary. With cross-tabulation data, we find the degrees of freedom using the following formula:

$$\text{df} = (r - 1)(c - 1) \qquad\qquad (14.3)$$

where

r = the number of rows
c = the number of columns

Thus, Table 14.1 with 2 rows and 2 columns has $(2 - 1)(2 - 1)$ or 1 degree of freedom. If the table had 3 rows and 2 columns it would have $(3 - 1)(2 - 1)$ or 2 degrees of freedom.

The degrees of freedom in a bivariate table can be interpreted as the number of cells in the table for which the expected frequencies are free to vary, given that the marginal totals are already set. Based on our data in Table 14.3, suppose we first calculate the expected frequencies for women who are not afraid to walk alone at night (f_e = 121.22). Because the sum of the expected frequencies in the first row is set at 213, the expected frequency of men who are not afraid has to be 91.78 (213 − 121.22 = 91.78). Similarly, all other cells are predetermined by the marginal total and are not free to vary. Therefore, this table has only 1 degree of freedom.

> **Learning Check.** *Degrees of freedom is sometimes a difficult concept to grasp. Review this section and if you don't understand the concept of degrees of freedom, ask your instructor for further explanation. How many degrees of freedom are there in your practice example?*

Appendix D shows values of the chi-square distribution for various df's. Notice how the table is arranged with the degrees of freedom listed down the first column and the level of significance (or P values) arrayed across the top. For example, with 5 df, the probability associated with a chi-square of 15.086 is .01. An obtained chi-square of 15.086 would occur only once in 100 samples.

> **Learning Check.** *Based on Appendix D, identify the probability for each chi-square value (df in parentheses):*
>
> ■ *22.307 (15)*
> ■ *20.337 (21)*
> ■ *54.052 (26)*

Let's get back to our example and find the probability of the chi-square we obtained. We can see that 36.07 does not appear on the first row (df = 1); in fact, it exceeds the largest chi-square value of 10.827 (P = .001). We can establish that the probability of obtaining a χ^2 of 36.07 is less than .001 if the null hypothesis were true. If our alpha was preset at .05, the probability of 10.827 would be well below this. Therefore, we can reject the null hypothesis that gender and fear are not associated in the population from which our sample was drawn. Remember the larger the χ^2 statistic, the smaller the P value, providing us with more evidence to reject the null hypothesis. We can be very confident of our conclusion that there is a relationship between gender and fear in the population because the probability of this result occurring due to sampling error is less than .001, a very rare occurrence.

Review

To summarize our discussion, let's apply the five-step process of hypothesis testing.

Making Assumptions   Our assumptions are as follows:

1. A random sample of $N = 369$ was selected.
2. The level of measurement of the variable *gender* is nominal.
3. The level of measurement of the variable *fear* is nominal.

Stating the Research and the Null Hypotheses and Selecting Alpha   The research hypothesis, H_1, is that there is a relationship between gender and fear (that is, gender and fear are statistically dependent). The null hypothesis, H_0, is that there is no relationship between gender and fear in the population (that is, gender and fear are statistically independent). Alpha is set at .05.

Selecting the Sampling Distribution and Specifying the Test Statistic   Both the sampling distribution and test statistic are chi-square.

Computing the Test Statistic   We should first determine the degrees of freedom associated with our test statistic:

$$df = (r - 1)(c - 1) = (2 - 1)(2 - 1) = (1)(1) = 1$$

Next, in order to calculate chi-square, we calculate the expected frequencies under the assumption of statistical independence. To obtain the expected frequencies for each cell, we multiply its row and column marginal totals and divide the product by N. The expected frequencies are displayed in Table 14.3.

Are these expected frequencies different enough from the observed frequencies presented in Table 14.1 to justify rejection of the null hypothesis? To find out, we calculate the chi-square statistic of 36.07. The calculations are shown in Table 14.5.

Making a Decision and Interpreting the Results   To determine the probability of obtaining our chi-square of 36.07, we refer to Appendix D. With 1 degree of freedom, the probability of obtaining 36.07 is less than .001 (less than our alpha of .05). We reject the null hypothesis that there are no differences in the level of fear among men and women. Thus, we can conclude that in the populations from which our sample was drawn, fear does vary by gender.

Based on our sample data, we know that women are more likely to report being afraid to walk alone in their neighborhood at night than men are.

> **Learning Check.** *What decision can you make about the association between preferred family size and support for abortion? Should you reject the null hypothesis at the .05 alpha level?*

■ ■ ■ ■ The Limitations of the Chi-Square Test: Sample Size and Statistical Significance

Although we found the relationship between gender and fear to be statistically significant, this in itself does not give us much information about the *strength* of the relationship or its *substantive significance* in the population. Statistical significance only helps us to evaluate whether the argument (the null hypothesis) that the observed relationship occurred by chance is reasonable. It does not tell us anything about the relationship's theoretical importance or even if it is worth further investigation.

The distinction between statistical and substantive significance is an important one in applying any of the statistical tests discussed in Chapter 13. However, this distinction is of particular relevance for the chi-square test because of its sensitivity to sample size. The size of the calculated chi-square is directly proportional to the size of the sample, independent of the strength of the relationship between the variables.

For instance, suppose that we cut the observed frequencies for every cell in Table 14.1 exactly in half—which is equivalent to reducing the sample size by one-half. This change will not affect the percentage distribution of fear among men and women; therefore, the size of the percentage difference and the strength of the association between gender and fear will remain the same. However, reducing the observed frequencies by half will cut down our calculated chi-square by exactly half, from 36.07 to 18.04. (Can you verify this calculation?) Conversely, had we doubled the frequencies in each cell, the size of the calculated chi-square would have doubled, thereby making it easier to reject the null hypothesis.

This sensitivity of the chi-square test to the size of the sample means that a relatively strong association between the variables may not be significant when the sample size is small. Similarly, even when the association between variables is very weak, a large sample may result in a statistically significant relationship. However, just because the calculated chi-square is large and we are able to reject the null hypothesis by a large margin does not imply that the relationship between the variables is strong and substantively important.

Another limitation of the chi-square test is that it is sensitive to small expected frequencies in one or more of the cells in the table. Generally, when the expected frequency in one or more of the cells is below 5, the chi-square statistic may be unstable and lead to erroneous conclusions. There is no hard and fast rule regarding the size of the expected frequencies. Most researchers limit the use of chi-square to tables that either (1) have no f_e values below 5 in value *or* (2) have no more than 20 percent of the f_e values below 5 in value.

Testing the statistical significance of a bivariate relationship is only a small step, albeit an important one, in examining a relationship between two variables. A significant chi-square suggests that a relationship, weak or strong, probably exists in the population and is not due to sampling fluctuation. However, to establish the strength of the association, we need to employ measures of association such as gamma, lambda, or Pearson's r (refer to Chapter 7). Used in conjunction, statistical tests of significance and measures of association can help determine the importance of the relationship and whether it is worth additional investigation.

Learning Check.   *If the total sample size of our practice data were reduced by 90 percent (we had only one-tenth the total respondents), could we reject the null hypothesis at the .01 alpha level?*

Box 14.1   Decision: Fail to Reject the Null Hypothesis

In Chapter 13 we learned how to test for differences between proportions. Tests between proportions can always be expressed as 2 × 2 (a bivariate table with 2 rows and 2 columns) chi-square tests. Let's take another look at the difference in educational attainment among men and women, this time in a 2 × 2 table.

We've collapsed educational level into two categories: high school or less versus some college or more. Results are presented in the following table.

Percentage of Men and Women by Educational Level

Educational Level	Men	Women	Total
High school or less	66.9%	71.6%	69.6%
	(168)	(235)	(403)
Some college or more	33.1%	28.4%	30.4%
	(83)	(93)	(176)
Total	100%	100%	100%
(N)	(251)	(328)	(579)

The research hypothesis is

H_1: There is a relationship between gender and educational level (that is, gender and educational level are statistically dependent).

Our null hypothesis is

H_0: There is no relationship between gender and educational level (that is, gender and educational level are statistically independent).

We set alpha at .05. The sampling distribution is chi-square; the test statistic is chi-square. The df for the table is

$$df = (r - 1)(c - 1)$$
$$= (2 - 1)(2 - 1)$$
$$= (1)(1)$$
$$= 1$$

To compute the obtained chi-square, we first determine the expected frequencies under the assumption of statistical independence. These expected frequencies are shown in the following table.

Expected Frequencies of Men and Women and Educational Level

Educational Level	Men (f_e)	Women (f_e)	Total (f_e)
High school or less	174.7	228.3	403
Some college or more	76.3	99.7	176
Total (N)	251	328	579

Next we calculate the obtained chi-square, as shown in the following table.

Calculating Chi-Square for Educational Level

Educational Level	f_o	f_e	$f_o - f_e$	$(f_o - f_e)^2$	$\dfrac{(f_o - f_e)^2}{f_e}$
Men/High school	168	174.7	−6.70	44.89	0.26
Men/College	83	76.3	6.70	44.89	0.59
Women/High school	235	228.3	6.70	44.89	0.20
Women/College	93	99.7	−6.70	44.89	0.45

$$\chi^2 = \sum \frac{(f_o - f_e)^2}{f_e} = 1.50$$

To determine if the observed frequencies are significantly different from the expected frequencies, we determine the P value of our calculated chi-square, 1.50. With 1 degree of freedom, our chi-square falls between critical chi-squares 1.074 (P = .30) and 1.642 (P = .20). We can say that the probability of 1.50 is between .30 and .20. Both P values indicate non-rare occurrences and are greater than our alpha level. Therefore, we fail to reject our null hypothesis that there are no differences in educational level between men and women. We have inconclusive evidence that there is a relationship between gender and educational attainment.

■ ■ ■ ■ **Statistics in Practice: Social Class and Health**

In Chapter 6 we examined the relationship between health condition and social class in a random sample of GSS respondents. We repeat the analysis here for 1996 data. Individuals were asked to identify their own social class (lower, working, middle, upper) and their level of health (poor/fair, good, excellent). These data are shown in Table 14.6. This bivariate table shows a clear pattern of positive association between social class (the independent variable) and health condition (the dependent variable). For instance, whereas 49 percent of individuals of lower social class reported fair or poor health, only 9 percent of those of upper class reported the same condition. Similarly, only 22 percent of respondents of lower class reported their health as excellent, whereas 58 percent of the respondents of upper class fell into that category.

The differences in the levels of health among the four social class groups seem sizable. However, it is not clear whether these differences are due to chance or sampling fluctuations, or whether they reflect a real pattern of association in the population. To answer these questions we perform a chi-square test following the five-step model of testing hypotheses.

Making Assumptions   Our assumptions are as follows:

1. A random sample of $N = 1{,}198$ is selected.
2. The level of measurement of the variable *class* is ordinal.
3. The level of measurement of the variable *health* is ordinal.

Stating the Research and the Null Hypotheses and Selecting Alpha   Our hypotheses are

H_1: There is a relationship between social class and health condition (that is, social class and health are statistically dependent).

H_0: There is no relationship between social class and health condition in the population (that is, social class and health are statistically independent).

For this test, we'll select an alpha of .01.

Table 14.6   **Health Condition by Social Class: A Positive Relationship**

Health	Class				
	Lower	Working	Middle	Upper	Total
Excellent	17 (22%)	135 (26%)	195 (35%)	26 (58%)	373 (31%)
Good	22 (29%)	268 (51%)	267 (48%)	15 (33%)	572 (48%)
Fair/Poor	38 (49%)	119 (23%)	92 (17%)	4 (9%)	253 (21%)
Total	77 (100%)	522 (100%)	554 (100%)	45 (100%)	1,198 (100%)

Selecting the Sampling Distribution and Specifying the Test Statistic The sampling distribution is chi-square; the test statistic is chi-square.

Computing the Test Statistic The degrees of freedom for Table 14.6 is:

$$df = (r - 1)(c - 1) = (3 - 1)(4 - 1) = (2)(3) = 6$$

To calculate chi-square, first we determine the expected frequencies under the assumption of statistical independence. To obtain the expected frequency for each cell, we multiply its row and column marginal totals and divide the product by N. Following are the calculations for all cells in Table 14.6.

For lower class/poor health

$$f_e = \frac{77 \times 253}{1,198} = 16.26$$

For lower class/good health

$$f_e = \frac{77 \times 572}{1,198} = 36.76$$

For lower class/excellent health

$$f_e = \frac{77 \times 373}{1,198} = 23.97$$

For working class/poor health

$$f_e = \frac{522 \times 253}{1,198} = 110.24$$

For working class/good health

$$f_e = \frac{522 \times 572}{1,198} = 249.24$$

For working class/excellent health

$$f_e = \frac{522 \times 373}{1,198} = 162.53$$

For middle class/poor health

$$f_e = \frac{554 \times 253}{1,198} = 117.00$$

For middle class/good health

$$f_e = \frac{554 \times 572}{1,198} = 264.51$$

For middle class/excellent health

$$f_e = \frac{554 \times 373}{1,198} = 172.49$$

For upper class/poor health

$$f_e = \frac{45 \times 253}{1,198} = 9.50$$

For upper class/good health

$$f_e = \frac{45 \times 572}{1,198} = 21.49$$

and finally, for upper class/excellent health

$$f_e = \frac{45 \times 373}{1,198} = 14.01$$

We next calculate the chi-square to determine whether these expected frequencies are different enough from the observed frequencies to justify rejection of the null hypothesis. These calculations are shown in Table 14.7.

Making a Decision and Interpreting the Results To determine if the observed frequencies are significantly different from the expected frequencies, we compare our calculated chi-square with Appendix D. With 6 degrees of freedom, our chi-square of 67.50 exceeds the largest listed chi-square value of 22.457 ($P = .001$). We determine that the probability of observing our ob-

Table 14.7 **Calculating Chi-Square for Health and Social Class**

Class/Health	f_o	f_e	$f_o - f_e$	$(f_o - f_e)^2$	$\dfrac{(f_o - f_e)^2}{f_e}$
Lower/poor	38	16.26	21.74	472.63	29.07
Lower/good	22	36.76	−14.76	217.86	5.93
Lower/excellent	17	23.97	−6.97	48.58	2.03
Working/poor	119	110.24	8.76	76.74	.70
Working/good	268	249.24	18.76	351.94	1.41
Working/excellent	135	162.53	−27.53	757.90	4.66
Middle/poor	92	117.00	−25.00	625.00	5.34
Middle/good	267	264.51	2.49	6.20	.02
Middle/excellent	195	172.49	22.51	506.70	2.94
Upper/poor	4	9.50	−5.50	30.25	3.18
Upper/good	15	21.49	−6.49	42.12	1.96
Upper/excellent	26	14.01	11.99	143.76	10.26

$$\chi^2 = \sum \frac{(f_o - f_e)^2}{f_e} = 67.50$$

tained chi-square of 67.50 is less than .001, and less than our alpha of .01. We can reject the null hypothesis that there are no differences in the level of health among the different social class groups. Thus, we conclude that in the population from which our sample was drawn, health condition does vary by social class.

■ ■ ■ ■ **Reading the Research Literature:**
Sibling Cooperation and Academic Achievement

In earlier chapters we have examined a number of examples showing how the results of statistical analyses are presented in the professional literature. We have learned that most statistical applications presented in the social science literature are a good deal more complex than those described in this book. The same can be said about the application and presentation of chi-square in the research literature. Rarely do research articles go through the detailed steps of reasoning and calculation that are presented in this chapter. In most applications, the calculated chi-square is presented together with the results of a bivariate analysis. Occasionally, an appropriate measure of association summarizes the strength of the relationship between the variables.

Such an application is illustrated in Table 14.8, which is taken from an article by Carl Bankston III (1998)[3] about the relationship between academic achievement and cooperative family relations among Vietnamese American high school students. In his study, the author examined sibling cooperation among Vietnamese American families: How does help given to or received from other siblings relate to grade achievement? Bankston hypothesized that among Vietnamese American families, there is a strong emphasis on the family as a "cooperative unit." He used 1994 survey data to test whether "cooperation among siblings is significantly associated with academic success."

Data were collected from two regular and one honors public high school near Vietnamese communities in New Orleans. Surveys were distributed during English classes to a total of 402 students.

Bankston relied on two primary measures for his analysis. Grade performance was based on self-reported letter grades: A, B, or C or less. He used an ordinal measure to determine how often the respondent provided his/her siblings with assistance and how often he/she received assistance from siblings: "never," "sometimes," or "often or always."

Table 14.8 shows the results cross-tabulated for the dependent variable, average letter grade, by the two independent variables: (a) reported help *received from* siblings and (b) reported help *given to* siblings. The table displays the percentage of cases in each cell (the observed frequencies) and row and column marginals.

[3]Carl L. Bankston III, "Sibling Cooperation and Scholastic Performance Among Vietnamese American Secondary Students: An Ethnic Social Relations Theory," *Sociological Perspectives* 41, no. 1 (1998): 41(1): 167–184.

Table 14.8 Average Letter Grade Received by Reported Sibling Cooperation on Schoolwork

a. Help Received from Siblings

	Never Help	Sometimes	Often or Always	Row Total (N)
C or less	35.0	13.7	6.8	19.6 (63)
B	50.5	58.3	52.3	55.0 (177)
A	14.6	28.0	40.9	25.5 (82)
Column Total (N)	32.0 (103)	54.3 (175)	13.7 (44)	100 (322)

Note: $\chi^2 = 29.33$; $p < .001$

b. Help Given to Siblings

	Never Help	Sometimes	Often or Always	Row Total (N)
C or less	54.3	18.3	8.7	19.9 (67)
B	32.6	57.7	60.0	55.1 (185)
A	13.0	24.0	31.3	25.0 (84)
Column Total (N)	13.7 (46)	52.1 (175)	34.2 (115)	100 (336)

Note: $\chi^2 = 44.32$; $p < .001$
Source: Adapted from Carl L. Bankston III, "Sibling Cooperation and Scholastic Performance Among Vietnamese American Secondary Students: An Ethnic Social Relations Theory," *Sociological Perspectives* 41, no. 1 (1998), pp. 174, 175. Used by permission of JAI Press, Inc.

Below each table Bankston reports the obtained chi-square and the actual significance (the P value) of the obtained statistic. This information indicates a very significant relationship between sibling cooperation and grade performance. A level of significance less than .001 means that a chi-square as high as 29.33 or 44.32 would have occurred less than once in 1,000 samples if the two variables were not related. In other words, the probability of the relationship occurring due to sampling fluctuations is less than 1 out of a 1,000.

Based on these tables, Bankston presents the following analysis:

Most of these students are good students, regardless of the amount of help they report receiving from siblings. 55% of all students in the survey who had siblings or who answered the question reported grades averaging to a "B." Among those whose brothers and sisters never helped, 51% received "B" averages; among those whose brothers and sisters sometimes helped, 58% achieved "B" averages; among those whose brothers and sisters often or always helped, 52% achieved "B" averages.

Despite this clustering at the level of the "B," however, those who reported a great deal of help from siblings were much more likely than others to make "A"s and those who reported no help from siblings were much more likely to receive "C"s or less. 35% of those who reported never receiving help from siblings had grades averaging "C" or less, while only 15% of those who reported never receiving help from siblings had grades averaging to "A." By contrast, only 7% of those who reported that their siblings often or always helped them had grades averaging to a "C" or less, and 41% of them had grades averaging to "A."

Table 2 [Table 14.8b] presents a cross tabulation of averaged grade received by the amount of help reported given to siblings. Comparing Tables 1 [Table 14.8a] and 2, it appears that there is a slight tendency among respondents to overreport help given to siblings and underreport help received. Despite this apparent slight bias in reporting, the very fact that help given and help received are systematically related to academic performance supports the view that these reported data are meaningful, if not necessarily precise, measures of amounts of sibling cooperation. Those who report more sibling cooperation do better in school than those who report less.

The greater the amount of help that respondents report giving to their siblings, the better the respondents tend to do in school. Over half of those who say they never help their siblings have grades averaging to "C" or less; fewer than one out of ten who say they often or always help their siblings have grades averaging to "C" or less.[4]

Learning Check. *The value of the chi-square is 5.97 with 1 degree of freedom. Based on Appendix D, we determine that its probability is between .01 and .02. Both probabilities are less than our alpha level of .05. We reject the null hypothesis of no relationship between preferred family size and support for abortion. If we reduce our sample size by half, the obtained chi-square is 2.99. Determine the P value for 2.99. What decision can you make about the null hypothesis?*

---

[4]Ibid., p. 175.

MAIN POINTS

- The chi-square test is an inferential statistics technique designed to test for a significant relationship between variables organized in a bivariate table. The test is conducted by testing the null hypothesis that no association exists between two cross-tabulated variables in the population and, therefore, the variables are statistically independent.

- The obtained chi-square (χ^2) statistic summarizes the differences between the observed frequencies (f_o) and the expected frequencies (f_e)—the frequencies we would have expected to see if the null hypothesis were true and the variables were not associated.

- The sampling distribution of chi-square tells the probability of getting values of chi-square, assuming no relationship exists in the population. The shape of a particular chi-square sampling distribution depends on the number of degrees of freedom.

KEY TERMS

chi-square (obtained)

chi-square test

expected frequencies (f_e)

independence (statistical)

observed frequencies (f_o)

SPSS DEMONSTRATION

Producing the Chi-Square Statistic for Cross-Tabulations [Module A]

The SPSS Crosstabs procedure was previously demonstrated in Chapters 6 and 7. This procedure can also be used to calculate a chi-square value for a bivariate table. We will continue our Chapter 7 analysis by calculating a chi-square value for the table of ABNOMORE and RELIG.

Click on *Analyze, Descriptive Statistics*, and *Crosstabs*, then on the *Statistics* button. You will see the dialog box shown in Figure 14.2.

To request the chi-square statistic click on the Chi-square box in the upper left corner. Notice that the chi-square choice is not grouped with the nominal or ordinal measures of association that we discussed in Chapter 7. SPSS separates it because the chi-square test is not a measure of association, but a test of independence of the row and column variables.

Click on *Continue*. Then place ABNOMORE in the Row(s) box and RELIG in the Column(s) box. Then click on *OK* to run the procedure.

The resulting table (Figure 14.3) is the same one we saw in Chapter 7, with the addition of the chi-square statistics on the bottom of the cross-tabulation. SPSS produces quite a bit of output, perhaps more than expected. We will concentrate on the first row of information, the Pearson chi-square.

The Pearson chi-square has a value of 13.948, with 4 degrees of freedom. SPSS calculates the significance of this chi-square with 4 degrees of freedom

Figure 14.2

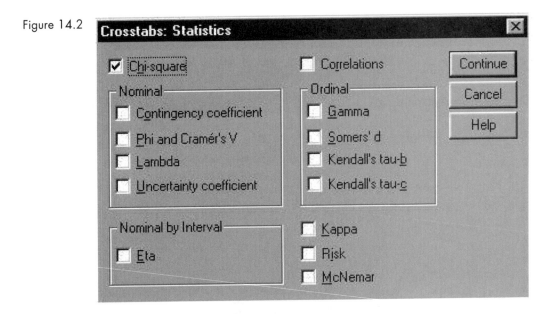

Figure 14.3

Chi-Square Tests

	Value	df	Asymp. Sig. (2-sided)
Pearson Chi-Square	13.948[a]	4	.007
Likelihood Ratio	14.076	4	.007
Linear-by-Linear Association	3.743	1	.053
N of Valid Cases	891		

a. 0 cells (.0%) have expected count less than 5. The minimum expected count is 9.59.

to be less than .007. The interpretation is that religious affiliation and support for abortion for women with children who don't want any more children are related. In particular, Jews, those who identify other religious affiliations, and those with no religious affiliation are more likely to support such abortions than are Protestants and Catholics.

The last portion of the output from SPSS allows us to check for the assumption that all expected values in each cell of the table are 5 or greater. The minimum expected value is 9.20, above the threshold of 5, so the use of chi-square is justified for this table.

As we calculated in Chapter 7, lambda, a measure of association for nominal variables, is .049 for this same table. (We do not repeat the lambda in the output shown here.) This indicates a weak relationship between religious affiliation and abortion attitude, yet the significance associated with the chi-square value for the table indicates there is little chance that the two variables are independent. These two statements are not contradictory. The magnitude of a relationship is not necessarily related to the statistical significance of that same relationship.

SPSS PROBLEMS

1. The 1996 GSS [Module A] contains a question about whether the respondent favors job preferences in the hiring of women (FEJOBAFF). It is very likely that the responses to this question vary by sex and race, but also by other variables as well.
   a. Use SPSS for Windows to investigate the relationship between SEX, RACE, and FEJOBAFF. Create Crosstab tables and ask for appropriate percentages and expected values. You should have two bivariate tables: SEX by FEJOBAFF, and RACE by FEJOBAFF. Does either table have a large number of cells with expected values less than 5? Are there any surprises in the data?
      [If a table has a problem with low expected numbers, recode the variable, collapsing categories or dropping a category with only a few respondents, and then recreate the table.]
   b. Have SPSS calculate chi-square for each table.
   c. Test the null hypothesis at the .05 significance level in each table. What do you conclude?
   d. Select another demographic variable and investigate its relationship with FEJOBAFF.
   e. Select a variable measuring an opinion or attitude and investigate its relationship with FEJOBAFF. Pick a variable you believe will be related to FEJOBAFF, and suggest the research hypothesis before creating the crosstab table.

2. Do men and women believe they have different job promotion opportunities because of their gender? In Module A, data from the GSS96 are presented for the variable SEXPROMO. Individuals were asked, "Do you think your being a man/woman makes your promotion opportunities better or worse?"
   a. Is there a relationship between SEX and SEXPROMO? Have SPSS calculate the cross-tabulation of both variables, along with chi-square. What can you conclude?
   b. Use RACE as a control variable. SPSS will calculate the chi-square for each subtable. (Remember to assess the minimum expected counts for each table to determine whether chi-square is appropriate for the table.) For all subtables, explain what relationship exists between race, sex, and the variable SEXPROMO.

3. Throughout the textbook we have been illustrating the use of SPSS with items measuring support for women's employment and women's rights. Many variables are predictors of abortion attitudes (whether they are a *cause* of the attitudes is a question that can't be answered directly by statistics). In this exercise we want you to explore the relationship of several variables and some of the abortion items. Good predictors to use are a general measure of political position (POLVIEWS), attitudes toward homosexuality (HOMOSEX), religious preference (RELIG), and religiosity (ATTEND).

    a. Create bivariate tables with some or all of these predictors and some of the GSS abortion items (Module A).

    b. Have SPSS calculate the appropriate percentages and chi-squares. You may have to recode some tables or drop some categories to complete the analysis.

    c. Summarize which variables are good predictors of which abortion attitudes. Did you find some general pattern?

    d. Add the demographic variables RACE, CLASS, SEX, or FEMINIST as control variables to a couple of these tables to see whether differences emerge between categories of respondents.

CHAPTER EXERCISES

1. In previous exercises we have examined the relationship between race and the fear of walking alone at night. In Chapter 6 we created a bivariate table with these two variables to investigate their joint relationship. Now we can extend the analysis by calculating chi-square for the same table.

    a. Use the data from Chapter 6, Exercise 1, to calculate chi-square for the bivariate table of *race* and *fear of walking alone at night*.

    b. What is the number of degrees of freedom for this table?

    c. Test the null hypothesis that race and fear of walking alone are independent (alpha = .05). What do you conclude? Is your conclusion consistent with your description of the percentage differences in Chapter 6, Exercise 1?

    d. It's always important to test the assumption that the expected value in each cell is at least 5. Does any cell fail to meet this criterion?

2. In Chapter 6, Exercise 5, we investigated the relationship between sex and the respondent's self-rated health in the GSS1996. Now we can calculate chi-square for this table to determine whether our previous advice to the neighborhood clinic was correct.

    a. Use the data from Chapter 6, Exercise 5. How many degrees of freedom does the table have?

    b. Calculate chi-square for the table. What is the expected number of females who rate their health as poor?

    c. Test the hypothesis that self-rated health and sex are independent. What is the P value of your obtained chi-square? What do you conclude (alpha = .05)? Is this consistent with your advice in Chapter 6?

3. The issue of whether and how minorities are able to advance in society is a complex matter, and people have diverse and conflicting ideas on these questions. Not surprisingly, blacks and whites often differ in their perception of how minorities can overcome the problems of bias and prejudice. The 1996 GSS contains several questions on these topics. The following bivariate table presents race and the variable WRKWAYUP, which asked whether blacks can overcome prejudice in society without favors. To make your task easier, the expected value (f_e) is also included as the second number in each cell.

RACE RACE OF RESPONDENT * WRKWAYUP BLACKS OVERCOME PREJUDICE WITHOUT FAVORS Crosstabulation

			WRKWAYUP BLACKS OVERCOME PREJUDICE WITHOUT FAVORS					
			1 AGREE STRONGLY	2 AGREE SOMEWHAT	3 NEITHER AGREE NOR DISAGREE	4 DISAGREE SOMEWHAT	5 DISAGREE STRONGLY	Total
RACE RACE OF RESPONDENT	1 WHITE	Count	136	97	38	36	14	321
		Expected Count	132.8	95.4	38.3	37.5	17.0	321.0
	2 BLACK	Count	20	15	7	8	6	56
		Expected Count	23.2	16.6	6.7	6.5	3.0	56.0
Total		Count	156	112	45	44	20	377
		Expected Count	156.0	112.0	45.0	44.0	20.0	377.0

a. What is the number of degrees of freedom for this table?
b. One cell has an expected value less than 5. Is this a serious violation of the expected value assumption?
c. Calculate chi-square for this table.
d. Test whether these two variables are independent (alpha = .01). What do you conclude?
e. To further specify the relationship, calculate an appropriate measure of association (refer to Chapter 7 if necessary).

4. In Chapter 6, Exercise 10, we studied whether there was a relationship between believing that pornography leads to a breakdown in morals and support for women working outside the home.
a. Calculate chi-square for the table in Chapter 6, Exercise 10.
b. Test the null hypothesis that the two variables are independent at the .01 alpha level. What did you find? Is this consistent with what you decided in Chapter 6?
c. Which cell has the greatest difference between the expected value (f_e) and the actual value (f_o)? What did you discover when you tried to answer this question?

5. Use the data from Chapter 7, Exercise 8, to investigate the relationship between educational attainment and attitudes toward premarital sex. In that earlier exercise, you used percentage differences and an appropriate measure of association for the table to study the relationships.

a. Calculate chi-square for the tables in Chapter 7, Exercise 8.

b. Based on an alpha of .01, test whether educational degree is independent of attitudes toward premarital sex. What do you find?

6. What is the relationship between education and attitude toward capital punishment? SPSS was used to create the following cross-tabulation output to study how educational attainment is related to support for capital punishment.

CAPPUN FAVOR OR OPPOSE DEATH PENALTY FOR MURDER * DEGREE RS HIGHEST DEGREE Crosstabulation

			DEGREE RS HIGHEST DEGREE					
			0 LT HIGH SCHOOL	1 HIGH SCHOOL	2 JUNIOR COLLEGE	3 BACHELOR	4 GRADUATE	Total
CAPPUN FAVOR OR OPPOSE DEATH PENALTY FOR MURDER	1 FAVOR	Count	144	585	67	153	69	1018
		Expected Count	140.8	564.9	66.6	160.7	85.0	1018.0
	2 OPPOSE	Count	40	153	20	57	42	312
		Expected Count	43.2	173.1	20.4	49.3	26.0	312.0
Total		Count	184	738	87	210	111	1330
		Expected Count	184.0	738.0	87.0	210.0	111.0	1330.0

Chi-Square Tests

	Value	df	Asymp. Sig. (2-sided)
Pearson Chi-Square	17.739[a]	4	.001
Likelihood Ratio	16.440	4	.002
Linear-by-Linear Association	13.616	1	.000
N of Valid Cases	1330		

a. 0 cells (.0%) have expected count less than 5. The minimum expected count is 20.41.

a. Based on the data and chi-square calculation, what do you conclude?

b. The table displays the observed and expected frequencies, in that order, in each cell. You can use the difference between these two values to get an idea of which categories contribute the most to the chi-square value of 17.74. Where are the greatest differences between the observed and expected frequencies? Which cells contribute the most to the chi-square value?

c. Use the information from (b) to describe further how education is related to support for the death penalty.

7. Refer to the table presented in Chapter 7, Exercise 3, the cross-tabulation of attitudes toward premarital sex and categories of age.

a. Complete the five-step model for these data, selecting an alpha of .01 to test whether attitudes toward premarital sex are independent of respondent's age.

b. Describe the relationship you found, using all available information.

c. Is the relationship between these two variables stronger or weaker than the relationship between educational degree and attitude toward premarital sex?

8. Gun ownership is quite prevalent in the United States, but those who own a gun are not necessarily a cross section of Americans. There are interesting differences in gun ownership between various groups. You suspect that there might be a difference by marital status and hypothesize that those individuals more likely to be living alone (those not married) are more likely to own guns (this is the alternative hypothesis H_1). You then use the 1996 GSS data to construct the following bivariate table and test your hypothesis.

OWNGUN HAVE GUN IN HOME * MARITAL MARITAL STATUS Crosstabulation

			MARITAL MARITAL STATUS					
			1 MARRIED	2 WIDOWED	3 DIVORCED	4 SEPARATED	5 NEVER MARRIED	Total
OWNGUN HAVE GUN IN HOME	1 YES	Count	99	16	12	4	15	146
		Expected Count	72.8	18.1	20.1	4.7	30.3	146.0
	2 NO	Count	86	30	39	8	62	225
		Expected Count	112.2	27.9	30.9	7.3	46.7	225.0
Total		Count	185	46	51	12	77	371
		Expected Count	185.0	46.0	51.0	12.0	77.0	371.0

a. Calculate the value of chi-square for this table. What is the number of degrees of freedom for this table?

b. Based on an alpha of .01, do you reject the null hypothesis?

c. If you rejected the null hypothesis, is your research hypothesis supported by the data in the table? What does your answer tell you about the implication of rejecting the null hypothesis in terms of support for a *specific* research hypothesis?

d. Can you suggest a substantive reason for the relationship you discovered between the variables?

9. Calculate chi-square statistics for data presented in Chapter 7, Exercises 12 and 13. For each set of data, use the five-step model (alpha = .05).

10. Are those who do not support affirmative action policies for blacks also likely to disagree with policies that support women in the workplace? We used the GSS96 data to create a cross-tabulation of AFFRMACT (favor preference in hiring blacks) and FEJOBAFF (for or against preferential hiring for women). (See table at top of page 539.)

a. Note that the table is so large that many cells have an expected value of less than 5. There are so many of these cells that this assumption of the chi-square test is most definitely violated. Because you still hope to test the relationship, group the cells in logical subsets for each variable to create a table with only three categories for each variable.

b. Check this table for the smallest expected frequency. What is its value?

c. Calculate chi-square for your new, collapsed table. How many degrees of freedom does this table have?

FEJOBAFF **for or against preferential hiring of women * AFFRMACT FAVOR PREFERENCE IN HIRING BLACKS Crosstabulation**

			AFFRMACT FAVOR PREFERENCE IN HIRING BLACKS				
			1 STRONGLY SUPPORT PREF	2 SUPPORT PREF	3 OPPOSE PREF	4 STRONGLY OPPOSE PREF	Total
FEJOBAFF for or against preferential hiring of women	1 Strongly for	Count	9	5	8	4	26
		Expected Count	2.1	1.8	7.0	15.1	26.0
	2 For	Count	3	4	6	5	18
		Expected Count	1.5	1.3	4.8	10.5	18.0
	3 Against	Count	2	2	19	24	47
		Expected Count	3.8	3.3	12.6	27.3	47.0
	4 Strongly Against	Count	0	1	13	67	81
		Expected Count	6.6	5.7	21.7	47.1	81.0
Total		Count	14	12	46	100	172
		Expected Count	14.0	12.0	46.0	100.0	172.0

d. Based on alpha = .05, test the null hypothesis that support of affirmative action for blacks and support of preferential hiring of women are independent. What do you conclude?

15   Reviewing Inferential Statistics

Box 15.6 Occupational Prestige of Male and Female Sociology Alumni: Another Example Using a *t* Test

Conclusion

SPSS DEMONSTRATION

SPSS PROBLEMS

CHAPTER EXERCISES

■ ■ ■ ■ Introduction[1]

The goal of this chapter is to provide a concise summary of the information presented in Chapters 10 through 14, to help sort out all that you've learned. Remember that it is a concise summary and it is not all-inclusive. If you are confused about any of the specific statistical techniques, please go back and review the relevant chapter(s).

■ ■ ■ ■ Normal Distributions

The normal distribution is central to the theory of inferential statistics. This theoretical distribution is bell-shaped and symmetrical, with the mean, the median, and the mode all coinciding at its peak and frequencies gradually decreasing at both ends of the curve. In a normal distribution, a constant proportion of the area under the curve lies between the mean and any given distance from the mean when measured in standard deviation units.

Although empirical distributions never perfectly match the ideal normal distribution, many are near normal. When a distribution is near normal and the mean and the standard deviation are known, the normal distribution can be used to determine the frequency of any score in the distribution regardless of the variable being analyzed. But to use the normal distribution to determine the frequency of a score, the raw score must first be converted to a standard or Z score. A Z score is used to determine how many standard deviations a raw score is above or below the mean. The formula for transforming a raw score into a Z score is

$$Z = \frac{Y - \overline{Y}}{S_Y}$$

where

Y = the raw score
$\overline{Y}$ = the mean score of the distribution
S_Y = the standard deviation of the distribution

[1]This chapter was co-authored with Pat Pawasarat.

A normal distribution expressed in Z scores is called a standard normal distribution and has a mean of 0.0 and a standard deviation of 1.0. The areas or proportions under the standard normal curve are summarized in the standard normal table in Appendix B.

The standard normal curve allows researchers to describe many characteristics of any distribution that is near normal. For example, researchers can find

- The area between the mean and a specified positive or negative Z score
- The area between any two Z scores
- The area above a positive Z score or below a negative Z score
- A raw score bounding an area above or below it
- The percentile rank of a score higher or lower than the mean
- The raw score associated with any percentile

Detailed explanations of the operations necessary to find any of these can be found in Chapter 10.

The standard normal curve can also be used to make inferences about population parameters using sample statistics. Later we will review how Z scores are used in the process of estimation and how the standard normal distribution can be used to test for differences between means or proportions (Z tests). But first let's review the aims of sampling and the importance of correctly choosing a sample, as discussed in Chapter 11.

■ ■ ■ ■ Sampling: The Case of AIDS

All research has costs to researchers in terms of both time and money, and the subjects of research may also experience costs. Often the cost to subjects is minimal; they may be asked to do no more than spend a few minutes responding to a questionnaire that does not contain sensitive issues. However, some research may have major costs to its subjects. For example, in the 1990s one of the focuses of medical research was on the control of, and a cure for, AIDS. Statistical hypothesis testing allows medical researchers to evaluate the effects of new drug treatments on the progression of AIDS by administering them to a small number of people suffering from AIDS. If a significant number of the people receiving the treatment show improvement, then the drug may be released for administration to all of the people who have AIDS. Not all of the drugs tested cause an improvement; some may have no effect and others may cause the condition to worsen. Some of the treatments may be painful. Because researchers are able to evaluate the usefulness of various treatments by testing only a small number of people, the rest of the people suffering from AIDS can be spared these costs.

Statistical hypothesis testing allows researchers to minimize all costs by making it possible to estimate characteristics of a population—population parameters—using data collected from a relatively small subset of the population, a sample. Sample selection and sampling design are an integral part of any research project, and you will learn much more about sampling when you take a methods course. However, two characteristics of samples must be stressed here.

First, the techniques of inferential statistics are designed for use only with probability samples. That is, researchers must be able to specify the likelihood that any given case in the population will be included in the sample. The most basic probability sampling design is the simple random sample; all other probability designs are variations on this design. In a simple random sample, every member of the population has an equal chance of being included in the sample. Systematic samples and stratified random samples are two variations of the simple random sample.

Second, the sample should—at least in the most important respects—be representative of the population of interest. Although a researcher can never know everything about the population he or she is studying, certain salient characteristics are either apparent or indicated by literature on the subject. Let's go back to our example of medical research on a cure for AIDS. We know that AIDS is a progressive condition, beginning when a person is diagnosed as HIV-positive and usually progresses through stages finally resulting in death. Some researchers are testing drugs that may prevent people who are diagnosed as HIV-positive from developing AIDS. When these researchers choose their samples, they should include only people who are HIV-positive, not people who have AIDS. Other researchers are testing treatments that may be effective at any stage of the disease. Their samples should include people in all stages of AIDS. AIDS knows no race, gender, or age boundaries, and all samples should reflect this. These are only a few of the obvious population characteristics researchers on AIDS must consider when selecting their samples. What you must remember is that when researchers interpret the results of statistical tests, they can only make inferences about the population their sample represents.

Every research report should contain a description of the population of interest and the sample used in the study. Carefully review the description of the sample when reading a research report. Is it a probability sample? Can the researchers use inferential statistics to test their hypotheses? Does the sample reasonably represent the population the researcher describes?

Although it may not be difficult to select, it is often difficult to implement a "perfect" simple random sample. Subjects may be unwilling or unable to participate in the study, or their circumstances may change during the study. Researchers may provide information on the limitations of the sample in their research report, as we will see in a later example.

■ ■ ■ ■ Estimation

The goal of most research is to provide information about population parameters, but researchers rarely have the means to study an entire population. Instead, data are generally collected from a sample of the population, and sample statistics are used to make estimates of population parameters. The process of estimation can be used to infer population means, variances, and proportions from related sample statistics.

When you read a research report of an estimated population parameter, it will most likely be described as a point estimate. A point estimate is a sample statistic used to estimate the exact value of a population parameter. But if we draw a number of samples from the same population, we will find that the sample statistics vary. These variations are due to sampling error. Thus, when a point estimate is taken from a single sample, we cannot determine how accurate it is.

Interval estimates provide a range of values within which the population parameter may fall. This range of values is called a confidence interval. Because the sampling distributions of means and proportions are approximately normal, the normal distribution can be used to assess the likelihood—expressed as a percentage or a probability—that a confidence interval contains the true population mean or proportion. This likelihood is called a confidence level.

Confidence intervals may be constructed for any level, but the 90, 95, and 99 percent levels are the most typical. The normal distribution tells us that

■ 90 percent of all sample means or proportions will fall between ±1.65 standard errors

■ 95 percent of all sample means or proportions will fall between ±1.96 standard errors

■ 99 percent of all sample means or proportions will fall between ±2.58 standard errors

The formula for constructing confidence intervals for means is

$$CI = \overline{Y} \pm Z(\sigma_{\overline{Y}})$$

where

$\overline{Y}$ = the sample mean
Z = the Z score corresponding to the confidence level
$\sigma_{\overline{Y}}$ = the standard error of the sampling distribution of the mean

If we know the population standard deviation, the standard error can be calculated using the formula

$$\sigma_{\overline{Y}} = \frac{\sigma_Y}{\sqrt{N}}$$

where

$\sigma_{\overline{Y}}$ = the standard error of the sampling distribution of the mean
σ_Y = the standard deviation of the population
N = the sample size

But since we rarely know the population standard deviation, we can estimate the standard error using the formula

$$S_{\overline{Y}} = \frac{S_Y}{\sqrt{N}}$$

where

$S_{\overline{Y}}$ = the estimated standard error of the sampling distribution of the mean
S_Y = the standard deviation of the sample
N = the sample size

When the standard error is estimated, the formula for confidence intervals for the mean is

$$CI = \overline{Y} \pm Z(S_{\overline{Y}})$$

The formula for confidence intervals for proportions is similar to that for means:

$$CI = p \pm Z(S_p)$$

where

p = the sample proportion
Z = the Z score corresponding to the confidence level
S_p = the estimated standard error of proportions

The estimated standard error of proportions is calculated using the formula

$$S_p = \sqrt{\frac{p(1-p)}{N}}$$

where

p = the sample proportion
N = the sample size

Interval estimation consists of the following four steps, which are the same for confidence intervals for the mean and for proportions.

1. Find the standard error.
2. Decide on the level of confidence and find the corresponding Z value.
3. Calculate the confidence interval.
4. Interpret the results.

When interpreting the results, we restate the level of confidence and the range of the confidence interval. If confidence intervals are constructed for two or more groups, they can be compared to show similarities or differences between the groups. If there is overlap in two confidence intervals, the groups are probably similar. If there is no overlap, the groups are probably different.

Remember, there is always some risk of error when using confidence intervals. At the 90 percent, 95 percent, and 99 percent confidence levels the respective risks are 10 percent, 5 percent, and 1 percent. Risk can be reduced by increasing the level of confidence. However, when the level of confidence is increased, the width of the confidence interval is also increased, and the estimate becomes less precise. The precision of an interval estimate can be increased by increasing the sample size, which results in a smaller standard error, but when $N \geq 400$ the increase in precision is small relative to increases in sample size.

■ ■ ■ ■ Statistics in Practice: The War on Drugs

If you pick up a newspaper, watch television, or listen to the radio, you will probably see the results of some kind of poll. Thousands of polls are taken in the United States every year, and the range of topics is almost unlimited. You might see that 75 percent of dentists recommend brand X or that 60 percent of all teenagers have tried drugs. Some polls may seem frivolous, whereas others may have important implications for public policy, but all of these polls use estimation.

The Gallup organization conducts some of the most reliable and widely respected polls regarding issues of public concern in the United States. In September 1995 a Gallup survey was taken to determine public attitudes toward combating the use of illegal drugs in the United States and public opinions about major influences on the drug attitudes of children and teenagers.[2]

The Gallup organization reported that 57 percent of Americans consider drug abuse to be an extremely serious problem. When asked to name the single most cost-efficient and effective strategy for halting the drug problem, 40 percent of Americans favor education; 32 percent think efforts to reduce the flow of illegal drugs into the country would be most effective; 23 percent favor convicting and punishing drug offenders; and 4 percent believe drug treatment is the single best strategy. The same poll found that 71 percent of Americans favor increased drug testing in the workplace, and 54 percent support mandatory drug testing in high schools. All of these percentages are point estimates.

Table 15.1 shows the percentage of Americans who think that peers, parents, professional athletes, organized religion, school programs, and television and radio messages have a major influence on the drug attitudes of

[2]*Gallup Poll Monthly*, December 1995, pp. 16–19.

Table 15.1 **Drug Attitudes of the Young: Major Influences**

	Peers	Parents	Pro Athletes	Organized Religion	School Programs	TV & Radio Messages	N
National	74%	58%	51%	31%	30%	26%	1,020
Sex							
Male	71	59	47	30	30	25	511
Female	76	57	55	32	30	27	509
Age							
18–29 years	72	55	54	26	23	26	172
30–49 years	79	62	48	30	32	24	492
50–64 years	74	57	54	39	31	27	187
65 & older	60	52	42	34	31	29	160
Region							
East	78	57	53	24	27	24	226
Midwest	73	56	46	28	31	26	215
South	73	61	56	42	33	31	363
West	72	57	48	27	29	21	216
Community							
Urban	70	57	53	32	32	27	420
Suburban	77	60	50	29	29	24	393
Rural	72	57	51	34	28	28	199
Race							
White	74	58	51	30	29	22	868
Nonwhite	73	56	54	42	37	47	143
Education							
College postgraduate	90	58	44	24	17	12	155
Bachelor's degree	79	58	44	29	25	21	151
Some college	76	60	53	30	32	26	308
High school or less	66	56	54	35	33	31	400
Income							
$75,000 & over	85	60	50	28	30	15	140
$50,000–74,999	81	61	52	26	27	14	323
$30,000–49,999	74	61	47	29	29	23	251
$20,000–29,999	75	59	56	34	30	34	158
Under $20,000	66	52	51	37	33	36	233
Family drug problem							
Yes	78	55	55	28	29	23	191
No	73	59	50	32	30	27	826

Source: Adapted from *The Gallup Poll Monthly,* December 1995, pp. 16–19. Used by permission.

children and teenagers. The table shows percentages for the total national sample and by subgroup for selected demographic characteristics. Notice that for most of the categories of influence, the percentages are similar across the subgroups, and the subgroup percentages are similar to the national percentage for the category. One exception is the Peers category. The Gallup Poll reports that 74 percent of Americans believe that peers are a major influence on the drug attitudes of young people (the highest percentage for any of the categories).

Many of the subgroups show percentages closely aligned with the national percentage. However, look at the subgroups under Education. The percentages for respondents with bachelor's degrees (79%) and some college (76%) are similar to each other and to the national percentage. The percentages for college postgraduates (90%) and high school or less (66%) differ more widely. The comparison of the point estimates leads us to conclude that education has an effect on opinions about peer influence on drug attitudes. However, remember that point estimates taken from single samples are subject to sampling error, so we cannot tell how accurate they are. Different samples taken from the populations of college postgraduates and people with a high school education or less might have resulted in point estimates closer to the national estimate, and then we might have reached a different conclusion.

A comparison of confidence intervals can make our conclusions more convincing because we can state the probability that the interval contains the true population proportion. We can use the sample sizes provided in Table 15.1 to calculate interval estimates. In Box 15.1 we followed the process of interval estimation to compare the national percentage of Americans who think peers are a major influence on drug attitudes with the percentages for college postgraduates and those who have a high school education or less.

Learning Check. *Use Table 15.1 to calculate 99 percent confidence intervals for opinions about the influence of television and radio messages on drug attitudes of the young for the national sample and by race (three intervals). Compare the intervals. What is your conclusion?*

The primary purpose of estimation is to find a population parameter, using data taken from a random sample of the population. Confidence intervals allow researchers to evaluate the accuracy of their estimates of population parameters. Point and interval estimates can be used to compare populations, but neither allows researchers to evaluate conclusions based on those comparisons.

The process of statistical hypothesis testing allows researchers to use sample statistics to make decisions about population parameters. Statistical hypothesis testing can be used to test for differences between a single sample and a population or between two samples. In the following sections, we will review the process of statistical hypothesis testing, using t tests, Z tests, and chi-square in two-sample situations.

Box 15.1 Interval Estimation for Peers as a Major Influence on the Drug Attitudes of the Young

To calculate the confidence intervals for peer influence we must know the point estimates and the sample sizes for all Americans, college postgraduates, and Americans with a high school education or less. These figures are shown in the following table.

Group	Point Estimate	Sample Size (N)
National	74%	1,020
College postgraduates	90%	155
High school or less	66%	400

We follow the process of estimation to calculate confidence intervals for all three groups.

1. **Find the standard error.** For all groups we use the formula for finding the standard error of proportions:

$$S_p = \sqrt{\frac{p(1-p)}{N}}$$

2. **Decide on the level of confidence and find the corresponding Z value.** We choose the 95 percent confidence level, which is associated with $Z = 1.96$.

3. **Calculate the confidence interval.** We use the formula for confidence intervals for proportions:

$$CI = p \pm Z(S_p)$$

4. **Interpret the results.** Summaries of the calculations for standard errors and confidence intervals and interpretations follow.

National	College Postgraduates	High School or Less
$S_p = \sqrt{\dfrac{(.74)(.26)}{1,020}}$	$S_p = \sqrt{\dfrac{(.90)(.10)}{155}}$	$S_p = \sqrt{\dfrac{(.66)(.34)}{400}}$
$= .014$	$= .024$	$= .024$
$CI = .74 \pm 1.96(.014)$	$CI = .90 \pm 1.96(.024)$	$CI = .66 \pm 1.96(.024)$
$= .74 \pm .027$	$= .90 \pm .047$	$= .66 \pm .047$
$= .713$ to $.767$	$= .853$ to $.947$	$= .613$ to $.707$

We can be 95 percent confident that the interval .713 to .767 includes the true population proportion.

We can be 95 percent confident that the interval .853 to .947 includes the true population proportion.

We can be 95 percent confident that the interval .613 to .707 includes the true population proportion.

We can use the confidence intervals to compare the proportions for the three groups. None of the intervals overlap, which suggests that there are differences between the groups. The proportion of college postgraduates who think peer pressure is a major influence on the drug attitudes of young people is probably higher than the national proportion, and the proportion of the population with a high school education or less who think this is probably lower than the national proportion. It appears that education has an effect on opinions about this issue.

■ ■ ■ ■ **The Process of Statistical Hypothesis Testing**

In Chapter 13 we learned that the process of statistical hypothesis testing consists of the following five steps:

1. Making assumptions
2. Stating the research and the null hypotheses and selecting alpha
3. Selecting a sampling distribution and a test statistic
4. Computing the test statistic
5. Making a decision and interpreting the results

Examine quantitative research reports, and you will find that all responsible researchers follow these five basic steps, although they may state them less explicitly. When asked to critically review a research report, your criticism should be based on whether the researchers have correctly followed the process of statistical hypothesis testing and if they have used the proper procedures at each step of the process. Others will use the same criteria to evaluate research reports you have written.

In this section we follow the five steps of the process of statistical hypothesis testing to review Chapter 13. We provide a detailed guide for choosing the appropriate sampling distribution, test statistic, and formulas for the test statistics. In the following sections we will present research examples to show how the process is used in practice.

Step 1: Making Assumptions

Statistical hypothesis testing involves making several assumptions that must be met for the results of the test to be valid. These assumptions include the level of measurement of the variable, the method of sampling, the shape of the population distribution, and the sample size. The specific assumptions may vary, depending on the test or the conditions of testing. However, all statistical tests assume random sampling, and two-sample tests require independent random sampling. Tests of hypotheses about means also assume interval-ratio level of measurement and require that the population under consideration is normally distributed or that the sample size is larger than 50.

Step 2: Stating the Research and the Null Hypotheses and Selecting Alpha

Recall that in Chapter 1 we learned that hypotheses are tentative answers to research questions, which can be derived from theory, observations, or intuition. As tentative answers to research questions, hypotheses are generally stated in sentence form. To verify a hypothesis using statistical hypothesis testing, it must be stated in a testable form called a research hypothesis.

We use the symbol H_1 to denote the research hypothesis. Hypotheses are always stated in terms of population parameters. The null hypothesis (H_0) is a contradiction of the research hypothesis and is usually a statement of no difference between the population parameters. It is the null hypothesis that researchers test. If it can be shown that the null hypothesis is false, researchers can claim support for their research hypothesis.

Published research reports rarely make a formal statement of the research and the null hypotheses. Researchers generally present their hypotheses in sentence form. In order to evaluate a research report, you must construct the research and the null hypotheses to determine whether the researchers actually tested the hypotheses they stated. Box 15.2 shows possible hypotheses for comparing the sample means and for testing a relationship in a bivariate table.

Statistical hypothesis testing always involves some risk of error because sample data are used to estimate or infer population parameters. Two types of error are possible—Type I and Type II. A Type I error occurs when a true null hypothesis is rejected; alpha (α) is the probability of making a Type I error. In social science research alpha is typically set either at the .05, .01, or .001 level. At the .05 level, researchers risk a 5 percent chance of making a Type I error. The risk of making a Type I error can be decreased by choosing a smaller alpha level—.01 or .001. However, as the risk of a Type I error decreases, the risk of a Type II error increases. A Type II error occurs when the researcher fails to reject a false null hypothesis.

How does a researcher choose the appropriate alpha level? By weighing the consequences of making a Type I or a Type II error. Let's look again at research on AIDS. Suppose researchers are testing a new drug that may halt the progression of AIDS. The null hypothesis is that the drug has no effect on the progression of AIDS. Now suppose that preliminary research has shown this drug has serious negative side effects. The researchers would want to minimize the risk of making a Type I error (rejecting a true null hypothesis) so people would not experience the negative side effects unnecessarily if the drug does not affect the progression of AIDS. An alpha level of .001 or smaller would be appropriate.

Alternatively, if preliminary research has shown the drug has no serious negative side effects, the researchers would want to minimize the risk of a Type II error (failing to reject a false null hypothesis). If the null hypothesis is false and the drug might actually help people with AIDS, researchers would want to increase the chance of rejecting the null hypothesis. In this case, the appropriate alpha level would be .05.

Box 15.2 Possible Hypotheses for Comparing Two Samples

When data are measured at the interval-ratio level, the research hypothesis can be stated as a difference between the means of the two samples in one of the following three forms:

1. $H_1: \mu_1 > \mu_2$
2. $H_1: \mu_1 < \mu_2$
3. $H_1: \mu_1 \neq \mu_2$

Hypotheses 1 and 2 are directional hypotheses. A directional hypothesis is used when the researcher has information that leads him or her to believe that the mean for one group is either larger (right-tailed test) or smaller (left-tailed test) than the mean for the second group. Hypothesis 3 is a nondirectional hypothesis, which is used when the researcher is unsure of the direction and can only assume that the means are different.

The null hypothesis always states that there is no difference between means:

$H_0: \mu_1 > \mu_2$

The form of the research and the null hypotheses for nominal or ordinal data is determined by the statistics used to describe the data. When the variables are described in terms of proportions, such as the proportions of elderly men and women who live alone, the research hypothesis can be stated as one of the following:

1. $\pi_1 > \pi_2$
2. $\pi_1 < \pi_2$
3. $\pi_1 \neq \pi_2$

The null hypothesis will always be

$H_0: \pi_1 = \pi_2$

When a cross-tabulation has been used to descriptively analyze nominal or ordinal data, the research and the null hypotheses are stated in terms of the relationship between the two variables.

H_1: The two variables are related in the population (statistically dependent).

H_0: There is no relationship between the two variables in the population (statistically independent).

Do not confuse alpha and P. Alpha is the level of probability—determined in advance by the investigator—at which the null hypothesis is rejected; P is the actual calculated probability associated with the obtained value of the test statistic. The null hypothesis is rejected when $P \leq$ alpha.

Step 3: Selecting a Sampling Distribution and a Test Statistic

The selection of a sampling distribution and a test statistic, like the selection of the form of the hypotheses, is based on a set of defining criteria. Whether you are choosing a sampling distribution to test your data or evaluating the use of a test statistic in a written research report, make sure that all of the criteria are met. Box 15.3 provides the criteria for the statistical tests for two-sample situations (Chapter 13) and for cross-tabulation (Chapter 14).

Box 15.3 Criteria for Statistical Tests When Comparing Two Samples

When the data are measured at the interval-ratio level, sample means can be compared using the t distribution and t test.

Criteria for using the t distribution and a t test with interval-ratio level data

■ Population variances unknown

■ Independent random samples

■ Population distribution assumed normal unless $N_1 > 50$ and $N_2 > 50$

When the data are measured at the nominal or ordinal level, either the normal distribution or the chi-square distribution can be used to compare proportions for two samples.

Criteria for using the normal distribution and a Z test with proportions (nominal or ordinal data)

■ Population variances unknown but assumed equal

■ Independent random samples

■ Combined sample size greater than 100 $(N_1 + N_2 > 100)$

For this test, the population variances are always assumed equal because they are a function of the population proportion (π), and the null hypothesis is $\pi_1 = \pi_2$.

Criteria for using the chi-square distribution and a χ^2 test with nominal or ordinal data

■ Independent random samples

■ Any size sample

■ Cross-tabulated data

■ No cells with expected frequencies less than 5, or not more than 20 percent of the cells with expected frequencies less than 5

The chi-square test can be used with any size sample, but it is sensitive to sample size. Increasing the sample size results in increased values of χ^2. This property can leave interpretations of the findings open to question when the sample size is very large. Thus, it is preferable to use the normal distribution if the criteria for a Z test can be met.

Step 4: Computing the Test Statistic

Most researchers use computer software packages to calculate statistics for their data. Consequently, when you evaluate a research report there is very little reason to question the accuracy of the calculations. You may use your computer to calculate statistics when writing a research report, but there may be times when you need to do manual calculations (such as during this course). The formulas you need to calculate t, Z, and χ^2 statistics are shown in Box 15.4.

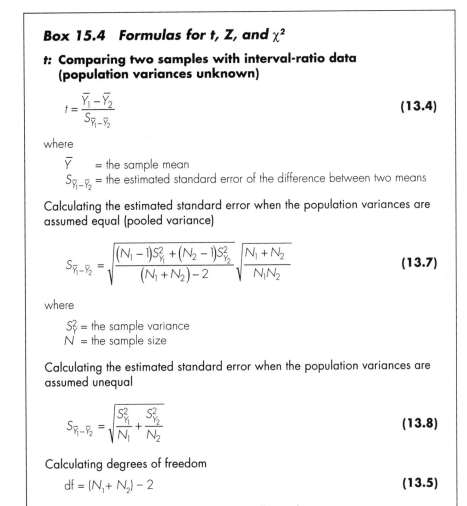

Box 15.4 Formulas for t, Z, and χ²

t: Comparing two samples with interval-ratio data (population variances unknown)

$$t = \frac{\overline{Y}_1 - \overline{Y}_2}{S_{\overline{Y}_1 - \overline{Y}_2}} \tag{13.4}$$

where

$\overline{Y}$ = the sample mean

$S_{\overline{Y}_1 - \overline{Y}_2}$ = the estimated standard error of the difference between two means

Calculating the estimated standard error when the population variances are assumed equal (pooled variance)

$$S_{\overline{Y}_1 - \overline{Y}_2} = \sqrt{\frac{(N_1 - 1)S_{Y_1}^2 + (N_2 - 1)S_{Y_2}^2}{(N_1 + N_2) - 2}} \sqrt{\frac{N_1 + N_2}{N_1 N_2}} \tag{13.7}$$

where

S_Y^2 = the sample variance

N = the sample size

Calculating the estimated standard error when the population variances are assumed unequal

$$S_{\overline{Y}_1 - \overline{Y}_2} = \sqrt{\frac{S_{Y_1}^2}{N_1} + \frac{S_{Y_2}^2}{N_2}} \tag{13.8}$$

Calculating degrees of freedom

$$df = (N_1 + N_2) - 2 \tag{13.5}$$

Adjusting for unequal variances (with small samples)

$$df = \frac{\left(S_{Y_1}^2 - S_{Y_2}^2\right)^2}{\left(S_{Y_1}^2\right)^2 (N_1 - 1) + \left(S_{Y_2}^2\right)^2 (N_2 - 1)} \tag{13.6}$$

(continued on next page)

Box 15.4   Formulas for t, Z, and χ^2 (continued)

where

S_Y^2 = the sample variance
N = the sample size

Z: Comparing two samples with nominal or ordinal data (population variances unknown but assumed equal; $N_1 + N_2 > 100$)

$$Z = \frac{p_1 - p_2}{S_{p_1 - p_2}}$$

(13.9)

$$S_{p_1 - p_2} = \sqrt{\frac{p_1(1 - p_1)}{N_1} + \frac{p_2(1 - p_2)}{N_2}}$$

(13.10)

where

p = the proportion of the sample
$S_{p_1 - p_2}$ = the estimated standard error
N = the sample size

χ^2: Comparing two samples with nominal or ordinal data (cross-tabulated data; any sample size; no cells or less than 20 percent of cells with expected frequencies < 5)

$$\chi^2 = \sum \frac{(f_o - f_e)^2}{f_e}$$

(14.2)

where

f_o = the observed frequency in a cell
f_e = the observed frequency in a cell

Calculating expected frequencies

$$f_e = \frac{(\text{column marginal})(\text{row marginal})}{N}$$

(14.1)

Calculating degrees of freedom

$$df = (r - 1)(c - 1)$$

(14.3)

where

r = the number of rows
c = the number of columns

Step 5: Making a Decision and Interpreting the Results

The last step in the formal process of statistical hypothesis testing is to determine whether the null hypothesis should be rejected. If the probability of the obtained statistic—t, Z, or χ^2—is equal to or less than alpha, it is considered to be statistically significant and the null hypothesis is rejected. If the null hypothesis is rejected, the researcher can claim support for the research hypothesis. In other words, the hypothesized answer to the research question becomes less tentative, but the researcher cannot state that it is absolutely true because there is always some error involved when samples are used to infer population parameters.

The conditions and assumptions associated with the two-sample tests are summarized in the flowchart presented in Figure 15.1. Use this flowchart to help you decide which of the different tests (t, Z, or chi-square) is appropriate under what conditions and how to choose the correct formula for calculating the obtained value for the test.

■ ■ ■ ■ ■ **Statistics in Practice: Education and Employment**

Why did you decide to attend college? Whether you made the decision on your own or discussed it with your parents, spouse, or friends, the prospect of increased employment opportunities and higher income after graduation probably weighed heavily in your decision. Although most college students expect that their major will prepare them to compete successfully in the job market and the workplace, undergraduate programs do not always meet this expectation.

In their introduction to a 1992 study of the efficacy of social science undergraduate programs, Velasco, Stockdale, and Scrams[3] note that sociology programs have traditionally been designed to prepare students for graduate school, where they can earn professional status. However, the vast majority of students who earn a B.A. in sociology do not attend graduate school and must either earn their professional status through work experience or find employment in some other sector. The result is that many people holding a B.A. in sociology are underemployed.

According to Velasco et al., certain foundational skills are critical to successful careers in the social sciences. The foundational skills include logical reasoning, understanding scientific principles, mathematical and statistical skills, computer skills, and knowing the subject matter of the major. In their study, the researchers sought to determine how well sociology programs develop these skills in students. Specifically, they focused on the following research questions:

[3]Steven C. Velasco, Susan E. Stockdale, and David J. Scrams, "Sociology and Other Social Sciences: California State University Alumni Ratings of the B.A. Degree for Development of Employment Skills," *Teaching Sociology* 20 (1992): 60–70.

Figure 15.1   **Flowchart of the Process of Statistical Hypothesis Testing: Two-Sample Situations**

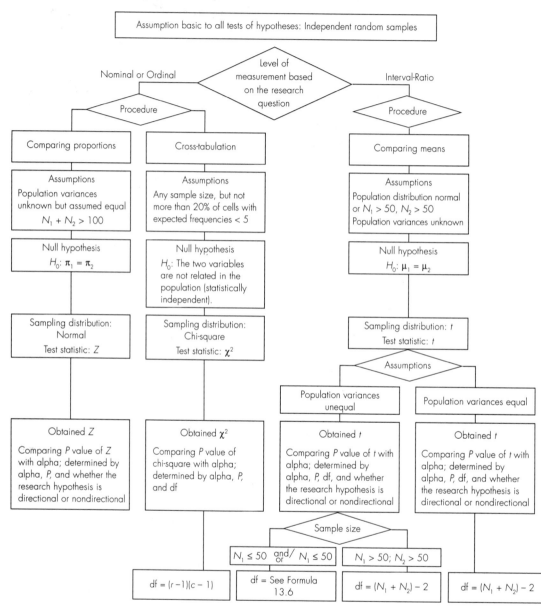

1. How do sociology alumni with B.A. degrees, as compared with other social science alumni, rate their major with respect to the helpfulness of their major in developing the "foundational skills"?

2. Has the percentage of sociology alumni who rate their major highly increased over time with respect to the development of these skills?

3. Do male and female alumni from the five social science disciplines differ in regard to ratings of the major in developing the foundational skills? Do male and female alumni differ with respect to occupational prestige or personal income?[4]

Clearly, surveying the entire population of alumni in five disciplines to obtain answers to these questions would be a nearly insurmountable task. To make their project manageable, the researchers surveyed a sample of each population and used inferential statistics to analyze the data. Their sampling technique and characteristics of the samples are discussed in the next section.

Sampling Technique and Sample Characteristics

Velasco et al. used the alumni records from eight diverse campuses in the California State University system to identify graduates of B.A. programs in anthropology, economics, political science, psychology, and sociology. The population consisted of forty groups of alumni (5 disciplines × 8 campuses = 40 groups). The researchers drew a random sample from each group.[5] Potential subjects were sent a questionnaire and, if necessary, a follow-up postcard. If after follow-up fewer than fifty responses were received from a particular group, random replacement samples were drawn and new potential subjects were similarly contacted.

The final response rate from the combined groups was about 28 percent. Such a low response rate calls into question the representativeness of the sample and, consequently, the use of inferential statistics techniques. The researchers caution that because the sample may not be representative, the results of the statistical tests they performed should be viewed as exploratory.

A total of 2,157 questionnaires were returned. Some of the responses were from people holding advanced degrees, and some of the respondents were not employed full-time. Because the researchers were interested in examining how undergraduate programs prepare students for employment, they limited their final sample to full-time employed respondents with only a B.A. degree, thereby reducing the total sample size to 1,194. Table 15.2 shows selected demographic characteristics for the total final sample and for each discipline.

---

[4]Ibid., p. 62.

[5]All members of groups with fewer than 150 members were included as potential subjects. Up to three questionnaire and follow-up mailings were made to each alumnus to maximize responses from these groups.

Table 15.2 **Selected Demographic Characteristics of the Sample Population with Bachelor's Degrees Who Are Employed Full-Time**

	All	Anthropology	Economics	Political Science	Psychology	Sociology
N	1,194	181	288	222	220	283
% sample in major	—	15.2	24.1	18.6	18.4	23.7
% female	48.7	64.1	26.4	31.5	66.4	61.1
% white	84.8	87.3	87.2	83.3	86.4	80.6
Mean age	35.5	37.6	34.7	33.4	34.1	37.9
SD age	9.11	10.1	9.27	8.22	8.45	8.75
Mean graduation age	27.2	29.9	26.0	25.5	26.6	28.3
SD graduation age	7.75	9.90	6.79	6.38	6.98	8.01

Source: Steven C. Velasco, Susan E. Stockdale, and David J. Scrams, "Sociology and Other Social Sciences: California State University Alumni Ratings of the B.A. Degree for Development of Employment Skills," *Teaching Sociology* 20 (1992): 60–70. Used by permission.

Comparing Ratings of the Major Between Sociology and Other Social Science Alumni

The first research question in this study required a comparison between sociology alumni ratings of their major on the development of foundational skills and the ratings given by alumni from other social science disciplines. To gather data on foundational skills, the researchers asked alumni to rate how well their major added to the development of each of the five skills, using the following scale: 1 = poor; 2 = fair; 3 = good; 4 = excellent. The mean rating for each of the foundational skills, by major, is shown in Table 15.3. The table shows that the skill rated most highly in all disciplines was subject matter of the major. Looking at the mean ratings, we can determine that economics alumni generally rated their major the highest, whereas sociology and political science alumni rated their majors the lowest overall.[6] The lowest rating in all disciplines was given to the development of computer skills.

Ratings of Foundational Skills in Sociology: Changes over Time

In recent years many sociology departments have taken steps to closely align undergraduate requirements with the qualifications necessary for a

[6]In this book we have limited our discussions to tests of differences between two sample means. Velasco et al. used a statistical test called the analysis of variance (ANOVA) to test for differences between several means (5 disciplines) and found that there were significant differences in the ratings given each foundational skill across majors.

Table 15.3 **Graduates' Mean Rating of Their Majors Regarding the Development of Foundational Skills**

	Anthropology	Economics	Political Science	Psychology	Sociology
Logical reasoning	2.99	3.30	3.16	3.13	2.94
Scientific principles	3.01	2.98	2.41	3.07	2.70
Mathematical and statistical skills	2.23	3.22	2.16	2.90	2.54
Computer skills	1.63	2.23	1.67	1.93	1.89
Subject matter of the major	3.36	3.36	3.20	3.26	3.14

Scale: 1 = poor; 2 = fair; 3 = good; 4 = excellent

Source: Adapted from Steven C. Velasco, Susan E. Stockdale, and David J. Scrams, "Sociology and Other Social Sciences: California State University Alumni Ratings of the B.A. Degree for Development of Employment Skills," *Teaching Sociology* 20 (1992): 60–70. Used by permission.

career in sociology. If these changes have been successful, then more recent graduates should rate program development of foundational skills higher than less recent graduates. This is the second research question addressed in this study. To examine the question of whether the percentage of sociology alumni who rate their major highly with respect to the development of foundational skills has increased over time, Velasco et al. grouped the sample of sociology alumni into three categories by number of years since graduation: 11+ years, 5–10 years, and 0–4 years. They grouped the ratings into two categories: "poor or fair" and "good or excellent." Table 15.4 shows percentage bivariate tables for each of the five foundation skills.

Cross-tabulation of the bivariate tables in Table 15.4 reveals the following relationship for all of the foundational skills: The percentage of alumni who rated the major as "good or excellent" in the development of the skill decreased as the number of years since graduation increased. For example, the bivariate table for scientific principles shows that 76.6 percent of the alumni who graduated 0 to 4 years ago rated the major as "good or excellent" compared with 64.1 percent of those who graduated 5 to 10 years ago and 46.4 percent of alumni who graduated 11+ years ago.

The researchers used the chi-square distribution to test for the significance of the relationship for each of the skills. (See Box 15.5 for an illustration of the calculation of chi-square for mathematical and statistical skills.) The chi-square statistic, degrees of freedom, and level of significance are reported at the bottom of each bivariate table in Table 15.4.

Table 15.4 **Sociology Alumni Ratings of the Major in Developing Foundational Skills by Number of Years Since Graduation**

	Number of Years Since Graduation		
	11+	**5–10**	**0–4**
Logical reasoning	$(N = 112)$	$(N = 93)$	$(N = 65)$
Poor or fair	31.3	23.7	17.4
Good or excellent	68.5	76.3	81.5
	chi-square = 3.802; 2 df; p = ns*		
Scientific principles	$(N = 110)$	$(N = 92)$	$(N = 64)$
Poor or fair	53.6	35.9	23.4
Good or excellent	46.4	64.1	76.6
	chi-square = 16.46; 2 df; $p < .001$		
Mathematical and statistical skills	$(N = 109)$	$(N = 92)$	$(N = 64)$
Poor or fair	59.6	46.7	34.8
Good or excellent	40.4	53.3	65.2
	chi-square = 10.41; 2 df; $p < .01$		
Computer skills	$(N = 52)$	$(N = 58)$	$(N = 64)$
Poor or fair	84.4	72.4	65.4
Good or excellent	15.6	27.6	34.6
	chi-square = 4.57; 2 df; $p < .10$		
Subject matter of the major	$(N = 116)$	$(N = 96)$	$(N = 66)$
Poor or fair	21.6	15.6	9.1
Good or excellent	78.4	84.4	90.9
	chi-square = 4.82; 2 df; $p < .10$		

*ns = not significant
Source: Adapted from Steven C. Velasco, Susan E. Stockdale, and David J. Scrams, "Sociology and Other Social Sciences: California State University Alumni Ratings of the B.A. Degree for Development of Employment Skills," *Teaching Sociology* 20 (1992): 60–70. Used by permission.

Look at the levels of significance. Remember that statistical software programs provide the most stringent level at which a statistic is significant, and researchers typically report the level indicated by the output. However, the alpha levels reported in Table 15.4 are somewhat deceptive. There is no problem with the levels reported for scientific principles ($p < .001$) or mathematical and statistical skills ($p < .01$) if we assume that the researchers set alpha at .05 or .01 because p is less than either of these levels for both skills. We can agree with their conclusion that there is a significant relationship between recency of graduation and alumni ratings of the major, and we can further conclude that sociology programs may be improving in the development of the two skills.

Box 15.5 Education and Employment: The Process of Statistical Hypothesis Testing, Using Chi-Square

To follow the process of statistical hypothesis testing, we will calculate chi-square for mathematical and statistical skills from Table 15.4.

Step 1. Making assumptions

A random sample of $N = 265$
Level of measurement of the variable *ratings:* ordinal
Level of measurement of the variable *years since graduation:* ordinal

Step 2. Stating the research and the null hypotheses and selecting alpha

H_1: There is a relationship between number of years since graduation and alumni ratings of the sociology major in developing mathematical and statistical skills (statistical dependence).

H_0: There is no relationship between number of years since graduation and alumni ratings of the sociology major in developing mathematical and statistical skills (statistical independence).

We select an alpha of .05.

Step 3. Selecting a sampling distribution and a test statistic

We will analyze cross-tabulated data measured at the ordinal level.

Sampling distribution: chi-square
Test statistic: χ^2

Step 4. Computing the test statistic

We begin by calculating the degrees of freedom associated with our test statistic:

$$df = 2 \, [(2 - 1)(3 - 1)] = 2$$

In order to calculate chi-square, we first calculate the observed cell frequencies from the percentage table shown in Table 15.4. The frequency table follows.

	NUMBER OF YEARS SINCE GRADUATION			
RATINGS	11+	5–10	0–4	
Poor or fair	65	43	22	130
Good or excellent	44	49	42	135
	109	92	64	265

(continued on next page)

Box 15.5 Education and Employment: The Process of Statistical Hypothesis Testing, Using Chi-Square (continued)

Next calculate the expected frequencies for each cell, based on Formula 14.1:

$$f_e = \frac{(\text{column marginal})(\text{row marginal})}{N}$$

Then calculate chi-square, as follows:

Calculating Chi-Square for Alumni Ratings

Rating	f_e	f_o	$f_o - f_e$	$(f_o - f_e)^2$	$\frac{(f_o - f_e)^2}{f_e}$
Poor or fair/11+	53.47	65	11.53	132.94	2.49
Good or excellent/11+	55.53	44	−11.53	132.94	2.39
Poor or fair/5–10	45.13	43	−2.13	4.54	.10
Good or excellent/5–10	46.87	49	2.13	4.54	.10
Poor or fair/0–4	31.40	22	−9.40	88.36	2.81
Good or excellent/0–4	32.60	42	9.40	88.36	2.71

$$\chi^2 = \sum \frac{(f_o - f_e)^2}{f_e} = 10.60$$

Step 5. Making a decision and interpreting the results

Referring to Appendix D, though 10.60 is not listed in the row for 2 degrees of freedom, we know that it falls between 9.210 and 13.815. We conclude that the probability of our obtained chi-square is somewhere between .01 and .001. Since the probability range is less than our alpha level of .05, we can reject the null hypothesis and conclude that there may be a relationship between the number of years since graduation and the rating given to the major. Sociology programs may have improved in the development of mathematical and statistical skills.

Notice that our calculation resulted in a χ^2 value of 10.60, which differs from that in Table 15.4 ($\chi^2 = 10.41$). The difference of .19 is probably due to rounding as the researchers undoubtedly used a statistical program to do their calculations.

The problem arises when we compare the values presented for logical reasoning (p = ns), computer skills ($p < .10$), and subject matter of the major ($p < .10$). None of the chi-square statistics for these skills is significant at even the .05 level, yet the researchers report the alpha levels differently. They clearly show that the chi-square statistic for logical reasoning skills is not significant (p = ns), but they report $p < .10$ for both of the other skills, thereby giving the impression that these chi-square statistics are significant. The reason for this bit of misdirection can be inferred from the text accompanying the table. The researchers state that "the increases in ratings for computer skills and for understanding the subject matter of the major approached statistical significance."[7] In other words, the researchers would like us to believe that these results were almost significant. Although statements like this are not rare in research reports, they are improper. There is no such thing as an almost significant result. The logic of hypothesis testing dictates that either the null hypothesis is rejected or it is not, and there is no gray area in between. The researchers should have reported "p = ns" for all three of the skills.

Does the lack of a significant result indicate that sociology programs are doing poorly in developing the skill in question? Does a significant finding indicate they are doing well? We need to analyze the results to answer these questions. For example, the chi-square statistic for subject matter of the major was not significant, indicating that the percentage of alumni who rate their major highly in this area has not increased. But let's look at the percentages shown in Table 15.4. Notice that a high percentage of the alumni graduating 11+ years ago (78.4%) felt their major did a good or excellent job of developing the skill. We would conclude that sociology programs have always performed well in developing this skill and would not expect to see significant improvement.

Learning Check. *Analyze the results for the remaining four skills. Where is improvement necessary? Where is it less critical?*

Gender Differences in Ratings of Foundational Skills, Occupational Prestige, and Income

The final research question explored by Velasco et al. concerned gender differences in alumni ratings of foundational skills, occupational prestige, and income. A foundational skills index was constructed by summing the responses for the five categories of skills for each alumnus. The index ranged from 5 to 20, and the mean index score was calculated for each of the disciplines by gender. Occupational prestige was coded using a recognized scale and job titles provided by respondents. Information on income was gathered by asking respondents to report their approximate annual income.

Table 15.5 shows the mean, standard deviation, and t for each of the variables by discipline and gender. The researchers used t tests for the difference between means because the variances were all estimated and the variables were measured at the interval-ratio or ordinal level. Significant t's are indicated by asterisks, with the number of asterisks indicating the highest level

---

[7] Velasco et al., p. 65.

Table 15.5   **Indicated Means and *t* Tests by Gender for Alumni from Each Major**

	Males		Females		
	Mean	**SD**	**Mean**	**SD**	***t***
Foundational skills index					
Anthropology	14.28	2.80	13.58	2.83	1.56
Economics	15.09	2.74	15.49	2.83	−1.08
Political science	12.98	3.08	12.67	3.36	.64
Psychology	15.23	2.84	14.42	2.22	2.06*
Sociology	13.67	2.74	13.52	3.19	.40
Occupational prestige					
Anthropology	49.83	14.01	48.75	11.04	.53
Economics	49.94	10.53	51.42	8.90	−1.08
Political science	48.19	10.18	49.54	9.05	−.93
Psychology	49.37	10.43	49.56	9.22	−.13
Sociology	47.27	10.32	48.81	9.45	−1.25
Income (in thousands of dollars)					
Anthropology	32.78	22.10	23.30	13.78	3.15**
Economics	40.09	22.73	31.43	15.44	3.53***
Political science	38.52	43.01	25.96	8.60	3.42***
Psychology	34.03	26.61	24.71	13.90	2.70**
Sociology	39.36	44.40	25.66	10.47	3.13**

*$p < .05$
**$p < .01$
***$p < .001$
Source: Adapted from Steven C. Velasco, Susan E. Stockdale, and David J. Scrams, "Sociology and Other Social Sciences: California State University Alumni Ratings of the B.A. Degree for Development of Employment Skills," *Teaching Sociology* 20 (1992): 60–70. Used by permission.

at which the statistic is significant. One asterisk indicates the .05 level, two asterisks indicate the .01 level, and three asterisks indicate the .001 level.

The mean ratings of foundational skills show that among males, psychology received the highest average rating (15.23), followed in order by economics (15.09), anthropology (14.28), sociology (13.67), and political science (12.98). Among females, economics received the highest average foundational skill rating (15.49) and political science received the lowest rating (12.67). Only one major, psychology, shows a significant difference between the mean ratings given by male and female alumni.

The mean occupational prestige scores are similar across disciplines within genders. They are also similar across genders within disciplines. The results of the *t* tests show no significant differences between the mean occupational prestige scores for male and female alumni from any major. In Box 15.6 we use the process of statistical hypothesis testing to calculate *t* for occupational prestige among sociology alumni.

Box 15.6   *Occupational Prestige of Male and Female Sociology Alumni: Another Example Using a t Test*

The means, standard deviations, and sample sizes necessary to calculate t for occupational prestige as shown in Table 15.5 are shown below:

	Mean	SD	N
Males	47.27	10.32	105
Females	48.81	9.45	162

Step 1. Making assumptions

Independent random samples

Level of measurement of the variable *occupational prestige*: interval-ratio

Population variances unknown but assumed equal

Because $N_1 > 50$ and $N_2 > 50$, the assumption of normal population is not required.

Step 2. Stating the research and the null hypotheses and selecting alpha

Our hypothesis will be nondirectional because we have no basis for assuming the occupational prestige of one group is higher than the occupational prestige of the other group:

$H_1: \mu_1 \neq \mu_2$
$H_0: \mu_1 = \mu_2$

Alpha for our test will be .05.

Step 3. Selecting a sampling distribution and a test statistic

We will analyze data measured at the interval-ratio level with estimated variances assumed equal:

Sampling distribution: t distribution
Test statistic: t

Step 4. Computing the test statistic

Degrees of freedom are:

$$df = (N_1 + N_2) - 2 = (105 + 162) - 2 = 265$$

The formulas we need to calculate t are

$$t = \frac{\overline{Y}_1 - \overline{Y}_2}{S_{\overline{Y}_1 - \overline{Y}_2}}$$

$$S_{\overline{Y}_1 - \overline{Y}_2} = \sqrt{\frac{(N_1 - 1)S_1^2 + (N_2 - 1)S_2^2}{(N_1 + N_2) - 2}} \sqrt{\frac{N_1 + N_2}{N_1 N_2}}$$

(continued on next page)

Box 15.6 Occupational Prestige of Male and Female Sociology Alumni: Another Example Using a t Test (continued)

First calculate the standard deviation of the sampling distribution:

$$S_{\bar{Y}_1 - \bar{Y}_2} = \sqrt{\frac{(104)(10.32)^2 + (161)(9.45)^2}{(105 + 162) - 2}} \sqrt{\frac{105 + 162}{(105)(162)}}$$

$$= \sqrt{\frac{11,076.25 + 14,377.70}{265}} \sqrt{\frac{267}{17,010}}$$

$$= 9.801(.125) = 1.23$$

Then plug this figure into the formula for t:

$$t = \frac{47.27 - 48.81}{1.23} = \frac{-1.54}{1.23} = -1.25$$

Step 5. Making a decision and interpreting the results

Our obtained t is -1.25, indicating that the difference should be evaluated at the left-tail of the t distribution. Based on a two-tailed test, with 265 degrees of freedom, we can determine the probability of -1.25 based on Appendix C. Recall that we will ignore the negative sign when assessing its probability. Our obtained t is less than any of the listed t values in the last row. The probability of 1.25 is less than .20, larger than our alpha of .05. We fail to reject the null hypothesis and conclude that there is no difference in occupational prestige between male and female sociology alumni.

Economics majors have the highest mean annual income for both males ($40,090) and females ($31,430); anthropology majors have the lowest mean incomes (males, $32,780; females, $23,300). The results of the t tests (for directional tests) show that the mean income of male alumni is significantly higher than the mean income of female alumni for each major. This finding is not surprising given that we know that women typically earn less than men. It is interesting, however, since no significant differences were found between the mean ratings of occupational prestige of male and female alumni. This may indicate that females are paid less than males for similar work.

■ ■ ■ ■ **Conclusion**

We hope that this book has increased your understanding of the social world and helped you to develop your foundational skills in statistics. As an undergraduate, you may need to use your statistics skills to complete a research project or to interpret research reports based on the techniques you have learned. If you choose to pursue a graduate degree, the principles and procedures you have learned here will serve as the basis for advanced graduate statistics classes. If you choose a career in the social sciences, you may be required to conduct research, analyze and report data, or interpret the research reports of others. Even if you are not required to use statistics in your educational or professional endeavors, your knowledge of statistics will help you to be a more knowledgeable consumer of the wide array of information we use in daily life.

SPSS DEMONSTRATION

Regression Revisited: An Application of Inferential Statistics

As presented in Chapter 8, regression is defined as a measure of association between interval (or ordinal) variables. In this demonstration we review regression models, this time introducing their relationship to statistical hypothesis testing.

As we've demonstrated with t, Z, and χ^2 statistics, each is part of a statistical hypothesis test procedure. In determining whether our findings are rare or unexpected, we are testing whether our obtained statistic could be based on chance or if something significant is indicated between the variables that we're investigating.

This same logic can be applied to the correlation coefficient, r, and the standardized slope, b. The appropriate distribution to assess the significance of r and b is the t distribution. In every SPSS Correlation output, SPSS automatically calculates the probability of r based on a two-tailed test. In each regression model, SPSS reports the corresponding t (obtained) and the probability of t for b.

For our demonstration, we'll estimate the correlation between occupational prestige and educational attainment (EDUC). The correlation output is reproduced in the following table.

Correlations

		EDUC HIGHEST YEAR OF SCHOOL COMPLETED	PRESTG80 RS OCCUPATIONAL PRESTIGE SCORE (1980)
EDUC HIGHEST YEAR OF SCHOOL COMPLETED	Pearson Correlation		
	Sig. (2-tailed)		
	N		
PRESTG80 RS OCCUPATIONAL PRESTIGE SCORE (1980)	Pearson Correlation	.559	
	Sig. (2-tailed)	.000	
	N	1364	

Note that the calculated r is .559. This indicates a moderate to strong positive relationship between occupational prestige and respondent's education. Listed under the Pearson Correlation is the significance level for a two-tailed test, .000. We usually assess the probability of a test statistic based on an alpha level, usually set at .05 or less. If p is less than alpha, we would reject the null hypothesis of no relationship between the variables. Though the t isn't reported, its probability is very rare, leading us to conclude that there is a significant relationship between prestige score and educational attainment.

Output for the regression model is reproduced below.

Model Summary

Model	R	R Square	Adjusted R Square	Std. Error of the Estimate
1	.559ª	.313	.312	11.90

a. Predictors: (Constant), EDUC HIGHEST YEAR OF SCHOOL COMPLETED

Coefficientsª

Model		Unstandardized Coefficients		Standardized Coefficients	t	Sig.
		B	Std. Error	Beta		
1	(Constant)	6.120	1.531		3.997	.000
	EDUC HIGHEST YEAR OF SCHOOL COMPLETED	2.762	.111	.559	24.886	.000

a. Dependent Variable: PRESTG80 RS OCCUPATIONAL PRESTIGE SCORE (1980)

In the Model Summary table, the correlation coefficient is reported. We know from the previous correlation output that .556 is significant at the .000 level. Let's consider the data in the Coefficients table.

We will not interpret the significance of the constant (a), but will analyze the information for EDUC. Both the unstandardized and standardized coefficients (Beta) for EDUC are reported. As indicated by b, for every additional year of education, occupational prestige is predicted to increase by 2.762 units. In the last two columns of the table, the t statistic and its significance level (or P value) are reported for the constant (a) and EDUC. Note that the t statistic for the EDUC coefficient is 24.886, significant at .000. Educational attainment has a very significant positive relationship with occupational prestige.

SPSS PROBLEMS

1. Two questions in the 1996 GSS file are concerned with respondent's support of government aid to the elderly (AIDOLD) and the unemployed (AIDUNEMP). Investigate the relationship between these questions and education (DEGREE). Identify the level of measurement for each variable. Calculate an appropriate statistical test, and describe the relationships you find. Also, describe any differences in the relationships between education and the two support variables. [Module B]

2. Test the null hypothesis that there is no difference in years of education between those who attended religious services at least one month and those who did not (ATTEND). Use the variable EDUC as the measure of educational attainment in years. Conduct your test at the .05 alpha level. [Module A]

3. The variable PARTNERS measures the number of self-reported sex partners the respondent had in the previous year. Use it to answer these questions. [Module A]
   a. Calculate a 90 percent confidence interval for PARTNERS, and provide an interpretation.
   b. Test the null hypothesis that there is no difference in the mean of sex partners for men and women. Conduct your test at the .05 alpha level.

4. In Chapter 13, SPSS Problems 2–4, we examined the relationship between hours of television watching (TVHOURS) and gender, age, and marital status. Some of the earlier analysis can be repeated, treating the variables as interval-ratio and conducting regression analyses. What is the relationship between educational attainment (EDUC) and respondent's age (AGE) as independent variables and hours of television viewing per week? Confirm how each of these variables is measured and scaled before beginning the exercise. [Module B]
   a. Construct scatterplots to relate TVHOURS to EDUC and AGE. (You should have two scatterplots.) Do the relationships appear to be linear? Describe the relationships.
   b. Calculate the correlation coefficient for each scatterplot, and the coefficient of determination. Describe the relationship between the variables.
   c. Calculate the regression equation for each scatterplot. Describe the relationship between the variables.
   d. Repeat a–c, this time computing separate scatterplots and statistics for men and women (SEX). What can you conclude?

CHAPTER EXERCISES

1. The 1987–1988 National Survey of Families and Households found, in a sample of 6,645 married couples, that the average length a marriage had

lasted was 205 months (about 17 years), with a standard deviation of 181 months. Assume that the distribution of marriage length is approximately normal.

a. What proportion of marriages lasts between 10 and 20 years?
b. A marriage that lasts 50 years is commonly viewed as exceptional. What is the percentile rank of a marriage that lasts 50 years? Do you believe this justifies the idea that such a marriage is exceptional?
c. What is the probability that a marriage will last more than 30 years?
d. Is there statistical evidence (from the data in this exercise) to lead you to question the assumption that length of marriage is normally distributed?

2. The 1994 National Election Study included a question on whether federal funding for AIDS research should be increased or decreased or stay about the same. Responses to this question are most likely related to many demographic and other attitudinal measures. The following table, shows the relationship between this item and the respondent's political preference in five categories.

Support for AIDS Research by Respondent's Political Preference

	Republican	Independent	No Preference	Other Party	Democrat	Total
Increased	199	284	59	2	338	882
Same	230	165	31	2	190	618
Decreased	111	73	11	0	44	239
Total	540	522	101	4	572	739

a. Describe the relationship in this table by calculating appropriate percentages.
b. Test at the .01 alpha level whether political preference and attitude toward AIDS funding are unrelated.
c. Are all the assumptions for doing a chi-square test met?

3. To investigate further the determinants for funding of AIDS research, the previous table is broken into the following two subtables for whites and blacks. Use them to answer these questions.

Support for AIDS Research by Respondent's Political Preference Whites Only

	Republican	Independent	No Preference	Other Party	Democrat	Total
Increased	182	228	52	1	245	708
Same	217	152	26	2	161	558
Decreased	106	62	9	0	32	209
Total	505	442	87	3	438	1,475

**Support for AIDS Research by Respondent's Political Preference
Blacks Only**

	Republican	Independent	No Preference	Other Party	Democrat	Total
Increased	5	45	5	1	84	140
Same	2	6	1	0	20	29
Decreased	1	8	2	0	7	18
Total	8	59	8	1	111	187

a. Test at the .05 alpha level the relationship between political preference and support for AIDS research funding in each table. Are the results consistent or different by race?

b. Is race an intervening control variable, or is it acting to specify the relationship between political preference and attitude toward AIDS funding?

c. If the assumptions of calculating chi-square are not met in these tables, how might you group the categories of political preference to do a satisfactory test? Do this, and recalculate chi-square for both tables. What do you find now?

d. Can you suggest substantive reasons for the differences between whites and blacks?

4. A large labor union is planning a survey of its members to ask their opinion on several important issues. The members work in large, medium, and small-sized firms. Assume that there are 50,000 members in large companies, 35,000 in medium-sized firms, and 5,000 in small firms.

a. If the labor union takes a proportionate stratified sample of its members of size 1,000, how many union members will be chosen from medium-sized firms?

b. If one member is selected at random from the population, what is the probability that she will be from a small firm?

c. The union decides to take a disproportionate stratified sample with equal numbers of members from each size of firm (to make sure a sufficient number of members from small firms are included). If a sample size of 900 is used, how many members from small firms will be in the sample?

5. The Census Bureau reported that in 1991, 66.9 percent of all Hispanic households were two-parent households. You are studying a large city in the Southwest and have taken a random sample of the households in the city for your study. You find that only 59.5 percent of all Hispanic households had two parents in your sample of 400.

a. What is the 95 percent confidence interval for your population estimate of 59.5 percent?

b. Do a test to determine whether the city's percentage of Hispanic two-parent households is significantly lower than the U.S. population figure. Set up appropriate research and null hypotheses, and test at the .01 level.

6. It is often said that there is a relationship between religious belief and education, such that belief declines as education increases. However, the recent revival of fundamentalism may have weakened this relationship. The 1994 National Election Study data can be used to investigate this question. One item asked whether religion was important to the respondent, with possible responses of either "Yes" or "No." We find that those who answered yes have 12.96 mean years of education, with a standard deviation of 2.61; those who answered no have 13.36 mean years of education, with a standard deviation of 2.44. Altogether, 775 answered yes and 252 no.

   a. Using a two-tailed test, test at the .05 level the null hypothesis that there is no difference in years of education between those who do and those who don't find religion personally important.

   b. Now do the same test at the .01 level. If the conclusion is different from that in (a), is it possible to state that one of these two tests is somehow better or more correct than the other? Why or why not?

7. Often the same data can be studied with more than one type of statistical test. The following table displays the relationship between gender and whether the respondent approves or disapproves of Congress, with data taken from the 1994 National Election Study.

Approval of Congress	Gender	
	Male	Female
Approve	249	303
Disapprove	564	540
Total	813	843

It is possible to study this table with both the chi-square statistic and a two-sample test of proportions.

   a. Conduct a chi-square test at the .01 level of the null hypothesis that females are more likely to disapprove of Congress than males.

   b. Conduct a two-sample proportion test at the .01 level to determine whether males and females differ in their disapproval of Congress.

   c. Construct a 95 percent confidence interval for the percentage of all respondents, both male and female, who disapprove of Congress.

   d. Were your conclusions similar or different in the two tests in (a) and (b)?

8. People who are self-employed are often thought to work more hours per week than those who are not self-employed. Study this question with data from 1994. Those who are self-employed (137 respondents) worked

an average of 44.80 hours per week, with a standard deviation of 20.70. Those not self-employed (1,035 respondents) worked an average of 41.69 hours per week, with a standard deviation of 12.21. Assume that the standard deviations are not equal.

a. Test at the .05 level with a one-tailed test the hypothesis that the self-employed work more hours than others.

b. The standard workweek is often thought to be 40 hours. Do a one-sample test to see whether those who are not self-employed work more than 40 hours at the .10 alpha level.

9. Ratings of the job being done by individuals often differ from ratings of the overall job done by the organization to which they belong. In an NBC/*Wall Street Journal* poll in October 1991, 60 percent of the respondents said that "in general, they disapprove of the job Congress is doing," whereas 40 percent approved. In an ABC/*Washington Post* poll done that same month, 70 percent of the respondents "approve of the way your own representative to the U.S. House in Congress is handling his or her job," whereas 30 percent disapproved. The first poll contacted 716 people, and the second contacted 1,398.

a. Test at the .01 level the null hypothesis that there is no difference in the approval ratings of Congress and individual representatives.

b. If you find a difference, suggest reasons why people can believe their own representative is doing a good job but not the Congress as a whole. Try to think of reasons why there might be a difference even if the individual representative is performing similarly to his or her colleagues.

10. The MMPI test is used extensively by psychologists to provide information on personality traits and potential problems of individuals undergoing counseling. The test measures nine primary dimensions of personality, with each dimension represented by a scale normed to have a mean score of 50 and a standard deviation of 10 in the adult population. One primary scale measures paranoid tendencies. Assume the scale scores are normally distributed.

a. What percentage of the population should have a Paranoia scale score above 70? A score of 70 is viewed as "elevated" or abnormal by the MMPI test developers. Based on your statistical calculation, do you agree?

b. What percentile rank does a score of 45 correspond to?

c. What range of scores, centered around the mean of 50, should include 75 percent of the population?

11. The 1994 National Election Study included a few questions that asked whether the respondent feels things are going to be better or worse next year, or have improved or gotten worse over the past year, for both the United States as a whole and the respondent himself or herself. The following table displays the relationship between answers to whether the respondent is doing better or worse than a year ago by marital status.

Better or Worse Off Than a Year Ago	Marital Status				
	Married	Never Married	Divorced	Separated	Widowed
Better off	414	151	42	21	21
Same	388	83	48	8	63
Worse off	276	90	57	16	40

a. Describe the relationship between marital status and belief that things have improved or not by calculating appropriate percentages.

b. Test whether these two characteristics are related at the .05 alpha level.

c. Offer some substantive reasons for the relationship you observe in the table.

12. Is there a relationship between smoking and school performance among teenagers? Data from Chapter 7, Exercise 13, are presented again in the following table. Calculate chi-square for the relationship between the two variables. Set alpha at .05.

School Performance	Nonsmokers	Former Smokers	Current Smokers	Total
Much better than average	753	130	51	934
Better than average	1,439	310	140	1,889
Average	1,365	387	246	1,998
Below average	88	40	58	186
Total	3,645	867	495	5,007

Source: Adapted from Teh-wei Hu, Zihua Lin, and Theodore E. Keeler, "Teenage Smoking: Attempts to Quit and School Performance," *American Journal of Public Health* 88, no. 6 (1998): 940–943. Used by permission of The American Public Health Association.

13. How different are users of alternative medicine from nonusers? Bivariate tables of age and household income with use of alternative medicine follow. Data are based on the Quebec Health Study (1987) and Quebec Health Insurance Board (QHIB) claims database. Calculate the chi-square for each table, setting alpha at .05.

Age (Yrs)	Users of Alternative Medicine	Nonusers of Alternative Medicine
0–29	39	51
30–44	72	44
45–64	42	52
65 and older	16	22

Household Income	Users of Alternative Medicine	Nonusers of Alternative Medicine
Less than $12,000	8	30
$12,000–19,999	27	22
$20,000–29,999	38	37
$30,000–39,999	26	23
$40,000 or over	53	41

Source: Adapted from Regis Blais, Aboubacrine Maiga, and Alarou Aboubacar, "How Different Are Users and Non-users of Alternative Medicine?" *Canadian Journal of Public Health* 88, no. 3 (1997): 159–162, Table 1. Used by permission of the publisher.

Appendix A
Table of Random Numbers

A Table of 14,000 Random Units

Line/Col.	(1)	(2)	(3)	(4)	(5)	(6)	(7)	(8)	(9)	(10)	(11)	(12)	(13)	(14)
1	10480	15011	01536	02011	81647	91646	69179	14194	62590	36207	20969	99570	91291	90700
2	22368	46573	25595	85393	30995	89198	27982	53402	93965	34095	52666	19174	39615	99505
3	24130	48360	22527	97265	76393	64809	15179	24830	49340	32081	30680	19655	63348	58629
4	42167	93093	06243	61680	07856	16376	39440	53537	71341	57004	00849	74917	97758	16379
5	37570	39975	81837	16656	06121	91782	60468	81305	49684	60672	14110	06927	01263	54613
6	77921	06907	11008	42751	27756	53498	18602	70659	90655	15053	21916	81825	44394	42880
7	99562	72905	56420	69994	98872	31016	71194	18738	44013	48840	63213	21069	10634	12952
8	96301	91977	05463	07972	18876	20922	94595	56869	69014	60045	18425	84903	42508	32307
9	89579	14342	63661	10281	17453	18103	57740	84378	25331	12566	58678	44947	05585	56941
10	85475	36857	43342	53988	53060	59533	38867	62300	08158	17983	16439	11458	18593	64952
11	28918	69578	88231	33276	70997	79936	56865	05859	90106	31595	01547	85590	91610	78188
12	63553	40961	48235	03427	49626	69445	18663	72695	52180	20847	12234	90511	33703	90322
13	09429	93969	52636	92737	88974	33488	36320	17617	30015	08272	84115	27156	30613	74952
14	10365	61129	87529	85689	48237	52267	67689	93394	01511	26358	85104	20285	29975	89868
15	07119	97336	71048	08178	77233	13916	47564	81056	97735	85977	29372	74461	28551	90707
16	51085	12765	51821	51259	77452	16308	60756	92144	49442	53900	70960	63990	75601	40719
17	02368	21382	52404	60268	89368	19885	55322	44819	01188	65255	64835	44919	05944	55157
18	01011	54092	33362	94904	31273	04146	18594	29852	71585	85030	51132	01915	92747	64951
19	52162	53916	46369	58586	23216	14513	83149	98736	23495	64350	94738	17752	35156	35749
20	07056	97628	33787	09998	42698	06691	76988	13602	51851	46104	88916	19509	25625	58104
21	48663	91245	85828	14346	09172	30168	90229	04734	59193	22178	30421	61666	99904	32812
22	54164	58492	22421	74103	47070	25306	76468	26384	58151	06646	21524	15227	96909	44592
23	32639	32363	05597	24200	13363	38005	94342	28728	35806	06912	17012	64161	18296	22851
24	29334	27001	87637	87308	58731	00256	45834	15398	46557	41135	10367	07684	36188	18510
25	02488	33062	28834	07351	19731	92420	60952	61280	50001	67658	32586	86679	50720	94953
26	81525	72295	04839	96423	24878	82651	66566	14778	76797	14780	13300	87074	79666	95725
27	29676	20591	68086	26432	46901	20849	89768	81536	86645	12659	92259	57102	80428	25280
28	00742	57392	39064	66432	84673	40027	32832	61362	98947	96067	64760	64584	96096	98253
29	05366	04213	25669	26422	44407	44048	37937	63904	45766	66134	75470	66520	34693	90449
30	91921	26418	64117	94305	26766	25940	39972	22209	71500	64568	91402	42416	07844	69618
31	00582	04711	87917	77341	42206	35126	74087	99547	81817	42607	43808	76655	62028	76630
32	00725	69884	62797	56170	86324	88072	76222	36086	84637	93161	76038	65855	77919	88006
33	69011	65797	95876	55293	18988	27354	26575	08625	40801	59920	29841	80150	12777	48501
34	25976	57948	29888	88604	67917	48708	18912	82271	65424	69774	33611	54262	85963	03547
35	09763	83473	73577	12908	30883	18317	28290	35797	05998	41688	34952	37888	38917	88050

Source: William H. Beyer, ed., *Handbook for Probability and Statistics*, 2nd ed. Copyright © 1966 CRC Press, Boca Raton, Florida. Used by permission.

Line/Col.	(1)	(2)	(3)	(4)	(5)	(6)	(7)	(8)	(9)	(10)	(11)	(12)	(13)	(14)
36	91567	42595	27958	30134	04024	86385	29880	99730	55536	84855	29080	09250	79656	73211
37	17955	56349	90999	49127	20044	59931	06115	20542	18059	02008	73708	83317	36103	42791
38	46503	18584	18845	49618	02304	51038	20655	58727	28168	15475	56942	53389	20562	87338
39	92157	89634	94824	78171	84610	82834	09922	25417	44137	48413	25555	21246	35509	20468
40	14577	62765	35605	81263	39667	47358	56873	56307	61607	49518	89656	20103	77490	18062
41	98427	07523	33362	64270	01638	92477	66969	98420	04880	45585	46565	04102	46880	45709
42	34914	63976	88720	82765	34476	17032	87589	40836	32427	70002	70663	88863	77775	69348
43	70060	28277	39475	46473	23219	53416	94970	25832	69975	94884	19661	72828	00102	66794
44	53976	54914	06990	67245	68350	82948	11398	42878	80287	88267	47363	46634	06541	97809
45	76072	29515	40980	07391	58745	25774	22987	80059	39911	96189	41151	14222	60697	59583
46	90725	52210	83974	29992	65831	38857	50490	83765	55657	14361	31720	57375	56228	41546
47	64364	67412	33339	31926	14883	24413	59744	92351	97473	89286	35931	04110	23726	51900
48	08962	00358	31662	25388	61642	34072	81249	35648	56891	69352	48373	45578	78547	81788
49	95012	68379	93526	70765	10593	04542	76463	54328	02349	17247	28865	14777	62730	92277
50	15664	10493	20492	38391	91132	21999	59516	81652	27195	48223	46751	22923	32261	85653
51	16408	81899	04153	53381	79401	21438	83035	92350	36693	31238	59649	91754	72772	02338
52	18629	81953	05520	91962	04739	13092	97662	24822	94730	06496	35090	04822	86772	98289
53	73115	35101	47498	87637	99016	71060	88824	71013	18735	20286	23153	72924	35165	43040
54	57491	16703	23167	49323	45021	33132	12544	41035	80780	45393	44812	12515	98931	91202
55	30405	83946	23792	14422	15059	45799	22716	19792	09983	74353	68668	30429	70735	25499
56	16631	35006	85900	98275	32388	52390	16815	69298	82732	38480	73817	32523	41961	44437
57	96773	20206	42559	78985	05300	22164	24369	54224	35083	19687	11052	91491	60383	19746
58	38935	64202	14349	82674	66523	44133	00697	35552	35970	19124	63318	29686	03387	59846
59	31624	76384	17403	53363	44167	64486	64758	75366	76554	31601	12614	33072	60332	92325
60	78919	19474	23632	27889	47914	02584	37680	20801	72152	39339	34806	08930	85001	87820
61	03931	33309	57047	74211	63445	17361	62825	39908	05607	91284	68833	25570	38818	46920
62	74426	33278	43972	10119	89917	15665	52872	73823	73144	88662	88970	74492	51805	99378
63	09066	00903	20795	95452	92648	45454	09552	88815	16553	51125	79375	97596	16296	66092
64	42238	12426	87025	14267	20979	04508	64535	31355	86064	29472	47689	05974	52468	16834
65	16153	08002	26504	41744	81959	65642	74240	56302	00033	67107	77510	70625	28725	34191
66	21457	40742	29820	96783	29400	21840	15035	34537	33310	06116	95240	15957	16572	06004
67	21581	57802	02050	89728	17937	37621	47075	42080	97403	48626	68995	43805	33386	21597
68	55612	78095	83197	33732	05810	24813	86902	60397	16489	03264	88525	42786	05269	92532
69	44657	66999	99324	51281	84463	60563	79312	93454	68876	25471	93911	25650	12682	73572
70	91340	84979	46949	81973	37949	61023	43997	15263	80644	43942	89203	71795	99533	50501
71	91227	21199	31935	27022	84067	05462	35216	14486	29891	68607	41867	14951	91696	85065
72	50001	38140	66321	19924	72163	09538	12151	06878	91903	18749	34405	56087	82790	70925
73	65390	05224	72958	28609	81406	39147	25549	48542	42627	45233	57202	94617	23772	07896
74	27504	96131	83944	41575	10573	08619	64482	73923	36152	05184	94142	25299	84387	34925
75	37169	94851	39117	89632	00959	16487	65536	49071	39782	17095	02330	74301	00275	48280
76	11508	70225	51111	38351	19444	66499	71945	05422	13442	78675	84081	66938	93654	59894
77	37449	30362	06694	54690	04052	53115	62757	95348	78662	11163	81651	50245	34971	52924
78	46515	70331	85922	38329	57015	15765	97161	17869	45349	61796	66345	81073	49106	79860
79	30986	81223	42416	58353	21532	30502	32305	86482	05174	07901	54339	58861	74818	46942
80	63798	64995	46583	09765	44160	78128	83991	42865	92520	83531	80377	35909	81250	54238
81	82486	84846	99254	67632	43218	50076	21361	64816	51202	88124	41870	52689	51275	83556
82	21885	32906	92431	09060	64297	51674	64126	62570	26123	05155	59194	52799	28225	85762
83	60336	98782	07408	53458	13564	59089	26445	29789	85205	41001	12535	12133	14645	23541
84	43937	46891	24010	25560	86355	33941	25786	54990	71899	15475	95434	98227	21824	19585
85	97656	63175	89303	16275	07100	92063	21942	18611	47348	20203	18534	03862	78095	50136

Line/Col.	(1)	(2)	(3)	(4)	(5)	(6)	(7)	(8)	(9)	(10)	(11)	(12)	(13)	(14)
86	03299	01221	05418	38982	55758	92237	26759	86367	21216	98442	08303	56613	91511	75928
87	79626	06486	03574	17668	07785	76020	79924	25651	83325	88428	85076	72811	22717	50585
88	85636	68335	47539	03129	65651	11977	02510	26113	99447	68645	34327	15152	55230	93448
89	18039	14367	61337	06177	12143	46609	32989	74014	64708	00533	35398	58408	13261	47908
90	08362	15656	60627	36478	65648	16764	53412	09013	07832	41574	17639	82163	60859	75567
91	79556	29068	04142	16268	15387	12856	66227	38358	22478	73373	88732	09443	82558	05250
92	92608	82674	27072	32534	17075	27698	98204	63863	11951	34648	88022	56148	34925	57031
93	23982	25835	40055	67006	12293	02753	14827	22235	35071	99704	37543	11601	35503	85171
94	09915	96306	05908	97901	28395	14186	00821	80703	70426	75647	76310	88717	37890	40129
95	50937	33300	26695	62247	69927	76123	50842	43834	86654	70959	79725	93872	28117	19233
96	42488	78077	69882	61657	34136	79180	97526	43092	04098	73571	80799	76536	71255	64239
97	46764	86273	63003	93017	31204	36692	40202	35275	57306	55543	53203	18098	47625	88684
98	03237	45430	55417	63282	90816	17349	88298	90183	36600	78406	06216	95787	42579	90730
99	86591	81482	52667	61583	14972	90053	89534	76036	49199	43716	97548	04379	46370	28672
100	38534	01715	94964	87288	65680	43772	39560	12918	86537	62738	19636	51132	25739	56947

Appendix B
The Standard Normal Table

The values in column A are Z scores. Column B lists the proportion of area between the mean and a given Z. Column C lists the porportion of area beyond a given Z. Only positive Z scores are listed. Because the normal curve is symmetrical, the areas for negative Z scores will be exactly the same as the areas for positive Z scores.

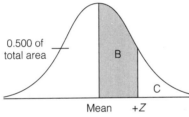

0.500 of total area

B

C

Mean   +Z

Areas between mean and Z and beyond Z (for positive Z)

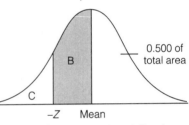

0.500 of total area

B

C

−Z   Mean

Areas between mean and Z and beyond Z (for negative Z)

A	B	C	A	B	C	A	B	C
	Area Between	Area Beyond		Area Between	Area Beyond		Area Between	Area Beyond
Z	Mean and Z	Z	Z	Mean and Z	Z	Z	Mean and Z	Z
0.00	0.0000	0.5000	0.11	0.0438	0.4562	0.21	0.0832	0.4168
0.01	0.0040	0.4960	0.12	0.0478	0.4522	0.22	0.0871	0.4129
0.02	0.0080	0.4920	0.13	0.0517	0.4483	0.23	0.0910	0.4090
0.03	0.0120	0.4880	0.14	0.0557	0.4443	0.24	0.0948	0.4052
0.04	0.0160	0.4840	0.15	0.0596	0.4404	0.25	0.0987	0.4013
0.05	0.0199	0.4801	0.16	0.0636	0.4364	0.26	0.1026	0.3974
0.06	0.0239	0.4761	0.17	0.0675	0.4325	0.27	0.1064	0.3936
0.07	0.0279	0.4721	0.18	0.0714	0.4286	0.28	0.1103	0.3897
0.08	0.0319	0.4681	0.19	0.0753	0.4247	0.29	0.1141	0.3859
0.09	0.0359	0.4641	0.20	0.0793	0.4207	0.30	0.1179	0.3821
0.10	0.0398	0.4602						

A Z	B Area Between Mean and Z	C Area Beyond Z	A Z	B Area Between Mean and Z	C Area Beyond Z	A Z	B Area Between Mean and Z	C Area Beyond Z
0.31	0.1217	0.3783	0.61	0.2291	0.2709	0.91	0.3186	0.1814
0.32	0.1255	0.3745	0.62	0.2324	0.2676	0.92	0.3212	0.1788
0.33	0.1293	0.3707	0.63	0.2357	0.2643	0.93	0.3238	0.1762
0.34	0.1331	0.3669	0.64	0.2389	0.2611	0.94	0.3264	0.1736
0.35	0.1368	0.3632	0.65	0.2422	0.2578	0.95	0.3289	0.1711
0.36	0.1406	0.3594	0.66	0.2454	0.2546	0.96	0.3315	0.1685
0.37	0.1443	0.3557	0.67	0.2486	0.2514	0.97	0.3340	0.1660
0.38	0.1480	0.3520	0.68	0.2517	0.2483	0.98	0.3365	0.1635
0.39	0.1517	0.3483	0.69	0.2549	0.2451	0.99	0.3389	0.1611
0.40	0.1554	0.3446	0.70	0.2580	0.2420	1.00	0.3413	0.1587
0.41	0.1591	0.3409	0.71	0.2611	0.2389	1.01	0.3438	0.1562
0.42	0.1628	0.3372	0.72	0.2642	0.2358	1.02	0.3461	0.1539
0.43	0.1664	0.3336	0.73	0.2673	0.2327	1.03	0.3485	0.1515
0.44	0.1700	0.3300	0.74	0.2703	0.2297	1.04	0.3508	0.1492
0.45	0.1736	0.3264	0.75	0.2734	0.2266	1.05	0.3531	0.1469
0.46	0.1772	0.3228	0.76	0.2764	0.2236	1.06	0.3554	0.1446
0.47	0.1808	0.3192	0.77	0.2794	0.2206	1.07	0.3577	0.1423
0.48	0.1844	0.3156	0.78	0.2823	0.2177	1.08	0.3599	0.1401
0.49	0.1879	0.3121	0.79	0.2852	0.2148	1.09	0.3621	0.1379
0.50	0.1915	0.3085	0.80	0.2881	0.2119	1.10	0.3643	0.1357
0.51	0.1950	0.3050	0.81	0.2910	0.2090	1.11	0.3665	0.1335
0.52	0.1985	0.3015	0.82	0.2939	0.2061	1.12	0.3686	0.1314
0.53	0.2019	0.2981	0.83	0.2967	0.2033	1.13	0.3708	0.1292
0.54	0.2054	0.2946	0.84	0.2995	0.2005	1.14	0.3729	0.1271
0.55	0.2088	0.2912	0.85	0.3023	0.1977	1.15	0.3749	0.1251
0.56	0.2123	0.2877	0.86	0.3051	0.1949	1.16	0.3770	0.1230
0.57	0.2157	0.2843	0.87	0.3078	0.1922	1.17	0.3790	0.1210
0.58	0.2190	0.2810	0.88	0.3106	0.1894	1.18	0.3810	0.1190
0.59	0.2224	0.2776	0.89	0.3133	0.1867	1.19	0.3830	0.1170
0.60	0.2257	0.2743	0.90	0.3159	0.1841	1.20	0.3849	0.1151

A	B Area Between	C Area Beyond	A	B Area Between	C Area Beyond	A	B Area Between	C Area Beyond
Z	Mean and Z	Z	Z	Mean and Z	Z	Z	Mean and Z	Z
1.21	0.3869	0.1131	1.51	0.4345	0.0655	1.81	0.4649	0.0351
1.22	0.3888	0.1112	1.52	0.4357	0.0643	1.82	0.4656	0.0344
1.23	0.3907	0.1093	1.53	0.4370	0.0630	1.83	0.4664	0.0336
1.24	0.3925	0.1075	1.54	0.4382	0.0618	1.84	0.4671	0.0329
1.25	0.3944	0.1056	1.55	0.4394	0.0606	1.85	0.4678	0.0322
1.26	0.3962	0.1038	1.56	0.4406	0.0594	1.86	0.4686	0.0314
1.27	0.3980	0.1020	1.57	0.4418	0.0582	1.87	0.4693	0.0307
1.28	0.3997	0.1003	1.58	0.4429	0.0571	1.88	0.4699	0.0301
1.29	0.4015	0.0985	1.59	0.4441	0.0559	1.89	0.4706	0.0294
1.30	0.4032	0.0968	1.60	0.4452	0.0548	1.90	0.4713	0.0287
1.31	0.4049	0.0951	1.61	0.4463	0.0537	1.91	0.4719	0.0281
1.32	0.4066	0.0934	1.62	0.4474	0.0526	1.92	0.4726	0.0274
1.33	0.4082	0.0918	1.63	0.4484	0.0516	1.93	0.4732	0.0268
1.34	0.4099	0.0901	1.64	0.4495	0.0505	1.94	0.4738	0.0262
1.35	0.4115	0.0885	1.65	0.4505	0.0495	1.95	0.4744	0.0256
1.36	0.4131	0.0869	1.66	0.4515	0.0485	1.96	0.4750	0.0250
1.37	0.4147	0.0853	1.67	0.4525	0.0475	1.97	0.4756	0.0244
1.38	0.4162	0.0838	1.68	0.4535	0.0465	1.98	0.4761	0.0239
1.39	0.4177	0.0823	1.69	0.4545	0.0455	1.99	0.4767	0.0233
1.40	0.4192	0.0808	1.70	0.4554	0.0446	2.00	0.4772	0.0228
1.41	0.4207	0.0793	1.71	0.4564	0.0436	2.01	0.4778	0.0222
1.42	0.4222	0.0778	1.72	0.4573	0.0427	2.02	0.4783	0.0217
1.43	0.4236	0.0764	1.73	0.4582	0.0418	2.03	0.4788	0.0212
1.44	0.4251	0.0749	1.74	0.4591	0.0409	2.04	0.4793	0.0207
1.45	0.4265	0.0735	1.75	0.4599	0.0401	2.05	0.4798	0.0202
1.46	0.4279	0.0721	1.76	0.4608	0.0392	2.06	0.4803	0.0197
1.47	0.4292	0.0708	1.77	0.4616	0.0384	2.07	0.4808	0.0192
1.48	0.4306	0.0694	1.78	0.4625	0.0375	2.08	0.4812	0.0188
1.49	0.4319	0.0681	1.79	0.4633	0.0367	2.09	0.4817	0.0183
1.50	0.4332	0.0668	1.80	0.4641	0.0359	2.10	0.4821	0.0179

A	B Area Between	C Area Beyond	A	B Area Between	C Area Beyond	A	B Area Between	C Area Beyond
Z	Mean and Z	Z	Z	Mean and Z	Z	Z	Mean and Z	Z
2.11	0.4826	0.0174	2.41	0.4920	0.0080	2.71	0.4966	0.0034
2.12	0.4830	0.0170	2.42	0.4922	0.0078	2.72	0.4967	0.0033
2.13	0.4834	0.0166	2.43	0.4925	0.0075	2.73	0.4968	0.0032
2.14	0.4838	0.0162	2.44	0.4927	0.0073	2.74	0.4969	0.0031
2.15	0.4842	0.0158	2.45	0.4929	0.0071	2.75	0.4970	0.0030
2.16	0.4846	0.0154	2.46	0.4931	0.0069	2.76	0.4971	0.0029
2.17	0.4850	0.0150	2.47	0.4932	0.0068	2.77	0.4972	0.0028
2.18	0.4854	0.0146	2.48	0.4934	0.0066	2.78	0.4973	0.0027
2.19	0.4857	0.0143	2.49	0.4936	0.0064	2.79	0.4974	0.0026
2.20	0.4861	0.0139	2.50	0.4938	0.0062	2.80	0.4974	0.0026
2.21	0.4864	0.0136	2.51	0.4940	0.0060	2.81	0.4975	0.0025
2.22	0.4868	0.0132	2.52	0.4941	0.0059	2.82	0.4976	0.0024
2.23	0.4871	0.0129	2.53	0.4943	0.0057	2.83	0.4977	0.0023
2.24	0.4875	0.0125	2.54	0.4945	0.0055	2.84	0.4977	0.0023
2.25	0.4878	0.0122	2.55	0.4946	0.0054	2.85	0.4978	0.0022
2.26	0.4881	0.0119	2.56	0.4948	0.0052	2.86	0.4979	0.0021
2.27	0.4884	0.0116	2.57	0.4949	0.0051	2.87	0.4979	0.0021
2.28	0.4887	0.0113	2.58	0.4951	0.0049	2.88	0.4980	0.0020
2.29	0.4890	0.0110	2.59	0.4952	0.0048	2.89	0.4981	0.0019
2.30	0.4893	0.0107	2.60	0.4953	0.0047	2.90	0.4981	0.0019
2.31	0.4896	0.0104	2.61	0.4955	0.0045	2.91	0.4982	0.0018
2.32	0.4898	0.0102	2.62	0.4956	0.0044	2.92	0.4982	0.0018
2.33	0.4901	0.0099	2.63	0.4957	0.0043	2.93	0.4983	0.0017
2.34	0.4904	0.0096	2.64	0.4959	0.0041	2.94	0.4984	0.0016
2.35	0.4906	0.0094	2.65	0.4960	0.0040	2.95	0.4984	0.0016
2.36	0.4909	0.0091	2.66	0.4961	0.0039	2.96	0.4985	0.0015
2.37	0.4911	0.0089	2.67	0.4962	0.0038	2.97	0.4985	0.0015
2.38	0.4913	0.0087	2.68	0.4963	0.0037	2.98	0.4986	0.0014
2.39	0.4916	0.0084	2.69	0.4964	0.0036	2.99	0.4986	0.0014
2.40	0.4918	0.0082	2.70	0.4965	0.0035	3.00	0.4986	0.0014

A Z	B Area Between Mean and Z	C Area Beyond Z	A Z	B Area Between Mean and Z	C Area Beyond Z	A Z	B Area Between Mean and Z	C Area Beyond Z
3.01	0.4987	0.0013	3.21	0.4993	0.0007	3.41	0.4997	0.0003
3.02	0.4987	0.0013	3.22	0.4994	0.0006	3.42	0.4997	0.0003
3.03	0.4988	0.0012	3.23	0.4994	0.0006	3.43	0.4997	0.0003
3.04	0.4988	0.0012	3.24	0.4994	0.0006	3.44	0.4997	0.0003
3.05	0.4989	0.0011	3.25	0.4994	0.0006	3.45	0.4997	0.0003
3.06	0.4989	0.0011	3.26	0.4994	0.0006	3.46	0.4997	0.0003
3.07	0.4989	0.0011	3.27	0.4995	0.0005	3.47	0.4997	0.0003
3.08	0.4990	0.0010	3.28	0.4995	0.0005	3.48	0.4997	0.0003
3.09	0.4990	0.0010	3.29	0.4995	0.0005	3.49	0.4998	0.0002
3.10	0.4990	0.0010	3.30	0.4995	0.0005	3.50	0.4998	0.0002
3.11	0.4991	0.0009	3.31	0.4995	0.0005	3.60	0.4998	0.0002
3.12	0.4991	0.0009	3.32	0.4995	0.0005			
3.13	0.4991	0.0009	3.33	0.4996	0.0004	3.70	0.4999	0.0001
3.14	0.4992	0.0008	3.34	0.4996	0.0004			
3.15	0.4992	0.0008	3.35	0.4996	0.0004	3.80	0.4999	0.0001
3.16	0.4992	0.0008	3.36	0.4996	0.0004			
3.17	0.4992	0.0008	3.37	0.4996	0.0004	3.90	0.4999	<0.0001
3.18	0.4993	0.0007	3.38	0.4996	0.0004			
3.19	0.4993	0.0007	3.39	0.4997	0.0003			
3.20	0.4993	0.0007	3.40	0.4997	0.0003	4.00	0.4999	<0.0001

Appendix C
Distribution of *t*

df	Level of Significance for One-Tailed Test					
	.10	.05	.025	.01	.005	.0005
	Level of Significance for Two-Tailed Test					
	.20	.10	.05	.02	.01	.001
1	3.078	6.314	12.706	31.821	63.657	636.619
2	1.886	2.920	4.303	6.965	9.925	31.598
3	1.638	2.353	3.182	4.541	5.841	12.941
4	1.533	2.132	2.776	3.747	4.604	8.610
5	1.476	2.015	2.571	3.365	4.032	6.859
6	1.440	1.943	2.447	3.143	3.707	5.959
7	1.415	1.895	2.365	2.998	3.499	5.405
8	1.397	1.860	2.306	2.896	3.355	5.041
9	1.383	1.833	2.262	2.821	3.250	4.781
10	1.372	1.812	2.228	2.764	3.169	4.587
11	1.363	1.796	2.201	2.718	3.106	4.437
12	1.356	1.782	2.179	2.681	3.055	4.318
13	1.350	1.771	2.160	2.650	3.012	4.221
14	1.345	1.761	2.145	2.624	2.977	4.140
15	1.341	1.753	2.131	2.602	2.947	4.073
16	1.337	1.746	2.120	2.583	2.921	4.015
17	1.333	1.740	2.110	2.567	2.898	3.965
18	1.330	1.734	2.101	2.552	2.878	3.922
19	1.328	1.729	2.093	2.539	2.861	3.883
20	1.325	1.725	2.086	2.528	2.845	3.850

df	Level of Significance for One-Tailed Test					
	.10	.05	.025	.01	.005	.0005
	Level of Significance for Two-Tailed Test					
	.20	.10	.05	.02	.01	.001
21	1.323	1.721	2.080	2.518	2.831	3.819
22	1.321	1.717	2.074	2.508	2.819	3.792
23	1.319	1.714	2.069	2.500	2.807	3.767
24	1.318	1.711	2.064	2.492	2.797	3.745
25	1.316	1.708	2.060	2.485	2.787	3.725
26	1.315	1.706	2.056	2.479	2.779	3.707
27	1.314	1.703	2.052	2.473	2.771	3.690
28	1.313	1.701	2.048	2.467	2.763	3.674
29	1.311	1.699	2.045	2.462	2.756	3.659
30	1.310	1.697	2.042	2.457	2.750	3.646
40	1.303	1.684	2.021	2.423	2.704	3.551
60	1.296	1.671	2.000	2.390	2.660	3.460
120	1.289	1.658	1.980	2.358	2.617	3.373
∞	1.282	1.645	1.960	2.326	2.576	3.291

Source: Abridged from R. A. Fisher and F. Yates, *Statistical Tables for Biological, Agricultural and Medical Research,* 6th ed. (London: Longman, 1974), Table III. Used by permission of Pearson Education Limited.

Appendix D
Distribution of Chi-Square

df	.99	.98	.95	.90	.80	.70	.50	.30	.20	.10	.05	.02	.01	.001 α
1	$.0^3157$	$.0^3628$	.00393	.0158	.0642	.148	.455	1.074	1.642	2.706	3.841	5.412	6.635	10.827
2	.0201	.0404	.103	.211	.446	.713	1.386	2.408	3.219	4.605	5.991	7.824	9.210	13.815
3	.115	.185	.352	.584	1.005	1.424	2.366	3.665	4.642	6.251	7.815	9.837	11.341	16.268
4	.297	.429	.711	1.064	1.649	2.195	3.357	4.878	5.989	7.779	9.488	11.668	13.277	18.465
5	.554	.752	1.145	1.610	2.343	3.000	4.351	6.064	7.289	9.236	11.070	13.388	15.086	20.517
6	.872	1.134	1.635	2.204	3.070	3.828	5.348	7.231	8.558	10.645	12.592	15.033	16.812	22.457
7	1.239	1.564	2.167	2.833	3.822	4.671	6.346	8.383	9.803	12.017	14.067	16.622	18.475	24.322
8	1.646	2.032	2.733	3.490	4.594	5.527	7.344	9.524	11.030	13.362	15.507	18.168	20.090	26.125
9	2.088	2.532	3.325	4.168	5.380	6.393	8.343	10.656	12.242	14.684	16.919	19.679	21.666	27.877
10	2.558	3.059	3.940	4.865	6.179	7.267	9.342	11.781	13.442	15.987	18.307	21.161	23.209	29.588
11	3.053	3.609	4.575	5.578	6.989	8.148	10.341	12.899	14.631	17.275	19.675	22.618	24.725	31.264
12	3.571	4.178	5.226	6.304	7.807	9.034	11.340	14.011	15.812	18.549	21.026	24.054	26.217	32.909
13	4.107	4.765	5.892	7.042	8.634	9.926	12.340	15.119	16.985	19.812	22.362	25.472	27.688	34.528
14	4.660	5.368	6.571	7.790	9.467	10.821	13.339	16.222	18.151	21.064	23.685	26.873	29.141	36.123
15	5.229	5.985	7.261	8.547	10.307	11.721	14.339	17.322	19.311	22.307	24.996	28.259	30.578	37.697
16	5.812	6.614	7.962	9.312	11.152	12.624	15.338	18.418	20.465	23.542	26.296	29.633	32.000	39.252
17	6.408	7.255	8.672	10.085	12.002	13.531	16.338	19.511	21.615	24.769	27.587	30.995	33.409	40.790
18	7.015	7.906	9.390	10.865	12.857	14.440	17.338	20.601	22.760	25.989	28.869	32.346	34.805	42.312
19	7.633	8.567	10.117	11.651	13.716	15.352	18.338	21.689	23.900	27.204	30.144	33.687	36.191	43.820
20	8.260	9.237	10.851	12.443	14.578	16.266	19.337	22.775	25.038	28.412	31.410	35.020	37.566	45.315
21	8.897	9.915	11.591	13.240	15.445	17.182	20.337	23.858	26.171	29.615	32.671	36.343	38.932	46.797
22	9.542	10.600	12.338	14.041	16.314	18.101	21.337	24.939	27.301	30.813	33.924	37.659	40.289	48.268
23	10.196	11.293	13.091	14.848	17.187	19.021	22.337	26.018	28.429	32.007	35.172	38.968	41.638	49.728
24	10.856	11.992	13.848	15.659	18.062	19.943	23.337	27.096	29.553	33.196	36.415	40.270	42.980	51.179
25	11.524	12.697	14.611	16.473	18.940	20.867	24.337	28.172	30.675	34.382	37.652	41.566	44.314	52.620
26	12.198	13.409	15.379	17.292	19.820	21.792	25.336	29.246	31.795	35.563	38.885	42.856	45.642	54.052
27	12.879	14.125	16.151	18.114	20.703	22.719	26.336	30.319	32.912	36.741	40.113	44.140	46.963	55.476
28	13.565	14.847	16.928	18.939	21.588	23.647	27.336	31.391	34.027	37.916	41.337	45.419	48.278	56.893
29	14.256	15.574	17.708	19.768	22.475	24.577	28.336	32.461	35.139	39.087	42.557	46.693	49.588	58.302
30	14.953	16.306	18.493	20.599	23.364	25.508	29.336	33.530	36.250	40.256	43.773	47.962	50.892	59.703

Source: R. A. Fisher & F. Yates: *Statistical Tables for Biological, Agricultural and Medical Research*, 6th ed. (London: Longman, 1974), Table IV. Used by permission of Pearson Education Limited.

Appendix E
How to Use a Statistical Package

by Joan Saxton Weber
with contributions by Thomas J. Linneman

The statistics and computers we use today have many roots, some of which originate in cryptography. In the decades between World War I and World War II, decoding encrypted messages evolved from something slightly more difficult than the Sunday *Times* crossword into a science based on statistics such as chi-square, which were used to compare observed versus expected frequencies of letters of the alphabet. Statistical methods such as measures of central tendency, dispersion, fit and skewness, probability, and sampling were all used to decode messages. These methods were limited because they were complex, cumbersome, and time consuming. Binary codes evolved, and many Rube-Goldberg–like devices—predecessors of today's computers—were invented to speed up the process.

In this appendix, you will not learn how to become a secret agent, but you will learn how to become a sleuth in search of social scientific meaning. To do this, you will use the Statistical Package for the Social Sciences (SPSS), one of the most popular and best-documented comprehensive social science software packages. You may have heard of other software such as Excel and Access, which also perform many statistical functions. SPSS was designed for the needs of social science research and is thus a better choice (and much simpler) for a course in this field. Because SPSS for Windows uses a graphic interface, SPSS is easier to use than to describe.

Before you begin to use SPSS, you will need to know some bare-bones aspects of the computer operating software known as Windows. Among other things, Windows is in charge of your computer's hardware, such as your CD player, your monitor, and your printer. Windows also assists software such as SPSS. So there's no way to avoid Windows. If you have had little contact with computers (let alone Windows), seek the guidance of someone who has more experience. A friend or a classmate can be very helpful. If your college offers free Windows crash courses, sign up fast! One of those "Windows in Plain English" guide books is invaluable; get a book for the version of Windows you will be using (for example, Windows 98). The first pitfalls you encounter with SPSS will probably be related to Windows. (A bonus: If you learn more about Windows for this course, it will be helpful in your other coursework.)

Don't be alarmed by the very idea of computers and statistics software. The computer is not going to self-destruct.[1] If you follow this brief summary of the basics of SPSS, you will see that the same elements are intrinsic to all SPSS procedures.

■ ■ ■ ■ Getting Acquainted with SPSS

Starting SPSS

SPSS gives you lots of hints about how to do things. (In computer jargon, the preceding sentence says: Like most programs, SPSS uses drop-down menus, Tool Bars, and dialog boxes.)

To start SPSS for Windows, click or double-click on the SPSS icon. (Clicking versus double-clicking depends on the version of Windows you are using and the way Windows was installed. Unsure? Just experiment.) If you do not see the SPSS icon (the little picture), click the *Start* button, then click on *Programs*, and then click on *SPSS*.

The program will begin to load, and soon you will see a new screen. The title in the brightly colored strip along the top of the screen says "Untitled – SPSS for Windows Student Version Data Editor."

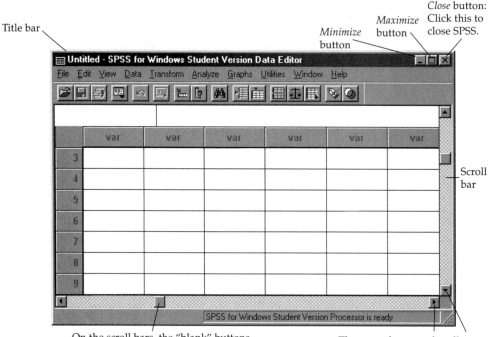

Title bar

Minimize button

Maximize button

Close button: Click this to close SPSS.

Scroll bar

On the scroll bars, the "blank" buttons enable you to move around the screen.

The arrow buttons also allow you to move around the screen.

[1]You should, however, be careful in shutting down the computer (which means shutting down Windows). If you don't know how to stop something the computer is doing—or if the computer is refusing to do something—don't just turn the computer off. If you get stuck, the best course of action is to leave your computer on until you find help, even if it takes a day or two. You should, however, turn off your monitor.

If you are using SPSS Version 8 or Version 9, a box will be superimposed on this screen. The box is titled "SPSS for Windows: What would you like to do?" For now, click on the *Cancel* button in the "What would you like to do?" box, to get it out of the way.

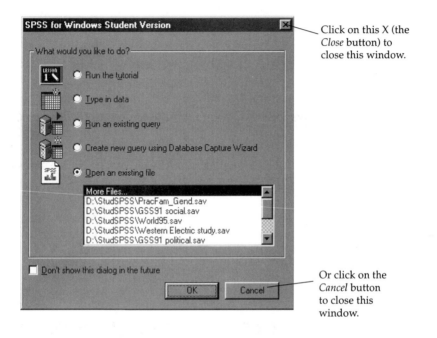

Click on this X (the *Close* button) to close this window.

Or click on the *Cancel* button to close this window.

Now you are looking at the SPSS for Windows Data Editor window. By convention, it is just called the "data window." As you can see, it is full of empty cells.

One element of most windows (such as the Data Editor window) is the symbols in the little boxes at the top right corner of your screen:

▫ is the *Minimize* button; click on it to temporarily shrink the screen so that only the program's name and icon appear at the bottom of your monitor.

▣ is the "medium-size" button; click on it to make the window fill just part of the screen.

▢ is the "full-size" button; click on it to make the window use the entire screen.

☒ is the *Close* button. Click on this to close a file or to close a program.

Getting Acquainted with Bars

Before you get your data, you should get to know the bars. These include the title bar, the scroll bars, the menu bar, and the tool bar.

You have already seen the title bar and the scroll bars in the first illustration. Scroll bars appear along the right side and bottom of the data window. Clicking on them allows you to move around the screen. Just below the title bar are the menu bar and the tool bar.

The menu bar offers you several alternatives. If you click on an item, additional selections tumble down. (The selections on the menu bar are referred to as "drop-down menus.") For example, if you click on *File*, you will see several choices for opening a new file, saving a file, closing a file, printing, and so on. If you want to create a graph, click on *Graphs*, and you will be offered many varieties.

The Menu Bar

Tool bar items also tell SPSS what to do. Tool bar items often provide quick ways to do things that would take several steps to do using the menu bar. To use a tool bar item, just click on it. (Items on the tool bar are usually called buttons—clearly a scheme to keep you confused; we will discuss a totally different kind of rectangular button below.)

The Tool Bar

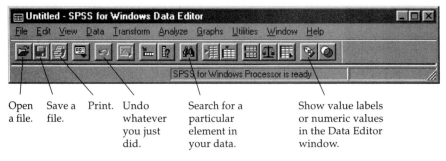

Open a file.   Save a file.   Print.   Undo whatever you just did.   Search for a particular element in your data.   Show value labels or numeric values in the Data Editor window.

Sometimes when you click on one of the tool bar items, SPSS responds with a dialog box. Dialog boxes are simply a collection of hints (or instructions) about what else SPSS needs to know in order to fulfill your command. Dialog boxes often appear when you select menu bar items or click on the tool bar buttons. For example, when you tell SPSS to print a document, a dialog box will appear.

To print via the menu bar, first select the *File* menu by clicking on File, then clicking on *Print*. The print dialog box appears. From now on, such a list of directions will appear as follows:

File → Print

Click the down arrow in the Printer
Name list box to change printers.

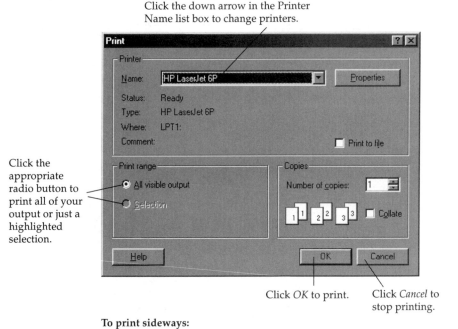

Click the
appropriate
radio button to
print all of your
output or just a
highlighted
selection.

Click *OK* to print. Click *Cancel* to
stop printing.

To print sideways:

Via SPSS:
Choose File → Page Setup. Click the radio button next to *Landscape*.

Via Printer Settings:
Click on *Properties*.
Select *Landscape*.

The print dialog box will let you print all of your output or only certain portions, and it will give you choices as to the type of printer and output you want. Decide what you want, and then click *OK*. If you know you want to print only part of the output, look for the "Print range" section of the dialog box, click on the radio button for *Selection*, and then click *OK*.

The other way to print is to click the "print" button on the tool bar. You can do this if you are certain that you want to print the entire output and if you know the printer you will be using has already been selected.

Remember that the computer will print whichever window is "active." Usually what you see on your monitor is the active window—it is what you want to print. The active window is usually the one in front of other windows, and its title is highlighted in a bright color (not "grayed out"). Printing will be discussed in more detail below; right now, just concentrate on getting used to SPSS.

Following are features that will regularly make an appearance on your monitor. Don't memorize this information. This box is a reference guide. Use it when you come upon puzzling choices as you explore SPSS.

Buttons (rectangular buttons, not the square ones on the tool bar or the scroll bar). Most buttons fall into one of the following two groups:

The "Make SPSS Do Something" buttons. Click on these to nudge SPSS into doing some actual work for you—for example:

Open
OK
Cancel
Continue
Save

The Ellipsis Buttons (buttons with three dots at the end). When you click on these, a dialog box appears.

Properties...
Options...
Statistics...

Text Boxes. These have a white "fill-in-the blank" area in which you type your choice. Press Enter on the keyboard when you are finished.

Drop-down list boxes. The same as text boxes, plus a downward-pointing arrow. Click the arrow and any available alternative choices will be shown. Highlight the choice you want. (You will see these when you open and close SPSS files.)

Check Boxes. ☐ These generally appear next to some cryptic word or phrase. You can usually check as many boxes as you like (by clicking in the box), but this is not very helpful unless you recognize what the check box is supposed to do. If you don't want SPSS to perform a certain task (especially ones you don't recognize), make sure the box beside the task is empty—that is, it doesn't have a check mark.

*Radio Buttons.** ⦿ Unlike check boxes, you can't choose more than one radio button in a group of radio buttons. If you click on one radio button, a dot will appear within the circle, and the dot from another radio button will disappear. However, on rare occasions SPSS puts several groups of radio buttons in one dialog box. So be alert when you are fooling around with the radio buttons.

*OK, we haven't seen radio buttons like these for decades, but this is what they are called in some of the more readable computer books. You can call them option buttons if you prefer.

Here is an illustration of some of these features in the print dialog box:

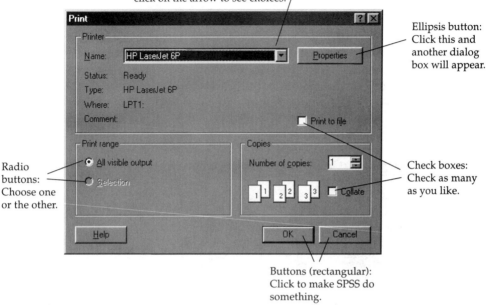

Text boxes and drop-down list boxes:
click on the arrow to see choices.

Ellipsis button:
Click this and
another dialog
box will appear.

Radio
buttons:
Choose one
or the other.

Check boxes:
Check as many
as you like.

Buttons (rectangular):
Click to make SPSS do
something.

Now that you have seen many different ways to make SPSS do things, you may be feeling a bit overwhelmed. When you first use SPSS, it may be easiest to use the menu bar items, because they display hints about how to proceed. When you become an SPSS wizard, you may like the tool bar shortcuts. SPSS is just trying to be helpful when it offers several different ways to do the same thing. You don't have to learn everything all at once.

Setting Options in SPSS

Setting options in SPSS are choices you need to make just once—and then you can forget about them. Different ways of displaying names and descriptions of variables can make using the GSS data and SPSS easier. (Remember that the term *variable* usually refers to a question in a survey.) Before Windows 95 variable names could be no more than 8 characters long. Now variables can have names or descriptions up to 25 characters long. The longer versions are often less convenient to work with, so the shorter variable names are used in this book. Short variable names can be thought of as nicknames. Nicknames such as EDUC for education are easy to spot. Others take a little getting used to, but they are descriptive and are easy to remember once you become accustomed to them. For example, ABNOMORE is pronounced Ab-No-More and is the name of a variable that asks about ABortion if a woman wants NO MORE children.

The following instructions aren't going to make much sense until you have used SPSS and the GSS data. Just make the changes below for the time being; you can always alter these choices later.

Video Display Options
You should make changes (or confirm settings) in the options that SPSS uses to display variable information on your computer monitor while you work with files. Click:

Edit → Options

General Tab showing   In Variable Lists, click the radio button next to *Alphabetical*.   In Variable Lists, click the radio button next to *Display names*.

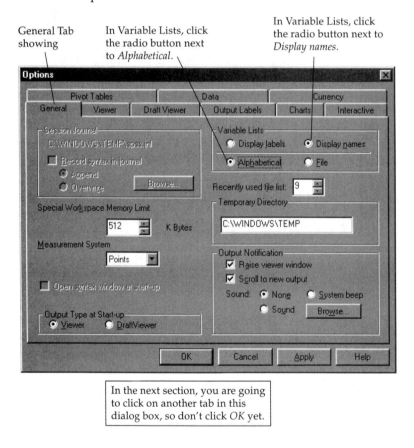

In the next section, you are going to click on another tab in this dialog box, so don't click *OK* yet.

You will see the Options window open, with a series of "tabs" along the top of the window. You should see the General options (if not, click on the "General" tab). On the upper right side of the "General" options screen you will see the *Variable Lists* options. Under Variable Lists, you must make two choices:

Choose between *Display Names* and *Display Labels*.

Choose between *Alphabetical* and *File*.

1. Click the radio button next to *Display Names* (to mark it with a small dot). This tells SPSS to display the short names of the variables.

2. Click the radio button next to *Alphabetical* (to mark it with a small dot). SPSS gives you the choice of seeing variables in alphabetic order or in the order in which they appear in the file. Listing variable names alphabetically is more convenient in most SPSS procedures.

By selecting these options, you will see the more convenient short variable names in alphabetical order on your computer monitor.

Viewer Output and Printing Options
The next change you should make in the Options dialog box is in Output labels—that is, how information will look on your monitor before it is printed and how it will look in the printed copy. If you have just finished changing the General options, you should still be looking at the Options dialog box. (If not, go to Options as you did above.) Click on the "Output Labels" tab. You will be using drop-down list boxes to make your choices.

First, click on the "Output Labels" tab.

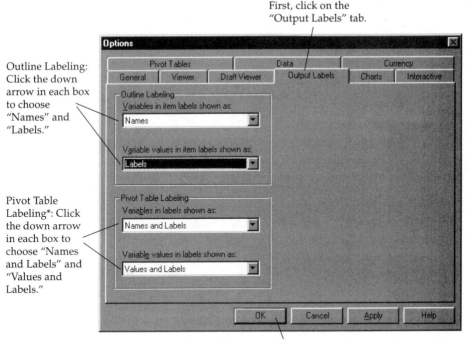

Outline Labeling: Click the down arrow in each box to choose "Names" and "Labels."

Pivot Table Labeling*: Click the down arrow in each box to choose "Names and Labels" and "Values and Labels."

Final step: Click *OK*.

*Pivot Table refers to the statistics output that you see in the Viewer and the tables that you print. The options shown above give you the maximum information about the variables in your printed output. However, long labels can be awkward in some output. When you become more familiar with SPSS and with the GSS variable names, you may prefer shorter information about the variables. If so, return to this Options dialog box and change the two options in Pivot Table Labeling to "Names" and "Labels."

The first two changes occur in the portion of the Output Labels screen called *Outline Labeling*.

1. The text box under "Variables in item labels shown as" should say "Names." If not, click on the down arrow. You will see three choices; click on the "Names" option to highlight it, and it will be placed in the text box.

2. Look at the settings for "Variable values in item labels shown as." The text box should say "Labels." If not, click on the down arrow and highlight this choice.

The next two changes occur in the portion of the Output Labels screen called "Pivot Table Labeling."

3. The text box below "Variables in labels shown as" should say "Names and Labels." If not, choose this option. This way, your output will contain both the short variable "nicknames" and the longer descriptions of the variables.

4. Look at the settings for "Variables values in labels shown as." "Values and Labels" should be shown in this rectangle. If not, choose this option. This tells SPSS that you want to see both the numeric codes (values) for your variables as well as the labels assigned to those numbers in all of the output you receive.

After you have made these Options changes, click *OK* and you will return to the SPSS data window. By selecting these options, your printed output will contain the maximum amount of information for each variable you use. This much information can, however, be cumbersome, especially in tables. When you have more experience with both SPSS and the GSS variable names, you may decide to change these options.

■ ■ ■ ■ **Data Files**

Getting Data

The General Social Survey (GSS) is a large annual survey of Americans. Excerpts from the 1996 GSS are included on the disk that accompanies this book. In this appendix, you will be using examples from a practice dataset that focuses on family and gender. The name of the dataset is PracFam_Gend.sav. If you have installed SPSS on your own computer, you will need to copy the data files onto your hard disk. SPSS dataset files always end with the extension .sav, so they are easy to recognize. SPSS will be much easier to use if you copy the dataset files into the SPSS subdirectory. (If the above all sounds like gibberish, seek help from anyone who is familiar with Windows; that person doesn't need to be familiar with SPSS.) Once the files are on the computer you are using, start SPSS and click

File → Open

This takes you to the Open File dialog box. If you see the name of the file you want, click on the name to highlight it, then click *OK*. (Search both your floppy drive and your hard disk drive if necessary.)

STEP ONE: Look for the data file you want. If the file is not shown, use the drop-down list arrow to tell the computer where to look for the file you want—on the floppy disk drive (drive A), the hard disk (drive C), etc.

STEP TWO: Click on the name of the file you want to open.

STEP THREE: Click the *Open* button to actually open the file.

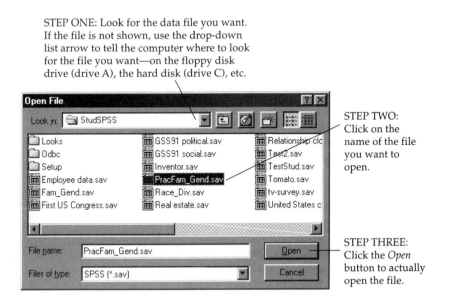

SPSS fetches the file and puts it into the data window. You will see many variables and their numeric values. Although the variables will be in a somewhat different order, the appearance of your screen will resemble the illustration below.

Data Window: Numeric Values

Is respondent 1 really 99 years old?

The data window is divided into rows and columns. Each row represents a person (a respondent to the survey, a "case"). The dataset you will be using has 1,427 rows, which means that there are 1,427 respondents. Each column represents a variable; at the top of each column is the name of the variable. The variable AGE is simply the age of the respondents. You can move around within the dataset in a number of ways. With the mouse, you can move the scroll buttons on the right-hand and lower margins of the window. With the keyboard, you can use the arrow keys and the Page Up and Page Down keys.

Looking at the Data

Is respondent 1 really 99 years old? Good question, and we will discuss the contextual meaning of these numbers.

All the answers to each GSS question have been coded. That is, for each variable (question), each possible response has been assigned a number (referred to as a value). For some variables, such as AGE and EDUC, the number is simply the number of years. For other variables (such as RACE and SEX), each number corresponds to a particular response or category. For example, for the variable SEX, men were coded with the number 1, and women were coded with the number 2. This works well for the computer, but it can be confusing to people who use the datasets. Therefore, SPSS allows space for a brief description of the meaning of the number. Each non-intuitive number is given a label. These labels are called value labels because, of course, they describe the value (the response category, the answer given for the question).

Value labels and numeric value codes are one of the most striking visual differences between SPSS version 9.0 and much earlier versions; however the meaning and categories of value labels and codes are the same in all versions of SPSS. There are several ways of locating the numeric codes, value labels, and other important information about the variables.

1. The usual way is to look up the variable in a codebook—in this case, the GSS Codebook. (Mini-codebooks for the datasets you will be using are included in the appendices of the GSS codebook.) A codebook is similar to a dictionary. It contains pertinent information about each variable in the dataset. You could look up, for example, the variable ABANY (approval of abortion for any reason) in the GSS codebook or the GSS web site. You would then discover how ABANY was coded by the GSS researchers (for example, a code of 2 means "No").

2. What if you don't have a codebook? Within SPSS it is quite easy to find out what the numeric values mean. One way to look up value labels in SPSS is by using the Variables dialog box. To get to this dialog box, click:

   Utilities → Variables

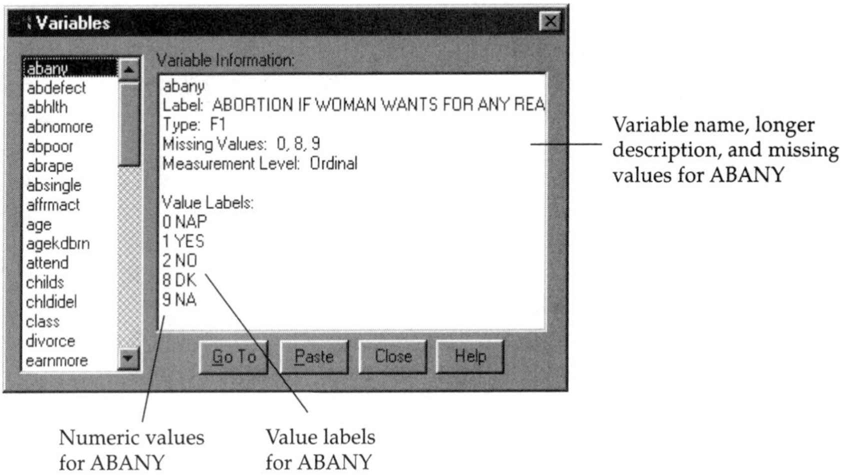

Variable name, longer
description, and missing
values for ABANY

Numeric values
for ABANY

Value labels
for ABANY

Respondent 4 has a code of 2 for ABANY. From the Utilities → Variables command illustrated above, you can see that 2 means "No." The respondent does not approve of abortion for any reason.

3. A way to see the value labels in the Data Editor window is to click on the tool bar button that looks like a price tag (it is the second-to-last button). When you click on this, the numbers in the Data Editor window will change to their respective labels. Or, if the labels are already displayed, clicking on this button will turn them back into numbers.

Data Window: Value Labels

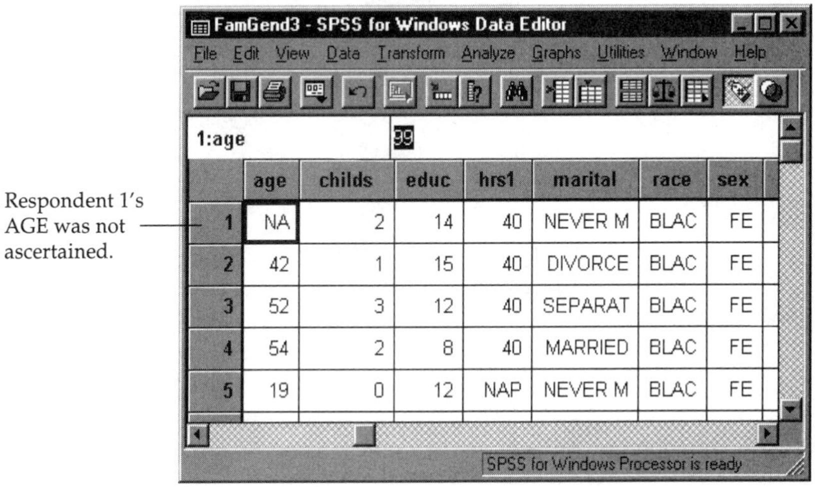

Respondent 1's
AGE was not
ascertained.

So, is respondent 1 really 99 years old? The value label says 99 means "NA." What is "NA"? The answer involves *missing values*, which are codes determined by the researcher to refer to missing data. The code 99 describes people who did not want to give their age. The value label "NA" stands for "Not Ascertained/No Answer."

What can the value labels tell us about the respondents? Respondent 4 is a married black woman who is 54 years old, works full-time (HRS1), and has 2 children (CHILDS). Respondent 5 is a 19-year-old unmarried black woman who is not currently employed (HRS1 is coded as NAP—not applicable) and has no children.

If you explore the characteristics of other respondents, you will quickly become familiar with what the numbers reveal.

Data Editor Window: Numeric Values and Value Labels

Respondent 9 was not asked these questions; this is coded as NAP (missing data).

Respondent 12 said "don't know"; this response is coded as missing data.

Let's compare the attitudes of respondents 9 through 12 on abortion. Respondents 10 and 11 each oppose abortion if a woman is poor (ABPOOR). However, their responses differ on the variable ABNOMORE (abortion if a woman doesn't want any more children). Respondent 10 agrees with abortion under these circumstances. Respondent 11 opposes abortion under these circumstances. Respondent 12's answer is recorded as 8 on ABNOMORE, and Respondent 9 was given a code of 0 for each question.

The numbers 0 and 8 again involve missing values. In a lengthy survey, not everyone will be willing to answer everything. To deal with such a situation, the GSS interviewer can pencil in responses such as "don't know" or "no answer," and for that one respondent, data on that question will be considered "missing." If you look at the value labels for ABNOMORE and ABPOOR (using the Utilities → Variables command) you will see that there are three missing values: 0, 8, and 9. All of the "missing" codes are known collectively as *missing data*.

The code 8 was given to people who said "don't know" (DK) when they were asked the question; and 9 was the code given to people who did not want to answer the question (NA).

The code 0 was given to people (such as respondent 9) who weren't even asked the question. Respondents to the GSS are asked only certain sets, or modules, of questions. Since they have no answer that can be coded, they are given a code of Not Applicable (NAP) for the questions that were not asked of them. Notice that 0 has a different meaning for these questions than it does in other questions. For variables such as CHILDS (number of children), 0 can be a valid response because it means zero children. For other questions—usually questions in which zero is not a valid response—0 is used to indicate NAP (inapplicable) or that these respondents were not asked the question.

What about the code of –1 on HRS1 (number of hours R worked last week) given to respondent 5 (shown in the earlier examples), which is also labeled NAP? Zero could be a valid answer for someone who is employed but is on vacation. So the GSS researchers use this distinctive code for respondents for whom this question is not appropriate (in this example, respondents who are not employed).

In general, you should watch for numbers that are often used for missing values and make sure missing values are not inadvertently included in your analysis. As you have seen, these numbers are –1, 0, 8, and 9. (If the variable is two digits wide, such as AGE, the missing codes are usually 97, 98, and 99.)

In calculations of statistics, SPSS takes those cases that had missing data "out of the equation" and calculates the statistics based on data for the respondents who answered the questions. Five respondents had missing data on AGE, so SPSS calculates the mean age of "all" the respondents—that is, for 1,422 of the 1,427 respondents. Some questions have more missing data than others, particularly questions that are of a sensitive nature or questions that were not asked of the entire sample.

To see and print responses of individual cases for a subset of variables, you could use the Summarize Cases procedure (formerly known as the List Cases procedure) by clicking Analyze → Reports → Case Summaries. This feature of SPSS was essential in earlier versions of SPSS where the variable values were encoded in hexadecimal (which was unreadable to users). If you want to have a printed copy of this information, the Case Summaries command is still useful. But for our purposes, it may be more instructive for you to view the data on your monitor in the manner illustrated above.

You have seen a number of SPSS procedures. Some instructors suggest that students make a short list of essential SPSS commands. Such a list should include the following commands:

Opening a file	File → Open → Filename
Closing a file	File → Close
Printing a file	File → Print

■ ■ ■ ■ Univariate Statistics

Frequencies

So far we have come dangerously close to doing actual research. Now you can tell SPSS to do some real work. The variable ABANY (approval or disapproval of abortion for any reason) will serve as a good example for a frequency distribution. A frequency distribution tells you how many people said yes, how many people said no, and so on. (You wouldn't want a frequency distribution for a variable such as age, which has a huge number of categories.) Click:

Analyze → Descriptive Statistics → Frequencies

STEP ONE: Click on the variable of interest (ABANY is chosen here).

STEP TWO: Click on the arrow button. This will move the highlighted variable into the Variable(s) box.

STEP THREE (optional): Click on *Statistics....*

FINAL STEP: Click *OK.* (The *OK* button is "grayed out" until there are items in the Variable(s) box.)

(Optional) STEP FOUR: Click to put checkmarks in the boxes next to the statistics you want—for example, Mean, Median, and Mode.

(Optional) STEP FIVE: When you are finished, click on *Continue*. The first dialog box will reappear.

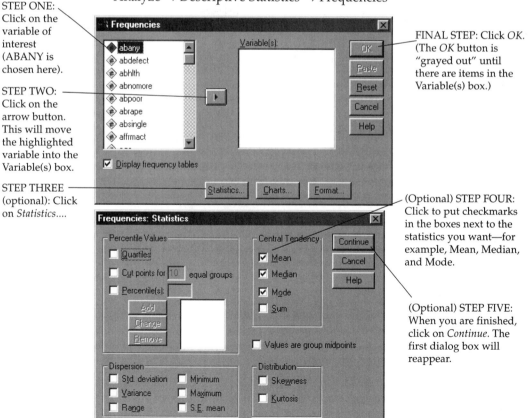

Here is what happens when you click *OK*. Scrolling down the output window below the Statistics (the *n* and missing data), you can see the Frequency Distribution.

Frequency Distribution for ABANY

The left side is called the Outline pane, a list of everything you have asked SPSS to do.

The right side shows the actual results of the computations SPSS made. It is the Output pane, but most people just call it the output.

Use the scroll bar to view all the output.

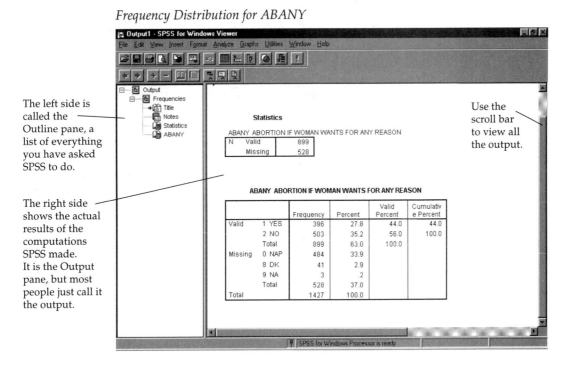

If you have requested the mean, median, and mode, these will appear in the statistics part of the window.

Statistics

ABANY ABORTION IF WOMAN WANTS FOR ANY REASON

N	Valid	899
	Missing	528
Mean		1.56
Median		2.00
Mode		2

After SPSS has processed the command, it whisks you into a new window called "Output 1 - SPSS for Windows Viewer." In the Output Viewer, you can see all of the results of the frequencies command. The Output Viewer has two panes. At the top of the left pane you can see the word *Outline*. To avoid confusion, this pane is referred to as the Outline pane. (Some of the default options for the outline, which you changed previously, affect the way variables and value labels are shown in this pane.) The Outline pane is simply a list of everything you have asked SPSS to do since you began the session. It is like a table of contents. Imagine you have been working for a couple of hours, and you have asked SPSS to do all kinds of things.

You have scores of tables, graphs, and statistical models to reconnoiter. To go to a specific graph, you find it in the Outline pane, click on it there, and the graph is displayed in the right pane.

The right pane is the actual output for the commands you gave SPSS. It is often just called the "output" or the "results." (Again, some of the options for output, which you changed previously, affect the way variables and value labels are shown in this pane.)

In the example above, SPSS calculated a frequency distribution for the variable ABANY. If the entire table is not visible, you can move around in the output window with the arrow keys and the scroll buttons. When you become more accustomed to SPSS, you may want to use the Zoom/Print Preview feature.

At this point, you could print your output—that is, the right pane. Remember, the computer will print whichever window is active. Therefore, if you want to print your output, make sure the output window is the active window. (If the data window is the active window, you will end up with a paper copy of the data window—a stack of paper that could be several inches high!)

If you want to save your output, refer to that section below.

The SPSS Version 9 output is a bit different from the output in some earlier versions, but the information about the distribution and statistical measures for the variables is the same.

■ ■ ■ ■ Data Transformation

Recoding Variables

There are times when variables with many categories are confusing to use and interpret. The variable EDUC (years of education) is a good example. You seldom want to know if people with 11 years of education are different from people who have 10 years of education. It might be better to examine differences of opinion among high school graduates versus college graduates.

To create the groups you want, you will need to *recode* the variable. Remember all the information earlier in this appendix about missing values? Here such information is important. Before you recode any variable, it is a good idea to check the missing values and to look at the minimum and maximum values for valid responses.

Let's say you want to recode EDUC (highest year of education completed) into the following five categories:

Those who have 0 to 11 years of school are put into category 1

Those who have exactly 12 years of school are put into category 2

Those who have 13 to 15 years of school are put into category 3

Those who have exactly 16 years of school are put into category 4

Those who have 16 to 20 years of school are put into category 5

This recode looks sensible for the United States, given our knowledge of education levels in the United States. If you did not know the distribution of a variable, you would first use the Frequencies command to determine the range of values and some other valuable information such as percentiles, quartiles, and so on. You don't want to reclassify some variable into five categories and then discover, for example, that there are only two people in the third category!

In this instance, you will collapse the original 21 categories (years of education from 0 to 20) into 5 categories. You are going to accomplish this task by creating a new variable called ED5CAT. Click:

Transform → Recode → Into Different Variables

This takes you to the first recode dialog box, in which you tell SPSS the name of the old variable you are recoding from (EDUC) and the name of your new recoded variable (ED5CAT).

STEP ONE: In the list of variables, highlight the variable you want to recode.

STEP TWO: Click the arrow to put it in the Numeric Variable → Output Variable box.

STEP THREE: Type the name of your new variable and give it a label.

STEP FOUR: Click the *Change* button.

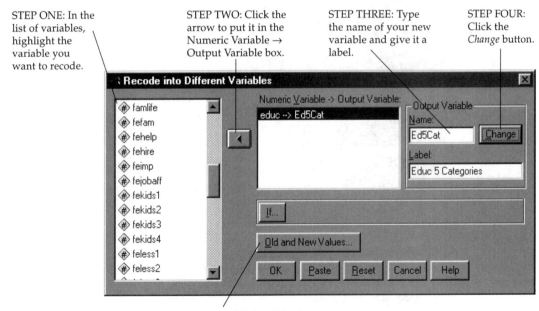

STEP FIVE: Click the *Old and New Values...* button to go to the second dialog box.

After doing this, you go to the second dialog box, in which you put the original ("old") values of the variable on the left side of the box, and you put the values for your new variable on the right side of the box. There is a lot going on in this dialog box, so we will take a close look at the steps involved.

STEP ONE:

Tell SPSS which old values are to be recoded (e.g., 12).

STEP TWO:

Tell SPSS a range of old values that are to be collapsed (recoded) into a single category.

STEP THREE: Recode missing values.

Click the radio button next to "All other values."

If you get this dire-looking error message, it just means you forgot to click the *Add* button next to the Old → New box.

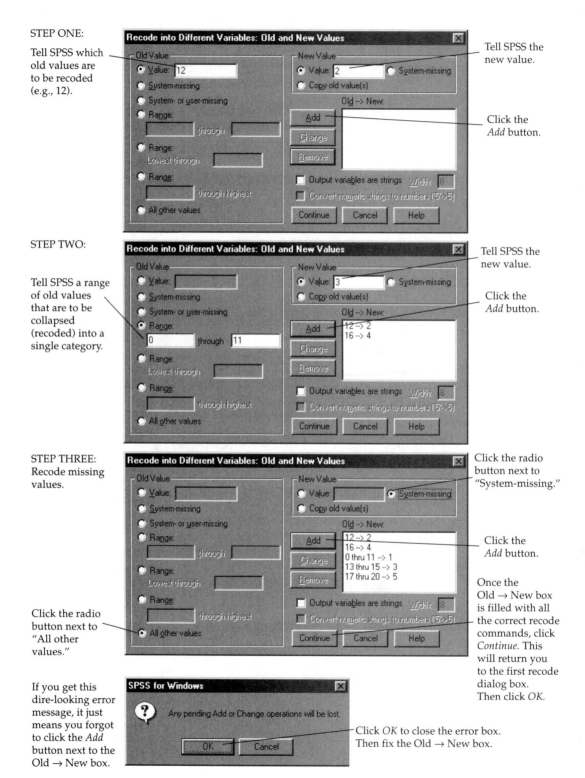

Tell SPSS the new value.

Click the *Add* button.

Tell SPSS the new value.

Click the *Add* button.

Click the radio button next to "System-missing."

Click the *Add* button.

Once the Old → New box is filled with all the correct recode commands, click *Continue*. This will return you to the first recode dialog box. Then click *OK*.

Click *OK* to close the error box. Then fix the Old → New box.

Here are results for the first five respondents for your recoded education variable.

GSS Respondent Number	Old Value on Variable EDUC	New Value on Variable ED5CAT
1	14	3
2	15	3
3	12	2
4	8	1
5	12	2

The recode appears to be working correctly. Respondent 1 attended school for 14 years and is put into the group that have some college education. Respondent 4 was put into the group of those who don't have a high school education.

Always check your recodes. Cross-tabulating the original variable with your new variable is the best way. If crosstabs are impractical (as is the case for variables with, for example, 100 categories), then looking at individual cases is especially necessary. You should check a sufficient number of cases to be convinced that your recode is working the way you expected. If the recoded variable will be the basis of your analysis, you don't want to discover a big mistake after you have been using the variable for hours!

Here is the frequency distribution generated by SPSS:

Frequency Distribution for Recoded Education (Ed5Cat)

ED5CAT Educ 5 Categories

		Frequency	Percent	Valid Percent	Cumulative Percent
Valid	1.00	232	16.3	16.3	16.3
	2.00	418	29.3	29.5	45.8
	3.00	399	28.0	28.1	73.9
	4.00	183	12.8	12.9	86.8
	5.00	187	13.1	13.2	100.0
	Total	1419	99.4	100.0	
Missing	System	8	.6		
Total		1427	100.0		

As you can see, there are no value labels. Someone who was handed this output would have no way of knowing what a 1 or 2 means. To make this output easier to read and remember, you should attach value labels to the numeric codes.

Defining Variables

Click over to the data editor window and go to the right-most column (this is where SPSS puts all the new variables you create). Click on the <u>name</u> of the new variable ED5CAT to select it. From the Menu Bar, click:

Data → Define Variable

The Define Variable dialog box appears:

Click the *Labels...* button to label the values in your new variable. This will take you to a second dialog box, shown below.

Define *Missing Values...* if necessary.

Click *Type...* and/or click *Column Format* to eliminate decimal values or change the format in other ways.

In this dialog box you can modify your variable in a number of ways. You can check the missing values SPSS has assigned to your variable and make sure SPSS did its work correctly. (You don't want SPSS thinking, for example, that someone is 99 years old!)

To describe the numeric values of your variable, click the *Labels...* button.

STEP ONE: Put the numeric value you want to label in the "Value" box.

STEP TWO: Put the value label you want in the "Value Label" box.

STEP THREE: Click the *Add* button.

STEP FOUR: When your list is complete, click the *Continue* button to return to the former dialog box, and then click *OK*.

Computing a New Variable

Recoding does not combine two variables. How can this be done? There are other commands that modify data. The Compute command works well for combining two variables and for performing mathematical transformations. The GSS data contain the educational levels (in years) of the respondents' mothers (MAEDUC) and fathers (PAEDUC). Suppose you wanted to calculate the average (mean) level of education for MAEDUC and PAEDUC. For example, if the father had a high school education and the mother had a college education, the mean for the two would be 14 years of education: $(12 + 16)/2 = 14$. Use the Compute command to create a new variable, PAREDUC, which combines two variables and computes the average. Click:

Transform → Compute

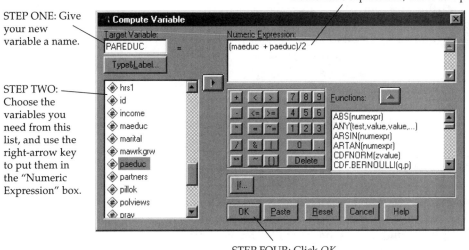

STEP THREE: Using the correct variables and mathematic expressions, create an equation.

STEP ONE: Give your new variable a name.

STEP TWO: Choose the variables you need from this list, and use the right-arrow key to put them in the "Numeric Expression" box.

STEP FOUR: Click *OK*.

Here are some examples of scores on the new variable.

GSS Respondent Number	MAEDUC	PAEDUC	New Variable: PAREDUC
1	13	97	•
2	11	10	10.5
3	9	97	•
4	98	98	•
5	11	16	13.5

The Compute command appears to be working correctly. PAREDUC has missing data for respondents 1, 3, and 4 because for one or both of the parents, the years of education was missing. (The symbol • is often used by SPSS for missing data.)

The mean was correctly calculated for respondents 2 and 5. Because you've really checked the only two cases that did not have any missing data, you would want to look at more cases before you spend hours on research using this variable.

■ ■ ■ ■ Saving and Retrieving Files in SPSS

Finding a Place to Put Your Files

In SPSS student version datasets (*.sav files), the maximum number of variables allowed is 50. Some of the datasets on the disk that accompanies this book are quite full. If you are trying to save a new variable (not output) and this problem arises, talk to your instructor.

Before you attempt to save your work, it is a good idea to be certain there is room on the hard disk or floppy disk you are using. When in doubt, have a spare floppy disk handy.

Saving SPSS Data Files ("System Files" with the Extension *.sav)

You will probably want to save the new variables you have created. Otherwise, the next time you use SPSS, you will have to start all over. On the menu bar, click:

File → Save As

The Save As dialog box opens. It should be similar to the File → Open dialog box shown at the beginning of this appendix.

If you are saving your file to your hard disk, your file should be saved in SPSS or in My Documents. If you have problems, talk to someone who is familiar with Windows. Be certain that your new variables have descriptive names and that they do not have the same names as any of the original, older variables. To minimize confusion if others are using the same file, give the dataset a new name and click the *Save As* button. If you get an error message saying that you have exceeded the maximum number of variables, talk to your instructor.

Saving SPSS Viewer Output Files ("Output Files" with the Extension *.spo)

What if you want to save your output? It is similar to saving SPSS data files. Be sure the output is the active window (if not, make it the active window). On the menu bar, click:

File → Save As

STEP ONE: On the SPSS menu bar, click File → Save As. The Save As dialog box will appear.

STEP TWO: Look in the "Save in" list box to see where your file is being saved. It should be in SPSS or My Documents.

STEP THREE: Output1.spo (or Output2, etc.) will appear as the file name in the "File name" text box. Change the file name to something you will remember, such as ABANYFreq.spo.

Save As

Save in: Spss

Acrobat
Looks
Odbc
sav_egs
Scripts
Setup

File name: Output1.spo

Save as type: Viewer Files (*.spo)

Save

Cancel

Don't change the "Save as type" list box. SPSS output should be saved as Viewer Files (with the *.spo extension).

STEP FOUR: Click the *Save* button.

If you are saving your output file to your hard disk, your file should be saved in SPSS or in My Documents. (If you have problems, talk to someone who is familiar with Windows.)

The bottom box—the "Save as type" drop-down list box—should now say "Viewer Files (*.spo)." This is correct for output files, so don't change this.

Output1.spo (or Output2.spo, and so on) will appear in the "File Name" text box. If you save the file with this name, the next time you open SPSS and you save your work, the new file will also be called Output1.spo by default; if you save new output under this name, you will erase the old file. To avoid this, always change the name of the file you are saving to something you will remember—for example, ABANYFreq.spo.

Click *Save* to actually save the file.

Opening Viewer Output Files

On the SPSS menu bar, click

File → Open

SPSS is trying to be helpful, and it wants to open a data file for you. Therefore the "File of type" drop-down list box says SPSS (*.sav). You need to locate Viewer output files. To do this, click on the arrow in the "Files of type" box, and a drop-down list will appear. Highlight "Viewer document (*.spo)."

STEP ONE: On the SPSS menu bar, click File → Open.
The Open File dialog box will appear.

STEP TWO: Check the "Look In" list box and make sure it shows the location where you saved your output files. If that location is not shown, click the arrow, and the drop-down list will show other possible locations for your file.

STEP THREE: To find Viewer output files, click on the arrow in the "Files of type" list box. From the drop-down list, highlight "Viewer document (*.spo)."

When you do this, the contents of the Open File dialog box will change. No data files (that is, files with the extension *.sav) will be shown. Instead, you will see a list of Viewer output files (files that end with *.spo). Highlight "ABANYFreq.spo" (or any other Viewer output file you want to open), and click the *Open* button. The Viewer output file will appear on your monitor.

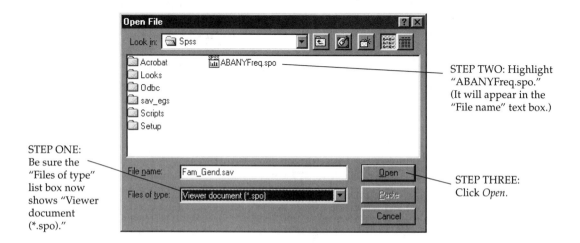

STEP ONE: Be sure the "Files of type" list box now shows "Viewer document (*.spo)."

STEP TWO: Highlight "ABANYFreq.spo." (It will appear in the "File name" text box.)

STEP THREE: Click *Open*.

■ ■ ■ ■ **Bivariate and Multivariate Statistics**

Cross-tabulation

Social scientists are interested in going beyond univariate statistics; they look for the explanations for relationships between two or more variables. Do you think that people's attitudes toward abortion depend on the number of years they attended school? And if so, do you think education increases or decreases support for abortion?

Let's examine the relationship between education (ED5CAT, the categorized measure of education you created earlier in this appendix with the Recode command) and attitudes toward abortion for single women (ABSINGLE). Run the Crosstabs procedure, with ED5CAT as the independent variable and ABSINGLE as the dependent variable. Click:

Analyze → Descriptive Statistics → Crosstabs

STEP TWO: Put your independent variable in the "Columns" box.

STEP ONE: Put your dependent variable in the "Row(s)" box.

STEP FIVE: When everything is the way you want it, click *OK*.

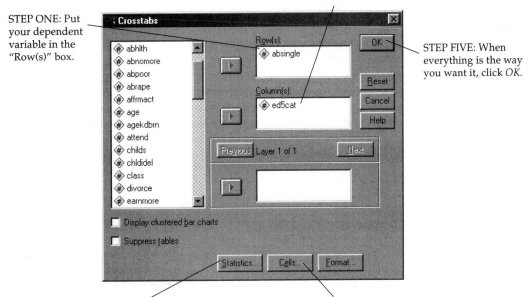

STEP THREE: Click *Statistics...* to choose which statistics you would like to run (in this case, we chose chi-square).

STEP FOUR: Click the *Cells...* button and choose column percentages.

In the dialog box, click on *Statistics...*, make appropriate choices (this example uses chi-square, which is discussed in Chapter 14), and click *Continue*. Also click on *Cells...* to get column percentages in addition to the cell counts. Then click *Continue* to return to the first dialog box. Click *OK* to obtain the output.

ABSINGLE NOT MARRIED * ED5CAT Educ 5 Categories Crosstabulation

			1 LT HS	2 High School	3 Some Col	4 Col Grad	5 Post Col	Total
			ED5CAT Educ 5 Categories					
ABSINGLE NOT MARRIED	1 YES	Count	54	89	107	73	68	391
		% within ED5CAT Educ 5 Categories	35.3%	34.1%	44.0%	59.3%	59.1%	43.7%
	2 NO	Count	99	172	136	50	47	504
		% within ED5CAT Educ 5 Categories	64.7%	65.9%	56.0%	40.7%	40.9%	56.3%
Total		Count	153	261	243	123	115	895
		% within ED5CAT Educ 5 Categories	100.0%	100.0%	100.0%	100.0%	100.0%	100.0%

Chi-Square Tests

	Value	df	Asymp. Sig. (2-sided)
Pearson Chi-Square	37.558[a]	4	.000
Likelihood Ratio	37.638	4	.000
Linear-by-Linear Association	32.163	1	.000
N of Valid Cases	895		

a. 0 cells (.0%) have expected count less than 5. The minimum expected count is 50.24.

Education appears to be a strong influence on attitudes toward abortion. As level of education increases, people are much more likely to support abortion. The chi-square test tells us that the relationship is unlikely to have occurred by chance.

Regression

What if you wish to examine the relationship between two (or more) variables that have ordered categories, that have too many categories to be suitable for crosstabs, and that you believe have a linear relationship? One good solution is to use the Regression command. Regression produces an equation in which scores on the X (independent) variable are used to predict scores on the Y (dependent) variable. Regression augments this algebraic formula with statistical techniques for adjusting the prediction line and assessing how good the prediction is; these techniques are explained in Chapter 8.

As one example, it seems very likely that parents' education influences the amount of child's education. Use the combined education of parents, PAREDUC (the variable you created earlier in this appendix using the Compute command) as the independent variable. Use EDUC (respondent's education) as the dependent variable. Click:

Analyze → Regression → Linear

STEP ONE: Put your
dependent variable here.

STEP TWO: Put your independent
variable(s) here.

STEP THREE:
Click *OK*.

Linear Regression

feless2	
feless3	
feminist	
fepres	
homosex	
hrs1	
id	
income	
maeduc	
marital	
mawrkgrw	
paeduc	
partners	
pillok	
polviews	
pray	

Dependent:
educ

Block 1 of 1

Previous   Next

Independent(s):
paeduc

Method: Enter

Selection Variable:

Rule...

Case Labels:

OK
Paste
Reset
Cancel
Help

WLS >>   Statistics...   Plots...   Save...   Options...

Here are some of the statistics SPSS calculates.

Child's Education by Parents' Education

Model Summary

The R², or
proportion
of variance
explained,
is .183.

Model	R	R Square	Adjusted R Square	Std. Error of the Estimate
1	.428[a]	.183	.182	2.49

a. Predictors: (Constant), PAREDUC

ANOVA[b]

The number of
cases is 939 + 1
+ 1 = 941.

Model		Sum of Squares	df	Mean Square	F	Sig.
1	Regression	1312.573	1	1312.573	210.857	.000[a]
	Residual	5851.445	940	6.225		
	Total	7164.018	941			

a. Predictors: (Constant), PAREDUC

b. Dependent Variable: EDUC HIGHEST YEAR OF SCHOOL COMPLETED

Coefficients[a]

Model		Unstandardized Coefficients		Standardized Coefficients	t	Sig.
		B	Std. Error	Beta		
1	(Constant)	9.792	.297		32.944	.000
	PAREDUC	.357	.025	.428	14.521	.000

a. Dependent Variable: EDUC HIGHEST YEAR OF SCHOOL COMPLETED

Using this information, we can
write the regression equation:
EDUC = 9.79 + .36(PAREDUC)

The slope for any PAREDUC
is highly significant.

Statistics run rampant! Parents' education explains about 18 percent of the variation in respondents' education. This is quite a respectable percentage in social science research, so it is clearly worth our attention. For each additional year of parental education, respondents' education is predicted to rise by over one-third of a year (.36). If each of your parents has a college degree, your predicted educational attainment is 15.6 years of education:

EDUC $= a + b$(PAREDUC)
15.55 $= 9.79 + .36(16)$

Comparing Means

The t test (discussed in Chapter 13) is a statistical procedure that is useful when the goal is to compare the means of two groups within categories of a predictor variable.

The dependent variable must have ordered categories.[2] Perhaps we are interested in the difference between incomes of women and men or in whether Catholics are more opposed to abortion than are Protestants.

Let's investigate how husbands and wives answer the following question: How much housework does the respondent's spouse do (SPHMEWRK)? Answers can range from 1 to 5, with 1 meaning "all" and 5 meaning "almost none." Click:

Analyze → Compare Means → Independent-Samples T test

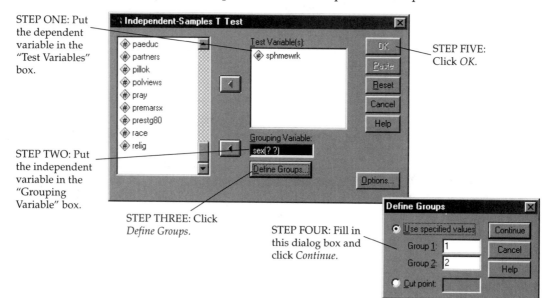

STEP ONE: Put the dependent variable in the "Test Variables" box.

STEP TWO: Put the independent variable in the "Grouping Variable" box.

STEP THREE: Click *Define Groups.*

STEP FOUR: Fill in this dialog box and click *Continue.*

STEP FIVE: Click *OK.*

[2]In an ideal situation, the dependent variable would have "ratio" or "interval" data. In the real world, the majority of survey and other research questions are ordinal-level data. After checking the distribution of ordinal variables, most social researchers treat them as interval data when they are using the t test for comparison of means. SPHMEWRK is an ordinal variable that is treated as interval for this example.

Directly after the variable you chose to put in the "Grouping Variable" box, you will see the following:

(? ?)

This is SPSS's way of asking you for more information. The information SPSS wants is which two groups you want to compare. With some variables, this will seem superfluous because there are only two possible groups (that is, the variable has only two categories, such as gender). But for other variables, this step is clearly necessary. If you choose the variable MARITAL, you still need to tell SPSS whether you want to compare people who are married versus those who are divorced, married versus never married, and so on. So fill in the groups, and click *OK*.

How Much Housework Does Your Spouse Do?

Group Statistics

	SEX RESPONDENTS SEX	N	Mean	Std. Deviation	Std. Error Mean
SPHMEWRK How much housework R's spouse does	1 MALE	175	2.83	.80	6.07E-02
	2 FEMALE	203	3.74	.92	6.46E-02

Independent Samples Test

		Levene's Test for Equality of Variances		t-test for Equality of Means						
									95% Confidence Interval of the Difference	
		F	Sig.	t	df	Sig. (2-tailed)	Mean Difference	Std. Error Difference	Lower	Upper
SPHMEWRK How much housework R's spouse does	Equal variances assumed	15.952	.000	-10.102	376	.000	-.90	8.95E-02	-1.08	-.73
	Equal variances not assumed			-10.205	375.944	.000	-.90	8.86E-02	-1.08	-.73

On average, men report that wives score 2.8 on housework (just under half). Women report that husbands average 3.7 ("some") on housework. So if women do just under half, and men do far less. . . . Use your imagination to come up with a research hypothesis that explains what is going on here.

In general, interviewers report that respondents try very hard to answer all of the questions. Detailed analyses of questionnaires supports this anecdotal evidence. But there are obviously a few situations in which (un)intentional bias influences the answers given by respondents.

■ ■ ■ ■ **Conclusion**

If you have reached this point, you should be able to understand a box or a bar, and you will not be intimidated by words such as *frequency distribution* and *crosstab*. Everyone runs into problems, but you can minimize these if you do the following.

- *Write down the error message you are getting.* That way the computer folks know you are trying, and they can actually help you.

- Try to determine whether your problem is a Windows problem or a computer hardware problem. Your college probably offers computer lab technical support. Introductory books for Windows are an excellent backup because they typically describe problems common to Windows and to many software programs.

- Try to determine whether your problem is an SPSS problem. There are many sources of assistance out there: the Help drop-down menu in SPSS, the SPSS manual, your campus computer lab consultant, your instructor, and quite possibly your classmates.

You won't learn SPSS overnight, but after you have done some of the exercises in this book by hand, you may begin to appreciate commanding SPSS to do the computations for you. Think of the computer as an accomplice who helps you complete your assignments. The computer will never, never add 2 + 2 and get 5!

When you use the General Social Survey and SPSS, have fun investigating questions about the diverse society in which we live.

Appendix F
The General Social Survey

The General Social Survey (GSS) has been conducted annually since 1972. Conducted for the National Data Program for the Social Sciences at the National Opinion Research Center at the University of Chicago, the GSS was designed to provide social science researchers with a readily accessible database of socially relevant attitudes, behaviors, and attributes of a cross section of the U.S. population.

The datasets accompanying this book are based on the 1996 GSS. A total of 2,904 surveys were completed for the 1996 survey. We randomly selected a sample of 1,427 cases for our datasets. The data, obtained through a sampling design known as multistage probability sample, are representative of Americans 18 years of age or older. This means that the GSS dataset allows us to estimate the characteristics, opinions, and behaviors of all non-institutionalized, English-speaking, American adults in a given year.

Selected survey items are divided into two subsets: family/gender and race/diversity. GSS items include attitudes toward the role of women, abortion, premarital sex, homosexuality, affirmative action, immigration, and racial/ethnic diversity.

Appendix G
A Basic Math Review

by James Harris

You have probably already heard that there is a lot of math in statistics and for this reason you are somewhat anxious about taking a statistics course. Although it is true that courses in statistics can involve a great deal of mathematics, you should be relieved to hear that this course will stress interpretation rather than the ability to solve complex mathematical problems. With that said, however, you will still need to know how to perform some basic mathematical operations as well as understand the meanings of certain symbols used in statistics. Following is a review of the symbols and math you will need to know to successfully complete this course.

Symbols and Expressions Used in Statistics

Statistics provides us with a set of tools for describing and analyzing *variables*. A variable is an attribute that can vary in some way. For example, a person's age is a variable because it can range from just born to over one hundred years old. "Race" and "gender" are also variables, though with fewer categories than the variable "age." In statistics, variables you are interested in measuring are often given a symbol. For example, if we wanted to know something about the age of students in our statistics class, we would use the symbol Y to represent the variable "age." Now let's say for simplicity we asked only the students sitting in the first row their ages—19, 21, 23, and 32. These four ages would be scores of the Y variable.

Another symbol that you will frequently encounter in statistics is Σ, or uppercase sigma. Sigma is a Greek letter that stands for summation in statistics. In other words, when you see the symbol Σ, it means you should sum all of the scores. An example will make this clear. Using our sample of students' ages represented by Y, the use of sigma as in the expression ΣY (read as: the sum of Y) tells us to sum all the scores of the variable Y. Using our example, we would find the sum of the set of scores from the variable "age" by adding each score together:

$$19 + 21 + 23 + 32 = 95$$

So, for the variable "age," $\Sigma Y = 95$.

Sigma is also often used in expressions with an exponent, as in the expression ΣY^2 (read as: the sum of squared scores). This means that we should first square all the scores of the Y variable and then sum the squared products. So using the same set of scores, we would solve the expression by squaring each score first and then adding them together:

$$19^2 + 21^2 + 23^2 + 32^2 = 361 + 441 + 529 + 1{,}024 = 2{,}355$$

So for the variable "age," $\Sigma Y^2 = 2{,}355$.

A similar, but slightly different, expression, which illustrates the function of parentheses, is $(\Sigma Y)^2$ (read as: the sum of scores, squared). In this expression, the parentheses tell us to first sum all the scores and then square this summed total. Parentheses are often used in expressions in statistics, and they always tell us to perform the expression within the parentheses first and then the part of the problem that is outside of the parentheses. To solve this expression, we need to sum all the scores first. However, we already found that $\Sigma Y = 95$, so to solve the expression $(\Sigma Y)^2$, we simply square this summed total,

$$95^2 = 9{,}025$$

So, for the variable "age," $(\Sigma Y)^2 = 9{,}025$.

You should also be familiar with the different symbols that denote multiplication. Most students are familiar with the times sign ($\times$); however, there are several other ways to express multiplication. For example,

$$3(4) \qquad (5)6 \qquad (4)(2) \qquad 7 \cdot 8 \qquad 9 * 6$$

all symbolize the operation of multiplication. In this text, the first three are most often used to denote multiplication. There are also several ways division can be expressed. You are probably familiar with the conventional division sign ($\div$), but division can also be expressed in these other ways:

$$4/6 \qquad \frac{6}{3}$$

This text uses the latter two forms to express division.

In statistics you are likely to encounter greater than and less than signs ($>$, $<$), greater than or equal to and less than or equal to signs ($\geq$, $\leq$), and not equal to signs ($\neq$). It is important you understand what each sign means, though admittedly it is easy to confuse them. Use the following expressions for review. Notice that numerals and symbols are often used together:

$4 > 2$ means 4 is greater than 2
$H_1 > 10$ means H_1 is greater than 10

$7 < 9$ means 7 is less than 9
$a < b$ means a is less than b

$Y \geq 10$ means that the value for Y is a value greater than or equal to 10
$a \leq b$ means that the value for a is less than or equal to the value for b

8 ≠ 10 means 8 does not equal 10

$H_1 \neq H_2$ means H_1 does not equal H_2

■ ■ ■ ■ Proportions and Percentages

Proportions and percentages are commonly used in statistics and provide a quick way to express information about the relative frequency of some value. You should know how to find proportions and percentages.

Proportions are identified by P; to find a proportion apply this formula:

$$P = \frac{f}{N}$$

where f stands for the frequency of cases in a category and N the total number of cases in all categories. So, in our sample of four students, if we wanted to know the proportion of males in the front row, there would be a total of two categories, female and male. Because there are 3 females and 1 male in our sample, our N is 4; and the number of cases in our category "male" is 1. To get the proportion, divide 1 by 4:

$$P = \frac{f}{N} \qquad P = \frac{1}{4} = .25$$

So, the proportion of males in the front row is .25. To convert this to a percentage, simply multiply the proportion by 100 or use the formula for percentaging:

$$\% = \frac{f}{N} \times 100 \qquad \% = \frac{1}{4} \times 100 = 25\%$$

■ ■ ■ ■ Working with Negatives

Addition, subtraction, multiplication, division, and squared numbers are not difficult for most people; however, there are some important rules to know when working with negatives that you may need to review.

1. When adding a number that is negative, it is the same as subtracting:

    $5 + (-2) = 5 - 2 = 3$

2. When subtracting a negative number, the sign changes:

    $8 - (-4) = 8 + 4 = 12$

3. When multiplying or dividing a negative number, the product or quotient is always negative:

    $6 \times -4 = -24, \qquad -10 \div 5 = -2$

4. When multiplying or dividing two negative numbers, the product or quotient is always positive:

    $-3 \times -7 = 21, \qquad -12 \div -4 = 3$

5. Squaring a number that is negative always gives a positive product because it is the same as multiplying two negative numbers:

$-5^2 = 25$ is the same as $-5 \times -5 = 25$

■ ■ ■ ■ **Order of Operations and Complex Expressions**

In statistics you are likely to encounter some fairly lengthy equations that require several steps to solve. To know what part of the equation to work out first, follow two basic rules. The first is called the rules of precedence. They state that you should solve all squares and square roots first, then multiplication and division, and finally, all addition and subtraction from left to right. The second rule is to solve expressions in parentheses first. If there are brackets in the equation, solve the expression within parentheses first and then the expression within the brackets. This means that parentheses and brackets can override the rules of precedence. In statistics, it is common for parentheses to control the order of calculations. These rules may seem somewhat abstract here, but a brief review of their application should make them more clear.

To solve this problem,

$$4 + 6 \cdot 8 = 4 + 48 = 52$$

do the multiplication first and then the addition. Not following the rules of precedence will lead to a substantially different answer:

$$4 + 6 \cdot 8 = 10 \cdot 8 = 80$$

which is incorrect.

To solve this problem,

$$6 - 4(6)/3^2$$

First, find the square of 3,

$$6 - 4(6)/9$$

then do the multiplication and division from left to right,

$$6 - \frac{24}{9} = 6 - 2.67$$

and finally, work out the subtraction,

$$6 - 2.66 = 3.33$$

To work out the following equation, do the expressions within parentheses first:

$$(4 + 3) - 6(2)/(3 - 1)^2$$

First, solve the addition and subtraction in the parentheses,

$$(7) - 6(2)/(2)^2$$

Now that you have solved the expressions within parentheses, work out the rest of the equation based on the rules of precedence, first squaring the 2,

$$(7) - 6(2)/4$$

Then do the multiplication and division next:

$$(7) - \frac{12}{4} = (7) - 3$$

Finally, work out the subtraction to solve the equation:

$$7 - 3 = 4$$

The following equation may seem intimidating at first, but by solving it in steps and following the rules, even these complex equations should become manageable:

$$\sqrt{(8(4-2)^2)/(12/4)^2}$$

For this equation, work out the expressions within parentheses first; note that there are parentheses within parentheses. In this case, work out the inner parentheses first,

$$\sqrt{(8(2)^2)/3^2}$$

Now do the outer parentheses, making sure to follow the rules of precedence within the parentheses—square first and then multiply:

$$\sqrt{\frac{32}{3^2}}$$

Now, work out the square of 3 first and then divide:

$$\sqrt{\frac{32}{9}} = \sqrt{3.55}$$

Last, take the square root:

$$1.88$$

Answers to Odd-Numbered Exercises

Chapter 1

3. a. Interval-ratio
   b. Nominal
   c. Interval-ratio
   d. Ordinal
   e. Nominal
   f. Ordinal
   g. Interval-ratio

Chapter 2

1. a. Race is a nominal variable. Class is an ordinal variable because the categories can be ordered from lower to higher status.
   b. Frequency table for Race:

Race	Frequency
White	17
Nonwhite	13

   Frequency table for Class:

Class	Frequency
Lower	3
Working	15
Middle	11
Upper	1

   c. The proportion nonwhite is $13/30 = .43$. The percentage white is $(17/30) \times 100 = 56.7\%$.
   d. The proportion middle class is $11/30 = .37$.

3. a. Relative frequency table for traumas experienced:

Number of Traumas	Percentage
0	50.0
1	36.7
2	13.3
Total $N = 30$	

Trauma is an interval-ratio level variable because it has a real zero point and a meaningful numeric scale.

   b. People in this survey are more likely to have experienced no traumas last year (50% of the group).

   c. The proportion who experienced one or more traumas is calculated by first adding 36.7 percent and 13.3 percent = 50 percent. Then divide that number by 100 to obtain .50, or half the group.

5. a. Men and women are divided in their belief that pornography leads to a breakdown in morals. About 50% of each gender think that it does and about half think that it does not. Slightly more females than males (52.2% to 48.9%) think that pornography does not have this effect.

Attitude Toward Pornography Leading to Moral Breakdown	Relative Frequencies
MALES	
No	48.9%
Yes	51.1%
FEMALES	
No	52.2%
Yes	47.8%

   b. No, the leaders' views are incorrect, as women are more likely than men to support this stance toward the effect of pornography.

7. a. Females are underrepresented on television. They compose 51 percent of the population but only 37 percent of characters on television.

   b. Young adults and those defined as middle-aged are over-represented.

   c. It appears that Hispanics are the most underrepresented. They compose 9 percent of the American population but only 2 percent of television characters, for a difference of 7 percent (another way to put it is that Hispanics are underrepresented by a ratio of 4.5 to 1—9%/2%).

d. No. Professional/executive workers are greatly overrepresented; labor, service, and clerical workers are greatly underrepresented; and law enforcement officers are included on programs at a rate about ten times greater than they are found in the U.S. population.

e. Blacks are represented at about the same relative frequency on television (13% to 12%).

f. Handicapped people are underrepresented, as are those who are overweight and those who wear glasses.

9. a.

EDUCATION	MALE		FEMALE	
	Percentage	Cum Percentage	Percentage	Cum Percentage
Some high school	24.0	24.0	26.4	26.4
High school grad	28.0	52.0	32.0	58.4
Some college	24.0	76.0	26.8	85.2
College graduate	24.0	100.0	14.8	100.0

EDUCATION	WHITE		BLACK	
	Percentage	Cum Percentage	Percentage	Cum Percentage
Some high school	21.2	21.2	31.5	31.5
High school grad	32.2	53.4	27.9	59.4
Some college	23.8	77.2	28.4	87.8
College graduate	22.7	100.0	12.2	100.0

b. 48% of males have more than a high school education, 41.6% of females have more than a high school education.

c. 53.4% of whites have completed high school or less. 59.4% of blacks have completed high school or less.

d. The cumulative percentages are not terribly different for either gender or race. That said, the greatest difference exists for blacks and whites. For example, there is a ten percentage point difference between whites and blacks (77.2% to 87.8%) for those who have completed some college or less. The comparable difference for males and females is about nine percentage points. But it should be emphasized that the cumulative percentage differences are still similar.

11. a. The majority of Americans (58%) believe that abortion should be legal in some circumstances.

b. Since 1975, the majority of those surveyed believe that abortion should be legal in some circumstances. Percentages in this category range from a low of 48% in 1992 to a high of 58% in 1998. Overall, while there is some variation in responses over the years, the pattern of support remains the same for 1992 and 1998. In these two years, the majority of Americans believe abortion should be legal in some circumstances, followed by always legal and always illegal.

Poll results for 1975 show a different pattern – 54% supporting legal in some circumstances, 22% always illegal, and 21% always legal.

13. a. The category with the highest percentages across all the reported survey years is Salesperson. The percentage of respondents who said yes ranged from 68% (1977) to 90% (1996).

b. There are two categories with the lowest percentages, elementary school teachers and clergy. The percentage of Americans who agreed that homosexuals should be hired as elementary school teachers ranged from a low of 27% in 1997 to a high of 55% in 1996. For clergy, the percentage who said yes ranged from 36% (1977) to 53% (1996).

c. Overall, Gallup Poll results indicate an increase in the level of support for hiring homosexuals in all listed occupations. Support has always been high for particular occupations (salesperson and armed forces). But even for the occupations with the lowest levels of support, elementary school teachers and clergy, in 1996 slightly over half of those surveyed agreed that homosexuals should be hired as both.

15. a. Overall, the number of homemakers declined by 3 million since 1976. In 1996, the age of homemakers increased in two age categories, 35–44 and 55–64 years. For both reported years, the majority of female homemakers were white and married. The percentage of women with some college education or more increased in 1996. While 56.0 of female homemakers reported having any child under 18 in 1976, only 45.3% reported the same in 1996.

b. Age – ordinal
Race/Ethnicity – nominal
Marital Status – nominal
Education – ordinal
Presence of related children – nominal

Chapter 3

1. a. A total of .945 of all inmates are men.
   b. A total of 55 percent had a full-time job before being imprisoned.
   c. A total of 65 percent of the inmates did not graduate from high school.
   d. A total of 65 percent of inmates are nonwhite.

3. a. A total of 14 percent of all inmates were using both alcohol and drugs.
   b. A total of 49 percent were using either alcohol or drugs, or both.
   c. A total of 33 percent of women were incarcerated for drug crimes. Only 21 percent of men were incarcerated for the same crime.
   d. About 12,692 women were incarcerated for drug crimes.

5. a.

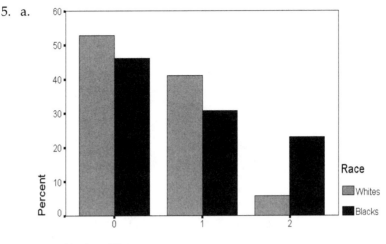

Number of Traumas

b. Blacks are more likely to have experienced two traumas. About 23 percent of blacks experienced two traumas as compared with about 6 percent of whites.

c. Frequencies don't control or adjust for the total number of people in each group. There are more whites (17) than blacks (13) in the survey, so percentages must be used to make the bars comparable.

7. a. Either bar charts or pie charts can be used to display these data, and either graph type will accurately represent these data. However, comparing values across graphs is usually easier when bar charts are used; this is especially true when the proportions change greatly (as in our comparison between whites and blacks).

9. a. Smoking by Grade Level

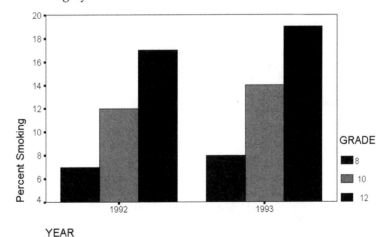

YEAR

b. Smoking by Year

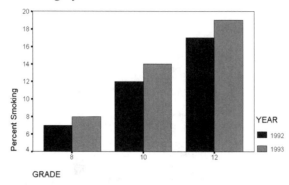

GRADE

c. The first bar graph shows that older students smoked more than younger students in each year, by placing the bars for the different grades next to each other. The second bar graph shows that students smoked more in 1993 than in 1992 at each grade level, by placing the bars for both years next to each other for each grade level. Either bar graph could be used to show that older teens smoked more and that smoking went up from 1992 to 1993, but it's easier to see the answer to each question with the data displayed in different formats.

11. a.

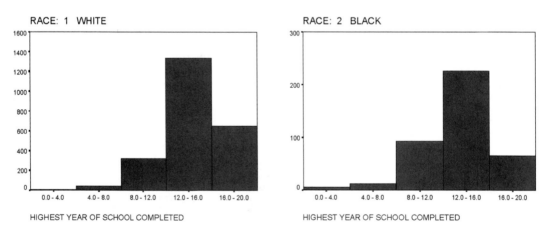

b. The histograms show that whites have more education than blacks. For example about 27.7% of whites have a college degree or higher, while only 16.2% of blacks in the GSS sample attained the same level of education. However, the shapes of the histograms are very similar, which indicates that whites and blacks have roughly the same type of distribution of educational attainment. For example, the largest category for both races are those who completed about a high school education.

13. a. The rate of first marriage has been as low as 80 per 1,000 single women in about 1931, at the beginning of the Depression, and as high as 140 per 1,000 just after World War II. It has been in a long-term decline since that period, reaching an all-time low of 76 per 1,000 in the late 1980s. The rate of first marriage was equal to 1 in 100 or 100 per 1,000 in 1935–1937 and 1971–1973. It would appear that there has been less fluctuation in the divorce rate, which rose slowly from the early 1920s until about 1970, then increased rapidly for ten years before beginning a slow decline. The absolute change in rate was from about 10 per 1,000 to 40 to 1,000, not as great as the absolute change in the first marriage rate. However, as a percentage change the divorce rate has changed by 400%, while the first mar-riage rate changed slightly less than 200%. Hence the concern a few years ago over the divorce rate in the United States.

    b. Notice that the remarriage rate follows the marriage rate very closely until the early 1960s. The first peak is easy to understand and is caused by the effects of World War II. Many women were widowed during the war. They remarried veterans returning after the war. Also, some veterans returned and got divorced. Many of the women they divorced in turn soon married again. The remar-riage peak in the 1960s is a bit more difficult to understand without more information. A tentative explanation is that the parents of some of the first baby boomers had stayed together for the sake of the children, but then got divorced when the children were in college. A second explanation is that people were simply more willing to get married to another partner in this period compared to other eras. In other words, values had changed and men and women, after being divorced or widowed, were more willing to get married again rather than live alone or with family.

15. Overall, the group with the highest rate of victimization is young adults, aged 18–21 years old. The rates of serious crime victimization for this age group is about 17 times higher than for persons aged 65 years or older. Americans 24 years of age or younger were more likely to be victims of serious crime; their rates ranging from about 35–50 per 1,000 persons. The rates of serious crime victimization steadily decline after age 24. The group with the lowest rates are persons 65 years or older.

Chapter 4

1. The mode for race is "white." The mode for class is "working class." The mode for traumas experienced is "0."

3. a. Education is measured on an interval-ratio scale. The mode is "12" years of education, or high school graduation. The median is "13" years of education.

b. The 25th percentile is 11 years; the 50th percentile is 13 years; the 75th percentile is 16 years. You don't need to calculate the 50th percentile because it is equivalent to the median.

5. a. For 18 to 29 year olds, the mode is once per day, and the median is several times a week. For 50 to 59 year olds, the mode is several times a day, and the median is once a day. For the 70- to 89-year-old group, the mode and median are both several times a day.

   b. Age is related to frequency of prayer because older people are more likely to pray. Thus, 50 to 59 year olds pray more than those in the 18 to 29 group, and those 50 to 59 year olds pray even more than 70 and above. The most common behavior in the two oldest groups is to pray several times a day. Moreover, over half of the oldest group prays at that frequency. On the other hand, the youngest respondents were most likely to pray once a day. At least as regards prayer behavior, there certainly seems to be a generation gap, as the differences between age groups are fairly large.

7. The mean number of persons per household is 1.85.

9.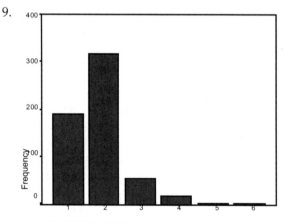

HOUSEHOLD MEMBERS 18 YRS AND OLDER

a. Household size appears to be positively skewed. A bar chart for this variable shows a tail trailing toward larger values of household size.

b. The median for household size is 2 and the mode is also 2. Both of these values are below the mean, offering further proof that household size is positively skewed, as the mean is pulled toward the tail for skewed distributions. The distribution is not symmetrical because there is a smallest minimum value of household size (1), but no theoretical maximum, so the distribution is cut off at the low end and trails out at large household sizes.

11. Yes, both of these politicians can be correct, at least in a technical sense. One politician can be referring to the mean; the other could be using the

median. The average or mean income of Americans can be greater than the median if income's distribution is positively skewed.

13. For the group of large cities, the mean is 23.39, the median is 22.10. Among small cities reported in the table, the mean murder rate is lower at 14.91, the median is also lower at 12.40. The murder rate is highest among large cities. The medians also show that larger cities have higher murder rates. For the 1995 data, the median is slightly lower than the mean for each group, the distributions are skewed in a positive direction.

15. a. The mean number of marriages is 6,681.40. The mean number of divorces is 3,648.30.
    b. Without the Florida numbers, the mean number of marriages drops to 6,122 and the mean number of divorces drops to 3,330.89. The higher number of marriages and divorces in Florida pulls each mean toward a higher value.

Chapter 5

1. a. The maximum possible number of differences is 132,963.86.
   b. The observed number of differences is 120,298.
   c. The IQV of .905 means that Americans are fairly diverse in their political position.

3. a. The range of poverty rates in the South is from 12.1 to 23.5 percent, or 11.4 percent. The range in the West is from 7.1 to 25.3 percent, or 18.2 percent. The West has a greater range.
   b. For the South, the 25th percentile is 13.6 percent and the 75th percentile is 20.0 percent, so the IQR is 6.4 percent. For the West, the 25th percentile is 10.43 percent and the 75th percentile is 16.25 percent, so the IQR is 5.82 percent. The IQR for the South is larger.
   c. The data support the idea that there is greater variability of poverty rates in western states. The range in poverty rates in western states is greater than in southern states by a factor of 18.2/11.4 or about 1.6. There is a difference in the IQRs for each region, though. The difference in variability seems to be because the West has more states with lower rates of poverty, such as Alaska, Utah, or Nevada.
   d. Standard deviation for the South is 3.86; for the West, 5.15.
   e. As indicated by the wider standard deviation, there is more variability in poverty rates among the western states listed in the table. This finding is consistent with previous results.

5. a. The range of percent increase in the elderly population for the Mountain states is 66.90 percent. The range of percent increase for the West North Central states is 5.5 percent. The Mountain states have a much larger range.
   b. The IQR for the Mountain states is 24.12 percent. The IQR for the West North Central states is 3.4 percent. Again, the value for the Mountain states is greater.

ANSWERS-10 ■ ANSWERS TO ODD-NUMBERED EXERCISES

    c. There is great variability in the increase in the elderly population in the Mountain states, chiefly caused by the large increases in Nevada, Arizona, and New Mexico, as measured by either the range or the IQR.

7. a. For Group 1, the standard deviation is 4.17. For Group 2, the standard deviation is 3.11.

    b. There is more cooperative behavior among children in Group 1. Over the fifteen days, the mean number of incidents of cooperative behavior in Group 1 was 9.13 versus 7.33 in Group 2. There is also more variability in Group 1, as its standard deviation of 4.17 is greater than the 3.11 for Group 2 children. A great deal of this difference is caused by the one day with 20 cooperative incidents in Group 1.

    c. If there were twice as many children in Group 1, then it would seem that there would be more chances for cooperative (and uncooperative) behavior, since there would be more children who would interact on average. This means that there might actually be less cooperative behavior in Group 1, on a per-person basis.

9. a. The mean for violent crime rate is 435.19. The mean for number of federal and state prisoners is 16,917.33.

    b. The standard deviation for crime rate is 246.05; the standard deviation for prisoners is 18,454.40.

    c. The IQR for violent crime rate is 387.50; the IQR for number of prisoners is 27,280.

    d. Because violent crime rate and number of prisoners are measured according to different scales (rate per 100,000 people versus the actual number), it isn't appropriate to directly compare the mean, standard deviation, or IQR for one variable with the other. It appears there is slightly more variability for the violent crime rate because the IQR and standard deviation are closer in size to the mean value. For violent crime rate, states like North Dakota, New York, Maine, New Hampshire, Vermont, and Illinois contribute more to its variability because they have values far from the mean. For number of prisoners, North Dakota, Vermont, Maine, New York, and Ohio contribute more to its variability.

11. a. The standard deviation is 17.98.

    c. Yes, the standard deviation would lead to about the same conclusion concerning the variability of the increase in elderly population as do the box plot and IQR. However, the box plot shows the actual range of values, which the standard deviation cannot (even though all values are used to calculate the statistic).

13. Since the variable is interval-ratio, according to our text, we should use variance (or standard deviation), range, or IQR. Among these three measures, variance and/or standard deviation is preferred. For measurements of central tendency, as discussed in Chapter 4, if we are

looking for the average percent of men and women who test positive for marijuana, we should rely on the mean. The mean percent of men who test positive is 32.52; while the mean percent for women is 17.38. On average, a higher percent of men, 15.14%, test positive for marijuana at the time of arrest. There is more variability in positive marijuana testing for men (standard deviation of 4.78) than for women (standard deviation is 3.67).

15. With a nominal variable (gender of employees), we are restricted to IQV as a measure of variability. IQV scores would indicate the degree to which a state's officer and/or civilian workforce is heterogeneous.

Chapter 6

1. a. The independent variable is race; the dependent variable is fear of walking alone at night.

RACE

FEAR	Black	White
No	4	7
Yes	6	4

b. A majority of whites (63.6%) are not afraid to walk alone in their neighborhoods at night, but a majority of blacks (60%) are afraid to walk alone. These differences undoubtedly have nothing to do with race as such, but rather with the fact that blacks are more likely to live in dangerous neighborhoods, where it is sometimes dangerous to be out at night alone.

RACE

FEAR	Black	White
No	40.0%	63.6%
Yes	60.0%	36.4%

3. a. Given how your classmate phrased the question, the dependent variable is belief about homosexuality, and the independent variable is whether or not someone has a gay or lesbian as a close friend or family member.
   b. A total of 254/1,154 = .22 have a close friend or family member who is gay or lesbian.
   c. No, they are more likely to believe that homosexuality is not a choice. Only 36 percent of those who are close to a gay or lesbian

think homosexuality is a choice, but 54.8 percent of those without close contact with gays or lesbians believe that homosexuality is a choice.

5. a. A total of 50 percent report being in good health, 30 percent in excellent health, and 20 percent think their health is fair or poor.

   b. There is very little difference in the perceived health of men and women. Slightly more men report being in excellent health (31.5% to 29.0%), but they are also very more likely to report being in fair or poor health (20.4% to 19.5%).

   c. It would seem that the clinic should focus equally on men and women, given that their perceived differences in health are so slight.

7. a. Residential segregation patterns have changed for both blacks and Hispanics between 1980 and 1990. Blacks increasingly live in more integrated neighborhoods; the percentage living in neighborhoods that are predominantly black has dropped from 35 to 13 percent. Hispanics have increasingly come to live in neighborhoods that are more segregated, as the percentages for them have increased from 17 to 27 percent.

   b. There is a relationship between ethnicity and changes in segregation, such that segregation increased for Hispanics and decreased for blacks. In fact, it would appear that Hispanics in 1990 live in more segregated neighborhoods than blacks, which was not true in 1980.

9. a. In three of the four cities, Los Angeles, Dallas, and Chicago, there has been little change in school segregation from 1968 to 1992, though in the two former cities, segregation did decrease slightly. In New York, segregation has increased over time; 23.2 percent of black students had predominantly white classmates in 1968, but that was true for only 8.4 percent of blacks in 1992.

   b. The trend is much stronger in New York. The percentage change was 14.8 percent in New York and only 3.7 percent in Dallas.

   c. In Los Angeles, segregation decreased from 1968 to 1986 (92.5% to 88.3%), then increased in 1992 (to 90.4 percent). In Dallas, segregation decreased from 1968 to 1974 (94.4% to 86.6%), but then increased in 1986 and 1992. In Chicago, segregation increased from 1968 to 1972 (94.6% to 98.8%), then decreased in 1986, then increased in 1992. In New York, segregation increased steadily from 1968 to 1992, so it was the only city to show a steady trend.

11. a. For whites, 72 percent of those who disapprove of women working outside the home think that pornography leads to a breakdown in morals, whereas only 60.6 percent of those who approve of women working think the same about the effect of pornography. For blacks, 71.4 percent of those who disapprove of women working outside the home think that pornography leads to a breakdown in morals, whereas only 56.3 percent who approve of women working think the same about pornography.

b. These differences between blacks and whites are small, so we conclude that race does not have a conditional effect on the relationship between attitudes toward women and the effect of pornography.

13. Suicide by firearm was more common than suicide by other methods across all the reported age groups. Overall, 11,424 or 76.74 percent of all reported suicides were committed with firearms. The use of firearms was more prevalent among males 65–84. In both age categories for white males, 65–74 and 75–84 years old, 78 percent were more likely to have used firearms. The percentage drops to 69 percent among men 85 years or older.

Chapter 7

1. a. We will make 496 errors, because we predict that everyone falls in the modal category that considers homosexuality to be a choice. The category of "Don't know" should be ignored.
   b. We will make 81 errors of prediction for those respondents who had a close friend or family member who is gay or lesbian. We will make 352 errors for those who don't have such a relationship.
   c. The proportional reduction in error is then $(496 - 433)/496 = 12.7$ percent, so we can increase our ability to predict attitude toward homosexuality by 12.7 percent (and equivalently, decrease the errors of prediction) by knowing whether or not someone has a close friend or family member who is lesbian or gay.

3. a. The number of same order pairs, Ns, is 127,248. The number of inverse order pairs, Nd, is 75,638.
   b. Gamma for the table is .254 and is positive. This implies that older people are more likely to believe that premarital sex is wrong. This value of gamma is moderate in size.

5. a. Time is the independent variable. Without knowing time, we would make 83 errors by predicting that all bills didn't pass. Knowing time period, we would make $19 + 64 = 83$ errors. Thus, lambda is $(83 - 83)/83 = 0$. This value seems to indicate that knowing the time period doesn't help to predict whether or not bills passed; put another way, it might be seen as evidence that there was no change in the proportion of bills passed on women's and family issues in the 1992/93 period versus 1990/91.
   b. Lambda has a value of exactly zero because the modal category for both time periods is the same—didn't pass—which is also the mode for the bill passage variable as a whole. The value of lambda does seem surprising, as 12.8 percent of the bills passed in 1992/93 versus 7.6 percent in 1990/91.

7. a. Gamma is equal to .28.
   b. That gamma is positive means that more bills passed in the later time period than in the earlier. That gamma is not close to 1 means

that the relationship is fairly weak, which we have also seen by looking at the percent difference in bill passage.

9. a. Somers' *d*, with attitude toward the death penalty the dependent variable, is .078.

   b. This is a very weak relationship because the value of Somers' *d* is so close to zero. People with a graduate education are more likely to oppose the death penalty. However, those most in favor of the death penalty are not people with the least education, but instead people with a high school education.

11. a. Gamma is −.277.

   b. Somers' *d* is −.128.

   c. The values of these two statistics are negative because disapproval of women working outside the home is associated with a greater belief that pornography leads to a breakdown in morals. In Chapter 6, Exercise 10, we found a 12.8 percent difference between belief about the effect of pornography, depending on whether one approved or disapproved of women working outside the home. Now we see that these differences translate into a modest statistical association between these two attitudes: both gamma and Somers' *d* are closer to 0 than to −1.

13. The appropriate measure of association for this table is lambda—with one ordinal variable (school performance) and a nominal variable (type of smoker). The lambda is .03, indicating a weak association between these two variables. If we had information about type of smoker, we would only improve our prediction about school performance by 3.0%.

Chapter 8

1. a. On the scatterplot the regression line has been plotted to make it easier to see the relationship between the two variables.

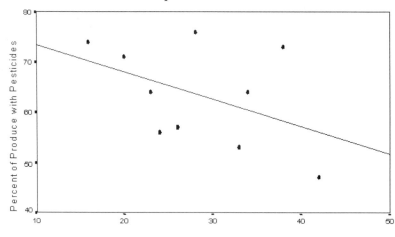

Number of Types of Pesticides

b. The scatterplot shows that there is a general linear relationship between the two variables, even though there is a fair amount of scatter about the straight line describing the relationship. As the number of pesticides detected increases, the percentage of a particular food type with pesticides decreases.

c. The Pearson correlation coefficient between the two variables is −.45. This seems rather odd, because it means that as the percent of produce with pesticides increases the different number of pesticides in that food decreases. It is hard to explain, but one possibility is that when a type of food is more often contaminated with pesticides, it is because of a few specific pesticides, rather than a large number.

3. a. The correlation coefficient between percent of unwed births is .954.

b. The correlation coefficient is very close to 1, which means that high values of unwed births for whites are closely associated with high values of unwed births for nonwhites, and vice versa.

5. a. Yes, as indicated by the negative r^2, the relationship between the variables is negative.

b. Based on the coefficient output, we can predict that an individual with 16 years of education will watch television 2.431 hours per week. In contrast, someone with 12 years of education (high school degree) is predicted to watch 4.039 hours of television per week. This is 1.608 hours more than someone with a college degree.

7. a.

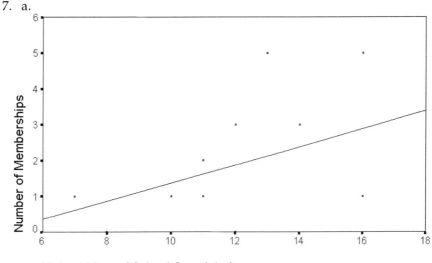

Highest Year of School Completed

b. The slope of the regression equation is .25. The intercept is −1.16. A straight line fits the data reasonably well, but it fits people with less education better. This is because the scatter about the line is less for people with less than 12 years of education than for those who have completed more education. In general, people with more education have more formal memberships.

c. The error of prediction for the second case is about 2.15 memberships. That is, we predicted that he or she would have 2.85 memberships, but this person actually belongs to 5 organizations. The error of prediction for a person with 10 years of education and 1 membership is –.34 membership. Here, we predicted a value of 1.34, but the person has less than that amount.

d. For someone with 14 years of education, we predict about 2.3 memberships. For someone with 4 years of education, we predict less than 0 memberships. This can't be right. This prediction illustrates the problem of making predictions beyond the range of the independent variable (also, the relationship between these two variables is most likely nonlinear at low values of education).

9. a. A straight line fits the data rather well, as shown in the scatterplot.

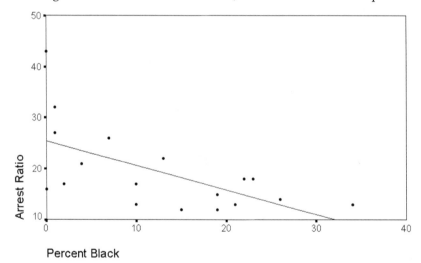

Percent Black

b. The slope of the regression equation is –.48. The intercept is 25.48. The city that falls farthest from the line is Livonia, Michigan, with no blacks and an arrest ratio of 43.

c. A value of 51 percent for the black population yields a predicted arrest ratio of 1.

Chapter 9

1. a. Nominal
   b. Interval-ratio
   c. Ordinal
   d. Nominal
   e. If an exact time, interval-ratio; if an approximate time, ordinal
   f. Ordinal

3.

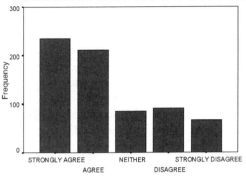

INCOME DIFFERENTIALS IN USA TOO BIG

INCOME DIFFERENTIALS IN USA TOO BIG

5. a. Yes, there appears to be some relationship. Those in the other race category are more likely to agree that income differentials in the United States are too big. When combining "strongly agree" and "agree" responses, 72.5% of other races responded versus 64.9% of whites and 61.1% of blacks. Black respondents were more likely to disagree with the statement. Thirty percent of blacks responded "disagree" or "strongly disagree," while 22.6 % of whites and only 10% of other races responded the same.

   b. Lambda = .000, indicating no association between race and responses to the variable INCGAP. Measures of association for nominal variables are not always effective at capturing what you can see in a table using percentages.

7. a. The median amount of sales tax paid is $338.84.

   b. The mean sales tax is $385.92 (based on the formula for calculating the mean from grouped data).

   c. The 25th percentile is $171.15 and the 75th percentile is $538.96, so the IQR is $367.81.

9. The bar chart uses a scale for the Y-axis that runs from only 2.60 to 2.75, thus exaggerating the difference in self-rated health between males and females. It would have been better to use a scale running from 1 to 7 on the Y-axis. Obviously, an unsuspecting reader might think that females rate their health much lower than do males.

11. a. The mean for feelings toward Clinton is 60.42. The mean feelings toward Dole is 51.67.

    b. The variance for feelings toward Clinton is 875.91, and the standard deviation is 29.60.

    c. The 25th percentile for Dole is 35. The 75th percentile is 70, so the IQR for feelings toward Dole is 35.

13. a. Working-class people are more likely to approve of Clinton: 22.3 percent of this group give the highest approval rating to Clinton,

compared with 15.9 percent of the middle class. Over one-third of the middle class (35.4%) "strongly disapprove" of Clinton, but only 22.5 percent of the working class do.

b. Somers' d for the table is –.164, with class as the predictor. The sign is negative because "working class" is coded lower than "middle class." This value of Somers' d indicates that there is a weak negative relationship between class and approval of President Clinton. It also indicates that by using the order of pairs of cases on class, we can improve the prediction of the order or relative ranking of pairs of cases on approval of President Clinton by 16.4 percent over a prediction without using class.

15. a. A statistical map would work best with the data from 1995. As discussed in Chapter 3, statistical maps can display geographic variations between different states.

b. A statistical map would also be appropriate to display the difference in poverty rates for each state.

c. For 1993: mean, 14.33; median, 13.30; standard deviation, 3.89. For 1995: mean, 13.03; median, 11.75; standard deviation, 3.69. The means indicate that there has been an overall decline in poverty rates since 1993 by 1.3 percent (14.33 – 13.03). Based on our calculations for the medians, we know that the distribution of poverty rates was negatively skewed for both years. A negatively skewed distribution indicates that there were some states with extremely low values (Connecticut and New Hampshire). There is not much of a change in the variability of poverty rates, based on the standard deviations of 3.89 (1993) and 3.69 (1995). The reported poverty rates for both years are relatively homogeneous.

Chapter 10

1. a. Yes, the applicant would be accepted because a score of 115 is at the 90.5th percentile.

b. The cutoff score is about 113 for the 88th percentile.

c. The Z score for 113 is 1.17.

3. a. China's Z score equivalent is 10.16.

b. For any normal distribution, about 15.9 percent of all cases should fall less than 1 standard deviation below the mean. For the distribution of population values, 1 standard deviation below the mean is less than zero (27.3 – 105.7), so clearly no countries have population values lower than 1 standard deviation below the mean.

c. The distribution of population must not be normal, because a true normal distribution would have cases well below the mean. Therefore, the median would be a better measure of central tendency.

5. a. About .17 of the males have incomes between $30,000 and $40,000. About .08 of the females have incomes in this range.

b. For a male drawn at random, the probability of having an income above $50,000 is .058. For females, the probability is only about .002.

c. The upper limit is $22,543.18; the lower limit is $6,119.24.

d. If income is positively skewed, then the proportion of cases at high incomes will be greater than for a normal distribution. This means, for example, that the probabilities of an income above $50,000 should be larger.

7. a. About 12 percent of whites should have scores above 60.

b. About 4 percent of blacks should have scores above 60.

c. About 82 percent of whites have prestige scores between 30 and 70.

d. Ten blacks have a prestige score between 50 and 60.

9. a. About .282 of all burglaries had dollar losses above $1,000.

b. About 5.5 percent of burglaries had dollar losses between $200 and $300.

c. The probability is about .69 that a dollar loss was above $400.

d. About 37 percent of all burglaries had losses below $500.

11. a. 3.14

b. .9015; this corresponds to 365 people.

c. .8 hour

d. .8523

13. For any Z distribution, the value of the mean is 0. The standard deviation of a Z distribution is 1. Z distributions are based on the mean of a variable and are centered around that value, so they have a mean of 0 by definition. A Z score of 1 or −1 is equivalent to a score in the original distribution that is 1 standard deviation above or below the mean, respectively. This direct mapping from the original distribution to a Z score means that the standard deviation of a Z distribution must be equal to 1.

15. a. Make sure that you understand the properties of the normal distribution, as we will apply its properties to other statistical techniques and procedures in the following chapters. Refer to the first part of Chapter 10 for a discussion of the normal distribution.

b. In general, standardized scores or Z scores provide a means to express the distance between the mean and a particular point or score. A Z score is the number of standard deviation units a raw score is from the mean. A negative Z score indicates that it is lower than the mean (or to the left of the mean), while a positive Z score indicates that it is higher than the mean (or to its right).

Chapter 11

1. a. Although there are problems with the collection of data from all Americans, the census is assumed to be complete, so the mean age would be a parameter.

b. A statistic because it is estimated from a sample

    c. A statistic because it is estimated from a sample

    d. A parameter because the bank has information on all employees

3. a. She is selecting a systematic random sample. The population might be defined as all persons shopping at that mall that day of the week. A more precise definition might limit it to all persons passing by the department store at the mall that day.

    c. This is neither a simple random sample nor a systematic random sample. It might be thought of as a sample stratified on last name, but even then, choosing the first twenty names is not a random selection process.

5. No.

7. a. They are not probability samples.

    b. The population is all those people who watch the television station and see the 800 advertised.

9. b. This is a systematic random sample because names are drawn systematically from the list of all enrolled students.

11. a. Mean = 5.3; standard deviation = 3.27

    b. Here are ten means from random samples of size 3: 6.33, 5.67, 3.33, 5.00, 7.33, 2.33, 6.00, 6.33, 7.00, 3.00.

    c. The mean of these ten sample means is 5.23. The standard deviation is 1.76. The mean of the sample means is very close to the mean for the population. The standard deviation of the sample means is much less than the standard deviation for the population. The standard deviation of the means from the samples is an estimate of the standard error of the mean we would find from one random sample of size 3.

Chapter 12

1. a. The best estimate at the 90 percent confidence level runs from 11.52 to 14.88 percent. This means that there are 90 chances out of 100 that the confidence interval we calculate in this manner will contain the percentage of victims in the U.S. population. The calculated interval for this particular sample either does or does not contain the population mean.

    b. The confidence interval at the 99 percent level is 13.2 percent $\pm$ 2.63 percent.

    c. Both confidence intervals will shrink by a factor of $1/\sqrt{2}$.

3. a. $.41 \pm .03$

    b. $.41 \pm .04$

5. The 95% confidence interval is $.52 \pm 4.37$. However, no matter whether the 90, 95, or 99 percent confidence interval is chosen, the calculated interval includes values below 50% for vote for George Johnson. Therefore, you should advise the newspaper to be cautious in declaring

Johnson the likely winner because it is quite possible that less than a majority of voters support him.

7. a. $42.36 \pm .90$

9. $1.68 \pm .07$

11. 41 percent ± 3.03 percent

13. a. $.31 \pm .02$; $.47 \pm .04$
   b. 95% C.I.: $.59 \pm .04$; $.35 \pm .02$. 99% C.I.: $.59 \pm .05$; $.35 \pm .03$

15. a. All calculated standard errors equal .01. (What can you attribute the small sampling error to?) For blacks: $44 \pm .02$; $.74 \pm .02$; $.53 \pm .02$; $.68 \pm .02$. For whites: $.51 \pm .02$; $.87 \pm .02$; $.72 \pm .02$; $.75 \pm .02$.

Chapter 13

1. Type II; beta; Type I

3. a. For research problem (a): Type I error would be to conclude that elderly women have lower income than all women when they do not. Type II error would be to conclude that elderly women have the same income as all women when they actually have lower incomes.
   b. Making a Type I error means rejecting the null hypothesis when it is, in fact, true. Since in most research the hypothesis of interest is not the null hypothesis, the implication of making a Type I error is to conclude that one's hypothesis is correct when it is not. Making a Type II error means accepting the null hypothesis when it is actually false. Thus, a Type II error leads you to reject a potentially valuable hypothesis or theory.
   c. Making a Type II error might seem to be a less serious error, but in studies of new drugs and procedures in medicine, rejecting a potentially valuable new procedure has serious consequences. If an illness is very severe and life threatening, then Type II error might be minimized for a drug or procedure that cures that illness. On the other hand, if a drug has serious side effects, it might be best to minimize Type I errors so that we are very sure that the drug has a positive side effect on an illness—the research hypothesis—before administering it to patients. Balancing Type I and Type II errors is an important consideration in any study based on samples.

5. a. $H_0: \mu_1 = \mu_2$. The research hypothesis is $\mu_1 > \mu_2$.
   b. We use the two-sample test with the variance in the two populations known, yielding a t value of 4.67. We estimate that the probability of t, based on a right-tailed test, is less than .0005 (a very rare occurrence). This leads to a rejection of the null hypothesis, and we conclude that the researcher is correct that city residents do experience more stressful events than suburban residents.

7. a. $H_0: \mu_1 = \mu_2$. The research hypothesis is $\mu_1 > \mu_2$.

    b. The obtained t is 1.16 (with 68 degrees of freedom). At the .05 level of significance, we cannot reject the null hypothesis. Based on our current data, length of time employed is not related to type of housing.

9. a. The variances cannot be assumed to be equal, so we must use a t test that meets these restrictions.
    b. The degrees of freedom for the test are approximately 59.8 (or about 60). The obtained t value is .767 (probability < .10), so we fail to reject the null hypothesis of no difference (probability > alpha). The friend was wrong.

11. a. The appropriate test is the Z test for differences in two proportions with large samples.
    b. The obtained value of Z is 1.13, so we fail to reject the null hypothesis. Males and females have an equal level of support for abortion for single women.

13. a. To compare Cuban with Puerto Rican men, we use the t test for unequal variances. But to compare Cuban with Mexican men, we can use the t test for equal variances.
    b. For the Cuban–Puerto Rican comparison, the obtained t is 28.59 (9.801 df), so we reject the null hypothesis. For the Cuban–Mexican comparison, the obtained t is 48.31, so we again reject the null hypothesis. The probability for both obtained t statistics is less than .0005. Both tests indicate significant differences in earnings.
    c. These results confirm the previous results from Chapter 12. These large obtained values of t make it very, very unlikely that the earnings of Cuban men are equal to those of Mexican or Puerto Rican men in the U.S. population.

15. Assuming a two-tailed test, the t-statistic of 1.035, with 728 degrees of freedom, is not very significant. Its probability is less than .20, above our alpha level of .05. Recall that we reject the null hypothesis only if the probability of our obtained t is less than (or equal to) alpha. In this case, we fail to reject the null hypothesis of no difference between men and women on their perceived success in their family life.

Chapter 14

1. a. $\chi^2 = 1.18$
    b. Degrees of freedom = 1
    c. Based on Appendix D, we estimate that the probability of 1.18 is between .20 and .30, above our .05 alpha level. Thus, we fail to reject the null hypothesis of independence between race and fear. This conclusion is not consistent with the description of the table in Chapter 6. Using percentages, it appeared that whites are less fearful than blacks of walking alone at night.
    d. The cell for blacks who are afraid of walking alone has an expected value of 4.76.

3. a. 4 degrees of freedom
    b. No, that one cell represents only 1/10th or 10% of the cells, which is below the rule-of-thumb of 20%.
    c. Chi-square is 4.73.
    d. The probability of 4.73 is between .30 and .50. This probability range is larger than our alpha of .01. We fail to reject the null hypothesis.

5. a. Chi-square equals 31.83, with 12 degrees of freedom.
    b. The probability of 31.83 is less than .001, below our alpha level of .01. We reject the null hypothesis and conclude that there is a relationship between educational degree and attitude toward premarital sex. In particular, respondents with a bachelor's degree or higher are more likely to indicate that premarital sex is "not wrong at all" than respondents with lower educational attainment.

7. a. Chi-square is 56.37. The probability of this chi-square is less than .000. Since $p <$ alpha, we reject the null hypothesis. We can conclude that respondent's age and attitudes toward premarital sex are dependent.
    b. There is a negative relationship between age and attitudes toward premarital sex—the younger a respondent, the higher his/her score on the PREMARSX scale. Younger respondents are more likely to believe that premarital sex is "not wrong at all." The distribution of responses for the first two age categories are nearly identical. Between 48% of 40–59 year olds and 50% of 20–39 year olds indicated that premarital sex was "not wrong at all." These percentages are larger than the 22% of 60+ year olds that indicated the same.

9. For the tables presented in Chapter 7, Exercise 12: For males, chi-square is 14.144; for females, chi-square is 50.26. For the table in Exercise 13: chi-square is 174.89.

Chapter 15

1. a. About .258 of all marriages last between 10 and 20 years.
    b. A marriage that lasts 50 years is at the 98.5th percentile. This is certainly a rare value, so it justifies the exceptional nature of such marriages.
    c. The probability is about .196.
    d. Yes, length of marriage is probably not normally distributed because there are few cases below 1 standard deviation from the mean.

3. a. The results are different by race. For whites, $\chi^2 = 64.25$ with 8 degrees of freedom, probability $< .001$. We reject the null hypothesis. For blacks, $\chi^2 = 6.99$ with 8 degrees of freedom, probability is between .50 and .70. Since $p >$ alpha, we fail to reject the null hypothesis.
    b. Race is a control variable that specifies the relationship between support for AIDS research funding and political preference. In other words, there is a conditional relationship between the latter two variables because of race.

c. The main problem comes about because of the small number of people who say they prefer a third, or other, party. Combine the "Other Party" category with the "Independent" and "No Prefer- ence" groups. These categories are not exactly comparable but this groups political preference into Republican, Democrat, and Other. In the table for whites, the relationship remains significant (chi- square 60.87). In the table for blacks, the relationship remains nonsignificant (chi-square 5.44).

5. a. The 95 percent confidence interval is 59.5 percent ± 4.8 percent.
   b. The upper end of the confidence interval is at 64.3 percent, which is below the census figure of 66.9 percent as a whole. This indicates that it is likely that the percentage of Hispanic two-parent house- holds is lower in the city.

7. a. $\chi^2 = 5.26$. At the .01 alpha level, we would fail to reject the null hypothesis. The probability of our obtained chi-square falls some- where between .02 and .05, both greater than our alpha level.
   b. The proportion of males disapproving is .694; for females it is .640. The obtained Z is 2.34, with a probability of .02 (.0095 × 2), so we fail to reject the null hypothesis. The probability of the obtained Z is slightly greater than our alpha of .01.
   c. The 95 percent confidence interval is 66.7 percent ± 2.3 percent.
   d. The conclusions are similar.

9. a. The 99 percent confidence interval for approval of Congress is .40 ± .47; for approval of their own representatives is .70 ± .03.

11. a. People who are divorced, separated, or widowed say they are worse off than they were a year ago. People who are married are the least likely to say this (25.6%). However, singles (46.6%) and those separated (46.7%) are the most likely to say they are better off. Only a majority of those widowed say they are about the same (50.8%).
    b. The chi-square for this table is 56.90 with 8 degrees of freedom. The probability of our chi-square is less than .001. We can confidently reject the null hypothesis and conclude that marital status and whether one is better off or not are related.

13. The chi-square for the first table is 10.36. With 3 degrees of freedom, we estimate its probability to be between .01 and .02. We reject the null hypothesis. We conclude that we have evidence that suggests a rela- tionship between a person's age and his/her use of alternative medi- cine. The chi-square for the second table is 12.84. With 4 degrees of freedom, the probability of our obtained chi-square is between .01 and .02. The range is less than our alpha of .05; thus, we reject the null hypothesis and conclude that a relationship also exists between a person's level of income and use of alternative medicine.

Index/Glossary

A

a (Y-intercept), 306, 307
  computing for prediction
    equation, 309–312, 329–331
  interpreting, 312–313
Abortion
  family size and, 227–230, 256–
    258, 515
  gender and, 498–499
  job security and, 210–212
  religion and, 225–231
  trauma and, 216–220
  worldview and, 280–283
Affirmative action, 254–256, 462–
  464
African American women, 127–
  128
Age
  demographic trends based on,
    82–84, 86–87
  educational level by race and,
    92–94
  female clerical and technical
    employees by, 178–179
  hours of reading per week and,
    316
AIDS
  sampling process and, 543–544
  study on risks among women,
    52–55
  See also HIV infections
Alpha (α) the level of probability
  at which the null hypothesis
  is rejected. It is customary to
  set alpha at the .05, .01, or
  .001 level, 485
Analysis of variance (ANOVA),
  560n
Analyzing data, 18. *See also* Descriptive data analysis

Asian Americans
  characteristics of women, 150–
    151
  earnings of men, 489–492
Association. *See* Measures of
  association
Assumptions
  of chi-square test, 515
  of statistical hypothesis testing,
    475, 551
**Asymmetrical measure of asso-
  ciation** a measure whose
  value may vary depending
  on which variable is considered the independent variable and which the dependent variable, 264
Average. *See* Mean
Axes, graphical, 89–90

B

b (slope), 306, 307
  calculating with a computa-
    tional formula, 313–315
  computing for prediction equa-
    tion, 309–312, 329–331
  interpreting, 312–313
Bar graph a graph showing the
  differences in frequencies or
  percentages among categories of a nominal or an ordinal variable. The categories
  are displayed as rectangles of
  equal width with their height
  proportional to the frequency
  or percentage of the category,
  75–77
  example of using, 92, 93
  producing with SPSS, 96–98
Bell-shaped curve, 380–381

Best-fitting line, 307–309
Bimodal distributions, 112, 113
Birth rate, 49
Bivariate analysis a statistical
  method designed to detect
  and describe the relationship
  between two variables, 202.
  See also Cross-tabulation
Bivariate procedure (SPSS for
  Windows), 336–338
Bivariate regression analysis, 297–
  348
  assessing prediction accuracy,
    318–327
  computational formulas, 309–
    315, 326–327
  interpreting coefficients, 312–
    313
  linear relationships and, 302–
    309
  main points about, 334
  nonlinear relationships and, 316
  Pearson's correlation coefficient
    and, 324–327
  practical examples of, 316–318,
    327–334
  prediction equation and, 309–
    312
  scatter diagram and, 299–302
  SPSS demonstrations and prob-
    lems on, 335–340
  study/review exercises related
    to, 340–348
Bivariate relationship, 216–234
  direction of, 218–220
  elaboration process and, 220–
    234
  existence of, 216–217
  properties of, 216–220
  SPSS output of, 368–370